Environmental Applications of Nanomaterials

Synthesis, Sorbents and Sensors

Environmental Applications of Nanomaterials

Synthesis, Sorbents and Sensors

editors

Glen E Fryxell
Pacific Northwest National Laboratory, USA

Guozhong Cao
University of Washington, USA

Imperial College Press

Published by

Imperial College Press
57 Shelton Street
Covent Garden
London WC2H 9HE

Distributed by

World Scientific Publishing Co. Pte. Ltd.
5 Toh Tuck Link, Singapore 596224
USA office: 27 Warren Street, Suite 401-402, Hackensack, NJ 07601
UK office: 57 Shelton Street, Covent Garden, London WC2H 9HE

British Library Cataloguing-in-Publication Data
A catalogue record for this book is available from the British Library.

ENVIRONMENTAL APPLICATIONS OF NANOMATERIALS
Synthesis, Sorbents and Sensors

ISBN-13 978-1-86094-662-2
ISBN-10 1-86094-662-3
ISBN-13 978-1-86094-663-9 (pbk)
ISBN-10 1-86094-663-1 (pbk)

Typeset by Stallion Press
Email: enquiries@stallionpress.com

Printed in Singapore.

Foreword

Nanotechnology has attracted a lot of attention recently, particularly in the research and industrial communities. It offers many unprecedented opportunities for advancing our ability to impact not only our day to day lives, but the very environment in which we live. The ability to design, synthesize and manipulate specific nanostructured materials lies at the very heart of the future promise of nanotechnology. Nanomaterials may have unique physical and chemical properties not found in their bulk counterparts, such as unusually large surface area to volume ratios or high interfacial reactivity. Such properties give hope for new chemical capabilities arising from exciting new classes of nanomaterials. Indeed, as this book summarizes, nanomaterials have been developed for specific applications that involve interfacial reactions and/or molecular transport processes.

The industrial revolution of the late 19th and early 20th century led to unprecedented economic growth in Europe and the United States. However, it also produced unprecedented environmental pollution. In those simpler, more naive times, contamination of the environment was largely ignored, a common sentiment being simply that "dilution is the solution to pollution". The economic benefits of increased industrial production outweighed the emerging environmental problems, and the vastness of the wilderness and ocean (along with the lower populations of the day) allowed this industrial contamination to dissipate to levels that made it relatively easy to ignore ... for a while. Today, we see other countries (e.g. China) going through similar growing pains, and experiencing similar environmental damage.

The 20th century also brought an unprecedented arms race, which in turn brought its own unique set of environmental concerns and needs. Of particular importance to the environment are the legacy wastes arising from 40 years of nuclear weapons production, as well as the vast stockpiles of chemical weapons, throughout the globe. Our parents devised and built these devastating weapons in order to fight back against the ruthless tyrants that threatened their world. They did not have the luxury of planning ahead for the eventual disposal of these deadly materials; they needed to fight, and they needed to fight NOW. Their success ultimately led to improved standards of living throughout much of Europe and the United States (eventually spreading to other parts of the world, as well). However, the issues raised by the presence of these difficult waste materials are still unresolved.

This new-found quality of life was starting to be threatened by industrial pollution in the 1960s and early 1970s, and society quickly realized that we must take a more active stance in terms of pollution management and prevention. In the years since, governments and industry have learned to work together (albeit awkwardly at times), monitoring industrial effluents and limiting new releases of toxic materials into the environment. Remediation methods have been developed to repair some of the damage that previously took place. We are learning. We have a responsibility, both to future generations, and to our global neighbors, to share these insights — both regulatory and remediatory.

The last ten to 15 years have seen a remarkable explosion of research in the design and synthesis of nanostructured materials–nanoparticles, nanotubes, nanorods, etc. Early work largely focused on making different shapes, or different sizes; then work started to focus on making a variety of compositions, and multicomponent materials. Tailoring the composition or interface of a nanomaterial is a key step in making it *functional*. This book is concerned with functional nanomaterials–materials containing specific, predictable nanostructure whose chemical composition, or interfacial structure enable them to perform a specific job — destroy, sequester or detect some material that constitutes an environmental threat. Nanomaterials have a number of features that make them ideally suited for this job — high surface area, high reactivity, easy dispersability, rapid diffusion, etc. The purpose of this book is to showcase how these features can be tailored to address some of the environmental remediation and sensing/detection problems faced by mankind today. A number of leading researchers have contributed to this volume, painting a picture of diverse synthetic strategies, structures, materials and methods. The intent of this book is to showcase the current state of environmental nanomaterials in such a way as to be useful either as a research resource, or as a graduate level textbook. We have organized this book into sections on nanoparticle-based remediation strategies, nanostructured inorganic materials (e.g. layered materials like the apatites), nanostructured organic/inorganic hybrid materials, and the use of nanomaterials to enhance the performance of sensors. The materials and methods described herein offer exciting new possibilities in the remediation and/or detection of chemical warfare agents, dense non-aqueous phase liquids (DNAPLs), heavy metals, radionuclides, biological threats, CO_2, CO and more. The chemistries captured by these authors form a rich and colorful tapestry. We hope the final result is both valuable and enjoyable to the reader.

Glen Fryxell, Ph.D.
Pacific Northwest National Laboratory, USA

Guozhong Cao, Ph.D.
University of Washington, USA

March 2006

CONTENTS

Nanoparticle Based Approaches

Chapter 1

Nanoparticle Metal Oxides for Chlorocarbon and Organophosphonate Remediation

Olga B. Koper*, Shyamala Rajagopalan*, Slawomir Winecki*
and Kenneth J. Klabunde*,†

*NanoScale Corporation, Inc., Manhattan, KS, USA

and

†Kansas State University, Manhattan, KS, USA

Introduction

The nanotechnology revolution affects many areas of science including chemistry and chemical engineering. Although nanotechnology advances in electronic or in manufacturing of "nano machines" are rather recent developments, nanoscale chemical structures are much older. Materials known and widely used over the past several decades, such as high surface area carbons, porous inorganic metal oxides, and highly dispersed supported catalysts all fall into the category of nanostructures. Although a few decades-old publications describing these materials rarely used the word "nano", they deserve proper credit for providing the foundation of current advances in nano-chemistry.

Since approximately the 1970s, enormous advances in the synthesis, characterization, and understanding of high surface area materials have taken place and this period can be justifiably described as the nanoscale revolution. Development and commercial use of methods like sol-gel synthesis, chemical vapor deposition, and laser induced sputtering allowed for manufacturing of countless new nanomaterials both at laboratory- and commercial-scale. Traditional chemical manufacturing methods, used in catalyst synthesis or in production of porous sorbents, were gradually improved by systematic development of methodologies, as well as quality control measures. In addition, characterization of nanostructures was revolutionized by an availability of new instruments. Various high-resolution

3

microscopic tools were developed and became available commercially. For the first time, pictures of nano-objects became easily obtainable. Today, they appear not only in scientific literature but also in popular magazines and textbooks. Measurement of specific surface area, once a tedious and lengthy laboratory procedure, is done today routinely using automatic yet affordable instruments. Numerous companies develop, manufacture and sell scientific equipment specifically for nano-research, which resulted in more instruments being available and at reduced costs. The understanding of nanostructures grows at a phenomenal rate due to efforts of countless research groups, academic institutions and commercial entities. Since approximately 1990, nano-science has been recognized as a distinct field of study with tremendous potential. As a result, nanotechnology research has proliferated, scientific journals dedicated to nano-science were created, nanotechnology research centers were established, and universities have opened faculty positions related to nano-research. These developments were, and continue to be, heavily supported by the federal and local governments in the form of research grants and other funding opportunities. Private organizations, including all major chemical manufacturers, started robust nanotechnology programs and investments.

This chapter is intended as a brief description of the technology and products developed in the laboratories of one of the authors (Kenneth J. Klabunde) at Kansas State University, Manhattan, KS; and by NanoScale® Corporation, Manhattan, KS. The description will start with a reminder of a few technical concepts related to all nanoscale materials intended for chemical uses.

Nano-materials have large specific surface areas and a large fraction of atoms are available for chemical reaction. Figure 1 presents a high-resolution transmission electron microscopy (TEM) image of aerogel-prepared nanocrystalline MgO. The image demonstrates the nano-morphology of this material with rectangular crystals, 2–4 nm in size, a length-scale that is only an order of magnitude larger than distances between atoms of this oxide. It is apparent that significant fractions of atoms are located on crystallite surfaces as well as on edges and corners. These atoms/ions are partially unsaturated (coordination less than 6) and can readily interact with chemical species that come in contact with the nanocrystallites. Such interaction can be a physical adsorption, if caused by van der Waals attraction, or a chemical reaction if chemical bonds are formed or modified. Simple estimations presented in Figure 2 demonstrate that reduction of size of nanocrystallites results in a sharp increase of specific surface areas, as well as fractions of atoms located on surfaces, edges and corners. For instance, a material with 2 nm crystallites has more than half of the atoms located on their surface, approximately 10% of atoms/ions

Figure 1. High-resolution Transmission Electron Microscopy (TEM) image of nanocrystalline MgO.

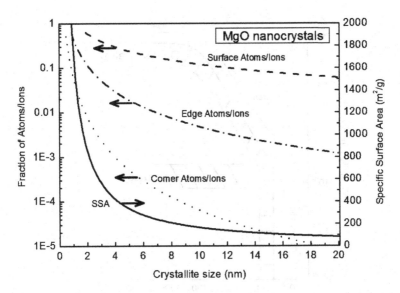

Figure 2. Specific surface area and fractions of atoms/ions located on surface, edges and corners of nanocrystalline magnesium oxide.

are on edges, 1% of atoms/ions are placed on corners, and with specific surface area approaching $1,000\,m^2/g$. It is important to recognize that the trends demonstrated in Figure 2 apply to all porous nanocrystalline and amorphous solids; and are the basis of the unique reactivity of nanocrystalline materials in chemical applications. Amorphous materials can be viewed as a limiting case of nanocrystalline materials with the crystallite size approaching inter-atomic distances.

One of the major important features of nanocrystalline materials is related to the size of pores between crystallites. It is apparent from Figure 1 that most pores between crystallites are similar in size as the crystallites themselves; therefore, nanocrystalline materials have a large fraction of pores below 10 nm that are traditionally described in literature as micropores. Fluids, and particularly gases, behave differently inside of micropores as compared to normal conditions. The first difference involves Knudsen diffusivity; conditions where collisions between fluid molecules are less frequent than collisions between fluid and pore-walls, and this is likely to dominate pores below 10–100 nm size-range, as shown in Figure 3. This mode of diffusivity is significantly slower than regular diffusivity for an unbound gas and will result in slower mass transfer within nanocrystalline materials. Gas diffusivities in micropores may be two- to four-orders of magnitude smaller

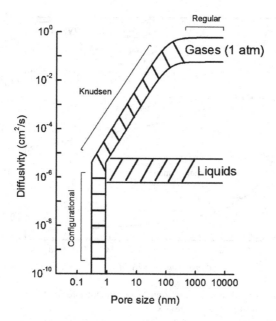

Figure 3.   Influence of the pore size on diffusivities of gases and liquids.

than diffusivities known for gases and may approach liquid diffusivities. The reduced diffusivities need to be taken into account in the engineering design. The second consequence of the micropores is the possibility of capillary condensation of vapors and gases and enhanced adsorption by nanocrystalline materials. The capillary condensation phenomenon is traditionally described by the Kelvin equation that relates reduction of the equilibrium vapor pressure inside of pores to pore-size and physical properties of a chemical. The reduction of the equilibrium vapor pressures for several common substances, predicted by the Kelvin equation, is shown in Figure 4. It is apparent that the equilibrium pressure reduction and capillary condensation effects become pronounced for micropores. On the other hand, nanocrystalline materials, when aggregated into microstructures or pellets, also possess numerous mesopores. Thus, both micropores and mesopores need to be considered in describing the behavior of this new class of materials (Table 1).

Physical properties of select nanocrystalline metal oxides (NanoActive® materials) manufactured by NanoScale are shown in Table 1. The surface areas range from 20 to over $600\,\text{m}^2/\text{g}$ for the NanoActive Plus metal oxides. Figure 5 shows the powder X-Ray Diffraction (XRD) spectra of commercial MgO, NanoActive MgO and NanoActive MgO Plus. The Plus material has the broadest peaks indicating the smallest crystallites. These nanometer-sized small crystallites, due to the high surface reactivity, aggregate into

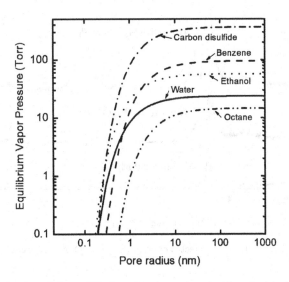

Figure 4. Reduction of the equilibrium vapor pressures inside cylindrical pores for several common substances as predicted by the Kelvin equation.

Table 1.   Physical properties of nanocrystalline metal oxides.

NanoActive Material	Powder Appearance	Surface Area $(m^2 g^{-1})$	Crystallite Size (nm)	Average Pore-Diameter (Å)	Mean Particle-Size (μm)
MgO Plus	White	≥ 600	≤ 4	30	12
MgO	White	≥ 230	≤ 8	50	3.3
CaO Plus	White	≥ 90	≤ 20	110	4
CaO	White	≥ 20	≤ 40	165	4
Al_2O_3 Plus	White	≥ 550	Amorphous	110	5
Al_2O_3	White	≥ 275	Amorphous	28	1.5
CuO	Black	≥ 65	≤ 8	85	6
ZnO	Off-White	≥ 70	≤ 10	170	4
TiO_2	White	≥ 500	Amorphous	32	5
CeO_2	Yellow	≥ 50	≤ 7	70	9.5

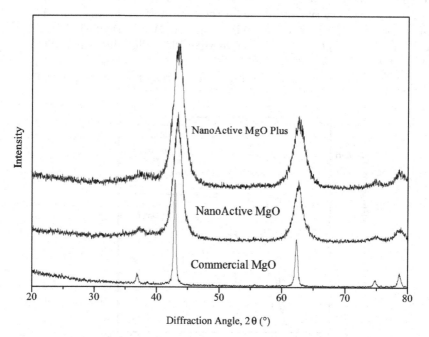

Figure 5.   Powder X-ray Diffraction spectra of commercial MgO, NanoActive MgO, and NanoActive MgO Plus.

larger particles (micron size) that preserve the high chemical reactivity, yet are much easier to handle.

Environmental Applications of NanoActive Materials

There are many areas where nanocrystalline materials can be used, but the environmental applications of these materials are particularly significant. Air and water pollution continues to be a challenge in all parts of the World. In many instances, the pollution control technology is limited by poor performance and/or high costs of existing sorbent materials; and considerable research activity is directed towards development of new sorbents. The following list gives a few examples of such cases:

- Control of chemical spills and accidental releases of harmful chemicals at various locations and settings: Accidental chemical releases are common occurrences in laboratories, industrial locations and all places where chemicals are used. Spills and accidental releases often result in environmental contaminations. NanoScale developed a specialized, safe sorbent formulation, FAST-ACT® (First Applied Sorbent Treatment — Against Chemical Threats), a powder capable of safe spill treatment and environmentally responsible disposal.
- Protection from chemical hazards: Nanocrystalline metal oxides can be incorporated into protective textiles or skin-creams to provide additional protection from chemical and biological hazards. The toxic agents are destroyed on nanoparticles, eliminating the threat of off-gassing and secondary contamination.
- Indoor air quality-control in buildings and vehicles: Currently, various air filtration technologies are available commercially; including particulate filters, electrostatic precipitators, and activated-carbon filters. There remains a large group of high volatility chemicals that cannot be effectively controlled by these approaches. Attempts to develop more effective sorbents or catalysts for these pollutants are on-going.
- Removal of elemental and oxidized mercury from combustion gases generated by electrical utilities using coal as a source of energy: Recent regulations enacted by the U.S. Environmental Protection Agency (EPA) spurred widespread development of new mercury sorbents, including activated-carbons and metal oxides.
- Control of arsenic, perchlorate, and methyl-*t*-butyl ether (MTBE) in drinking water: The presence of these compounds in water causes serious health concerns; and has resulted in mandatory maximum contaminant levels (MCL) for arsenic and may trigger new EPA regulations.

Development of new sorbents for economical and efficient removal of these pollutants is an active area of research.

The above applications, as well as numerous others, can benefit from the use of nanocrystalline sorbents owing to enhancement in reaction kinetics, increased removal capacities, and permanent destruction of harmful chemicals.

Nanocrystalline materials exhibit a wide array of unusual properties, and can be considered as new materials that bridge molecular and condensed matter.[1] One of the unusual features is enhanced surface chemical reactivity (normalized for surface area) toward incoming adsorbates.[2] For example, nanocrystalline MgO, CaO, TiO_2 and Al_2O_3 adsorb polar organics such as aldehydes and ketones in very high capacities, and substantially outperform the activated-carbon samples that are normally utilized for such purposes.[3] Many years of research at Kansas State University, and later at NanoScale, have clearly established the destructive adsorption capability of nanoparticles towards many hazardous substances, including chlorocarbons, acid gases, common air-pollutants, dimethyl methylphosphonate (DMMP), paraoxon, 2-chloroethylethyl sulfide (2-CEES), and even military agents such as GD, VX, and HD.[4–9] The enhanced chemical reactivity suggests a two-step decomposition mechanism of the adsorbates on nanoparticles (first step — adsorption of toxic agent on the surface by means of physisorption, followed by the second step — chemical decomposition). This two-step mechanism substantially enhances the detoxification abilities of nanoparticles because it makes the decomposition less dependent on the rate (speed) of chemical reaction. The rate of chemical reaction depends on the agent-nanoparticle combination; therefore, for some agents the rate may be quite low. In addition, the reaction rate strongly decreases at lower temperatures. For these reasons, any detoxification method that relies only on chemical reactivity would not work for many toxic agents and would not be effective at low temperatures. Reactive nanoparticles do not have this drawback because the surface adsorption sites remain active even at very low temperatures. In fact, the physisorption of the potential toxic agents is enhanced at low temperatures. In this way, the toxins are trapped and eventually undergo "destructive adsorption."

Destructive Adsorption of Hazardous Chemicals by Nanocrystalline Metal Oxides

A wide variety of chlorinated compounds such as cleaning-solvents, plasticizers, lubricants, and refrigerants are used by society in many beneficial functions. While some of these chlorinated compounds are being replaced by less harmful chemicals, many continue to be used because of the lack of suitable replacements or as a result of economic considerations. Therefore, considerable interest exists in developing methods for the safe disposal of chlorinated and other problematic wastes. It has been shown that nanocrystalline metal oxides are particularly effective decontaminants for several classes of environmentally problematic compounds at elevated temperatures; enabling complete destruction of these compounds at considerably lower temperatures than that required for incineration.[10]

Koper *et al.*,[11-15] have examined the reaction of aerogel-prepared CaO with carbon tetrachloride and other chlorinated hydrocarbons. The primary carbon-containing product was CO_2, with the CaO converted to $CaCl_2$. Conventionally-prepared CaO, however, produced little CO_2. A three-step mechanism was proposed. Initially, CO_2 and $CaCl_2$ are produced from metathesis of CaO and CCl_4; CO_2 then combines with CaO in a second step to yield $CaCO_3$; finally, metathesis of $CaCO_3$ and CCl_4 generates $CaCl_2$ and CO_2. Phosgene, $COCl_2$, is an intermediate, which reacts with the remaining CaO to form calcium chloride and carbon dioxide. Low-temperature infrared studies of CCl_4 monolayers on CaO demonstrated that CCl_x intermediates begin to form at temperatures above 113 K; at 200 K, CCl_4 and CCl_x are no longer observed.[13] Although the reactions of chlorocarbons with metal oxides are often thermodynamically favorable, liquid-solid or gas-solid reactions are involved only on the surface of the metal oxide. Kinetic parameters involving ion (Cl^-/O^{2-}) migration inhibit complete reaction. Therefore, the use of ultrahigh surface area metal oxides allows reasonably high capacities for such chlorocarbon destructive adsorption processes[16]: Li *et al.*,[17,18] have studied the reactions of aerogel-prepared CaO and MgO with chlorinated aromatics at 500–900°C. Upon destruction of chlorobenzene, a biphenyl was formed as the reaction by-product. Destruction of chlorinated aromatics occurred to a much greater extent, and at significantly lower temperatures, relative to destruction in the absence of the nanoscale metal oxides. Trace toxins such as dibenzo-*p*-dioxins were not observed under any conditions when aerogel-prepared MgO was used, nor when aerogel-prepared CaO was used in the presence of oxygen in the carrier gas. However, when low surface area CaO was used in the presence of oxygen, small amounts of dibenzo-*p*-dioxins were produced. Hooker and Klabunde[19] have studied the reaction of iron(III) oxide with carbon tetrachloride in a fixed-bed pulse reactor, from 400 to 620°C. The main carbon-containing product was CO_2; other carbon-containing products included C_2Cl_4, and graphite. It was found that the extent of reaction was greater than what could be accounted for by reaction only on surfaces of Fe_2O_3; this suggested that Fe_2O_3 on the surface was regenerated. The apparent regeneration of Fe_2O_3 on the surface of the particles suggested that coating other metal oxide nanoparticles with iron(III) oxide might enable more complete utilization of the core metal oxide. Accordingly, aerogel-prepared MgO overlaid with Fe_2O_3 (designated as $[Fe_2O_3]MgO$) was prepared, and its reactions with carbon tetrachloride were examined by Klabunde et al.,[20] Khaleel and Klabunde,[21] and Kim et al.[22] It was found that when $[Fe_2O_3]MgO$ reacted with CCl_4, nearly all of the metal oxides (MgO) were consumed, in contrast, Fe_2O_3 treatment of conventionally-prepared MgO did not give this enhancement. Since the iron oxide coating is regenerated by the core MgO, the Fe_2O_3 coating can be considered to be catalytic. The scheme below indicates the reaction of carbon tetrachloride with nanocrystalline MgO and iron-coated nanocrystalline MgO.

Further work, using X-ray Photoelectron Spectroscopy (XPS), showed that the best catalysts were those where Fe_2O_3 was coated on the surface of the MgO,[23] and was highly dispersed (as shown by Mossbauer spectroscopy[24] and XRD). Simple mixtures of Fe_2O_3 with MgO were not as effective and solid state mixtures of Fe_2O_3-MgO were also less effective. Indeed, a layered $[Fe_2O_3]MgO$ structure worked best. Figure 6 shows the destructive adsorption of carbon tetrachloride on two forms of nanocrystalline CaO coated with iron oxide and vanadium oxide. Again, the same trends were observed, where the sol-gel-based form of the CaO exhibited higher chemical reactivity as compared to the nanocrystalline CaO (nano), whose reactivity was much higher than the commercially available CaO. Furthermore, a coating of the transition metal oxide imparted additional reactivity driving the reaction closer to its stoichiometric limit.

Other metal oxides were also found to be effective coatings for enhancing the utilization of aerogel-prepared MgO. Indeed, Fe_2O_3 enhanced the catalytic properties of both MgO and SrO.[25,26] In an attempt to gain more detailed information on the chemical state of the $Fe_2O_3/FeCl_x$ before, during, and after CCl_4 reaction, a series of $[Fe_2O_3]SrO$ nanocrystals and microcrystals were studied by Jiang et al., using Extended X-ray Absorption Fine-Structure Spectroscopy (EXAFS).[27] Strontium was chosen as the base oxide because of available synchrotron energies. The results of these experiments were quite surprising. First of all, the data showed that SrO

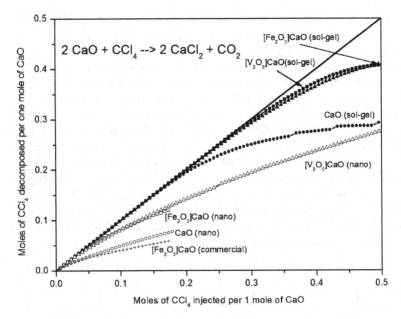

Figure 6. Decomposition ability of CaO and $[V_2O_5]$CaO towards CCl_4. In a stoichiometric reaction 1 model of CaO can decompose 0.5 moles of CCl_4.

itself was more reactive than CaO, which is known to be more reactive than MgO (per unit surface area, although MgO can be prepared in the highest surface area). Also, it was found that the Cl^-/O^{2-} exchange was extremely facile, and even after near stoichiometric exposure to CCl_4, Fe_2O_3 remained as Fe_2O_3. In other words, if any SrO was still available, it readily gave up its O^{2-} to $FeCl_x$ to reform Fe_2O_3.[27]

Besides CCl_4, other chlorocarbons have been examined. The decomposition of CCl_3F by several vanadium oxides- and vanadium oxide-coated aerogel-prepared MgO was studied by Martyanov and Klabunde.[28] V_2O_3 was consumed in the reaction, with some vanadium-halogen species being produced. Capture of the volatile vanadium-halogen species by MgO was thought to be responsible for the catalytic effect of the vanadium compounds on MgO. Furthermore, Fe_2O_3 exchange catalysis was found to work well for 1,3-dichlorobenzene, but not for trichloroethylene. Obviously, the catalytic effect of transition metal oxides depends a great deal on the intimate mechanistic details that are, of course, different for each chlorocarbon under study. Interestingly, this type of catalytic action is not restricted to O^{2-}/Cl^- exchange, but also operates in O^{2-}/SO_x^{n-}, and O^{2-}/PO_x^{n-} systems as well.[24]

Chemical decomposition of dimethyl methyl phosphonate (DMMP), trimethyl phosphate (TMP), and triethyl phosphate (TEP) on MgO was studied by Li *et al.*[29-31] The agents were allowed to adsorb on the MgO samples both in vacuum environment and in a helium stream. Substantial amounts of strongly chemisorbed agents were observed at room and at elevated temperatures (500°C). At low temperatures, the main volatile reaction products were formic acid, water, alcohols, and alkenes. At higher temperatures CO, CH_4, and water predominated. Phosphorous-containing products remained immobilized at all temperatures. Interestingly, addition of water enhanced decomposition abilities of nanocrystalline MgO.

Detailed studies using infrared photoacoustic spectroscopy and isotope-labeling confirmed that the chemisorption occurs through the P=O bond destruction. For instance, nanocrystalline MgO is able to hydrogen bond with DMMP at room temperature; with hydrolysis of DMMP occurring to produce surface bound species as shown below.

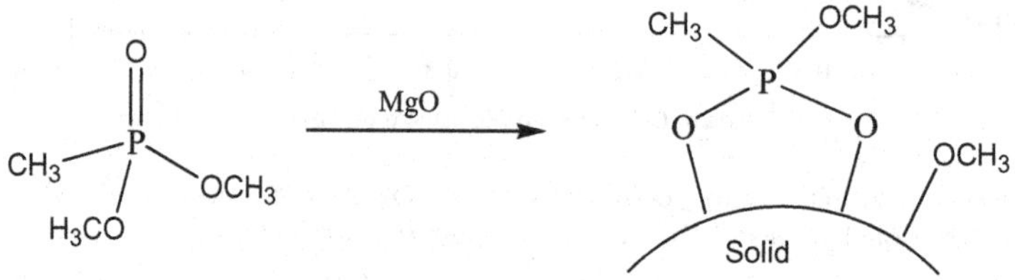

Further studies on the decomposition of phosphonate esters $(RO)_2IP{=}O$ identified two important reactions, nucleophilic substitution at P and nucleophilic substitution at the alkyl carbon of an alkoxy group.[32] Treatment of DMMP with MgO yielded formic acid (HCOOH) as the major volatile product; other volatile products included methanol (CH_3OH), dimethyl ether (CH_3OCH_3), and ethane (CH_3CH_3). Carbon dioxide, carbon monoxide, water, hydrogen, and phosphoric acid were not observed as products. All phosphorus-containing products were immobilized on the MgO surface; elemental analysis was consistent with O_{solid}-Mg-O-P(OMe)(Me)-O-Mg-O_{solid}, with an O_{solid} between the two Mg atoms. For this structure, the PO bond order of the O-P-O linkage is 1.5, i.e. the two P-O bonds share one double bond.[33] Proton abstraction of a β-hydrogen of an ethoxy group can lead to ethylene production, which is sometimes observed in reactions of ethyl esters.[34] Reactivity of nanocrystalline MgO towards a range of organophosphates $(RO)_3P{=}O$, organophosphites $(RO)_3P$, and organophosphines R_3P was studied by Lin *et al.*[34] Phosphorous compounds were allowed to adsorb on thermally activated MgO at room temperature and at elevated temperatures reaching 175°C. Most of phosphorous compounds were adsorbed and chemically decomposed in large quantities, and in some cases the reactions were essentially stoichiometric. In the most favorable cases, approximately one phosphorus molecule was decomposed for every two MgO molecules present in the bulk. Infrared studies performed on spent sorbents indicated that phosphates adsorb very strongly through the P=O bond (P=O bond is destroyed upon adsorption) accompanied by net electron loss from the RO groups. Phosphites adsorb through the phosphorous atom with a net electron density gain by RO groups. Phosphines adsorb less strongly through the phosphorous atom. Decomposition of the phosphorous compounds yielded volatile hydrocarbons, ethers, and alcohols. Products containing phosphorous were strongly retained by the MgO surface which prevented product desorption and release into the gas phase:

$$3DMMP + 6MgO \rightarrow CH_3OH + H_2O + 2HCOOH + [OH]_{adsorbed}$$
$$+ 3[-P(OCH_3)CH_3-]_{adsorbed}.$$

As another example of the extremely high reactivity of nanocrystalline metal oxides, ambient temperature destructive adsorption of paraoxon is illustrated. Paraoxon is an insecticide and is a suitable simulant for nerve warfare agents. Figure 7 illustrates the rate of removal of paraoxon from a pentane solution over time employing various adsorbents. The removal was monitored using UV-Vis spectroscopy and reduction of the paraoxon (265–270 nm) peak was observed over time. Nanocrystalline magnesium oxide nanoparticles achieved complete adsorption within two-minutes of

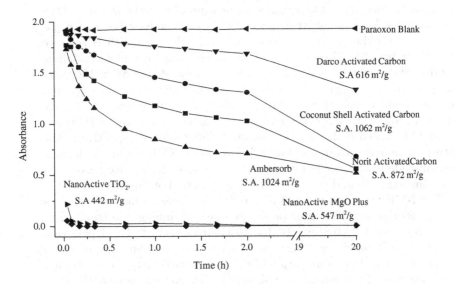

Figure 7. Rate of Adsorption of Paraoxon by various samples.

exposure to paraoxon while all the activated carbon and ion-exchange-resin (IER) samples were significantly less adsorptive and unable to adsorb paraoxon completely, even after 20 hours of exposure. In addition, with activated-carbons the agent was merely absorbed, whereas with nanoparticles it was first adsorbed onto the metal oxide surface and then decomposed into phosphate and p-nitrophenol:

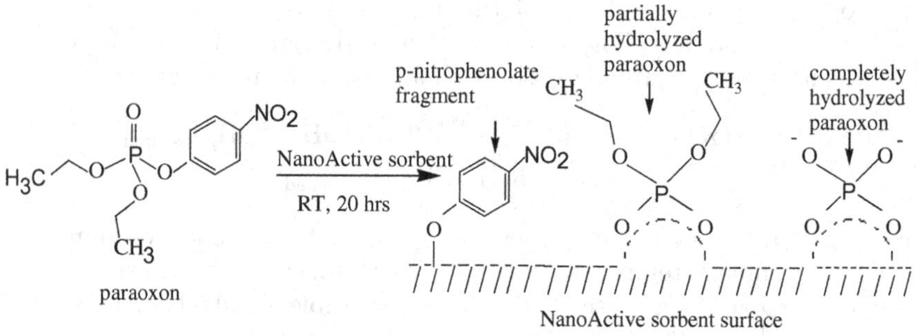

Adsorption and hydrolysis of paraoxon is readily visible on magnesium oxide nanoparticles due to the change in color of the powder from white to bright yellow. Note that paraoxon is a light yellow, oily substance and its color does not duplicate the bright yellow observed after contact with the nanoparticles. The anion $O_2NC_6H_4O^-$ is bright yellow, and the rapid color formation clearly shows that the anion is formed quickly on the surface of the nanoparticles. More definitive evidence for the destructive adsorptive capability of metal oxide nanoparticles comes from Nuclear Magnetic Resonance (NMR) studies. ^{31}P NMR spectra of intimately mixed dry nanocrystalline metal oxide/paraoxon mixtures after 20 hours are displayed in Figure 8. Paraoxon in deuterochloroform solvent exhibits a signal around -6.5 ppm and the product derived via complete hydrolysis of paraoxon, namely, the phosphate ion (PO_4^{3-}), is expected to show a signal around 0 ppm. However, it should be noted that the exact chemical shift values are sensitive to the nature of the metal oxide employed. Results from NMR studies with magnesium oxide nanoparticles over time indicate that destructive adsorption starts immediately and continues over a long period of time.

The NMR spectrum of MgO nanoparticles/paraoxon mixture contains at least four peaks. The outer peaks are due to spinning side bands. The major peak at $\delta = 0.8$ ppm is due to completely hydrolyzed paraoxon and the peak around $\delta = -8.2$ ppm is due to adsorbed but unhydrolyzed paraoxon. Upon close examination of the NanoActive TiO_2/paraoxon mixture spectrum, three signals were observed; a sharp peak around -8.4 ppm attributed to unhydrolyzed paraoxon, a broad peak around -5 ppm due

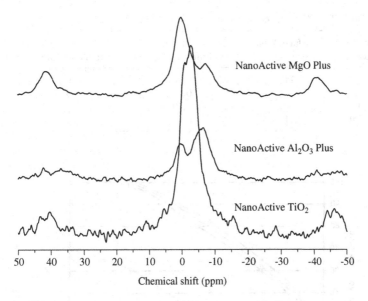

Figure 8. ^{31}P NMR spectra of nanocrystalline metal oxide/Paraoxon Mixture after 20-hours.

to partially hydrolyzed paraoxon, and a shoulder peak between 0 and −2 ppm was attributed to completely hydrolyzed paraoxon. The spectrum of NanoActive Aluminum Oxide Plus/paraoxon mixture displayed two major peaks. The upfield peak at $\delta = -6.7$ ppm is attributed to partially hydrolyzed paraoxon, and the peak at 0 ppm is due to completely hydrolyzed paraoxon. In short, adsorption of paraoxon by nanoparticles occurs instantly with the destruction of the agent as an ongoing active process.

Nanocrystalline materials can be utilized in a powdered or a compacted/granulated form. Compaction of the nano crystals into pellets does not significantly degrade surface area or surface reactivity when moderate pressures are employed, ensuring that these nano structured materials can be utilized as very fine powders or as porous, reactive pellets.[35] As shown in Table 2, pressure can be used to control pore structure. It should be noted that below about 5,000 psi the pore structure remains relatively unchanged, and these pellets behave in adsorption processes essentially identical to the loose powders. For example, both NanoScale MgO and Al$_2$O$_3$ (Figure 9) vigorously adsorb acetaldehyde with much higher rates and capacities than activated-carbon. This is the case whether the samples are loose powders or compressed pellets. As noted in Table 2, only when very high compression pressure was used, the adsorption process was hindered.

Table 2.   Effects of compaction pressure on nanocrystalline MgO.

Load (psi)	Surface Area (m^2g^{-1})	Total Pore Volume (cm^3g^{-1})	Average Pore Diameter (nm)
0	443	0.76	6.9
5,000	434	0.57	5.3
10,000	376	0.40	4.2
20,000	249	0.17	2.7

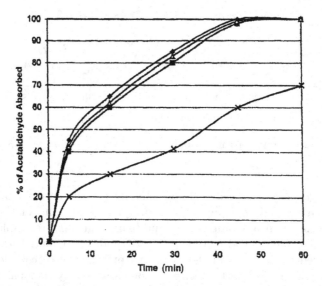

Figure 9.   Rate of adsorption of acetaldehyde by nanocrystalline Al_2O_3 in powder or pellet form. The following pressures were used to compact and form the pellets at room temperature: ◆ = powder, ■ = 5,000 psi, △ = 10,000 psi, X = 20,000 psi.

The application of nanocrystalline materials as destructive adsorbents for acid gases such as HCl, HBr, CO_2, H_2S, NO_X, and SO_X has been investigated by Klabunde,[36] Stark,[37] and Carnes.[38] These materials are much more efficient than commercially-available oxides. This is seen with nanocrystalline ZnO, which reacts in a stoichiometric molar ratio of 1 mol of hydrogen sulfide to 2.4 mol of ZnO, whereas the commercial ZnO reacts in a molar ratio of 1 mol hydrogen sulfide to 32 mol ZnO. This indicates that nanoparticles are chemically more reactive at room temperatures than their commercial counterparts. For example, nanocrystalline ZnO stays at near stoichiometric ratio at room temperature, whereas the commercial ZnO does not.[39]

In addition to high-surface-area pure-metal-oxide nanoparticles, the development of intermingled metal oxide nanoparticles has been reported which yield special advantages.[40] The adsorption properties of intermingled metal oxide nanoparticles were found to be more superior than, when comparing their reactivity to, that of individual metal oxide nanoparticles and physical mixtures. For example, the nanoparticles of Al_2O_3/MgO mixed product have enhanced chemical reactivity over pure metal oxides of Al_2O_3 or MgO for the adsorption of SO_2. In comparison, the intermingled Al_2O_3/MgO showed an adsorption capacity of 6.8 molecules for SO_2 (of SO_2/nm^2 adsorbed), while the nanoparticles of Al_2O_3 and MgO was observed at 3.5 molecules and 6.0 molecules, respectively.[41]

Destructive Adsorption of Chemical Warfare Agents (CWAs) by Nanocrystalline Metal Oxides

Nanocrystalline metal oxides not only neutralize toxic industrial chemicals, but also destroy chemical warfare agents, V-, G- and H-series, through hydrolysis and/or dehydrohalogenation. The G-series tend to be volatile and highly toxic by inhalation, while the V-agents are relatively non-volatile, persistent and highly toxic by the percutaneous route. HD is an acronym for mustard gas and it belongs to the vesicant class of chemical warfare agents.

Wagner *et al.*,[7–9] reported that NanoScale metal oxide nanoparticles are very effective in destructive adsorption of VX, GD and HD. These studies were done at room temperature using the pure agent on a column of dry metal oxide nanoparticles. It was found that the products formed in reactions with HD were the less-toxic thioglycol (TG) and divinyl (DVHD) compounds while the nerve agents afforded the surface bound hydrolyzed species. Destruction of nerve agents (GD and VZ) and the blistering agent (HD) was studied using solid-state NMR, as well as extraction, followed by GC-MS analysis. The destruction products of GD are pinacolyl methylphosphonic acid (GD-acid) that converts to surface-bound methylphosphonic acid (MPA), and HF, which is acidic and is neutralized by the metal oxide surface as well.

| GD | GD-acid | MPA |

The phosphonate destruction product of VX is ethyl methylphosphonic acid (EMPA) which converts into surface-bound methylphosphonic acid (MPA). It should be noted that during this reaction the toxic EA-2192 (S-(2-diisoppropyloamino)ethyl methylphosphonothioate) does not form, in contrast to basic VX hydrolysis in solution:

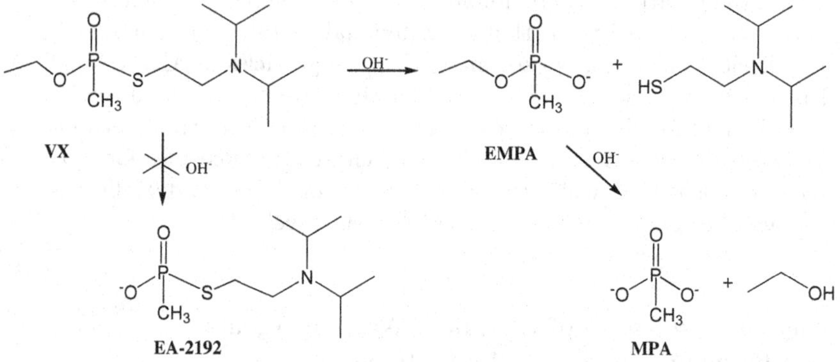

The decomposition product of HD is promoted by nucleophilic substitution on the β-carbon of a 2-chloroethyl group, resulting in a 2-hydroxyethyl group – thiodiglycol (TG), whereas decomposition products of HD *via* elimination are 2-chloroethyl-vinyl sulfide (CEVS) and divinyl sulfide (DVS). Both processes convert mustard to considerably less toxic compounds:

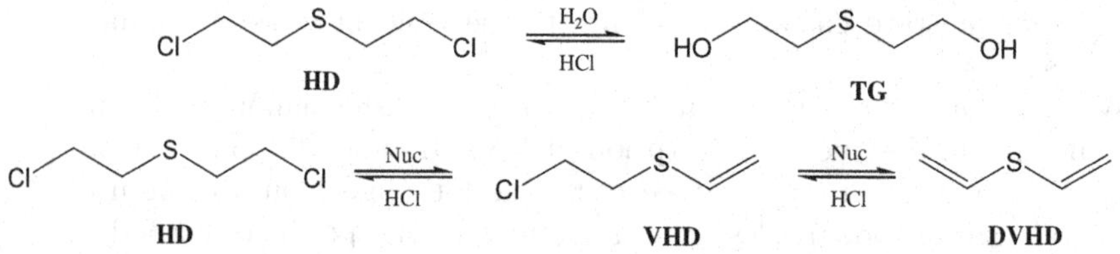

The off-gassing experiments for reactive nanoparticles after exposure to HD, GD and VX indicated, through the absence of any major chromatographable peaks, that the reaction products are either low molecular weight (less than 60 amu), small gaseous materials that were eluted with the air peaks, or compounds that are tightly bound to the nano materials.

NanoScale has developed a nanocrystalline metal-oxide-based product, FAST-ACT, that is utilized as a chemical hazard containment and neutralization system. This product can be applied manually as a dry powder for treating liquid spills, or in an aerosol form for treating toxic vapors, decontamination of vertical surfaces, or contaminated suits (Figure 10). Nanocrystalline metal oxides have low bulk density and remain suspended in the air for a prolonged period of time.

Figure 10.   Decontamination of a First Responder's suit utilizing FAST-ACT.

The effectiveness of FAST-ACT has been verified against chemical warfare agents GD (Soman), VX and HD (Distilled Mustard) by two independent laboratories: Battelle Memorial Institute (Battelle), Columbus, OH; and the United States Soldier and Biological Chemical Command (SBCCOM), Edgewood, MD. All studies were conducted using controlled protocols at ambient room-conditions (temperature, pressure and humidity). The agent was placed on a glass surface and FAST-ACT was applied. Within 90 seconds FAST-ACT was validated to remove over 99.9% of HD and GD; and over 99.6% (detection limit) of VX from the surface (as determined by surface extraction followed by GC-MS analysis). Upon contact with FAST-ACT, the agent was quickly adsorbed and then destroyed. The destruction was confirmed by changes in the NMR spectra (SBCCOM) and by inability to extract the agent from the powder (Battelle). In 10 minutes 99% of GD and over 99.9% of VX is destroyed, while in about 60 minutes 70–80% of HD is destroyed.

Safety of NanoActive Materials

Considering the high chemical reactivity of nanocrystalline metal oxides, a question to pose is, how safe are humans or animals if exposed to these materials? NanoScale has conducted rigorous toxicity testings at independent laboratories studying the oral, dermal, pulmonary and ocular effects of NanoActive metal oxides. The testings have revealed that there are no safety-hazards associated with the nano nature of these materials. Dermal LD_{50} (rabbit) was $> 2\,g/kg$ and oral LD_{50} (rabbit) was $> 5\,g/kg$. The formulation has also been tested for inhalation-toxicity and proven to be non-toxic to rats.

In addition, it was determined that in excess of 99.9% of the particles are captured by standard NIOSH particle filters (Flat media, Model 200 Series, N95 NIOSH, 30981J and Pleated filter with flat media, NIOSH Pro-Tech respirator, PN G100H404 OVIP-100). The non-toxic behavior of these particles in the inhalation testings, as well as the high removal efficiency can be explained by the formation of weak aggregates of nanomaterials. The nanocrystalline metal oxides manufactured at NanoScale aggregate into micron-sized particles due to their high surface reactivity. Micron-sized particles do not penetrate into the alveoli, but are stopped in the bronchia. NanoScale's products utilized for environmental applications are made from inherently non-toxic materials (magnesium oxide, calcium oxide, titanium dioxide, aluminum oxide) and possess higher solubility than their non-nano counterparts. Therefore, even if introduced into a human or animal body, they will dissolve and be expelled. Overall, the nanocrystalline materials produced at NanoScale were found to be no more toxic than the respective commercially available metal oxides.[42]

Conclusions

Nanocrystalline metal oxides are highly effective adsorbents towards a broad range of environmental contaminants ranging from acids, chlorinated hydrocarbons, organophosphorus and organosulfur compounds to chemical warfare agents. These materials do not merely adsorb, but actually destroy many chemical hazards by converting them to much safer byproducts under a broad range of temperatures. Metal oxides produced by NanoScale were proven to be no more toxic than their non-nano commercial counterparts and continue to be a great choice for abating environmental pollutants.

References

1. L. Interrante and M. Hampden-Smith (eds.), *Chemistry of Advanced Materials*, Chap. 7 (Wiley-VCH, 1998), pp. 271–327.
2. K. J. Klabunde, J. V. Stark, O. Koper, C. Mohs, D. G. Park, S. Decker, Y. Jiang, I. Lagadic and D. Zhang, *J. Physical Chemistry B* **100**, 12142–12153 (1996).
3. A. Khaleel, P. N. Kapoor and K. J. Klabunde, *NanoStruct. Materials.* **11**, 459–468 (1999); E. Lucas and K. J. Klabunde, *NanoStruct. Materials.* **12**,

179–182 (1999); O. Koper and K. J. Klabunde, U. S. Patent 6, 057, 488; May 2 (2000).

4. O. Koper, K. J. Klabunde, L. S. Martin, K. B. Knappenberger, L. L. Hladky, Decker P. Shawn, U.S. Patent 6, 653, 519; November 25 (2003).

5. K. J. Klabunde, U.S. Patent 5,990,373; November 23 (1999).

6. S. Rajagopalan, O. Koper, S. Decker and K. J. Klabunde, *Chem. Eur. J.* **8**, 2602 (2002).

7. G. W. Wagner, P. W. Bartram, O. Koper and K. J. Klabunde, *J. Phys. Chem. B* **103**, 3225 (1999).

8. G. W. Wagner, O. B. Koper, E. Lucas, S. Decker and K. J. Klabunde, *J. Phys. Chem. B* **104**, 5118 (2000).

9. G. W. Wagner, L. R. Procell, R. J. O'Connor, S. Munavalli, C. L. Carnes, P. N. Kapoor and K. J. Klabunde, *J. Am Chem. Soc.* **123**, 1636 (2001).

10. S. Decker, J. Klabunde, A. Khaleel and K. J. Klabunde, *Enviro. Sci. Technol.* **36**, 762–768 (2002).

11. O. Koper, Y. X. Li and K. J. Klabunde, *Chem. Mater.* **5**, 500 (1993).

12. O. Koper and K. J. Klabunde, *Nanophase Materials*, eds. G. C. Hadjipanayis and R. W. Siegel (Kluwer Academic Publishers, 1994), pp. 789–792.

13. O. B. Koper, E. A. Wovchko, J. A. Glass Jr., J. T. Yates Jr. and K. J. Klabunde, *Langmuir* **11**, 2054–2059 (1995).

14. O. Koper, I. Lagadic and K. J. Klabunde, *Chem. Mater.* **9**, 838–848 (1997).

15. O. Koper and K. J. Klabunde, *Chem. Mater.* **9**, 2481–2485 (1997).

16. R. J. Hedge and M. A. Barteau, *J. Catal.* **120**, 387–400 (1989).

17. Y. X. Li, H. Li and K. J. Klabunde, *Nanophase Materials*, eds. G. C. Hadjipanayis and R. W. Siegel (Kluwer Academic Publishers, 1994), pp. 793–796.

18. Y. X. Li, H. Li and K. J. Klabunde, *Environ. Sci. Technol.* **28**, 1248–1253 (1994).

19. P. D. Hooker and K. J. Klabunde, *Environ. Sci. Technol.* **28**, 1243–1247 (1994).

20. K. J. Klabunde, A. Khaleel and D. Park, *High Temp. Mater. Sci.* **33**, 99–106 (1995).

21. A. Khaleel and K. J. Klabunde, *Nanophase Materials*, eds. G. C. Hadjipanayis and R. W. Siegel (Kluwer Academic Publishers, 1994), pp. 785–788.

22. H. J. Kim, J. Kang, D. G. Park, H. J. Kweon and K. J. Klabunde, *Bull. Korean Chem. Soc.* **18**, 831–840 (1997).

23. K. J. Klabunde, A. Khaleel and D. Park, *High Temperatures and Material Science*, **33**, 99–106 (1995).

24. S. Decker and K. J. Klabunde, *J. Amer. Chem. Soc.* **118**, 12465–12466 (1996).

25. Y. Jiang, S. Decker, C. Mohs and K. J. Klabunde, *J. Catal.* **180**, 24–35 (1998).

26. S. Decker, I. Lagadic, K. J. Klabunde, J. Moscovici and A. Michalowicz, *Chem. Mater.* **10**, 674–678 (1998).

27. Y. Jiang, S. Decker, C. Mohs and K. J. Klabunde, *J. Catalysis.* **180**, 24–35 (1998).

28. I. N. Martyanov and K. J. Klabunde, *J. Catal.* **224**, 340–346 (2004).

29. Y. X. Li and K. J. Klabunde, *Langmuir.* **7**, 1388–1393 (1991).
30. Y. X. Li, J. R. Schulp and K. J. Klabunde, *Langmuir.* **7**, 1394–1399 (1991).
31. Y. X. Li, O. Koper, M. Atteya and K. J. Klabunde, *Chem. Mater.* **4**, 323–330 (1992).
32. M. K. Templeton and W. H. Weinberg, *J. Am. Chem. Soc.* **107**, 774–779 (1985).
33. Y. X. Li and K. J. Klabunde, *Langmuir.* **7**, 1388–1393 (1991).
34. S. T. Lin and K. J. Klabunde, *Langmuir* **1**, 600–605 (1985).
35. E. Lucas, S. Decker, A. Khaleel, A. Seitz, S. Fultz, A. Ponce, W. Li, C. Carnes and K. J. Klabunde, *Chem. Eur. J.* **7**, 2505–2510 (2001).
36. K. J. Klabunde, J. V. Stark, O. B. Koper, C. Mohs, D. G. Park, S. Decker, Y. Jiang, I. Lagadic and D. Zhang, *J. Phys. Chem.* **100**, 12142 (1996).
37. J. V. Stark and K. J. Klabunde, *Chem. Mater.* **8**, 1913–1918 (1996).
38. C. L. Carnes, J. Stipp, K. J. Klabunde and J. Bonevich, *Langmuir.* **18**, 1352–1359 (2002).
39. C. L. Carnes and K. J. Klabunde, *Chem. Mater.* **14**, 1806 (2002).
40. G. M. Medine, V. Zaikovskii and K. J. Klabunde, *J. Mater. Chem.* **14**, 757–763 (2004).
41. C. L. Carnes, P. N. Kapoor, K. J. Klabunde and J. Bonevich, *Chem. Mater.* **14**, 2922–2929 (2002).
42. USCHPPM Reports: 85-XC-01BN-03; 85-XC-5302-01; 85-XC-5302-03; 85-XC-03GG-11-05-01-02c; 85-XC-03GG-05; 85-XC-03GG-10-02-09-03. (Publication in preparation)

Chapter 2

Nanoscale Zero-Valent Iron (nZVI) for Site Remediation

Daniel W. Elliott*, Hsing-Lung Lien† and Wei-xian Zhang*

**Department of Civil and Environmental Engineering
Lehigh University, Bethlehem, PA 18015, USA*

*†Department of Civil and Environmental Engineering
National University of Kaohsiung
Kaohsiung 811, Taiwan, ROC*

Introduction

Over the span of little more than a decade, the multi-disciplinary nanotechnology boom has inspired the creation and development of powerful new tools in the ongoing challenge of addressing the industrialized world's legacy of contaminated sites.[1] These might include improved analytical and remote-sensing methodologies, novel sorbents and pollution-control devices, as well as superior soil- and groundwater-remediation technologies. The nanoscale zero-valent iron (nZVI) technology described in this chapter represents an important, early stage achievement of the burgeoning environmental nanotechnology movement.

Since 1996, our research group at Lehigh University has been actively engaged in the developing new nanometal materials, improving the synthetic schemes, performing both bench-scale and field-scale assessments, and extending the technology to increasing numbers of amenable contaminant classes. We have tested the nZVI technology with more than 75 different environmental contaminants from a wide variety of chemical classes. These include chlorinated aliphatic hydrocarbons (CAHs), nitroaromatics, polychlorinated biphenyls (PCBs), chlorinated pesticides like the lindane and DDT, hexavalent chromium, and perchlorate.

The nZVI technology may prove to be useful for a wide array of environmental applications including providing much-needed flexibility for both *in situ* and *ex situ* applications. Successful direct *in situ* injection of nZVI particles, whether under gravity-fed or pressurized conditions, to remediate chlorinated hydrocarbon-contaminated groundwater has already been demonstrated.[2,3] In addition, nZVI particles can be deployed in slurry reactors for the treatment of contaminated soils, sediments, and solid wastes; and can be anchored onto a solid matrix such as activated-carbon and/or zeolite for enhanced treatment of water, wastewater, or gaseous process streams (Figure 1).

In this chapter, an overview of the nZVI technology is provided beginning with a description of the process fundamentals and applicable kinetics models. This is followed by a discussion of the synthetic schemes for the nZVI types developed at Lehigh University. Next, a summary of the major research findings is provided, highlighting the key characteristics and remediation-related advantages of the nZVI technology versus the granular/microscale ZVI technology. A discussion of fundamental issues related to the potential applications of the nZVI technology and economic hurdles facing this technology is also included.

Overview and ZVI General Process Description

Zero-valent iron (Fe^0) has long been recognized as an excellent electron donor, regardless of its particle size. Even to the non-scientifically inclined, people have observed the familiar corrosion (oxidation) of iron-based structures, implements, and art objects for millennia. In the rusting process, Fe^0 is oxidized to various ferrous iron (Fe^{+2}) and ferric iron (Fe^{3+}) salts (i.e. various oxidation products collectively known as rust) while atmospheric oxygen is reduced to water. Interestingly, while iron corrosion has been observed and known for millennia, the potential application of the iron corrosion process to environmental remediation remained largely unknown until the mid-1980s.[4]

Applications of ZVI to Environmental Remediation

Gillham and co-workers serendipitously observed that chlorinated hydrocarbon solvents in contaminated groundwater samples were unstable in the presence of certain steel and iron-based well casing materials.[5] This discovery triggered enormous interest in the potential applications of relatively inexpensive and essentially non-toxic iron materials in the remediation of contaminated groundwaters. Since that time, exhaustive research on various aspects of the topic has been undertaken by numerous research groups around the world.

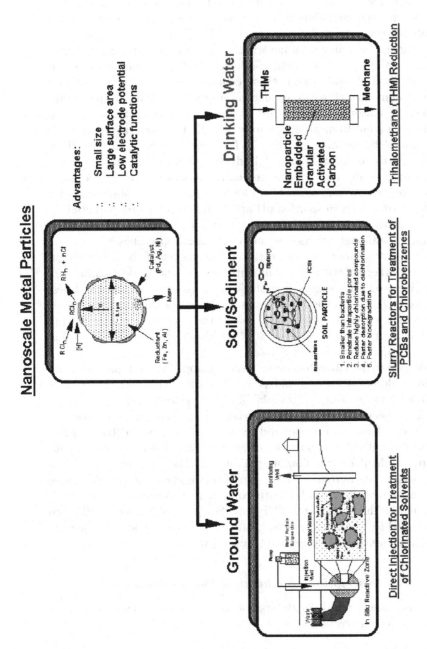

Figure 1. Applications of nanoscale zero-valent iron technology for environmental remediation.

The majority of the ZVI research conducted thus far has focused on the bench-scale degradation of relatively simple one and two carbon chlorinated-hydrocarbon contaminants in either batch aqueous systems or column studies. Including such notorious solvents and feedstock chemicals as perchloroethene (PCE), trichloroethene (TCE), carbon tetrachloride (CT), and vinyl chloride (VC), these contaminants pose a considerable and well-documented threat to groundwater quality in many of the world's industrialized nations. In batch-test studies, aqueous contaminant concentrations ranging from approximately 1 mg/L to near saturation levels were degraded by variable granular ZVI dosages of about 20–250 g/L in timescales on the order of hours.[4–7] Reductive dehalogenation of the contaminant to typically non-halogenated end-products (e.g. ethene, ethane, methane, etc.) was accompanied by profound changes in the observed water chemistry, particularly in solution pH and standard reduction potential. In many studies, the experimental data was modeled according to pseudo first-order kinetics although others used more complex approaches.[8,9]

Other investigators have demonstrated the efficacy of the ZVI technology towards other notorious contaminants including hexavalent chromium, nitrate, nitroaromatics, energetic munitions compounds, azo dyes and pesticides like DDT.[10–15] One of the common themes of this rich and extensive body of research is the essential role of the iron surface in mediating the contaminant degradation process[5,16] although other factors including sorption[17,18] also play an integral role.

In addition to the laboratory-scale research, applied investigations at the field pilot-scale has also been underway since the early-1990s. More than 100 ZVI-field-deployments of granular ZVI in permeable reactive barriers (PRBs) have been completed at industrial, commercial, and governmental sites in North America, Europe, and in Asia.[19] In essence, PRBs consists of sub-surface walls of granular iron; often constructed in a funnel-and-gate design, perpendicular to the flow of groundwater. The funneling mechanism directs contaminated groundwater through the reactive gate(s). The gates function as a plug flow reactor with contaminant degradation occurring throughout the thickness of the ZVI zone. For potable aquifers, PRBs are often designed to theoretically treat groundwater contamination to acceptable USEPA Maximum Contaminant Levels (MCLs).

ZVI Process Description — Chemical Fundamentals

As previously mentioned, Fe^0 exhibits a strong tendency to donate electrons to suitable electron acceptors:

$$Fe^{2+} + 2e^- \rightarrow Fe^0 \tag{1}$$

Zero-valent iron and dissolved ferrous iron form a redox couple with a standard reduction potential ($E°$) of $-0.440 \, \text{V}$.[5] Metallic ZVI can couple with several environmentally significant and redox-amenable electron acceptors including hydrogen ions (i.e. protons), dissolved oxygen, nitrate, sulfate, and carbonate.[5] Under aerobic conditions typical of vadose zone soils or shallow, oxygenated groundwaters, ZVI can react with dissolved oxygen (DO) as follows:

$$2Fe^0_{(s)} + 4H^+_{(aq)} + O_{2(aq)} \rightarrow 2Fe^{2+}_{(aq)} + 2H_2O_{(l)} \tag{2}$$

yielding ferrous iron and water. The overall $E°$ for this reaction ($E°_{rxn}$) is $+1.71 \, \text{V}$ at $25°C$, indicating a strongly favorable reaction from a thermodynamics perspective.[20] Implicit in the stoichiometry is the transfer of four electrons from the iron surface and associated increase of solution pH (based on the consumption of protons). Assuming that residual DO levels remain, the ferrous iron would be expected to undergo relatively facile oxidation to ferric iron, Fe^{3+}. The increasing pH favors formation of one or more iron hydroxide; or carbonate-based precipitates, and can have the effect of passivating the reactivity of the metal surface.[5]

In anaerobic or low DO groundwater environments, which are more characteristic of deeper aquifers and contaminated plumes, ZVI also forms an effective redox couple with water yielding ferrous iron, hydroxide, and hydrogen gas:

$$Fe^0_{(s)} + 2H_2O_{(l)} \rightarrow 2Fe^{2+}_{(aq)} + H_{2(g)} + 2OH^-_{(aq)} \tag{3}$$

Unlike the reduction of DO, the iron-mediated reduction of water is not thermodynamically favored as indicated by the $E°_{rxn}$ of $-0.39 \, \text{V}$ at $25°C$.[20] Thermodynamics considerations notwithstanding, the kinetics of these and other reactions in natural waters tend to be rather sluggish and; consequently, chemical equilibrium is generally not attained.[21]

In addition to the common environmentally relevant electron acceptors, ZVI also readily reacts with a wide variety of redox-amenable contaminants. Using a generalized chlorinated hydrocarbon, RCl, as an example, the ZVI-mediated transformation of RCl to the corresponding hydrocarbon, RH can be represented as:

$$RCl + H^+ + Fe^0 \rightarrow RH + Fe^{2+} + Cl^- \tag{4}$$

From a thermodynamics perspective, the large positive $E°_{rxn}$ values imply that a spontaneous reaction with ZVI should occur. In terms of the chlorinated hydrocarbons, the degree of favorability increases with the number of chlorine substituents. For many chlorinated hydrocarbons, the $E°_{rxn}$ is on the order of $+0.5$ to $+1.5 \, \text{V}$ at $25°C$.[5,22]

ZVI Process Description — Mechanistic Aspects

Examination of Equations (2) through (4) reveals that multiple species exist in the systems under investigation which are capable of serving as reducing agents. The electron donating potential of ZVI has already been discussed. However, ferrous iron can serve as a reductant by donating an electron to a suitable electron acceptor yielding ferric iron, Fe^{3+}. Hydrogen gas, too, is a well known reducing agent, particularly in the realm of microbiology where it is often characterized as the "universal electron donor". The roles of these three electron donors in the reduction of chlorinated hydrocarbons were studied. For example, Matheson and Tratnyek[5] proposed three possible mechanisms: (1) direct reduction at the metal surface, (2) reduction by ferrous iron, and (3) reduction by hydrogen with catalysis. They studied the potential roles of these reductants and found that ferrous iron, in concert with certain ligands, can slowly reduce the chlorinated hydrocarbons; and that dissolved hydrogen gas, in the absence of a suitable catalytic surface, failed to reduced the chlorinated hydrocarbon.

In the presence of granular ZVI, Matheson and Tratnyek[5] and numerous other research groups, observed the rapid transformation of various contaminants confirming the validity of the direct surface reduction model. Weber[16] elegantly confirmed these findings in a study using 4-aminoazobenzene (4-AAB), an aromatic azo-dye which readily undergoes ZVI-mediated reduction. In this work, 4-AAB which was immobilized by electrophilic derivatization to a solid support was not reduced by ZVI because it could not associate with the iron surface while the control, non-derivatized 4-AAB, was rapidly transformed.[16] Thus, these studies clearly demonstrated the fact that the degradation of contaminants by ZVI is a surface-mediated process via one or more heterogeneous reactions. The nature of these reactions, whether occurring sequentially or in concert depends upon the particular degradation pathways involved; which in turn are a function of the specific contaminant(s).

Kinetics Models to the ZVI Process

Although many investigations have invoked more-complex models recognizing the heterogeneous nature of the ZVI degradation process,[9] first-order kinetics was the most common model used to explain the experimental datasets from batch degradation studies. As has been previously described, these datasets covered a wide variety of chemical classes (e.g. chlorinated solvents, chlorinated aromatics, pesticides, PCBs, nitroaromatics, and metals such as hexavalent chromium). To the feasible, extent first-order kinetics is presumed to be applicable to a wide variety of reactions of environmental significance primarily because of the relative simplicity of the mathematical treatment. Thus, it is often the first model utilized to fit the experimental data for a new environmental process. As a result, our discussion mainly focused on the first-order kinetics model.

In a system in which first-order kinetics is determined to prevail, the rate of contaminant loss, or degradation, is proportional to its concentration in solution as expressed in Equation (5):

$$\frac{dC}{dt} = -kC \qquad (5)$$

where C is the contaminant concentration at time t, and k is the constant of proportionality known as the first-order rate constant. In order for the units to balance, k is reported in units of reciprocal time, $1/t$.

It stands to reason that ZVI-mediated transformations should not only depend upon the contaminant concentration but also on the iron concentration as well. After all, the reaction involves two chemical entities: the reductant and the reductate. In this case, the first-order kinetics expression must be modified as follows:

$$\frac{dC}{dt} = -k[C][\text{Fe}^0] \qquad (6)$$

where the rate of contaminant transformation now depends on both the contaminant and iron concentrations in solution. Thus, Equation (6) is applicable to reactions characterized by second-order kinetics. However, in the vast majority of cases involving the ZVI-mediated degradation of contaminant(s), the concentration of iron is appreciably larger than that of the aqueous contaminant. That is, $[\text{Fe}^0] \gg [C]$. For the vast majority of the peer-reviewed ZVI literature regarding granular, microscale, or nanoscale iron, this simplifying assumption can be made. If $[\text{Fe}^0]$ is large enough such that it does not change meaningfully over the course of the observed changes in $[C]$, it can be said to remain constant and Equation (7) can now be represented as:

$$\frac{dC}{dt} = -k_{\text{obs}}[C] \qquad (7)$$

The observed first-order rate constant, k_{obs}, is related to k from Equation (6) in that $k_{\text{obs}} = k[\text{Fe}^0]$. Equation (7) is referred to as a pseudo first-order expression and, as has just been described, results from applying a valid simplifying assumption to a system formally characterized by second order kinetics.

Although Equation (7) is indeed applicable to the ZVI-mediated degradation process, investigators such as Johnson *et al.*,[8] noted that k_{obs} data from the literature for specific contaminants in batch degradation studies under differing experimental conditions exhibited variability of up to three-orders of magnitude. In an effort to account for the majority of the experimental factors accounting for this variability, they expanded Equation (8) as follows:

$$\frac{dC}{dt} = -k_{SA}\rho_a C \tag{8}$$

where k_{SA} is defined as the surface area normalized rate constant and is reported in units of liters per hour per meter squared (L/hr-m^2) and ρ_a is the surface area concentration of ZVI in square meters per liter (m^2/L) of solution.[8] The iron surface area concentration is related to the specific surface area of the ZVI, a_s, and mass concentration of iron, ρ_m, by Equation (9):

$$\rho_a = a_s \rho_m \tag{9}$$

where the units of a_s are square meters per gram (m^2/g) and those of ρ_m are grams per liter of solution (g/L). The specific surface area concentration is also related to k_{obs} from Equation (10) as follows:

$$k_{\text{obs}} = k_{SA}\rho_m a_s \tag{10}$$

In using Equations (8) through (10) describe the kinetics of the ZVI-mediated transformation of various chlorinated hydrocarbon contaminants, Johnson et al.,[8] found that most of the variability could be eliminated. After replotting the literature-derived dataset using these expressions, one-order of magnitude variability was still observed, probably associated with uncertainty in the methodology and/or measurement of ZVI surface area.

Overview of Major Methodologies for Synthesizing nZVI

At least three distinct methods have been used to prepare the nanoscale ZVI (nZVI). All involved the reduction and precipitation of zero-valent-iron from aqueous iron salts using sodium borohydride as the reductant. These included one approach in which the iron salt was iron(III) chloride (referred to as the "chloride method"), and two schemes where iron(II) sulfate was the principal ZVI precursor salt (termed the "sulfate method"). Each of these methodologies is discussed in detail in the following paragraphs. It is important to note that the laboratory-scale synthetic protocols were not optimized and that potential significant batch-to-batch variability in terms of characteristics, behavior, and performance is to be expected.

Type I nZVI Using the Chloride Method

The chloride method synthesis represents the original means of producing nanoscale ZVI at Lehigh University (Zhang et al., 1998).[23–25] Consequently, the chloride-method iron, also referred to as Type I nZVI, was the earliest generation of nanoscale iron used in experimental and in fieldscale work.

In this synthesis, 0.25 molar (M) sodium borohydride was slowly added to 0.045 M ferric chloride hexahydrate in aqueous solution under vigorously mixed conditions such that the volumes of both the borohydride and ferric salt solutions were approximately equal (i.e. a 1:1 volumetric ratio). The mixing time utilized was approximately one hour. This reaction is shown in Equation (11) as follows[25]:

$$4Fe^{3+}_{(aq)} + 3BH^-_{4\ (aq)} + 9H_2O_{(l)} \rightarrow 4Fe^0_{(s)} + 3H_2BO^-_{3\ (aq)} + 12H^+_{(aq)} + 6H_{2(g)}$$
$$(11)$$

The ratio between the borohydride and ferric salt exceeded the stoichiometric requirement by a factor of approximately 7.4.[25] This excess is thought to help ensure the rapid and uniform growth of the nZVI crystals. The harvested nano-iron particles were then washed successively with a large excess of distilled water, typically $> 100\,mL/g$. The solid nanoparticle mass was recovered by vacuum filtration and washed with ethanol. The residual water content of the nZVI mass was typically on the order of 40–60%. A Transmission Electron Microscopy (TEM) micrograph of the Type I nZVI is shown in Figure 2.

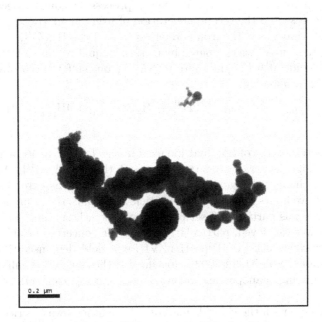

Figure 2.　TEM image of Type I nZVI aggregates. Note the size bar represents 100 nm.

If bimetallic particles were desired, the ethanol-wet nZVI mass was soaked in an ethanol solution containing approximately 1% palladium acetate as indicated in Equation (12):

$$Pd^{2+} + Fe^0 \rightarrow Pd^0 + Fe^{2+} \tag{12}$$

The chloride method proved to be readily adapted to any standard chemical-laboratory assuming that fume-hoods and adequate mixing was provided to dissipate and vent the significant quantities of hydrogen gas produced. The synthetic scheme was expensive, due to the high cost of the sodium borohydride and lack of production scale.

Type II and III nZVI Using the Sulfate Method

The development of the sulfate method for producing nZVI arose from two fundamental concerns associated with the chloride method: (1) potential health-and-safety concerns associated with handling the highly acidic and very hygroscopic ferric chloride salt; and (2) the potential deleterious effects of excessive chloride levels from the nZVI matrix in batch-degradation tests where chlorinated hydrocarbons are the contaminant of concern. In addition, the reduction of the iron feedstock from Fe(II) requires less borohydride than the chloride method, in which Fe(III) is the starting material which may favorably enhance overall process economics. Because this method represented the second generation of iron nanoparticles developed at Lehigh University; the iron is referred to as Type II nZVI.

Sulfate-method-nZVI was prepared by metering equal volumes of 0.50 M sodium borohydride at 0.15 L/min into 0.28 M ferrous sulfate according to the following stoichiometry:

$$2Fe^{2+}_{(aq)} + BH_4^-{}_{(aq)} + 3H_2O_{(l)} \rightarrow 2Fe^0_{(s)} + H_2BO_3^-{}_{(aq)} + 4H^+_{(aq)} + 2H_{2(g)} \tag{13}$$

The stoichiometric excess of borohydride used in the Type II nZVI synthesis was about 3.6, considerably less than that with the Type I nZVI. Owing to this process change, the rate of borohydride addition was extended to approximately two hours to help control particle size. Thus, the reduction in production costs was partially off-set by the longer synthesis time. Process improvements were achieved, particularly during the concerted production of 10 kg, dry weight basis, of Type II nZVI for a field demonstration in Research Triangle Park, NC in 2002.[3] Specifically, these included enhancements to the mixing, nanoparticle recovery, and storage elements of the synthesis.

As was the case with the chloride method, the synthesis was carried out in a fume-hood in open five-gallon (18.9 L) polyethylene containers fitted with variable speed, explosion-resistant mixers. No attempt was made to exclude air from the reaction mixture. The freshly prepared nZVI particles were allowed to settle for approximately one hour and were then harvested by vacuum filtration. The harvested nanoparticles were washed with copious amounts of distilled water ($> 100\,\text{mL/g}$), then by ethanol, purged with nitrogen, and refrigerated in a sealed polyethylene container under ethanol until needed. The residual water content of the Type II nZVI was typically on the order of 45–55%, very similar to that observed for the Type I iron. A representative TEM image of the Type II nZVI is shown in Figure 3.

The Type III nZVI was also synthesized using the sulfate method. As the procedure for synthesizing the Type III iron was very similar to that for the Type II nZVI, the details are not repeated here. It represented the latest generation of nZVI and exhibited an average particle size of approximately 50–70 nm, very similar to that observed for the Type II nZVI. However, the moisture content was appreciably lower than observed for the previous nZVI types: 20–30% versus 40–60% for Types I and II. The basis for this difference is not known. A TEM image of the Type III nZVI is shown in Figure 4.

200 nm

Figure 3.   TEM image of Type II nZVI aggregates. Note the size bar represents 200 nm.

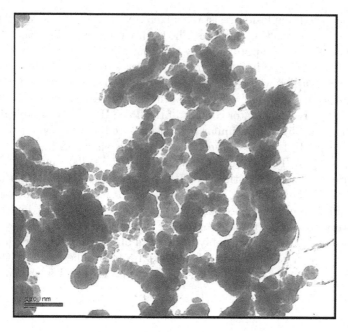

Figure 4.   TEM image of Type III nZVI aggregates. Note the size bar represents 100 nm.

Characterization of n-ZVI

The average particle size of the chloride method nZVI was of the order of 50–200 nm and the specific surface area was measured by mercury porosimetry to be approximately 33.5 m^2/g.[24] X-ray diffraction of the Type I palladized nZVI (nZVI/Pd) surface composition indicated that the major surface species of freshly prepared nanoscale Pd/Fe is Fe^0 (44.7°) with lesser quantities of iron(III) oxide, Fe$_2$O$_3$ (35.8°), and Pd^0 (40.1°) (Figure 5). The presence of iron(III) oxide, which results from air exposure of the nZVI, visually appeared as a surficial rust patina on the surface of the iron and typically did not extend into the bulk iron mass. Not surprisingly, the XRD diffractogram for nZVI/Pd aged for 48 hours shows relatively larger peaks for the iron oxide but still an appreciable zero-valent-iron peak (Figure 5).

The average particle size of the sulfate method nZVI was on the order of 50–70 nm. Analysis of over 150 individual particles and clusters yielded a mean diameter of 66.6 ± 12.6 nm with more than 80% being smaller than 100 nm and fully 30% being smaller than 50 nm.[26] Moreover, the BET specific surface area of the iron was 35 ± 2.7 m^2/g, approximately equivalent to that of the Type I nZVI.[26]

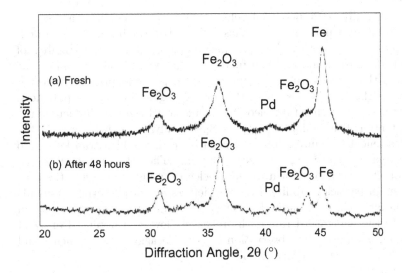

Figure 5. X-ray diffractogram of nZVI/Pd particles.

Using an electroacoustic spectrometer (Dispersion Technologies DT 1200), the zeta potential of a 0.85% (by weight) slurry of Type III nZVI in water was $-27.55\,\mathrm{mV}$ at a pH of 8.77 (Figure 6). According to the Colloidal Science Laboratory, Inc. (Westhampton, NJ), colloidal particles with zeta potential values more positive than $+30\,\mathrm{mV}$ or more negative than $-30\,\mathrm{mV}$ are considered stable with maximum instability (i.e. aggregation) occurring at a zeta potential of zero. Thus, using this benchmark, the Type III nZVI would be considered meta-stable.

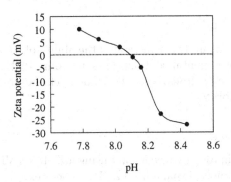

Figure 6. Titration of zeta potential versus pH using 2.0 N sulfuric acid for a 10 g/L suspension of nZVI in water.

Theoretically, zeta potential refers to the potential drop across the mobile or diffuse portion of the classical electric double layer surrounding a particle in solution.[27] It provides an indication of the stability of the colloidal suspension and is a function of many variables including the nature of the particle surface, ionic strength, pH, and presence of other substances that can interact with the surface (e.g. surface-active polymers, etc.). It has been well established that solution pH strongly influences the zeta potential of the colloidal particles. Specifically, as the pH increases, the particles tend to acquire additional negative charge which translates into a decreased (i.e. more negative) zeta potential. The pH corresponding to a zeta potential of zero, known as the isoelectric point, represents the area of minimum particle stability. Figure 6 depicts a zeta potential versus pH titration for a chloride method nZVI of 50–70 nm average particle size in water. In the neutral-pH range, the relatively low zeta potential of nZVI in solution supports the observation of particle aggregation which could adversely impact sub-surface mobility.

Summary of the nZVI Research and Applications

As demonstrated in the Zhang group research, the nZVI technology exhibits enhanced reactivity and superior field deployment capabilities as compared with microscale and granular iron as well as other *in situ* approaches. The enhanced reactivity stems from the appreciably greater specific surface area of the iron. The colloidal size of the iron nanoparticles, their amenability to direct subsurface injection via gravity feed conditions (or under pressure, if desired), and need for substantially less infrastructure all contribute to the technology's portability and relative ease of use.

Laboratory Batch Studies

Since 1996, the Zhang group has investigated the ability of nZVI to degrade a wide variety of environmental contaminants including PCBs, chlorinated aliphatic and aromatic hydrocarbons, hexavalent chromium, chlorinated pesticides, and perchlorate.

Chlorinated hydrocarbons

Degradation of chlorinated hydrocarbons using nZVI, mZVI (microscale ZVI) and nZVI/Pd (nanoscale palladized ZVI) has been extensively studied. A wide array of chlorinated hydrocarbons including aliphatic compounds (e.g. carbon tetrachloride, trichloroethylene, and hexachloroethane), alicyclic compounds (e.g. lindane), and aromatic compounds (e.g. PCB and hexachlorobenzene) have been tested. In general, nZVI/Pd showed the best overall performance followed by nZVI and then mZVI. Degradation of carbon tetrachloride by different types of ZVI represents a typical example. The surface area normalized rate constant (k_{SA}) data derived from the experimental datasets followed the order nZVI/Pd > nZVI > mZVI. The surface area normalized rate constant of nZVI/Pd was two-order magnitudes higher than that of mZVI. Furthermore, same reaction products including chloroform, dichloromethane, and methane were observed in the use of mZVI, nZVI, and nZVI/Pd; however, product distributions were significantly different. The highest yield of methane (55%) and lowest production of dichloromethane (23%) was found in the use of nZVI/Pd. In comparison, the accumulation of dichloromethane, a more toxic compound than the parent, accounted for more than 65% of the initial carbon tetrachloride and a yield of methane of less than 25% (Figure 7).

The superiority of nZVI/Pd can be attributed to two key factors. First and foremost, a significantly increased surface area of nZVI. As compared to mZVI, the enhanced reactivity of the nZVI can be attributed to the smaller average particle size which translates into a much larger specific surface area, $33.5\,m^2/g$ versus $< 0.9\,m^2/g$, for the irons studied.[24] Second, the presence of palladium resulting in the catalytic reducing power of the nZVI/Pd. It should be pointed out that palladium not only enhances the reactivity but also alters the product distribution, implying that different reaction mechanisms are likely involved.[28,29]

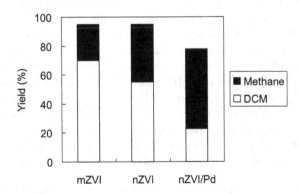

Figure 7. Comparison of yields of ethane and dichloromethane for dechlorination of carbon tetrachloride with nZVI/Pd, nZVI and mZVI. The dose of nZVI/Pd, nZVI and mZVI was $5\,g/L$, $5\,g/L$ and $400\,g/L$, respectively.

In comparison, chlorinated alicyclic and aromatic compounds have received less attention by investigators thus far. Very few systematic studies have been reported in the peer-reviewed literature, partially due to complicated transformation processes as they possess relatively large molecular structures. The fact that numerous degradation-related intermediates are often produced is another factor. Nevertheless, Xu and Zhang[30] demonstrated that nZVI or nZVI/Ag (iron-silver nanoparticles) were quite effective in transforming chlorinated aromatics through a sequential dechlorination pathway. For example, nZVI/Ag, with a measured specific surface area of $35.0\,m^2/g$, transformed hexachlorobenzene (HCB) to a series of lesser chlorinated benzenes including the following principal products: 1,2,4,5-tetrachlorobenzene, 1,2,4-trichlorobenzene, and 1,4-dichlorobenzene (Xu and Zhang, 2000). No chlorobenzene or benzene was observed as reaction products. HCB concentrations were reduced below the detection limit ($< 1\,\mu g/L$) after four days.

However, conventional mZVI powder at an iron to solution ratio of $25\,g/100\,mL$ produced little reaction with HCB under similar experimental conditions. After 400 hours of elapsed time, the total conversion of HCB to products was approximately 12% with primarily 1,2,4,5-tetrachlorobenzene and 1,2,4-trichlorobenzene detected as intermediates.[30] Clearly, nZVI/Ag is more well-suited than mZVI in degrading polychlorinated aromatics like HCB.

Hexavalent chromium

Hexavalent chromium, Cr(VI), is a highly toxic, very mobile, and quite common groundwater contaminant. The efficacy of the nZVI technology was evaluated in batch aqueous systems containing soils and groundwater impacted by chromium ore processing residuals (COPR) at a former manufacturing site in New Jersey. The average Cr(VI) concentration in groundwater samples from the site were measured to be $42.83 \pm 0.52\,mg/L$ while the concentration in air-dried soils was $3,280 \pm 90\,mg/kg$.[31] The total chromium concentration, that is Cr(III) plus Cr(VI), was determined to be $7,730 \pm 120\,mg/kg$. Due to the presence of lime in the COPR-contaminated media, the pH of groundwater typically exceeded 10–11. The basis for this reaction involves the very favorable reduction of Cr(VI) to Cr(III), a relatively non-toxic, highly immobile species which precipitates (i.e. Cr(III) oxyhydroxides) from solution at alkaline pH.

$$\frac{3}{2}Fe^0_{(s)} + CrO_4^{2-}{}_{(aq)} + 5H^+_{(aq)} \rightarrow \frac{3}{2}Fe^{2+}_{(aq)} + Cr(OH)_{3(s)} + H_2O_{(l)} \qquad (14)$$

Batch solutions containing 10 g of COPR-contaminated soils and 40 mL groundwater were exposed to 5–50 g/L (89.5–895 mM) Type I

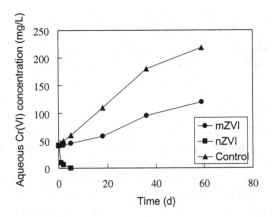

Figure 8. Results of batch Cr(VI) reduction by mZVI and nZVI. The dose of mZVI and nZVI was 150 g/L and 5 g/L, respectively.

nZVI under well-mixed conditions (Figure 8). Not surprisingly, once the COPR-contaminated soils were placed in the reactors, the Cr(VI) concentration increased from 42 mg/L to approximately 220 mg/L, demonstrating that substantial desorption and dissolution of hexavalent chromium occurs during the course of the experiment. Within a timeframe of up to six days, Cr(VI) concentrations in solution containing Type I nZVI were generally less than the detection limit, $< 10 \mu g/L$ (Figure 8). In contrast, the Cr(VI) concentration was observed to increase substantially in the reactors containing microscale iron (Fisher, $10 \mu m$) during the course of the reaction.

The reductive capacity of the Type I nZVI was found to be on the order of 84–109 mg Cr(VI) per gram of iron, approximately two-orders of magnitude greater than the reaction with mZVI. Given the highly heterogeneous nature of the COPR materials and the highly alkaline pH, it is likely that Cr(VI) desorbing from the matrix is reduced and rapidly precipitated as chromium(III) hydroxide, $Cr(OH)_3$. This precipitate enmeshes the COPR soils, forming a shell that tends to encapsulate any Cr(VI) remaining in the potentially still reactive core region.

Perchlorate

Perchlorate (ClO_4^-), emerged as a high profile environmental contaminant in the late-1990s when improved analytical techniques revealed widespread and previously undetected contamination in water supplies, particularly in the Western U.S.[32] Perchlorate concentrations in excess of $100 \mu g/L$ have been detected in Nevada's Lake Mead, well beyond USEPA guidance levels of $1 \mu g/L$.[33]

While the iron-mediated reduction of perchlorate, shown below in Equation (15), is a strongly thermodynamically favored process (having a large, negative value for the standard Gibbs Free Energy of the reaction, ΔG_{rxn}, of $-1,387.5\,kJ/mol$), one recent study reported that it is generally not reactive with ZVI.[34]

$$ClO_4^- + 4Fe^0 + 8H^+ \rightarrow Cl^- + 4Fe^{2+} + 4H_2O \qquad (15)$$

In general, ZVI transformations require either direct contact with the reactive iron surface or through suitable bridging groups. The structure of perchlorate makes this difficult because its reactive chlorine central atom, Cl(VII), is shielded by a tetrahedral array of bulky oxygen substituents, which also fully delocalize the oxyanion's negative charge.

Representative results of the reaction at $25°C$ are shown in Figure 9. Batch reactors containing nitrogen-purged deionized water spiked with 1–$200\,mg/L$ sodium perchlorate and 1–$20\,g/L$ Type II sulfate-method nZVI. After 28 days of elapsed time, negligible reaction with $20\,g/L$ microscale iron (Aldrich, $4.95\,\mu m$) was observed while significant removal of perchlorate was measured in the reactor containing $20\,g/L$ Type II nZVI. As depicted in Figure 10, progressively better removal of perchlorate was observed as the ambient temperature was increased to $40°C$ and finally to $75°C$. The most interesting feature of these plots is the dramatic reduction in the transformation timescale achieved with increasing temperature from $25°C$ to $75°C$. The half-life for perchlorate declined from 18–20 days at $25°C$ to about 80–100 hours at $40°C$ and finally to approximately eight hours at $75°C$. Thus, temperature was clearly an important factor in catalyzing the reduction of perchlorate in aqueous solution.

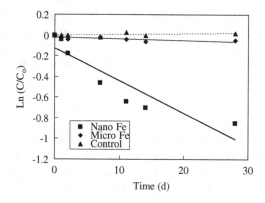

Figure 9.   Degradation of $200\,mg/L$ aqueous perchlorate by nZVI ($20\,g/L$) at $25°C$.

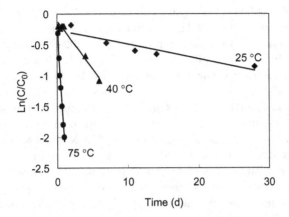

Figure 10. A plot of Ln (C/C_o) vs time for the reduction of perchlorate by nZVI (20 g/L) at various temperature conditions.

The appearance of chloride as the major end-product demonstrated that perchlorate transformation was occurring and not due to mere sorption or other surface-associated sequestration. Besides chloride, the only intermediate identified was chlorate, ClO_3^-, and only at trace levels.[32] Values of 0.013, 0.10, 0.64, and 1.52 mg perchlorate per gram of nZVI per hour $(mg\text{-}g^{-1}\text{-}hr^{-1})$ were calculated at temperatures of 25°C, 40°C, 60°C, and 75°C, respectively. The experiments conducted at different temperatures enabled calculation of the activation energy for the nZVI-mediated degradation process. The activation energy was determined to be 79.02 ± 7.75 kJ/mol, a relatively formidable value for aqueous reactions.

Field Testing Demonstration

Several *in situ* field demonstrations of the nZVI technology have been conducted at sites with contaminated groundwater since 2000, including the first field pilot-test in Trenton, New Jersey.[2] A subsequent field demonstration in North Carolina involved the injection of approximately 12 kg of chloride-method-nZVI/Pd into a fractured bedrock aquifer. The test area was situated approximately 38 meters downgradient of a former waste disposal area, which had groundwater contaminated with chlorinated solvents.[26] The site hydrogeology is quite complex, with much of the contaminated groundwater traversing through fractures in the sedimentary sandstones and siltstones. In addition to the injection well, the test area infrastructure included three monitoring wells at distances of 6.6, 13, and 19 meters downgradient of the injection well.

Approximately 6,056 L (1,600 gal) of a 1.9 g/L nZVI/Pd slurry in tap water was injected at a rate of 2.3 L/min (0.6 gal/min) over a nearly 2-day period from September 13 through September 15, 2002.[26] The slurry was prepared on-site in a 1,500 L (400 gal) heavy-duty polyethylene tank and mixed during the course of the injection process. Within approximately seven days from the injection, a reduction of more than 90% of the pre-injection TCE concentration of 14,000 μg/L was observed at the injection well and nearest monitoring well. These results are comparable to those from the first nZVI field demonstration conducted in New Jersey two years earlier.[2] Groundwater quality standards for TCE, PCE, and cis-DCE were generally achieved within six weeks of the injection at these two monitoring wells without increases in VC being observed. Although pre-injection groundwater ORP levels were in the range of +50 to −100 mV, indicative of iron-reducing conditions, post-injection ORP readings plunged to −700 mV in the injection well and −500 mV in nearby monitoring wells. The radius of influence measured at the injection well was approximately 6–10 meters.[26]

This demonstration successfully showed that nZVI can travel distances of more than 20 meters in groundwater, that the longevity of reactivity can reach periods of greater than 4–8 weeks in the field, and that very high degrees of contaminant removal can be achieved. However, due to the significant costs of fieldwork of this nature, the test was not conducted long enough to observe significant recovery of the contaminant levels within the test area.

Challenges Ahead

As the nZVI technology moves rapidly from laboratory research to real world implementation, a number of fundamental issues relating to the potential applications of iron nanoparticles for environmental remediation remain. These include: (1) mobility, (2) environmental impact and (3) cost-effectiveness. An important attribute of nZVI is their potential mobility in porous environmental media, especially in the sub-surface environment. Due partly to their very small surface charge, empirical evidence suggests that iron nanoparticles tend to undergo strong aggregation and thereby form much larger chains, rings, or assemblages of nZVI particles. Stable dispersions of iron particles can be achieved by modifying the iron surface or using microemulsion mixtures to protect nanoparticles. The surface modified (i.e. charged) iron nanoparticles can potentially remain in suspension for extended periods.[35] More recently, He and Zhao[36] reported the use of a new starch-based stabilizer which facilitated the suspension of nZVI/Pd in water for days without precipitation or agglomeration. Only a stable dispersion of iron particles offers the possibility of facile injection, mixing and rapid transport within the slow-flowing groundwater.

Thus far, no studies on the eco-toxicity of low-level iron in soil and water have been reported in the peer-reviewed literature. However, it is the authors' opinion that systematic research on the environmental transport, fate and eco-toxicity of nanomaterials is needed to help overcome increasing concerns in the environmental use of nanomaterials; so as to minimize any undesirable consequences. The use of iron nanoparticles in such studies could actually provide a valuable opportunity to demonstrate their expected positive overall effect on environmental quality. Iron is the fifth most used element; only hydrogen, carbon, oxygen and calcium are consumed in greater quantities during our daily activities. Iron is also typically found at the active center of many biomolecules and plays an key role in oxygen transfer and other life-essential biochemical processes. Iron in the body is mostly present as iron porphyrin or heme proteins, which include hemoglobin in the blood, myoglobin and the heme enzymes.[37] The heme enzymes permit the reversible combination with molecular oxygen. By this mechanism, red blood cells carry oxygen from one part of the body to another. Nature has evolved highly organized systems for iron uptake (e.g. the siderophore-mediated iron uptake), transport (via transferrins) and storage of iron (e.g. in the form of ferritin). The intracellular concentration of free iron (i.e. iron not-bound to organic ligands) is tightly controlled in animals, plants, and microbes. This is partly a result of the poor solubility of iron under aerobic conditions in aqueous solution: the solubility production of $Fe(OH)_3$ is 4×10^{-38}. Within the cell, free iron generally precipitates as polymeric hydroxides. The challenge for us is to determine the transport, aquatic, and intra-organism biochemical fate of manufactured iron nanoparticles.

In the early phase of our work (prior to 2000), no commercial vendors or manufacturing techniques existed for the production of iron nanoparticles. Small quantities ($< 10\,kg$) were prepared on a batch-wise basis in laboratory for bench- and pilot-scale tests. Not surprisingly, the production cost was rather high ($> \$200/kg$). Over the past few years, several vendors have emerged and are in the process of developing and scaling up different nZVI manufacturing techniques. These include both bottom-up chemical synthetic schemes similar to the borohydride method used at Lehigh as well as more advance top-down particle size reduction approaches. Consequently, the price is coming down rapidly (e.g. $\ll \$50/kg$ as of mid-2005). Even with the increased-supply and reduced-price, the general perception is that the nanoparticles are still too expensive for real world applications. A careful analysis suggests the contrary, that iron nanoparticles potentially offer superior economics based on their large specific surface area. The specific surface area (SSA) can be calculated according to Equation (16) as follows:

$$SSA = (Surface\ Area/Mass) = \frac{\pi d^2}{\rho \frac{\pi}{6} d^3} = \frac{6}{\rho d} \qquad (16)$$

where ρ is the density of the particle, $7.8\,\text{kg/m}^3$. With diameters on the order of $0.5\,\text{mm}$, the theoretical SSA of granular iron used in typical PRBs can be calculated to be approximately $1.5\,\text{m}^2/\text{g}$. In contrast, $50\,\text{nm}$ nZVI particles would exhibit SSA values on the order of $15{,}000\,\text{m}^2/\text{kg}$. Although granular iron is considerably less expensive that nZVI, approximately \$0.50/kg versus \$100/kg, the cost-effectiveness expressed in terms of surface area per dollar is dramatic: $3\,\text{m}^2/\text{dollar}$ for granular iron versus $150\,\text{m}^2/\text{dollar}$ for nZVI.

Research has demonstrated that iron nanoparticles have much higher surface activity per unit of surface area. For example, the surface area normalized reaction rates for TCE and DCEs dechlorination for iron nanoparticles are one to two orders of magnitude higher than bulk ($\sim 10\,\mu\text{m}$) iron powders.[38] Accordingly, nZVI particles have much lower cost per unit surface area and consequently should offer much higher return in terms of performance per unit iron-mass. The key hurdle for the application of nZVI particles is perhaps not the cost, but demonstrating the field efficacy of the technology at field scales.

References

1. T. Masciangioli and W.-X. Zhang, Environmental technologies at the nanoscale, *Environ. Sci. Technol.* 102A–108A (2003).
2. D. W. Elliot and W.-X. Zhang, Field assessment of nanoscale bimetallic particles for groundwater treatment, *Environ. Sci. Technol.* **35**, 4922–4926 (2001).
3. R. Glazier, R. Venkatakrishnan, F. Gheorghiu and W.-X. Zhang, Nanotechnology takes root, *Civil Engineering* **73**, 64–69 (2003).
4. R. W. Gillham and S. F. O'Hannesin, Enhanced degradation of halogenated aliphatics by zero-valent iron, *Ground Water* **32**, 958–967 (1994).
5. L. J. Matheson and P. G. Tratnyek, Reductive dehalogenation of chlorinated methanes by iron metal, *Environ. Sci. Technol.* **28**, 2045–2053 (1994).
6. B. Deng, D. R. Burris and T. J. Campbell, Reduction of vinyl chloride in metallic iron-water systems, *Environ. Sci. Technol.* **33**, 2651–2656 (1999).
7. J. Farrell, M. Kason, N. Melitas and T. Li, Investigation of the long-term performance of zero-valent iron for reductive dechlorination of trichloroethylene, *Environ. Sci. Technol.* **34**, 514–521 (2000).
8. T. L. Johson, M. M. Scherer and P. G. Tratnyek, Kinetics of halogenated organic compound degradation by iron metal, *Environ. Sci. Technol.* **30**, 2634–2640 (1996).

9. W. F. Wüst, O. Schlicker and A. Dahmke, Combined zero- and first-order kinetic model of the degradation of TCE and cis-DCE with commercial iron, *Environ. Sci. Technol.* **33**, 4304–4309 (1999).

10. R. M. Powell, R. W. Puls, S. K. Hightower and D. A. Sabatini, Coupled iron corrosion and chromate reduction mechanisms for subsurface remediation, *Environ. Sci. Technol.* **29**, 1913–1922 (1995).

11. D. P. Siantar, C. G. Schreier, C.-S. Chou and M. Reinhard, Treatment of 1,2-dibromo-3-chloropropane and nitrate-contaminated water with zero-valent iron or hydrogen/palladium catalysts, *Water Res.* **30**, 2315–2322 (1996).

12. J. F. Devlin, J. Klausen and R. P. Schwarzenbach, Kinetics of nitroaromatic reduction on granular iron in recirculating batch system, *Environ. Sci. Technol.* **32**, 1941–1947 (1998).

13. J. Cao, L. Wei, Q. Huang, S. Han and L. Wang, Reducing degradation of azo dye by zero-valent iron in aqueous solution, *Chemosphere* **38**, 565–571 (1999).

14. J. Singh, S. D. Comfort and P. J. Shea, Iron-mediated remediation of RDX-contaminated water and soil under controlled Eh/pH, *Environ. Sci. Technol.* **33**, 1488–1494 (1999).

15. G. D. Sayles, G. You, M. Wang and M. J. Kupferle, DDT, DDD, and DDE dechlorination by zero-valent iron, *Environ. Sci. Technol.* **31**, 3448–3454 (1997).

16. E. J. Weber, Iron-mediated reductive transformations: Investigation of reaction mechanism, *Environ. Sci. Technol.* **30**, 716–719 (1996).

17. D. R. Burris, T. J. Campbell and V. S. Manoranjan, Sorption of trichloroethylene and tetrachloroethylene in a batch reactive metallic iron-water system, *Environ. Sci. Technol.* **29**, 2850–2855 (1995).

18. R. M. Allen-King, R. M. Halket and D. R. Burris, Reductive transformation and sorption of cis- and trans-1,2-dichloroethene in a metallic iron-water system, *Environ. Sci. Technol.* **16**, 424–429 (1997).

19. EnviroMetal Technologies, Inc., http://www.eti.ca (2005).

20. V. L. Snoeyink and D. Jenkins, *Water Chemistry*, John Wiley & Sons, New York, NY (1980).

21. D. Langmuir, *Aqueous Environmental Geochemistry*, Prentice Hall (1997).

22. T. M. Vogel, C. S. Criddle and P. L. McCarty, Transformation of halogenated aliphatic compounds, *Environ. Sci. Technol.* **21**, 722–736 (1987).

23. G. N. Glavee, K. J. Klabunde, C. M. Sorensen and G. C. Hadjipanayis, Chemistry of borohydride reduction of iron(II) and iron(III) ions in aqueous and nonaqueous media, Formation of nanoscale Fe, FeB, and Fe_2B powders. *Inorg. Chem.* **34**, 28–35 (1995).

24. C.-B. Wang and W.-X. Zhang, Synthesizing nanoscale iron particles for rapid and complete dechlorination of TCE and PCBs, *Environ. Sci. Technol.* **31**, 2154–2156 (1997).

25. H.-L. Lien and W.-X. Zhang, Dechlorination of chlorinated methanes in aqueous solutions using nanoscale bimetallic particles, *J. Environ. Eng.* **125**, 1042–1047 (1999).

26. W.-X. Zhang, Nanoscale iron particles for environmental remediation: an overview, *J. Nanoparticle Res.* **5**, 323–332 (2003).
27. W. Stumm and J. J. Morgan, *Aqueous Chemistry*, 3rd edn. (John Wiley & Sons, Inc., 1996).
28. J. H. Brewster, Mechanisms of reductions at metal surfaces. I. A general working hypothesis, *J. Am. Chem. Soc.* **76**, 6361–6363 (1954).
29. T. Li and J. Farrell, Reductive dechlorination of trichloroethene and carbon tetrachloride using iron and palladized-iron cathodes, *Environ. Sci. Technol.* **34**, 173–179 (2000).
30. Y. Xu and W.-X. Zhang, Subcolloidal Fe/Ag particles for reductive dehalogenation of chlorinated benzenes, *Ind. Eng. Chem. Res.* **39**, 2238–2244 (2000).
31. J. Cao and W.-X. Zhang, Reduction and immobilization of hexavalent chromium by nanoscale iron particle, *J. Hazard. Mater.* **2005** (*in press*).
32. J. Cao, D. W. Elliott and W.-X. Zhang, Perchlorate reduction by nanoscale iron particle, *J. Nanoparticle Res.* **7**, 499–506 (2005).
33. E. T. Urbansky, *Perchlorate in the Environment*. New York: Kluwer Academic/Plenum Publishers, 2000.
34. B. E. Logan, Assessing the outlook for perchlorate remediation, *Environ. Sci. Technol.* **35**, 482 A–487 A (2001).
35. B. Schrick, B. W. Hydutsky, J. L. Blough and T. E. Mallouk, Delivery vehicles for zerovalent metal nanoparticles in soil and groundwater, *Chem. Mater.* **16**, 2187–2193 (2004).
36. F. He and D. Zhao, Preparation and characterization of a new class of starch-stabilized bimetallic nanoparticles for degradation of chlorinated hydrocarbons in water, *Environ. Sci. Technol.* **39**, 3314–3320 (2005).
37. D. P. Ballou (ed.), *Essays in Biochmemistry: Metalloproteins* (Princeton University Press, Princeton, NJ, USA, 1999).
38. H.-L. Lien and W.-X. Zhang, Nanoscale iron particles for complete reduction of chlorinated ethenes, *Colloids and Surfaces A: Physicochemical & Eng. Aspects* **191**, 97–106 (2001).
39. J. Cao and W.-X. Zhang, Nanoporous zero-valent iron, *J. Mater. Res.* (in press) (2005).
40. H.-L. Lien and W.-X. Zhang, Hydrodechlorination of chlorinated ethanes by nanoscale Pd/Fe bimetallic particles, *J. Environ. Eng.* **131**, 4–10 (2005).

Chapter 3

Synthesis, Characterization, and Properties
of Zero-Valent Iron Nanoparticles

D. R. Baer[*,§], P. G. Tratnyek[†,¶], Y. Qiang[‡], J. E. Amonette[*],
J. Linehan[*], V. Sarathy[†], J. T. Nurmi[†], C.-M. Wang[*] and J. Antony[‡]

[*]*Pacific Northwest National Laboratory, Richland, WA, USA*

[†]*Department of Environmental Science and Engineering*
Oregon Health and Sciences University, Beaverton, OR 97006, USA

[‡]*Department of Physics, University of Idaho, Moscow Idaho, USA*
[§]*don.baer@pnl.gov*
[¶]*tratnyek@ebs.ogi.edu*

Introduction

The chemical reactivity of nanometer-sized materials can be quite different
from that of either bulk forms of a material or the individual atoms and
molecules that comprise it. Advances in our ability to synthesize, visualize,
characterize and model these materials have created new opportunities to
control the rates and products of chemical reactions in ways not previously
possible. One application of this sort of tuning of materials for optimum
properties is in remediation of environmental contamination. In fact, it has
been suggested that nanotechnology applications for energy production and
storage; and water treatment and remediation will prove to be the number
one and three most important applications of nanotechnology for developing
countries.[1]

Nanoparticles of various types (metal, oxide, semiconductor or polymer) can be applied in a variety of ways to deal with environmentally-
related issues. They may be used for their sorption properties,[2,3] or for
their ability to facilitate chemical reactions, including those involving catalytic or photocatalytic behaviors.[4] Particles may be distributed by transport in solution[5] or immobilized in a polymer composite[2] or some type of

membrane.[6] Nanoparticles that rely on surface sorption sites have a high but finite storage capacity because of the large surface area of nanoparticles. For particles with catalytic properties, the desirable behaviors may continue indefinitely, assuming that some type of poisoning process does not occur. A third-type chemistry is where the particle is a reactant that reacts with contaminants, transforming them to more desirable products. The main example of this, is zero-valent iron which reduces many contaminants, resulting in oxidation of the particle. In general, reaction occurs until all of the accessible metal is oxidized, and this determines the useful lifetime of such materials. For nanoparticles, all of the metal contained in the particles can be oxidized, so its capacity to reduce contaminants is limited by the total quantity of metal.[7] In other cases, especially for larger particles, they may become passivated by a shell of unreactive material which prevents the metal core from reacting further. In this case, the capacity of iron to reduce contaminants will be limited by rate and degree of passivation.

Zero-valent iron (ZVI), including non-nanoparticle forms for iron, is one of the most promising remediation technologies for the removal of mobile chlorinated hydrocarbons and reducible inorganic anions for groundwater.[8-10] Suggestions that iron nanoparticles have an enhanced reactivity and may be relatively easily delivered to deep contamination zones[5] have great appeal to the engineering community and summaries of field demonstrations have already been completed and described in the literature.[11,12] In addition to enhanced reactivity and particle mobility in the sub-surface, iron nanoparticles potentially offer an ability to select the reaction products formed. The prospect for better remediation technologies using nanoparticles of iron, iron oxides, and iron with catalytic metals (i.e. "bimetallics") has potentially transformative implications for environmental management of contaminated sites around the world. Of particular interest is the potential to avoid undesirable products from the degradation of organic contaminants by taking advantage of the potential selectivity of nanoparticles to produce environmentally benign products.

Although ZVI nanoparticles may have great potential to assist environmental remediation, there are significant scientific and technological questions that remain to be answered. Work in our research groups has confirmed that different types of nanoparticulate iron do lead to different reaction pathways.[13] However, the factors controlling the pathway are not yet understood and therefore not readily controlled. We need to know more about the characteristics of the nanoparticles and what properties most influence the reaction pathways. Because of the high reactivity of iron in many environments, iron nanoparticles are either precoated with some type of protecting layer or such a layer forms upon exposure to air or water. This outer layer is sometimes called the shell leading to a core-shell structure with the metal being in the core. Many of the properties of these core-shell nanoparticles will be controlled or at least moderated by the transport and chemical nature of the shell. Thus core-shell nanoparticles containing ZVI are inherently more complex than the "pure" single-material nonreactive nanoparticles that are often studied. Among the complications are the time-dependence and environmental-dependence of the particle composition, structure and properties.

Understanding of reactive metal core-shell nanoparticles requires use of particles that are as well-characterized and understood as possible. We have, therefore, undertaken a series of studies that include synthesis of particles, analytical characterization of the particles and finally measurements of their chemical properties. Current efforts include understanding how the particles evolve in time during storage and when reacting with contaminants in aqueous solution. The remainder of this chapter describes the approaches we (and others) have used and are extending to determine and ultimately control properties of reactive metal nanoparticles for environmental remediation.

Synthesis

The terms top-down and bottom-up[14] have become relatively common descriptions for an engineering approach of making small objects from larger ones (top-down) or the building up of larger objects from smaller ones (bottom-up). Although most nanoparticles are made by self-assembly processes, it is possible to use grinding and ball-milling processes to produce nanostructured particles including metallic iron nanoparticles.[15-18] Because they are more common, we focus here on a limited set of self-assembly methods.

Many different self-assembly approaches can be used to produce metallic iron nanoparticles including wet chemical and gaseous or vacuum routes. The different approaches often produce particles with variations in physical shapes and coatings structure. Vacuum-based methods can allow a high level of control of the particle composition and some ability to alter the particle coatings. The coatings on vacuum- or gaseous-formed particles are often relatively well-defined oxide structures. Such particles are well-suited for fundamental mechanistic studies. However, they are difficult to produce in large quantities that would be required for actual remediation work. In all cases, a variety of materials and shape may appear and impurities may be introduced. A close coupling of synthesis and characterization is required to determine the true nature of the particles actually produced.

Solution synthesis and the hydrogen reduction of oxide nanoparticles have been used to produce enough nanoparticles for in-ground remediation tests. These particles may be less well-defined in several ways, including shape and size uniformity and many have coatings that contain a residue from or organics deliberately added during the synthesis process. The oxides formed on solution produce particles that often appear to be less crystalline and may contain elements in addition to iron and oxygen. We have used different types of synthesis processes to allow studies that enable us to address fundamental questions as well as understanding the important properties of the types of particles used in field studies.

Vapor-Phase Nanoparticle Synthesis

Nanoparticle synthesis is an important component of many rapidly growing research efforts in nanoscale science and engineering. They are the starting point for many "bottom-up" approaches for preparing nanostructured materials and devices. Nanoparticles of a wide range of materials can be prepared by a variety of methods. This section describes methods for preparing nanoparticles in the vapor phase. There are significant variations in the nature and quantities that can be prepared by each method. The production method must be matched to the need. Mechanistic studies often require very well-defined and uniform materials while applications need an adequate supply of material that functions in the appropriate manner. The detailed information can found in the book edited by Granqvist.[19]

In vapor-phase synthesis of nanoparticles, conditions are created where the vapor-phase mixture is thermodynamically unstable relative to formation of the solid material to be prepared in nanoparticulate form. This usually involves generation of a supersaturated vapor. If the degree of supersaturation is sufficient and the reaction condensation kinetics permits, particles will nucleate homogeneously. Once nucleation occurs, remaining supersaturation can be relieved by condensation or reaction of the vapor-phase molecules on the resulting particles, and particle growth will occur rather than further nucleation. Therefore, to prepare small particles, one wants to create a high degree of supersaturation, thereby inducing a high nucleation density, and then immediately quench the system, either by removing the source of supersaturation or slowing the kinetics, so that the particles do not grow. In most cases, this happens rapidly in about milliseconds in a relatively uncontrolled fashion, and lends itself to continuous or quasi-continuous operation.

Flame synthesis

The most commercially successful approach to nanoparticle synthesis — producing millions of tons per year of carbon black and metal oxides is flame synthesis. One carries out the particle synthesis within a flame, so that the heat needed is produced *in situ* by the combustion reactions. However, the coupling of the particle production to the flame chemistry makes this a complex process that is rather difficult to control. It is primarily useful for making oxides since the flame environment is quite oxidizing. Recent advances are expanding flame synthesis to a wider variety of materials and providing greater control over particle morphology. Janzen and Roth[20] recently presented a detailed study of flame synthesis of Fe_2O_3 nanoparticles, including comparison of their results to a theoretical model.

Chemical vapor synthesis

In analogy to the chemical vapor deposition (CVD) processes used to deposit thin solid films on surfaces, vapor-phase precursors during chemical vapor synthesis are brought into a hot-wall reactor under conditions that favor nucleation of particles in the vapor-phase rather than deposition of a film. This method has tremendous flexibility in producing a wide range of materials and can take advantage of the huge database of precursor chemistries that have been developed for CVD processes. The precursors can be solid, liquid or gas at ambient conditions, but are delivered to the reactor as a vapor (from a bubbler or sublimation source, as necessary). There are many good examples of the application of this method in the recent literature.[21,22]

Physical vapor synthesis

Four different physical vapor deposition processes are described, inert gas condensation, pulsed laser ablation, spark discharge generation, and sputtering gas-aggregation.

Inert gas condensation

This straightforward method is well suited for production of metal nanoparticles, since many metals evaporate at reasonable rates at attainable temperatures. By heating a solid to evaporate it into a background gas, one can mix the vapor with a cold gas to reduce the temperature, and produce nanoparticles. By including a reactive gas, such as oxygen, in the cold gas stream, oxides or other compounds of the evaporated material can be prepared. Wegner *et al.*,[23] presented a systematic experimental and modeling study of this method, as applied to preparation of bismuth nanoparticles, including both visualization and computational fluid dynamics simulation of the flow fields in their reactor. They clearly showed that they could control the particle size distribution by controlling the flow field and the mixing of the cold gas with the hot gas carrying the evaporated metal.

Pulsed laser ablation

Rather than simply evaporating a material to produce supersaturated vapor, one can use a pulsed laser to vaporize a plume of material that is tightly confined, both spatially and temporally. This method can generally only produce small amounts of nanoparticles. However, laser ablation can vaporize materials that cannot readily be evaporated.

Spark discharge generation

Another means of vaporizing metals is to charge electrodes made of the metal to be vaporized in the presence of an inert background gas until the breakdown voltage is reached. The arc (spark) formed across the electrodes then vaporizes a small amount of metal. This produces very small amounts of nanoparticles, but does so relatively reproducibly. Weber et al.,[24] recently used this method to prepare well-characterized nickel nanoparticles for studies of their catalytic activity in the absence of any support material. By preparing the nanoparticles as a dilute aerosol they were able to carry out reactions on the freshly prepared particles while they were still suspended. Metal oxides or other compounds can be prepared by using oxygen or another reactive background gas.

Sputtering gas-aggregation

A schematic view of a newly developed sputter-gas-aggregation (SGA) nanocluster deposition apparatus[25-27] that has been used to produce Fe nanoparticles important to our research is shown in Figure 1.[29] We label materials produced by this process as Fe^{SP}. The system uses a combination of magnetron-sputtering and gas-aggregation techniques. The cluster beam deposition apparatus is mainly composed of three parts: a cluster source, an e-beam evaporation chamber and a deposition chamber. The sputtered Fe atoms from a high-pressure magnetron-sputtering gun are decelerated by collisions with Ar gas (the flow rate: 100–500 sccm) injected continuously into the cluster growth gas-aggregation chamber, which is cooled by chilled water. The clusters formed in this chamber are ejected from a small nozzle by differential pumping and a part of the cluster beam is intercepted by a skimmer, and then deposited onto a sample holder in the deposition chamber. The mean size of clusters, from 1 nm to 100 nm, is easily varied by adjusting the aggregation distance, the sputter power, the pressure in the aggregation tube, and the ratio of He to Ar gas flow rate. The aggregation distance and the ratio of He to Ar gas flow rate are important parameters for getting a very high intensity (> 5 mg/h), monodispersed nanocluster beam.

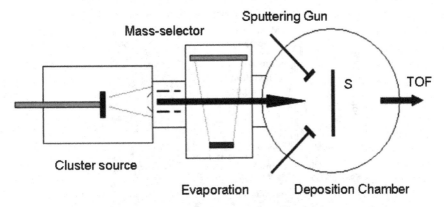

Figure 1.   Schematic drawing of three-chamber nanocluster deposition system. This is a 3rd generation particle source with increased particle production rates. Particles are produced in the cluster source and mass selector areas. They can be coated or react with a gas in the evaporation chamber and can be exposed to gas and collected on a substrate in the deposition chamber. After Ref. 28.

A major advantage of this type of system is that the clusters have much smaller size dispersion than grains obtained in a typical vapor deposition system. Studies with Transmission Electron Microscopy (TEM) and time-of-flight (TOF) mass spectrometer have shown that the observed lognormal size distribution has a standard deviation of below 10%. When oxygen gas is introduced into the deposition chamber during processing, uniform iron oxide shells covering the Fe clusters are formed. For a constant flow of Ar gas, the gas pressure in the deposition chamber can be adjusted by changing the flow rate of oxygen gas (O_2). In this circumstance, the iron clusters have an oxide shell before they are collected on a substrate. The clusters land softly on the surface of substrates at room temperature and retain their original shape. This process ensures that all Fe clusters are uniformly oxidized before the cluster films are formed.

Transmission Electron Microscopy (TEM) images of the core-shell Fe cluster assemblies deposited on a TEM grid are shown in Figure 2. In this example the mean cluster diameter (D) is 6.65 nm and the standard deviation is less than 7%. High-Resolution TEM (HRTEM) images of four Fe nanoparticles with sizes ranging from 3 nm to 85 nm deposited on a carbon film of microgrids are shown in Figure 3. The smaller nanoparticles appear almost spherical while crystal shape begins to appear for larger particles. The smallest particle is fully oxidized as are all of the air-exposed particles with a diameter of 7 nm or less. The dark center is an iron core region while the light-gray shell is oxide. To the extent we can measure by TEM, the oxide shells on these particles do not significantly grow when exposed to air.

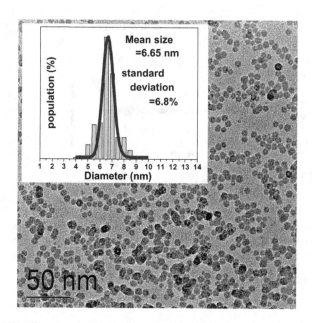

Figure 2.   TEM image and size distribution of iron/iron oxide nanoclusters deposited on a TEM grid by the SGA nanocluster source. The inset shows the size distribution of the particles as determined from TEM measurements. After Ref. 29.

Solution Phase Synthesis

Incipient wetness is one the oldest and most time-honored methods of producing nanoparticles on surfaces, especially for catalytic purposes.[30] In a typical procedure, a solid support is "wetted" with a metal salt solution, the solvent is removed and the metal ions are treated to produce the desired materials (metals, metal oxides, metal sulfides). The morphology of the particles produced depends upon the metal salts, solvents, solution ionic strength, co-precipitation agents (if any), temperature, and supports. The typical particles can be up to micron-sized and polydisperse, but for precious metals with low loadings the particles can be nanometer-sized and surprisingly monodisperse.

Micellar techniques

Micellar techniques, which include normal oil-in-water microemulsions and reverse water-in-oil systems, are intuitively attractive methods of controlling particle size by controlling the size of the microemulsion droplet in which the particles are formed.[31,32] These nanoreaction vessels are tailored

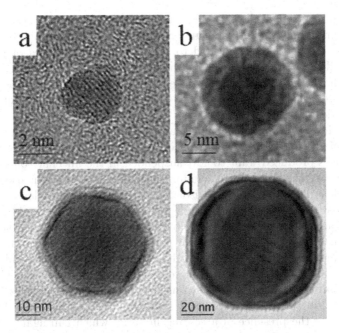

Figure 3.   HRTEM micrographs of the Fe clusters with diameters from around 3 nm to 85 nm prepared on carbon microgrids by SGA cluster source. The smaller particles are round while crystallographic faces begin to appear at larger sizes. Particles below 8 nm are fully oxidized. All particles have an oxide shell. After Ref. 28.

for core sizes up to 50 nm. In micellar and reverse micellar systems, a surfactant (or amphiphilic compound) is dissolved in a continuous phase and the second phase is mixed with the dissolved surfactant to form a thermodynamically stable, optically transparent, homogeneous solution. This solution actually consists of monodisperse droplets of the second phase in the nanometer- to hundred nanometer-size range. The size of the droplets and the shape of the droplets are controlled by the ratio of the second phase to the surfactant.

Reverse micelles allow the use of aqueous soluble reagents and aqueous reaction chemistries to produce nanoparticulates. The water-to-surfactant ratio actually determines the shape and the size of micellar droplet. Ternary phase diagrams, such as those found in Darab *et al.*,[33] are maps of the phase stability and can be maps of the shape and size of the micelle core. Working in the non-spherical region of the phase diagram allows researchers to produce mesoporous templates as described in other chapters. Here, we are concerned primarily with spherical aqueous droplets in an organic continuous phase.

Standard reverse micelle systems have worked well for the production of milligram amounts of quantum crystallites.[34,35] Unfortunately the high ionic strength solutions needed to produce gram quantities of particles destabilize the reverse micelles. To counter this negative effect, co-surfactants have been successfully utilized to stabilize reverse micelles at high ionic strengths.[33] This method has allowed the production of multigram quantities of nanoparticles.[36]

Nanoparticles produced using micellar methods are usually monodisperse. One of the problems with the micellar method is particle collection as the particles produced want to stay suspended in the micellar solution. Methods for particle collection, including micelle destabilization, usually lead to particle agglomeration. The isolated particles from micellar methods are almost always coated to some degree with the surfactants. Almost any type of aqueous reaction can adapted to the reverse micellar system to produce nanoparticles. The types of particles which can be produced include metals, metal oxides, metal sulfides, core-shell particles, mixed metal-containing materials, and organic coated particles. Supercritical CO_2 reverse micellar systems have also been used to produce nanoparticles.[37]

A variant on the micellar particle synthesis method is to produce nanoparticles in the presence of a coating material which can again be a surfactant, a polymer or another reagent which attaches itself to the surface of the particle. This technique has been successfully exploited using thermal or sonochemistry to produce metallic particles from metallic carbonyl or strained organometallic precursors.[38,39] In these syntheses, milligrams to gram quantities of coated nanoparticles can be produced. The coatings can be beneficial in that they can stabilize the small particle size and they stabilize the thermodynamically unstable species, i.e. nanometallics in the presence of oxygen. In some cases it is not necessary to add a surfactant as the particles produced are self-protecting. This is the case for the iron particles produced by Zhang and co-workers using an aqueous/ethanol borohydride reduction of ferrous salts.[40,41] In this case, a boron-rich layer is formed on the iron metal core which passivates the particles to growth and toward oxidation. We have examined some of this material produced by the Zhang group and material we have synthesized at Pacific Northwest National Laboratory (PNNL). This material is labeled as Fe^{BH} in our studies.

Hydrothermal synthesis

Hydrothermal synthesis either in batch or in a continuous mode can be useful in producing multi-gram quantities and nanocrystalline materials. As the name implies, these reactions are run with metal salts dissolved in water at high temperatures in pressure vessels. These reactions produce metaloxides by hydrolysis of the metal ions. Batch reactions can yield highly crystalline materials but the long heat-up and cool-down times of large pressure vessels preclude the production of nanoparticulates. The ionic strength of the solution does determine to some extent the size of the crystals and particulates produced.

An alternative to the batch hydrothermal method is the Rapid Thermal Decomposition of Solutes (RTDS) method in which the metal ions in solution under pressure are subjected to hot reaction zones for very short times (seconds) and then immediately cooled as the superheated liquid is passed through a nozzle.[42] The RTDS method allows one to control the particle size of metal oxide products through temperature, reaction time, pressure, and by adding co-reactants to the reactant stream. This is an especially good method to produce mixed metal or doped species. For example one can add 1% of a nickel salt to a ferric salt solution and obtain a 1% uniform nickel oxide doping of the iron oxide produced.

Post synthesis processing

In all of the above methods (solution and vapor-phase) one can produce particles which may or may not be reduced. Further processing, either through gas-phase reduction with hydrogen or another solution reduction step can lead to reduced particles, even to the point of producing metallic particles. During the subsequent steps, care must be taken to ensure minimal particle growth which usually entails mild reduction conditions. Residues from earlier synthesis or processing steps can have a significant influence on product shape and surface chemistry; increasing the need for particle characterization. In unique cases, the reduction quickly can follow the oxide particle formation. A commercial product produced by Toda Kogyo Corp. (Schaumberg, IL) known as RNIP is produced by the reduction of goethite and hematite particles with H_2 (200–600°C).[43] We have conducted some experiments on this type of material which are indicated by the label Fe^{H2} in our work.

Particle Handling and Characterization

In this section and the following one, we first show by example the application of some methods for *ex situ* characterization and then provide a summary of methods used to characterize reactions in solution. Before looking at specific techniques or examples, it is important to consider some of the general issues and challenges related to characterizing and measuring the properties of reactive metal nanoparticles. Readers unfamiliar with some of the basics of the methods described are referred to the "Encyclopedia of Materials Characterization" by Brundle, Evans and Wilson.[44]

Challenges for Nanoparticle Handling and Characterization

General issues

Sample handling, environmental stability, and contamination issues that are sometimes problems for micron-scale or bulk materials, are of significantly greater importance for nanoparticles. It is possible, for example, for nanopores to retain solvents that would evaporate from surfaces.[45] This solvent retention, which could occur during a standard solvent cleaning precedure before analysis, could alter sample analysis in several ways and lead to confusion in interpreting results. Surface contamination may be a minor issue for a "bulk" material but is of significant concern to nanoparticles. Even relatively stable (in comparison to metallic iron) nanoparticles may change in different environments. For example, the structure of ZnS nanoparticles has been show by Zhang *et al.*,[46] to change due to the presence or absence of water in the surrounding environment. This environmental effect is not an accident, but a characteristic of small systems, "the thermodynamics of small systems is highly dependent on the environment".[47] The size, structures and even composition of nanoparticles will significantly depend on the history and processing of the particles. For nanoparticles, any set of physical and chemical measurements is incomplete without a relatively complete history of the sample handling and processing.

In addition to environmental effects and specimen history, there is a more general instability or dynamic behavior of nanoparticles. The existence of many local minima for nanostructures, the shape and physical structure of nano-sized objects may be readily altered by any incident radiation including incident electron[48] or X-ray flux.[49] TEM images of Au nanoparticles show that the structure is not constant but varies with time.[48] This dynamic behavior of nanoparticles can be important and it has been shown, as one example, that the changes in the structure of catalyst nanoparticles as a function of time (at somewhat higher than ambient temperatures) are an important property of catalyst particles during the growth of carbon nanotubes (CNTs).[50] Particles large enough to minimize the shape changes are not useful catalysts for the CNTs.

It is also useful to recognize that some electronic and magnetic characteristics of individual nanoparticles are altered by close proximity to other nanoparticles. Proximity effects might be expected for magnetic interactions. As one example, inter-particle interactions are observed to significantly change the superparamagnetic relaxation behavior of NiO nanoparticles.[51] However, proximity effects also influence electronic properties of nanoparticles. For instance, Au nanoparticles in dilute solution have a size-dependent surface plasmon resonance that can be observed in UV-Vis absorption measurements. When Au particles reversibly aggregate in solution, an additional extended plasmon band is observed in addition to the size-dependent surface plasmon peak.[52] Packing or dispersing nanoparticles for measurement purposes may in some circumstances alter the properties being measured.

Special issues for reactive metals

Many of the issues noted above are present for most nanoparticles. However, because of their chemical reactivity, iron and other reactive metal nanoparticles add an additional complexity as they react to solutions (or other environments) to which they are exposed, including the chlorinated hydrocarbons of particular interest. Conceptually it is sometimes useful to recognize that passive corrosion films may react or evolve with time. This might be called an intra-particle effect since it does not necessarily involve extensive interaction with the surrounding environment (e.g. the so called "autoreduction" between core and shell material[53]). In addition, the particles will interact with the environment, including both major components of the environment (e.g. water and water components) and contaminants (usually relatively minor species in the water). These might be called extra-particle interactions. These different types of interactions may occur at different rates and with different sensitivities to environmental conditions, so their distinctions can be useful, even if they are all — ultimately — interrelated.

To a significant degree, time is an additional variable not independent of environmental effects. Since metallic iron is thermodynamically unstable in many environments, it is likely that with long term storage there will be additional oxidation of the iron or other changes in the material. Therefore, properties of the nanoparticles will likely vary somewhat as a function of time. Although a small amount of corrosion may not significantly alter the bulk behavior of a piece of iron, the same amount of corrosion may significantly alter the nature of nanoparticles. Although we generally store and handle specimens in relatively inert environment (anoxic, dry, and/or vacuum), we often see time-dependent changes in the particles.

Although the oxides that formed on the sputtering produced iron nanoparticles, as mentioned earlier, can be described as stable, stable is only a relative term. It is nearly impossible to produce a metallic particle with no coating and the particles have a tendency to continue to oxidize over time. We have examined solution formed particles (with their oxide coatings) using surface sensitive techniques with and without exposure to air; and found an additional amount of oxidation for samples exposed to air for even a few minutes. Environmental-stability questions frequently require careful sample preparation and handling so as to minimize the effects of air-exposure or time on the analysis results. Control specimens and other test procedures may be needed to provide some bounds on the environmental effects on the results.

Sample preparation and handling

It is safe to say that any characterization tool that has been applied to any type of material is currently (or soon will be) applied to nanostructured materials. Many of the tools useful for the most detailed characterization of materials, surfaces and nanoparticles require removing the material from the environment of interest to an atmosphere useful for the analysis. For example, electron microscopy and spectroscopy methods usually require moving the material into vacuum. Although there should always be a concern about the impacts of the change in environment, much critical useful information is obtained from these *ex situ* methods. A few methods, such as optical spectroscopy can more easily be applied in the environment of interest (or something similar). These *in situ* methods are very useful, but often provide only a subset of the information needed. Ideally it would be possible to collect all of the information we would like to obtain about a nanoparticle in the environment of interest as a function of time. Since this is not possible we must combine the use of *in situ* and *ex situ* methods to piece together information about the particles. Since each technique has limitations and no one method provides all the desired information it is helpful to know the strengths and limitations of each experimental tool, including the possible effects of sample handling and probe induced damage. It is hard to over emphasize the importance of application of multiple and complementary methods for obtaining an accurate understanding of nanoparticles.

The reactivity of metallic iron makes sample preparation and handling an important challenge. In general, we presume that these materials will be most stable when stored dry and under an inert (anoxic) atmosphere. However, some samples are received (or synthesized) in water or under air. The wet samples must be dried for analysis by some characterization methods (e.g. TEM) and the dry samples must be immersed to allow for other characterization methods (e.g. electrochemistry). All of these manipulations will affect the results of characterization and not necessarily in a consistent or (at this time) predictable way. To manage this issue, we have adopted a protocol for removing nanoparticles from solutions that we call "flash drying", and applied this to most of the wet samples we have received or synthesized.[54] Then, when necessary, the flash-dried powder can be weighed and re-immersed in solution — along with samples that were received or synthesized dry. This process made the samples easier to handle and process and enhanced the reproducibility of the measurements without altering the properties (physical or chemical) significantly.[13] Additional comments about specific specimen handling methods associated with specific methods are discussed below.

Ex Situ Examples Iron Nanoparticle Characterization

The techniques to be used and the order in which they are used depend upon the questions being asked. When characterizing nanoparticles, natural questions include: size, size distribution and surface area; surface and "bulk" composition and chemical state; electronic and physical structure, including the presence of defects. General characteristics of these materials and other samples used in much of our work are included in Table 1. In Table 2 we list a set of the techniques applied for the characterization of iron nanoparticles and a rough indication of the type of information they can be used to obtain. The lighter symbols indicate that the information is not a primary result of the method. The "X" indicates that we routinely apply the technique in our work and that examples of the use will be presented in the text.

To demonstrate some of the similarities and differences between different types of nanoparticles and how these differences can be identified, several examples are provided from three types of iron nanoparticles that we have examined most extensively.[13] Although the methods have not been described in great detail, each of these materials was mentioned in the synthesis section. For ease of identification, they are labeled in the following ways. Materials produced by the cluster sputter process are identified as Fe^{sp}; nanoparticles produced by the borohydride-reduction process are identified as Fe^{BH}; and particles produced by the H_2 reduction of iron oxides are labeled Fe^{H2}.

Transmission Electron Microscopy (TEM)

Transmission Electron Microscopy is an essential tool for characterization of the natural or synthesized nanoparticles.[55] When the particles are deposited at relatively low density on a TEM grid, it is possible to gain information about the size, shape, and structure of the particles. TEM images of different sized sputter-grown particles were shown earlier. With careful work, it is possible to examine the structure of the particles and particle coatings.

Table 1. Nanoparticles and characteristics.

Name	Source	Method	Particle Size (dia.)	Surface Area	Major Phase	Minor Phase
Fe^0/ Disc		High purity polished Fe^0			Fe^0	
Fe^{H2}	Toda Americas, Inc.	High temp. reduction of oxides with H_2	70 nm	29 m^2/g	α-Fe^0	Magnetite
Fe^{BH}	W.-X. Zhang, Lehigh Univ.	Precip. w/NaBH$_4$	10–100 nm	33.5 m^2/g	Fe^0	Goethite, Wustite
Fe^{EL}	Fisher Scientific	Electrolytic	150 μm	0.2–1 m^2/g	99% Fe^0	
Fe^{SP}	Y. Quiang, Univ. Idaho	Sputter Deposition	2–100 nm		Fe^0	Magnetite, Maghemite
Fe_3O_4	PNNL	Precip from $FeSO_4$ w/KOH	30–100 nm	4–24 m^2/g	Fe_3O_4	
Fe_2O_3	Nanotek, Corp.	Physical Vapor Synthesis (PVS)	23 nm	50 m^2/g	γ-Fe_2O_3	

Adapted from Ref. 13.

Figure 4 shows a high-resolution TEM (HRTEM) image of a relatively large Fe^{SP} nanoparticle where crystal structure has an effect on the particle shape. At these larger sizes, the nanoparticles are transforming from the round shapes (characteristic of smaller particles) towards a more crystallographic structure. It is relevant to remember that the TEM images are a projection of the particle shape on one plane. Unless tomography measurements are taken or the diffraction patterns are modeled, the relationship of the 2d projection and the 3d shape is not necessarily certain. In this case, the diffraction image has been modeled and the data is consistent with a slightly truncated cube defined by {100} planes. The particle is also covered by an iron oxide shell. In other work, we have identified vacancies in the particles[56] that result from their oxidation.

TEM images of Fe^{H2} and Fe^{BH} materials removed from their as-received storage conditions (dry inert gas for this particular Fe^{H2} and in an alcohol solution for the Fe^{BH}) and quickly[57] inserted into the TEM to minimize sample changes are shown in Figure 5. The shapes of the Fe^{H2} particles, one example of which is shown in Figure 5(a), vary significantly and the particles are covered by a crystalline oxide. The Fe^{BH} particles appear (Figure 5b) almost as bubbles, usually round but of varying size. The coating on these particles does not appear to be crystalline.

Table 2. Partial listing techniques and information that can be obtained about nanoparticles or collections of nanoparticles.

Techniques	Information Obtained								
	Size & Shape	Surface Area	Grain Size	Physical Structure	Composition	Surface Chemistry	Electronic Structure	Molecular Structure	Chemical Reactivity
• Scanning Electron Microscopy + EDS	✓				✓				
• Transmission Electron Microscopy + EDS + EELS	✗	✗	✓	✓	✓		✓		
• X-ray Diffraction			✓	✓					
• Small Angle X-ray Scattering	✓								
• Gas Sorption (BET)		✗							
• X-ray Photoelectron Spectroscopy					✗	✗	✓		
• Auger Electron Spectroscopy					✓	✓			
• X-Ray Adsorption Spectroscopy and Microscopy					✓		✓		
• Scanning Probe Microscopy	✗								

(*Continued*)

Table 2. (Continued)

Techniques	Information Obtained								
	Size & Shape	Surface Area	Grain Size	Physical Structure	Composition	Surface Chemistry	Electronic Structure	Molecular Structure	Chemical Reactivity
• Secondary Ion Mass Spectrometry						✓		✓	
• Raman Spectroscopy								✓	
• Fourier Transform Reflection Infrared Spectroscopy								✓	
• Light Scattering	✓								
• Mossbauer Spectroscopy				✓					
• Electron Paramagnetic Resonance									✗
• Electrochemical Methods									✗
• Batch Reaction Studies									✗

The symbols (✓ and ✗) indicate information that can be obtained by the techniques listed. Lighter symbols indicate information that is not a primary result of the method. The ✗'s indicate techniques that we routinely apply in our work and for which examples of use of the technique are presented in this chapter.

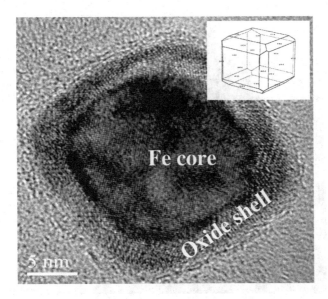

Figure 4. HRTEM image of a core-shell structured Fe-iron oxide nanoparticle. The inset shows the particle morphology following the Wulf construction. The particle is defined by 6 {100} planes and truncated slightly at the edges by the {110} type planes.

Hydrogen Reduced – Iron Nanoparticles –Dry Stored
As received

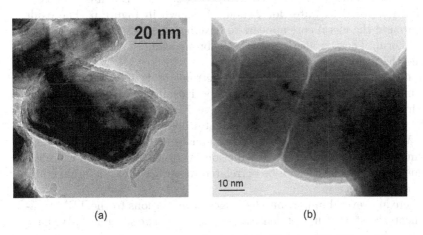

(a)   (b)

Figure 5. TEM images of dry, stored as-received (a) Fe^{H2} and (b) Fe^{BH} nanoparticles. Both types of particles show a core-shell structure.

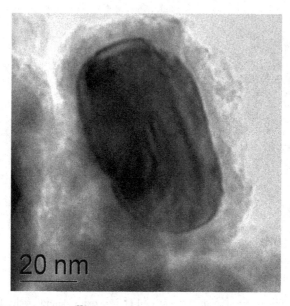

Figure 6.   TEM image of Fe^{H2} nanoparticle after 24 hours exposure to DI water. The core shell structure remains, but the shell appears thicker and less uniform. Fibrous iron oxide is observed around the metal particles.

The TEM images in Figure 5 are from material before exposure to water or water containing a chlorinated hydrocarbon. The TEM image of Fe^{H2} specimen exposed to water for 24 hours is shown in Figure 6. The oxide shell around the metal core appears to have become somewhat thicker and less uniform. There is also a growth of fibrous oxide — likely goethite — around the particles.

When an Electron Energy Loss spectrometer is incorporated in an instrument, it is possible to obtain compositional and chemical information from the particles. Energy Dispersive X-ray Spectroscopy (EDS or EDX) can also be used to gather elemental information about the particles. When many images are collected and analyzed, it is possible to obtain information about particle size distribution. As discussed below, TEM measurements of particle size can be correlated to surface areas determined by other methods.

There are several important challenges or limitations to the TEM measurements. First, the current densities used can damage specimens. We have observed that some particles shells can be damaged and evaporated by the electron beam while in other circumstances the shells can grow. There can also be a sample analysis bias or selectivity in TEM because TEM measurements can only be made with the specimen that is transparent to the electrons. Consequently, it can be useful to compare TEM results with those obtained by other methods to verify consistency. Inconsistent results may not be an indication of error, but provide useful information related to particle size distributions.

X-ray Diffraction (XRD)

X-ray Diffraction is also an important tool for analysis of the crystal structure and grain size of the nanoparticles. Specimens of Fe^{H2} paste or Fe^{BH} slurry were prepared for XRD analysis by spreading ~ 25 mg of material on a zero-background slide and allowing it to dry for 2–3 days in an anoxic glove-box that was continuously purged with N_2. In an attempt to protect them from oxidation in air during transport and the diffraction measurements, the dried specimens were coated with glycerol by applying a few drops of 10% glycerol in 95% ethanol, and allowed to dry for 1–2 additional days. The coated specimens were analyzed in ambient air using Cu-K_α radiation in a Philips X'Pert MPD diffractometer (PW3040/00) operated at 40 KVP and 50 mA. Continuous scans from 2–75 °2θ were collected at a scan rate of about 2.4 °2θ min^{-1}.

The XRD patterns for as-received Fe^{BH} and Fe^{H2} specimens are shown in Figures 7(a) and (b) respectively.[58] Both samples have a major peak at a 2-Theta value of approximately 44° characteristic of metallic iron. The difference in peak width is readily apparent. Based on the Scherrer equation, after correction for instrumental broadening, the relatively wide main diffraction peak from the Fe^{BH} samples is consistent with Fe^0 crystallites with an average grain size of 1.5 nm. There is also a broad diffraction peak that may be from Goethite. The narrower peaks from the Fe^{H2} samples indicate Fe^0 with a 30 nm average grain size and magnetite with a 50 nm grain size.

To test for sample stability, successive scans were taken 30 minutes, 60 minutes and 24 hours of exposure to ambient air. No change in the structural properties of the glycerol-coated specimens was observed, verifying either that the protective effect of the glycerol film with respect to oxidation or that the drying process itself was sufficient to slow the reactivity of the nanoparticles over the time frame of the XRD measurement. Mean crystallite dimension was estimated using the Scherrer equation, after correction for instrumental broadening.

Useful information can be obtained by combining information about the two types of material from TEM and XRD data. The Fe^{H2} sample shown in Figure 5(a) has a particle size observed by TEM quite similar to the grain sizes estimated from the XRD peak width. This implies that the Fe^{H2} nanoparticles are highly crystalline with each particle made up of one

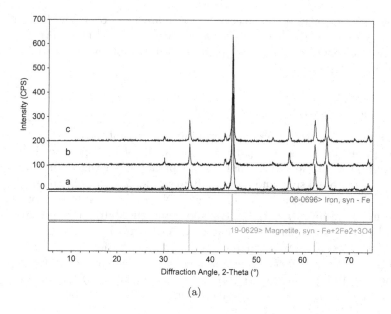

(a)

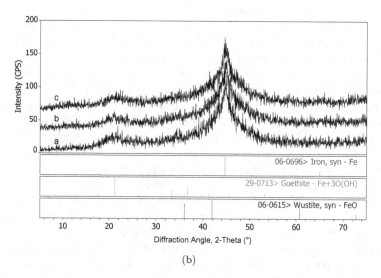

(b)

Figure 7. (a) X-Ray diffraction pattern from Fe^{H2} particles received in a slurry. The patterns show the presence of iron metal with an average grain size of approximately 30 nm and magnetite with grain sizes of approximately 50 nm. (b) X-Ray diffraction pattern from Fe^{BH} particles received in a solution. The patterns show the presence of iron metal with an average grain size of ~1.5 nm and possibly some goethite (after Ref. 13 figure S1). The different scans (a)–(c) in the figure were collected a) 30 minutes, b) 60 minutes, and c) 24 hours after removal from the glove box as described in the text.

or a few metal grains. In contrast, the Fe^{BH} particles as shown in Figure 5b are approximately the same size as observed in TEM, but they are made up of 1.5 nm metallic grains.

It is also possible to estimate the ratio of metal to oxide in a collection of material. We have made XRD and TEM measurements on a series of Fe^{H2} samples (in collaboration with Greg Lowry at Carnegie Mellon University) that have been exposed to water for different periods of time. The fraction of Fe in the magnetite phase is plotted for both TEM and XRD data in Figure 8. Several approximate assumptions are needed in both TEM and XRD to determine the fraction of Fe in magnetite. However, up to about 70 days, the trends in both types of data are very consistent. The longer time difference between XRD and TEM data is likely due to growth in size of the oxide grains to the point that they become opaque to the electron beam from the TEM.

The major limitations of XRD include sensitivity (the amount of material needed) and limitation to detection of crystalline phases. Micro-XRD can help minimize the amount of material needed for analysis. Small Angle X-ray Scattering (SAXS) can provide information about particle size, shape and pore size.

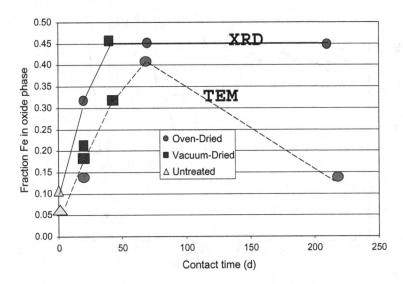

Figure 8. Fraction of measured Fe in magnetite phase as a function of time exposed to water. Up to about 70 days, the TEM and XRD data are relatively consistent. The differences at longer times are likely due to selective sampling of the TEM. In particular, larger oxide particles are not transparent to the electron beam.

X-ray Photoelectron Spectroscopy (XPS)

X-ray Photoelectron Spectroscopy is particularly useful for examining shell composition and the presence of impurities in or on nanoparticles. Because of the surface sensitivity of XPS, extra care is required to minimize changes in analysis results due to contamination or air exposure. A glove-bag has been connected to the spectrometer to allow transfer from a glove-box to the glove-bag from which the sample is inserted into the spectrometer. The Fe 2p photoelectron peaks from a Fe^{BH} sample transferred without air exposure and after air exposure are shown in Figure 9. The as-received material had a strong Fe^0 peak with indication of Fe^{+2}. After air exposure the Fe^0 was still present, but significantly decreased in size. The surface iron was oxidized and a significant amount of Fe^{+3} was present.

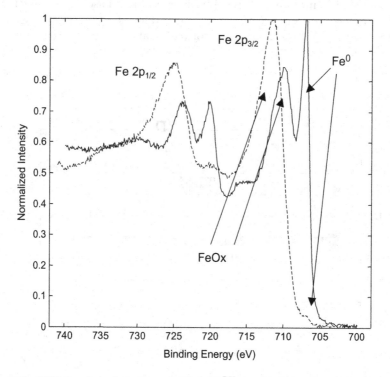

Figure 9. XPS 2p photoelectron peaks from Fe^{BH} material after anaerobic transfer (solid line) from the shipping container and the same sample after exposure to air for five minutes (dashed line).

In addition to looking for changes in surface chemical state, the XPS measurements provide an overall composition. The initial composition (in atomic %) of the as-received Fe^{BH} sample was 20% Fe, 49% O, 16% B, 15% Na, 0.5% S; while the initial composition of the Fe^{H2} was 51% Fe, 44% O, 3.0% Na, 1.9% S. The significant differences in the Fe are due to the presence of B in the coating of the Fe^{BH} material. The B is mostly oxidized indicating that the outer shell of this material has a significant BO_3 component.

The combination of TEM, XRD and XPS data indicate that even though both the Fe^{H2} and Fe^{BH} are made up of mostly Fe^0 particle 10s of nanometers in size, their grain size, shell structures and shell compositions are totally different.

Significant limitations of XPS include the need to place the samples in vacuum and to assemble a collection of nanomaterials to facilitate analysis. Usually, quantitative information extracted from XPS assumes that the sample is ideally flat. It is possible to combine TEM or other shape and size data to extract more elemental distribution information from XPS data through the use of particle and signal intensity models.[49,59]

Surface area determination

Particle size (and distribution) and the surface area associated with a collection of particles (or other nano-structured materials) are highly important and useful parameters regarding nanomaterials. The standard method for determining surface area is referred to as BET (Brunauer-Emmett-Teller); and involves N_2 (or Kr) physisorption and desorption as a function of heating a known weight of material. Although a common method, the procedures have sensitivity, pore volume and other complications that can limit accuracy. Because heating is involved, it is possible for nanoparticles to sinter or grow during heating. For many types of materials, the final temperature of the BET analysis needs to be lower than that used for routine analysis. The most widely-used alternative method of determining specific surface area is calculation — usually assuming spherical geometry — of surface areas from particle size distributions determined from TEM images. This approach also has advantages and disadvantages and rarely gives specific surface areas that are identical to those determined by BET. We compared the results obtained by these two methods in Ref. 13.

Consistent with the earlier discussion about particle stability, sample handling is an important issue. Damp powders place on filter paper in air, can start to burn the filter paper, clearly destroying the powder and the ability to obtain a surface area. We have found that surface area measurements for Fe^{BH} and Fe^{H2} stored as a paste or slurry appeared to have smaller surface areas per gram than reported in the literature. In particular, the nominal Fe^{H2} surface areas are reported to be about $29\,m^2/g$ with a range from 4 to $60\,m^2/g$. Our BET measurements on this material determined a value of $3\,m^2/g$. For the Fe^{BH}, the nominal literature value is $33\,m^2/g$ while our BET measured value was approximately $5\,m^2/g$.

As shown in Figure 2, TEM measurements are frequently taken to obtain particle size and particle size distribution. Because of the importance of obtaining good statistics to get reliable data, accurately obtaining surface area data from TEM measurements is very tedious. It is important to note that using the average size particle is not adequate to determine surface area data from TEM measurements, as particle size distribution can make a large difference. In spite of these significant uncertainties, our BET and TEM estimates of surface areas between one set of Fe^{BH} and Fe^{H2} material were relatively consistent. TEM surface area was approximately $7.5\,m^2/g$ in comparison to the BET value of $5\,m^2/g$. For Fe^{H2} the TEM value was $3.5\,m^2/g$ in comparison to a value of $3\,m^2/g$ from the BET measurement.

It is also possible to get particle size information from light scattering measurements, although there is uncertainty and some disagreements regarding the accuracy at lower particle sizes. We found that our iron particles tended to aggregate in the solution used for the light scattering measurements; and we primarily learned about the size of the aggregates. Small angle X-ray scattering can be used to get average particle size, particle shape, and pore size information. Because of differences in the measurement process and as well as sample handling, it is strongly recommended that multiple methods be used to obtain particle size, aggregation, size distribution and surface area information.

Other methods

Several other techniques are potentially important and have been used for iron nanoparticle characterization. We note particularly the use of Mossbauer spectroscopy to characterize the core shell structures of iron nanoparticles by Kuhn et al.,[60] who were able to show that the oxide shell structure changed depending the oxidation conditions. Synchrotron-based X-ray Absorption Spectroscopy has also proven very useful for characterizing the particle size dependence of the shell structure for iron core-shell particles.[61]

Scanning probe microscopy, particularly Atomic Force Microscopy (AFM) and Scanning Tunneling Microscopy (STM) have been major tools for characterizing nanostructures. AFM is a useful, but not primary tool for our characterization of nanoparticles or films produced by collecting

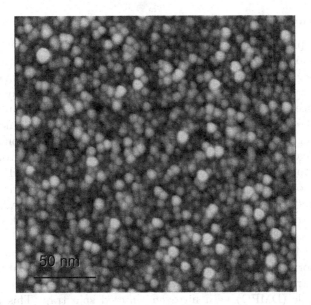

Figure 10. Atomic Force Microscopy image of Fe^{SP} particles for same material as shown in Figure 2.

nanoparticles on a substrate. AFM image of a particle film — for the same conditions as the TEM image in Figure 2 — is shown in Figure 10.

Although we have not applied them frequently, laser Raman Spectroscopy,[62] Fourier-Transform Infrared Spectroscopy (FTIR) and Time-of-Flight Secondary Ion Mass Spectrometry (TOF-SIMS) are each sensitive to molecular species and can sometimes be very useful for characterizing molecular species on particles surfaces.

Methods for Characterizing Reactivity in Solution

Reactions between nanoparticles and an aqueous medium can be viewed — and studied — from the perspective of the particles or that of the reactants in solution. *In situ* spectroscopies, such as Electron Paramagnetic Resonance (EPR), Raman, and Fourier-Transform Infrared (FTIR), as well as electrochemical methods can be used to monitor reactivity in near-real time and their results are used to infer changes in the properties of the particles. When the focus is on changes in solution composition such as the destruction of contaminants, periodic sampling and analysis of the solution from well-mixed batch reactors containing slurries of the particles or from flow-through reactors containing supported nanoparticles can be used.

Electron Paramagnetic Resonance (EPR) Spectroscopy

Chemical and physical properties of the shells surrounding metallic iron cores of the nanoparticles are likely to play an important role in determining the chemical reaction pathway, as noted in the introduction. The availability of electrons to transfer from the nanoparticle to a chlorinated hydrocarbon and the presence of hydrogen are two important factors that may influence the pathway. We have used EPR to monitor the changes in the mode of electron transfer by iron nanoparticles as they age in aqueous solutions.[63] In particular, we can determine when there is direct electron transfer from the particle to a spin-trap agent and when there is atom-transfer via H atoms generated by corrosion of the iron. The electron transfer mode is associated with production of relatively harmless reduction products whereas the hydrogen transfer is associated with greater production of undesirable products such as chloroform.

The approach involves the use of spin-trapping agents that react with reductants to form adducts having distinctive EPR signatures. Initial experiments demonstrating this technique were conducted using 5,5-dimethyl-1-pyrroline-N-oxide (DMPO) as a nitrogen-centered spin trap. This agent yields no EPR signal until it is reduced, and if reduced by a radical species, an adduct is formed that yields a distinctive EPR signal. In these experiments, portions of two solutions (1) a suspension of initially dry Fe^{H2} nanoparticles in buffered (pH 7 bis-tris propane), de-oxygenated de-ionized (DODI) H_2O, and (2) a solution of DMPO in DODI water mixed and injected into a flow-through flat cell located in the sample cavity of the EPR. The intensity and nature of the EPR signal from the spin-trap agent is then monitored over time periods of up to 24 hours, before additional solution is inserted into the cell. The initially dry Fe nanoparticles evolve with time (Figures 6 and 8) while suspended in the buffered DODI H_2O. Thus, the experiment can be used as an *in situ* monitor of the successive changes in the reactivity of the Fe.

The results of this experiment demonstrate that Fe^{H2} nanoparticles reduce DMPO in two ways. First, direct electron transfer to the DMPO occurs. At a later time atomic H is generated and H-associated electron transfer takes place. These produce distinct EPR signals that can be used to reveal the change in the reductive process (between the particles and DMPO) as a function of the time the particles have aged in solution. During the first 24 hours essentially all reduction occurs by direct electron transfer. After this period, H production begins and reduction by H transfer surpasses direct electron transfer after about 60 hours.

The particular nanoparticles used in this experiment consist of two phases, a magnetite phase associated with an oxide-coated zero-valent-Fe phase. Our interpretation of the results suggest that the oxide shell initially present on the zero-valent Fe phase shields the zero-valent Fe core for about 24 hours and all electron transfer occurs at the surface of the magnetite phase. Thereafter, breaches in the oxide shell occur to allow direct reaction of solution with zero-valent Fe and production of atomic H (see Figure 6). Over the next several days, total reductive capacity doubles and the H-transfer assumes dominance as oxide coatings form on the magnetite surface and decrease the contribution from the direct-electron transfer mechanism.

Electrochemical Methods

Electrochemical methods offer a unique perspective on the properties and reactivity of nano-structured materials. For the present purpose, key advantages include: (1) applicability to small quantities of material, such as is obtained by the gas-phase vacuum deposition methods described above, (2) control of mass transport so its contribution to reaction kinetics can be quantified, and (3) control over the formation of the oxide shell(s) that form on the particle's metal core.

To apply electrochemical methods to the characterization of nanoparticles, it is necessary to make working electrodes from the particles with minimal alteration of the particle structure and composition. One way to do this involves molding the nanoparticles into a disk electrode using binders and other additives. For example, by adapting the "soft-embedded" rotating disk electrode (RDE) described by Cha[64]; Rolison[65] made a "sticky carbon" electrode specifically for studying the corrosion of nano-sized catalysts. Subsequently, this electrode design was used to study the aqueous corrosion of nanoparticulate Fe metal and Fe-Ni bimetallics.[66,67]

In addition to the soft-embedded electrodes, there is a range of alternative electrode designs. For example, it is possible to make "powder" electrodes of nanoparticulate metals and metal oxides by compaction (without binders) in cavities on microelectrode supports.[64,68] Although most applications of powder electrodes have been to noble metals and semiconductors, one early study reported success with powder electrodes of iron metal.[69] An advantage of these methods for electrode fabrication is that they are used to support nanoparticles that are prepared separately, by whatever means and for whatever purpose. In other cases, it is desirable to form nanoparticles directly on electrodes, and this can be done by chemical vapor deposition,[70] electrochemical deposition[70] or *in situ* precipitation.[71] The latter was used in making thin film wall jet electrodes of microcrystalline iron oxides for studies of contaminant reduction by Fe(II) surface sites.[72,73]

The former — chemical vapor deposition — we have tested using the method described above.

We have applied the above methods in order to make electrodes out of a variety of nano-sized iron powders (Table 1), so that electrochemical methods can be applied to the characterization of these materials. Selected results obtained by linear sweep voltammetry are shown in Figure 12. The data show that electrodes made by gas-phase deposition of iron nanoparticles directly onto the electrode (Fe^{SP}) gives anodic polarization curves that are similar to what is obtained with electrodes packed with nano-sized iron (Fe^{H2} and Fe^{BH}), and this is indicative of the higher reactivity that is expected for nano-sized particles.[13,41,73–76] Ultimately, we hope to model the polarization curves obtained using these electrodes in the presence and absence of dissolved contaminants, and thereby, extract kinetic data for contaminant reduction on iron powders, as we have done previously with polished iron disk electrodes.[77,78]

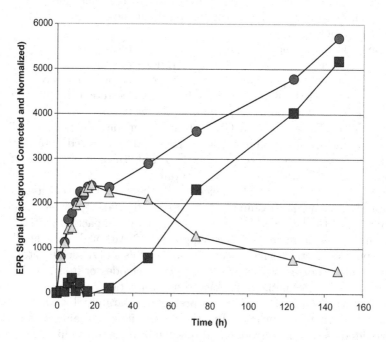

Figure 11. Changes in the relative proportion of direct electron-transfer and hydrogen-transfer products with aging time in H_2O. The triangles (△) are for direct electron transfer. The squares (■) are for atom (Hydrogen) transfer and the circles (●) are the total spin transfer.

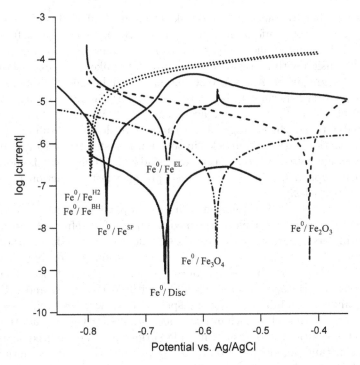

Figure 12. Log absolute current versus potential plot from anodic polarization of obtained with electrodes made from nano-sized iron and iron oxides. All data are for powder disk electrode (PDE) except for Fe^{SP}, where the particles were applied directly by vapor-deposition and conventional polished Fe^0 disk. All data at scan rate $= 0.1\,mV/s$ in anoxic aqueous borate buffer (pH 8.4).

Batch and Column Methods

Continuous and well-mixed batch reactors are the most common way to study the reaction of solutes with (nano)particle suspensions. This experimental configuration offers a number of potential advantages, including long contact times, high particle loads and therefore high total surface areas, controlled mixing to reduce external mass transfer effects, and sampling for external analysis. The latter is essential for characterization of environmentally-relevant concentrations of contaminants and their degradation products (by, for example, gas chromatography).

While the implementation of batch experiments to study the reactivity of nanoparticles with solutes is straightforward, rigorous consideration of the design and interpretation of such experiments has shown that they are fraught with many ambiguities and complications. A major issue of this sort is whether to control pH and risk secondary buffer effects[79,80] or avoid buffers and risk confounding the results with the effects of drifting pH.[13,81] This dilemma applies to batch studies of all particle sizes, but has an added dimension with nanoparticles because of the likely effect of pH on nanoparticle aggregation. Another example of the challenges in designing good quality batch experiments is whether to perform analysis of organic residues on headspace[79,81] or the aqueous phase.[13,80] While the former is more amenable to gas chromatography, it can give misleading results when the rate of reaction with the particles in solution is fast relative to rate of analyte exchange between the solution and headspace. This effect is more likely with nanoparticles, in so far as they produce faster reaction rates than coarser particles.

As part of extensive investigation of iron nanoparticle reactivity with aqueous contaminants, we have performed a series of carefully designed batch experiments using nanoparticles of zero-valent iron to reduce carbon tetrachloride (CT). These experiments were done without buffer, but with 24 hours of "equilibration" between deionized water and the particles which produces a stable pH during contact with CT; without headspace and with aggressive mixing to minimize mass transport effects; with frequent sampling and global fitting of a mechanistic kinetic model for CT degradation and product formation. The results include surface-area normalized first-order rate constants for degradation of CT (k_{SA}) and yields of the major product chloroform (Y_{cf}).

Figure 13 provides a summary of our data for CT degradation by three types of zero-valent iron, two nano-scale materials (Fe^{BH} and Fe^{H2}) and one micron-sized granular iron (Fe^{EL}) for comparison. Clearly, the material produced by reduction of nano-sized iron oxides with H_2 (Fe^{H2}) has the lowest Y_{cf}, which is advantageous because CF is a persistent and toxic by-product. Contrary to the popular perception that nano particles are intrinsically more reactive than bulk particles, surface normalized rate constants are higher for Fe^{EL} than for the two types of nano iron studied here. We think this result is definitive for CT, but we do not yet know how it applies to other contaminants because our preliminary analysis of k_{SA} data for trichloroethylene suggest more differences between nano- and micro-sized Fe^0 particles.[82]

Summary

The field represented by subject of this chapter is largely new and therefore its standard practices and fundamental principles are only just beginning to take shape. Nevertheless, we have tried to provide a comprehensive snapshot

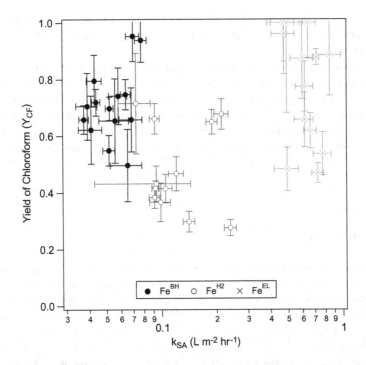

Figure 13. Yield of CF (toxic by-product of CT degradation by iron) versus surface normalized pseudo first-order rate constant for CT reduction by zero-valent iron. Fe^{H2} and Fe^{BH} are nano-sized iron synthesized by H_2 and $NaBH_4$ reduction of iron oxides respectively, and Fe^{EL} is a micro-sized iron obtained from Fisher Scientific. For more details on these materials, see Table 1. Conditions for these experiments were similar to those described in Nurmi *et al.*[13]

of the current state of theory and experimental practice in the area of nanoparticle applications to decontamination of environmental materials.

In principle, there are fundamental differences between nanoparticles and other types of materials, and these differences make the materials scientifically and technologically interesting. However, the most interesting fundamental properties of nanomaterials tend to be unsustainable under environmental conditions and this makes it difficult to characterize or utilize those properties in environmental contexts.

The examples of the synthesis and characterization provided in this chapter show the variety of ways that metal iron nanoparticles can be produced. Because of the variety of synthesis methods used, the likely presence of contamination and coatings, it is important to apply tools (surface spectroscopies and optical methods) that can sense molecular or elemental surface impurities. It is clear that adequate understanding of the behavior of these particles requires a more detailed characterization than often has been reported.

Two major elements in proper characterization of these materials are proper sample handling and the application of multiple characterization methods. Although many techniques are routinely applied to characterization of nanomaterials, it is important to understand the strengths and limitations of each method. It is often important to combine information from the different methods to produce a "smart" analysis. Modeling of the experimental data is important for many techniques including TEM, XRD and XPS.

Acknowledgments

In addition to samples synthesized as part of the project, samples of nano-iron were provided by Greg Lowry (Carnegie Mellon University), K. Okinaka (Toda Kogyo Corp.), and W.-X. Zhang (Lehigh University). We thank EMSL staff Mark Engelhard, David McCready, and Cathy Chin for assistance with the XPS, XRD, and BET analysis, respectively; and Prof. Paul Davis of Pacific Lutheran University for discussions on Au particle aggregation. This work was supported by the U.S. Department of Energy (DOE) Office of Science, Offices of Basic Energy Sciences and Biological and Environmental Research. Parts of the work were conducted in the William R. Wiley Environmental Molecular Sciences Laboratory (EMSL), a DOE User Facility operated by Battelle for the DOE Office of Biological and Environmental Research.

References

1. F. Salamanca-Buentello, D. L. Persad, E. B. Court, D. K. Martin, A. S. Daar and P. A. Singer, Nanotechnology and the developing world, *PLoS Medicine* **2**, 300–303 (2005).
2. L. Cumbal and A. K. Sengupta, *Environ. Sci. Technol.* **39**, 6508–6515 (2005).
3. W. Tungittiplakorn, L. W. Lion, C. Cohen and J.-Y. Kim, *Environ. Sci. Technol.* **38**, 1605–1610 (2004).
4. G. Vissokov and T. Tzvetkoff, *Eurasian Chemico-Technological Journal* **5**, 185–191 (2003).
5. N. Saleh, T. Phenrat, K. Sirk, B. Dufour, J. Ok, T. Sarbu, K. Matyjaszewski, R. D. Tilton and G. V. Lowry, *Nano Lett.* **5**(12), 2489–2494 (2005).
6. J. Xu and D. Bhattacharyya, *Environmental Progress* **24**, 358–366 (2005).

7. Y. Q. Liu, S. A. Majetich, R. D. Tilton, D. S. Sholl and G. V. Lowry, *Environ. Sci. Technol.* **39**(5), 1338–1345 (2005).

8. R. M. Powell, R. W. Puls, D. W. Blowes, J. L. Vogan, R. W. Gillham, P. D. Powell, D. Schultz, R. Landis and T. Sivavic, *Permeable Reactive Barrier Technologies for Contaminant Remediation. U.S. Environmental Protection Agency*, EPA/600/R-98/125 (1998).

9. P. G. Tratnyek, M. M. Scherer, T. J. Johnson and L. J. Matheson, in *Chemical Degradation Methods for Wastes and Pollutants: Environmental and Industrial Applications*, ed. M. A. Tarr (Marcel Dekker: New York, 2003), pp. 371–421.

10. D. W. Blowes, C. J. Ptacek, S. G. Benner, C. W. T. McRae, T. A. Bennett and R. W. Puls, *J. Contam. Hydrol.* **45**, 123–137 (2000).

11. J. Quinn, C. Geiger, C. Clausen, K. Brooks, C. Coon, S. O'Hara, T. Krug, D. Major, W. S. Yoon, A. Gavaskar and T. Holdsworth, *Environ. Sci. Technol.* **39**(5), 1309–1318 (2005).

12. D. W. Elliott and W. X. Zhang, Field assessment of nanoscale bimetallic particles for groundwater treatment, *Environ. Sci. Technol.* **35**(24), 4922–4926 (2000).

13. J. T. Nurmi, P. G. Tratnyek, V. Sarathy, D. R. Baer, J. E. Amonette, K. H. Pecher, C. Wang, J. C. Linehan, D. W. Matson, R. L. Penn and M. D. Driessen, *Environ. Sci. and Technol.* **39**(5), 1221–1230 (2005).

14. X. Zhang, C. Sun and N. J. Fang, *Nanopart. Res.* **6**, 125–130 (2004).

15. J. Z. Jiang, Y. X. Zhou, S. Moeup and C. B. Koch, *Nanostructured Materials* **7**(4), 401–410 (1996).

16. H. Hermann, W. Gruner, N. Mattern, H. D. Bauer, F. Fugaciu and T. Schubert, Evolution of nanostructure and chemical reactivity of carbon during ball-milling, in *Mechanically Alloyed, Metastable and Nanocrystalline Materials: Part 1*, Material Science Forum, Vols. 269–272 (Trans Tech Publications, 1998), pp. 193–198.

17. D. B. Vance, *Pollution Engineering* **37**(7), 16–18 (2005).

18. S. S. Suthersan, D. Vance, P. Palmer and U. S. Patent, *Application* **890**, 066 (2005).

19. C. Granquist, L. Kish and W. Marlow (eds.), *Gas Phase Nanoparticle Synthesis* (Springer: New York, 2004).

20. C. Janzen and P. Roth, *Combust Flame* **125**, 1150 (2001).

21. H. Hahn, *NanoStruct. Mater* **9**, 3–12 (1997).

22. M. L. Ostraat, J. W. De Blauwe, M. L. Green, L. D. Bell, H. A. Atwater and R. C. Flagan, *J. Electrochem. Soc.* **148**, G265–G270 (2001).

23. K. Wegner, B. Walker, S. Tsantilis and S. E. Pratsinis, *Chem. Eng. Sci.* **57**, 1753 (2002).

24. A. P. Weber, M. Seipenbusch and G. Kasper, *J. Phys. Chem. A* **105**, 8958 (2001).

25. Y. Qiang, Y. Thurner, Th. Reiners, O. Rattunde and H. Haberland, *Surface and Coatings Technol.* **101**(1–3), 27–32 (1998).

26. Y. Qiang, R. F. Sabiryanov, S. S. Jaswal, Y. Liu, H. Haberland and D. J. Sellmyer, *Phys. Rev. B* **66**, 064404 (2002).

27. Y. Xu, Z. Sun, Y. Qiang and D. J. Sellmyer, *J. Appl. Phys.* **93**, 8289 (2003).
28. Y. Qiang, J. Antony, A. Sharma, S. Pendyala, J. Nutting, D. Sikes and D. Meyer, *J. Nanoparticle Res.* **8**, 489 (2006).
29. J. Antony, Y. Qiang, D. R. Baer and C. Wang, *J. Nanosci. Nanotechnol.* **6**, 568–572 (2006).
30. J. R. Anderson and M. Boudart, *Catalysis, Science & Technology* (Springer-Verlag, Berlin, 1996).
31. T. Dwars, E. Paetzold and G. Oehme, *Angew. Chem. Int. Ed.* **44**, 7174–7199 (2005).
32. V. Uskokovic and M. Drofenik, *Surface Review and Letters* **12**, 239–277 (2005).
33. J. G. Darab, J. L. Fulton, J. C. Linehan, M. Capel and Y. Ma, Characterization of a water-in-oil microemulsion containing a concentrated ammonium ferric sulfate aqueous core, *Langmuir* **10**, 135–141 (1994).
34. A. R. Kortan, R. Hull, R. L. Opila, M. G. Bawendi, M. L. Steigerwald, P. J. Carroll and L. E. Brus, *J. Am. Chem. Soc.* **112**, 1327–1332 (2005).
35. M. L. Steigerwald, A. P. Alivistos, J. M. Gibson, T. D. Harris, R. Kortan, A. J. Muller, A. M. Thayer, T. M. Duncan, D. C. Douglass and L. E. Brus, *J. Am. Chem. Soc.* **110**, 3046–3050 (1988).
36. D. W. Matson, J. C. Linehan, J. G. Darab and M. F. Buehler, *Energy & Fuels* **8**, 10–18 (1994).
37. J. L. Fulton, M. Ji, C. M. Chen and J. Wai, *J. Am. Chem. Soc.* **121**, 2631 (1999).
38. N. A. D. Burke, H. D. H. Stover and F. P. Dawson, *Chem. Mater.* **14**, 4752 (2002).
39. K. S. Suslick and G. J. Price, *Ann. Rev. Mater. Sci.* **29**, 295–326 (1999).
40. C.-B. Wang and W.-X. Zhang, *Environ. Sci. Technol.* **31**, 2154–2156 ((1997).
41. H.-L. Lien and W.-X. Zhang, *Colloids and Surfaces. A. Physicochemical and Engineering Aspects* **191**(1–2), 97–105 (2001).
42. D. W. Matson, J. C. Linehan, J. G. Darab, M. R. Buehler, M. R. Phelps and G. G. Neuenschwander, in *Advanced Techniques in Catalyst Preparation*, ed. W. R. Moser (Academic Press, New York, 1996), pp. 259–283.
43. M. Uegami, J. Kawano, T. Okita, Y. Fujii, K. Okinaka, K. Kakuya and S. Yatagai, Iron particles for purifying contaminated soil or groundwater. Process for producing the iron particles, purifying agent comprising the iron particles, process for producing the purifying agent and method of purifying contaminate soil or groundwater. Toda Kogyo Corp. U.S. Patent Application (2003).
44. C. R. Brundle, C. A. Evans and S. Wilson (eds.), *Encyclopedia of Materials Characterization* (Butterworth-Heinemann, Stoneham, MA, 1992).
45. D. J. Gaspar, M. H. Engelhard, M. C. Henry and D. R. Baer, *Surf. Interface Anal.* **37**, 417–423 (2005).
46. H. Zhang, B. Gilbert, F. Huang and J. F. Banfield, *Nature* **424**, 1025 (2003).
47. T. L. Hill, *Nano Lett.* **1**, 273–275 (2001).
48. M. J. Yacaman, J. A. Ascencio, H. B. Liu and J. Gardea-Torresdey, *J. Vac. Sci. Technol. B.* **19**, 1091 (2001).

49. D. R. Baer, M. H. Engelhard, D. J. Gaspar, D. W. Matson, K. H. Pecher, J. R. Williams and C. M. Wang, *J. Surf. Anal.* **12**, 101–108 (2005).

50. P. M. Ajayan, *Nature* **427**, 402–403 (2004).

51. F. Bodker, M. F. Hansen, C. B. Koch and S. Morup, *J. Magnetism Magnetic Mat.* **221**(1–2), 32–36 (2000).

52. T. J. Norman, C. D. Grant, D. Magana, J. Z. Zhang, J. Liu, D. Cao, F. Bridges and A. V. Buuren, *J. Phys. Chem. B.* **106**, 7005–7012 (2002).

53. K. Ritter, M. S. Odziemkowski and R. W. Gillham, *J. Contaminant Hydrology* **554/15/02**, 87–111 (2002).

54. R. Miehr, P. G. Tratnyek, J. Z. Bandstra, M. M. Scherer, M. Alowitz and E. J. Bylaska, *Environ. Sci. Technol.* **38**, 139–147 (2004).

55. D. J. Burleson, M. D. Driessen and R. L. Penn, *J. Environ. Sci. Health, Part A: Toxic/Hazard. Subst. Environ. Eng.* **39**(10), 2707–2753 (2004).

56. C. M. Wang, D. R. Baer, L. E. Thomas, J. E. Amonette, J. Anthony, Y. Qiang and G. Duscher, *J. App. Phys.* **98**, 094308 (2005).

57. Although the "quick" sample transfer from glove-box to microscope appears to provide reproducible and useful results, we are currently implementing a controlled atmosphere transfer system to further minimize any changes that might occur during transfer.

58. These XRD patterns are from the supplementary material Figure S1 for Ref. 13.

59. Multiquant is an example of one program that includes sample shape information in the analysis www.chemres.hu/aki/XMQpages/XMQhome.htm.

60. L. T. Kuhn, A. Bojesen, L. Timmermann, M. M. Nielsen and S. Morup, *J. Phys. Condens. Matter.* **14**, 13551–13567 (2002).

61. L. Signorini, L. Pasquini, L. Savini, R. Carboni, F. Boscherini, E. Bonetti, A. Giglia, M. Pedio, N. Mahne and S. Nannarone, *Phys Rev B.* **68**, 195423 (2003).

62. L. Guo, Q. Huang, X.-y. Li and S. Yang, *Phys. Chem. Chem. Phys.* **3**(9), 1661–1665 (2001).

63. J. E. Amonette, V. Sarathy, J. C. Linehan, D. W. Matson, C. Wang, J. T. Nurmi, K. Pecher, R. L. Penn, P. G. Tratnyek and D. R. Baer, *Geochim. Cosmochim. Acta.* 69:A263 Supplement S. Presented at 15th Annual Goldschmidt Conference, Moscow, ID, May 21–25, (2005).

64. C. S. Cha, C. M. Li, H. X. Yang and P. F. Liu, *J. Electroanal. Chem.* **368**(1–2), 47–54 (1994).

65. J. W. Long, K. E. Swider, C. I. Merzbacher and D. R. Rolison, *Langmuir* **15**(3), 780–785 (1999).

66. S. M. Ponder, J. G. Darab, J. Bucher, D. Caulder, I. Craig, L. Davis, N. Edelstein, W. Lukens, H. Nitsche, L. Rao, D. K. Shuh and T. E. Mallouk, *Chem. Materials* **13**(2), 479–486 (2001).

67. B. Schrick, S. M. Ponder and T. E. Mallouk, *Extended Abstracts, Division of Environmental Chemistry*, Vol. 42, No. 2 (American Chemical Society, Washington, DC, 2000), pp. 639–640.

68. V. Vivier, C. Cachet-Vivier, B. L. Wu, C. S. Cha, J. Y. Nedelec and L. T. Yu, *Electrochem. Solid-State Lett.* **2**(8), 385–387 (1999).

69. G. V. Zhutaeva, N. D. Merkulova, M. R. Tarasevich and V. V. Surikov, *J. Appl. Chem. USSR* **61**(1), 56–59 (1988).
70. H. O. Finklea, *Studies in Physical and Theoretical Chemistry Semiconductor Electrodes* (Elsevier, Amsterdam, 1988).
71. J. S. Gao, T. Arunagiri, J. J. Chen, P. Goodwill, O. Chyan, J. Perez and D. Golden, *Chemistry of Materials* **12**(11), 3495–3500 (2000).
72. B. A. Logue, Ph.D. thesis, Oregon State University, Corvallis, OR, (2000).
73. B. A. Logue and J. C. Westall, *Environ. Sci. Technol.* **37**, 2356–2362 (2003).
74. S. Choe, Y. Y. Chang, K. Y. Hwang and J. Khim, *Chemosphere* **41**(8), 1307–1311 (2000).
75. H.-L. Lien and W.-X. Zhang, *Journal of Environmental Engineering* **125**(11), 1042–1047 (1999).
76. B. Schrick, J. L. Blough, A. D. Jones and T. E. Mallouk, *Chemistry of Materials* **14**(12), 5140–5147 (2002).
77. M. M. Scherer, J. C. Westall, M. Ziomek-Moroz and P. G. Tratnyek, Kinetics of carbon tetrachloride reduction at an oxide-free iron electrode, *Environ. Sci. and Technol.* **31**(8), 2385–2391 (1997).
78. M. M. Scherer, K. Johnson, J. C. Westall and P. G. Tratnyek, Mass transport effects on the kinetics of nitrobenzene reduction by iron metal, *Environ. Sci. Technol.* **35**(13), 2804–2811 (2001).
79. S. H. Joo, A. J. Feitz, D. L. Sedlak and T. D. Waite, *Environ. Sci. Technol.* **39**(5), 1263–1268 (2005).
80. M. M. Scherer, J. C. Westall, M. Ziomek-Moroz and P. G. Tratnyek, *Environ. Sci. Technol.* **31**, 2385–2391 (1997).
81. T. E. Meyer, C. T. Przysiecki, J. A. Watkins, A. Bhattacharyya, R. P. Simondsen, M. A. Cusanovich and G. Tollin, *Proc. Natl. Acad. Sci. USA* **80**, 6740–6744 (1983).
82. P. G. Tratnyek, V. Sarathy and B. Bae, Nano effects on the kinetics of contaminant reduction by iron and iron oxides, *Extended Abstracts, Division of Environmental Chemistry*, American Chemical Society, 230th National Meeting, Washington, D.C. August 28–September 1, Vol. 45, No. 2, (2005), p. 4.

Nanostructured Inorganic Materials

Chapter 4

Formation of Nanosize Apatite Crystals in Sediment for Containment and Stabilization of Contaminants

Robert C. Moore*, Jim Szecsody†, Michael J. Truex†,
Katheryn B. Helean*, Ranko Bontchev* and Calvin Ainsworth†

*Sandia National Laboratory, Albuquerque, NM, USA

†Pacific Northwest National Laboratory
Richland, WA, USA

Introduction

Apatite is a calcium phosphate mineral that has a strong affinity for sorption of many radionuclides and heavy metals. A large number of laboratory studies and field applications have focused on the use of apatite for remediation of contaminated sites.[1] Recently, numerous studies have appeared in the literature focused on the development of apatite with a nanoporous structure. The majority of these studies are focused on biomedical applications. However, the development of a high surface area, nanoporous apatite sorbent with an enhanced capacity for contaminant uptake is highly significant for environmental applications. In this chapter, we describe a method for forming nanosize apatite crystals in sediment using aqueous solutions containing calcium, phosphate and the citrate ligand (Figure 1). The method has applications in construction of *in situ* reactive barriers for groundwater remediation and stabilization of contaminated sites.

Apatite, or hydroxyapatite, $Ca_{10}(PO_4)_6(OH)_2$, in its purest form, has been the focus of many laboratory studies and has been used as a radionuclide sorbent in reactive barriers and for groundwater remediation.[1-3] Apatite has been demonstrated to be a strong sorbent for uranium,[4-6] neptunium,[7] plutonium,[8] strontium[9-11] and lead.[12-14] Apatite can be synthesized through precipitation reactions from supersaturated solutions of calcium phosphate,[15] high temperature solid state reactions[16] and high

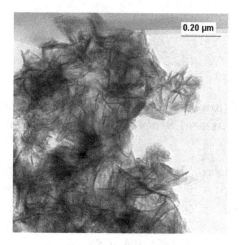

Figure 1.   Nano-sized apatite crystals formed in sediment through precipitation of aqueous solution. Transmission electron microscope taken in bright field mode.

temperature treatment of animal bone.[17] An excellent source of information on calcium phosphates including apatite is given by LeGeros.[18] Mineral apatite, phosphate rock, is mined through out the world, but typically is associated with significant amounts of impurities including radionuclides and heavy metals.

Permeation grouting, high pressure jet grouting and excavation followed by backfilling with sorbent are methods typically used for placing solids, including apatite, in sediment for construction of reactive barriers and contaminant immobilization. For certain sites, it may not be practical or even possible (i.e. deep placement below 50 feet) to use these conventional methods. For example, the presence of obstacles such as buried large waste tanks or geological formations can make solids placement difficult. Additionally, if the sorbent can not be homogeneously dispersed the formation of a continuous barrier can not be assured and may result in preferential flow around the reactive material.

An alternative to using a solid sorbent for contaminant immobilization is to inject a reagent in aqueous solution into the sediment that converts the contaminant to a less toxic form or reacts with the contaminant to form an insoluble precipitate. The use of phosphoric acid for lead stabilization in sediment through the formation of sparingly-soluble lead phosphates is an example. A novel approach for forming apatite in sediment using an aqueous solution of calcium chloride, sodium phosphate, sodium citrate and nutrients for microbial activity has been reported by Moore and others,[7,19] allowing sufficient time to inject the solution into an aquifer (0.5–4 days) before citrate is degraded and apatite precipitates. Calcium phosphates are sparingly soluble in water and precipitation of calcium phosphates from a supersaturated solution is rapid. However, citrate forms strong complexes with calcium and can be used to prevent precipitation or slow precipitation kinetics of calcium phosphates. When the aqueous solution containing calcium-citrate and phosphate is then mixed with sediment, the indigenous soil microorganisms biodegrade the citrate freeing up the calcium, so amorphous calcium phosphate solids precipitate. Over time (1–3 weeks), these solids change into the thermodynamically favored crystalline apatite structure. The method has been used to form $100\,\mu$m apatite crystals in sediment and studied for immobilization of ^{90}Sr.[7,19]

Calcium Phosphate Precipitation

The formation of a particular calcium phosphate solid through precipitation from aqueous solution is strongly dependant on solution conditions and the rate of solution mixing. Pure hydroxyapatite cannot be synthesized at ambient temperatures through precipitation. However, apatite or apatitic calcium phosphates with varying degrees of crystallinity and purity can be easily prepared by precipitation. Certain species are known to promote or inhibit the formation of calcium phosphate crystalline phases. Known promoters are fluoride and chloride ions whereas citrate, aluminum, tin and carbonate are known to inhibit apatite formation. Additionally, various substitutions can be made into the apatite structure. These include Na^+ and Sr^{2+} for Ca^{2+}, CO_3^{2-} and HPO_4^{2-} for PO_4^{3-} and F^- for OH^-. The extent of substitution of a specific ion into apatite is dependant on the concentration of that ion in solution (LeGeros, 1990). Substitutions into the apatite structure can significantly affect physical and chemical properties. Carbonate substitution, for example, increases apatite aqueous solubility whereas fluoride decreases apatite solubility.[20,21]

In sediment and groundwater, significant amounts of ionic species are available for substitution into apatite forming through precipitation *in situ*. The composition of natural sediment/water systems can widely vary. Typical arid region sediments tested are Ca/Mg-carbonate dominated, with groundwater containing Ca (53 mg/L), Mg (13 mg/L), Na (25 mg/L), and K (8.2 mg/L) that are held on oxide and clay surfaces (in somewhat higher concentrations than in solution) primarily by ion exchange. Anion composition is carbonate dominated (166 mg/L) with some sulfate (67 mg/L) and chloride (24 mg/L) present. Even in a high Ca-carbonate system, the aqueous concentration of Ca (1.3 mM) and surface Ca concentration (1.8 mM/g) is an insufficient amount to complex with the injected phosphate to form apatite, therefore Ca must be injected in a stoichiometric ratio with phosphate to form apatite.

Apatite Formation in Sediment

Citrate is a tetraprotic organic ligand with three carboxylate groups and forms strong complexes with calcium and is easily biodegraded by soil microorganisms (T. Rosswall, 1997).[22,23] The products of aerobic degradation of citrate by soil microorganisms are reported to be acetate, lactate, propionate and formate.[24] Equilibrium conditions for a metal, M, bonding with an organic ligand, L, is expressed by the complexation constant given by:

$$\beta = [M - L]/[M][L] \tag{1}$$

The stability constant for citrate is two-orders of magnitude higher than for the ligands with a single carboxylate group. For example, the logarithm of the complexation constants for citrate and acetate are 3.5 and 0.35, respectively.[25] Van der Houwen and Valsami-Jones[26] have reported that citrate significantly increases the degree of supersaturate of calcium phosphate solutions whereas acetate has little effect. The method for forming apatite in sediment takes advantage of this chemistry. By chelating calcium with citrate, calcium phosphate precipitates do not form in solutions containing phosphate.

The formation of apatite in sediment using solutions of sodium citrate, calcium chloride and sodium phosphate have been shown to be stable for weeks without any observed formation of precipitates.[7] At pH 7.0 to 10, a stable solution requires a ratio of 2.5 citrate molecules for each calcium ion to keep all of the calcium complexed (i.e. some free citrate present). A typical formulation for apatite formation in sediment is given in Table 1. Higher concentrations of reagents, > 100 mM citrate, result in the formation of precipitates almost immediately. The ratio of reagents in the formulation is significant. The ratio of sodium citrate to ammonium nitrate or another nitrate source is set at approximately 10:1 to support microbial growth. Hoagland's solution, as modified by Johnson et al.[27] contains the necessary

Table 1. Composition of apatite forming solution. Solution pH is adjusted to 7.2 with HCl and NaOH.

Component	Concentration (mM)
trisodium citrate	50
calcium chloride	25
disodium phosphate	18
ammonium nitrate	5
Hoagland's solution	1 mL added to 1 L of solution

micronutrients for microorganisms. The ratio of phosphate to calcium is set at approximately 3:5, the same as apatite. A critical parameter is solution pH that must be adjusted to between 7 and 8. At pH lower than 7.0 and higher than 9.5, precipitates other than amorphous calcium phosphate and apatite can form.

The solutions can be injected into sediment or allowed to infiltrate from the surface. Over time, approximately one to four weeks depending on the initial reagent concentrations, temperature and sediment, the soil microorganisms convert the citrate to acetate, lactate and formate under anaerobic conditions or carbon dioxide under oxic conditions. Oxic aquifers are driven to reducing conditions quickly even with oxygen saturated (8.4 mg/L) due to low oxygen solubility in water. The calcium, chelated with the single carboxylated ligands, reacts with phosphate and precipitation of calcium phosphate occurs. Initially, amorphous calcium phosphate forms. Over time, the amorphous material changes into apatite. An application utilizing the process is illustrated in Figure 2 for the formulation of an *in situ* reactive barrier. The formulation is injected into the sediment in the migration path of the contaminant. Apatite forms in the sediment and as the metal-contaminated groundwater passes through the barrier the contaminant is initially sorbed by the apatite, then over weeks incorporated into the apatite structure.

Phosphate in the formulation not only serves as a reagent in the reaction to form apatite, but also buffers the pH at a value where conditions are reported to be very favorable for apatite formation. The use of a constant composition method for precipitation of apatite is described by Nancollas and coworkers.[28] For our system, the pH of the solution is buffered at 7.2 as given by the reactions:

$$H_2PO_4^- + OH^- \rightarrow HPO_4^{2-}$$
$$HPO_4^{2-} + H^+ \rightarrow H_2PO_4^-$$

$$(2)$$

Other than preventing the premature formation of calcium phosphate precipitates before the solution can be injected into sediment, citrate may play an additional role in apatite formation. Citrate is reported to strongly interact with apatite.[29,30]

In a recent study, Van der Houwen *et al.*[30] indicate citrate affects apatite crystal size and can increase the amount of impurities in the apatite structure. This effect may be significant when precipitation of apatite occurs in the presence of sediment.

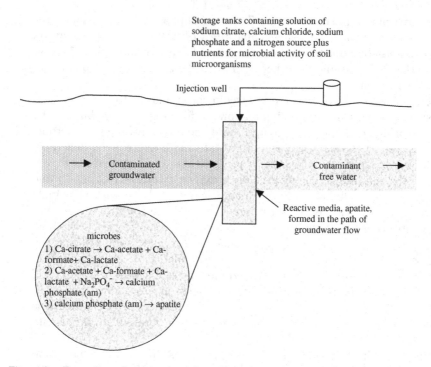

Figure 2. Formation of apatite in sediment to create a reactive barrier for removing contaminants in groundwater. Apatite is formed through the biodegradation of citrate resulting in the precipitation of amorphous calcium phosphate that changes to a poorly crystallized apatite over time.

Biodegradation of Citrate

The mechanism of apatite formation in sediment using the described method is driven by the biodegradation of citrate by the indigenous soil microorganisms. Citrate can be microbially metabolized under both aerobic and anaerobic conditions. Typically, citrate is added well in excess of the amount that can be degraded by any dissolved oxygen present in an aquifer, but not in the vadose zone. Additionally, aerobic degradation of the citrate portion of a calcium citrate complex results in the formation of calcium carbonate (Figure 3). Aerobic biodegradation of citrate shows 108.5 mM citrate generating a maximum of 70 mM carbonate, so we have accounted for only 11% of the carbon mass from the citrate. Anaerobic degradation processes for citrate are most relevant to *in situ* apatite formation in aquifers where oxygen supply is limited. Under anaerobic conditions, fermentation of citrate is a primary means of citrate degradation.

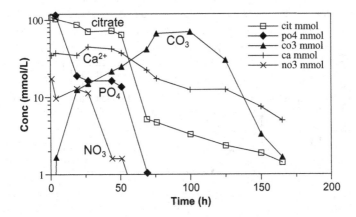

Figure 3. Citrate biodegradation in an aerobic sediment/water system forming predominantly carbonate.

Citrate Fermentation

Citrate ferments to form acetate and formate as has been observed in unamended sediments from the Hanford site and in these same sediments amended with a known citrate-fermenting microorganism (Figures 4a and b). A lag time of about 3 days prior to fermentation observed in the unamended sediments was likely due to low initial populations of microorganisms. While acetate and formate are the primary fermentation products, the amended sediments showed small amounts of lactate and propionate produced in addition to the acetate and formate. Citrate concentration remained constant in the sterile control over the duration of the incubation period. In these experiments (Figures 4a and b), citrate fermentation followed the reaction stoichiometry given by:

$$C_6H_8O_7 + H_2O + OH^- \rightarrow 2C_2H_4O_2 + CH_2O_2 + HCO_3^- \qquad (3)$$

Biomass in the pre-test unamended sediment sample was 1.51×10^5 cells/g-soil and an average of 2.25×10^8 cells/g-soil after the 16-day incubation period. Using a generic cell formula of $C_5H_7O_2N$, the molecular weight of cells is 113 mg/mmole. These anaerobic citrate biodegradation experiments show that 73% of the carbon from citrate can be accounted for in formate and acetate, and it is likely that the other 25% is carbon dioxide, as shown in Equation (3). With this cell molecular weight, the estimated yield from citrate is 0.06 mmole-cells/mmole-citrate (0.04 mg-cells/mg-citrate). The biomass yield is important to consider for *in situ* citrate degradation as part of determining whether the added citrate will cause enough biomass growth to significantly decrease the permeability of the aquifer.

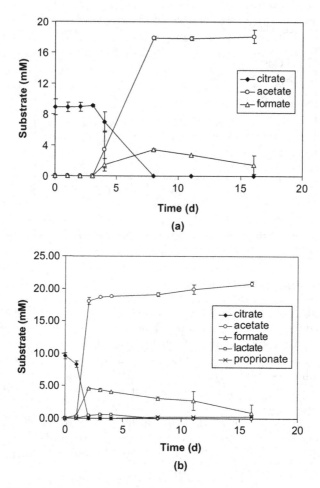

Figure 4.   Citrate fermentation in a composite sediment from the 100 North Area of the Hanford Site. (a) without inoculation (unamended); and (b) inoculated with approximately 10^7 cells/mL of *K. pneumonia* (amended). Error bars are for results from three replicate treatments. Initial solution composition for each experiment was 10 mM calcium citrate, 1 mM ammonium chloride, and 5 mM phosphate buffer adjusted to pH 7.3. The soil to solution ratio was 10:3.

Modeling Citrate Biodegradation and Temperature Effect

Citric acid is utilized by many organic systems as part of the TCA (Krebs) photosynthetic process, where the citrate (a C6 organic acid) is converted to C6, C5, and C4 organic acids producing CO_2 and H^+, then cycled from oxaloacetic acid (C4) to citric acid.[31] Citrate can also be further degraded to acetic acid (C2), formaldehyde, formic acid (C1) and CO_2. For the purpose of this study, citrate is used to complex Ca, so only the decrease in citrate concentration (by biodegradation) is of significance, as the lower molecular weight organic acids only form weak complexes with Ca. Two different modeling approaches were considered to quantify citrate biodegradation, a first-order model and a Monod model. A first-order model is an empirical approach that describes citrate removal with a single reaction rate coefficient. A Monod model is also an empirical approach that describes citrate removal externally to microbial organisms with a similar mathematical form of enzyme degradation of a compound (Michaelis-Menton kinetics). Monod kinetics are utilized when the observed data clearly show a considerable slowing of reaction rate at low concentration that cannot be accounted for using the simpler first-order kinetic model.

Citrate biodegradation experiments showed a slower rate at colder temperature and a slower rate at higher citrate concentration (Figure 5). At 10 mM citrate concentration, citrate was completely gone by 200 h (21°C) to 300 h (10°C, Figure 3a). At 50 mM citrate concentration, citrate was gone by 250 h (21°C) to 450 h (10°C, Figure 3b). At 100 mM citrate concentration, a small amount of citrate remained at 300 h (21°C) to 600 h (10°C, Figure 5c). At each concentration, duplicate experiments showed similar results. A first-order model (lines, Figure 5) showed good fits, and indicated that in some cases, citrate biodegradation may be somewhat more rapid at lower concentration than a first-order approximation. For example the 100 mM citrate data at 10°C (Figure 5c) showed a good first-order fit to 500 h, but then citrate more rapidly degraded. This effect is observed for all citrate concentrations at 10°C, but not at 21°C. A Monod kinetic model would describe the data equally as well with small half-saturation constants, but would describe the data more poorly with higher concentration half-saturation constants, which would slow citrate biodegradation at low concentration, the opposite effect of that observed. Therefore, a pseudo first-order model was used to quantify the rate data (Table 2). The citrate biodegradation rate was 3.0 times slower (10°C data) to 3.3 times slower (21°C data) as the citrate concentration increased from 10 mM to 100 mM. The citrate biodegradation rate averaged 3.3 times slower as the temperature decreased from 21°C to 10°C. The activation energy estimated from the reaction rate change with temperature is 35 kJ/mol for the 10 mM citrate data, 16 kJ/mol for the 50 mM citrate and 32 kJ/mol for the 100 mM citrate (Figure 6). These activation energies indicate the rate is controlled by the biochemical reaction and not diffusion, which is expected.

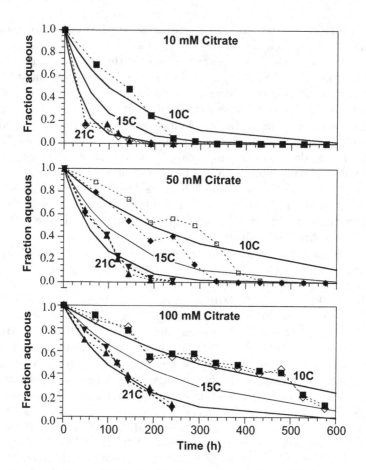

Figure 5. Citrate biodegradation by Hanford 100N sediment at different temperature for citrate concentrations of: (a) 10 mM, (b) 50 mM, and (c) 100 mM.

Table 2. Citrate biodegradation rates (1/h) as a function of temperature and initial citrate concentration.

Concentration (mM)	Rate Constant (10°C)	Rate Constant (15°C)	Rate Constant (21°C)
10	0.0071	0.013	0.025
50	0.0074	0.0101	0.013
100	0.0024	0.0042	0.0075

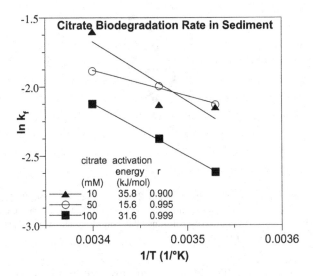

Figure 6. Arrhenius plot of citrate biodegradation rate change with absolute temperature with calculated activation energy.

Analysis of Apatite Precipitates Formed in Sediment

The mineral apatite, $Ca_5(PO_4)_3(F, Cl, OH)$, is the most abundant, ubiquitous phosphorous bearing mineral on Earth and is used broadly by Earth scientists in the study of igneous, metamorphic and sedimentary petrogenetic processes.[32] In addition to its geologic utility, apatite is also used in industry for such diverse applications as: a source for fertilizer, a component of fluorescent lamps and lasers, as nuclear waste forms and in biomedical applications.[33-35] In fact, hydroxyapatite is the primary mineral component in bone and teeth.[35] The apatite structure and its chemical facility provide the basis for its broad and varied application.

Apatite can refer to three specific end-member minerals, fluorapatite, chloroapatite and hydroxyapatite. All can be viewed as slight modifications of the $P6_3/m, Z = 2$ structure. For a complete description and structural details see Hughes *et al.*[36] Several features of the structure are noteworthy. PO_4-tetrahedra are found in a hexagonal arrangement within the $(00l)$ planes defining columns parallel to the z-axis (Figure 7). The PO_4 oxygens are coordinated with Ca in two different sites. The Ca1 site is intercalated between the $(00l)$ planes and coordinated to nine oxygens. The Ca2 site is coordinated to six oxygens and the column anions (typically F, Cl, or OH). The Ca2 site shows the greatest degree of structural distortion

Figure 7.   The crystal structure of apatite ($P6_3/m$, $Z = 2$) projected on the (001) plane.
PO_4 tetrahedra lay parallel to the plane and are coordinated to two different calcium
sites. The anions F, Cl and OH occupy the columns that run along the z-axis. (structure
rendered using CrystalMaker® 7.0 software).

upon chemical substitution. The apatite structure exhibits extensive solid
solution with respect to both cations and anions. Metal cations including
actinides, K, Na, Mn, Ni, Cu, Co, Zn, Sr, Ba, Pb, Cd and Fe substitute for
Ca; and oxyanions (e.g. AsO_4^{3-}, SO_4^{2-}, CO_3^{2-}, SiO_4^{4-}, CrO_4^{2-}) replace PO_4^{3-}
through a series of coupled substitutions to preserve electro neutrality.[32]

Multiple characterization techniques were employed to assess the crystal chemistry of the apatite formed by the microbial digestion of Ca-citrate
in sediment from the US DOE Hanford reservation. The apatite crystals
are shown in Figure 1. High-resolution Transmission Electron Microscopy
(HRTEM) and powder X-ray diffraction (XRD) were used to assess apatite
crystallinity and to document the transformation from an amorphous calcium phosphate to nanocrystalline apatite. Energy dispersive spectroscopy
(EDS) and Fourier-Transform Infrared (FTIR) spectroscopy were used to
analyze the chemical constituents. The apatite was formed in sediment collected from the 100 North Site on the U.S. DOE Hanford Reservation by
treatment with a solution of 50 mM sodium citrate, 25 mM calcium chloride, 20 mM sodium phosphate and nutrients for microbial activity at pH
7.4. The blade-like crystals are in an amorphous matrix and are approximately 0.1 μm in size. Figure 8a is the XRD pattern for the apatite crystals
shown in Figure 1. The observed broad overlapping peaks in the pattern
at 2θ of approximately 32° are characteristic of poorly crystallized apatite.
The remaining peaks in the XRD pattern correspond to components of
the sediment. The FTIR spectra is given in Figure 8c along with the spectra for a pure hydroxyapatite produced by precipitation and heat treatment at 700°C. The lower resolution of the PO_4^- bands confirms the lower
crystallinity of the sample, as observed by both HRTEM and XRD. The

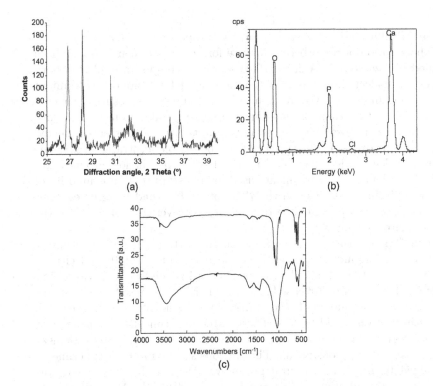

Figure 8. Chemical characterization of apatite crystals by: (a) XRD, (b) EDS, and (c) FTIR of nanocrystalline apatite formed in Hanford sediment by microbially mitigated Ca-citrate degradation in the presence of aqueous phosphate.

bands at 1455 cm^{-1} and 879 cm^{-1} indicate the presence of carbonate in the apatite structure. The TEM-EDS, Figure 8c, spectrum identifies calcium and phosphate as the major components with a stoichiometric apatite ratio of approximately 5:3.

Sorption of ^{90}Sr by Apatite Formed in Sediment

Moore *et al.*[19] studied the sorption of ^{90}Sr by apatite precipitates in sediments. The study consisted of equilibrating apatite-treated sediment, collected from the Albuquerque desert, with solutions containing a range of ^{90}Sr concentrations. Untreated soil was also tested. The untreated soil sorbed from 6.4 to 38% of the ^{90}Sr from the solutions whereas the apatite treated soil sorbed 76 to 99.2% of the ^{90}Sr. Possibly even more significant, the untreated soil released 20.4 to 41% of the sorbed ^{90}Sr back into solution when mixed with fresh water. The treated soil released only 1.7 to 11.2% of the ^{90}Sr illustrating the strong sorptive properties of the apatite.

The sorption of Sr sorbed to apatite formed in sediment has also been studied by Szecsody.[37] A sub-surface sediment composite used in experiments was collected from the ^{90}Sr contaminated 100N site on the Hanford Reservation in eastern Washington. At the 100N site, sub-surface Sr contamination is the result of once-through cooling water from the Hanford 100N Reactor disposed into the two disposal trenches. Sediments collected in a bore-hole drilled in 2004 showed that the ^{90}Sr concentration was greatest in the deep vadose zone (12′ to 15′ depth) and shallow groundwater (15′ to 22′ depth). Two sediments used in this study were a composite of sediments from all depths and the 13′ depth, which had the greatest 90Sr concentration (280 pCi/g). The ratio of Sr/^{90}Sr averages 1.4×10^6 in the contaminated sediment.

The ^{90}Sr adsorption rate was measured for Hanford sediments under both natural groundwater conditions and in a solution of 10 mM citrate, 4 mM Ca, and 2.4 mM mixed with the Hanford groundwater. Because ^{90}Sr is retained by Hanford 100N sediments primarily by ion exchange,[38] it was expected that the amount of ^{90}Sr adsorption by sediments in the higher ionic strength solution would be less. These batch experiments were conducted using 116 g of sediment and 70 mL of water to achieve a soil/water ratio near that in saturated porous media. The aqueous solution in both experiments contained 0.2 mg/L Sr (natural groundwater), and the experiment was initiated by the addition of ^{90}Sr to achieve 45,000 pCi/L. At specified time intervals, 0.5 mL of water was removed, filtered with a 0.45 micron PTFE filter (13 mm diameter), and placed in a scintillation vial with scintillation fluid. These samples were counted after 30 days after ^{90}Sr/^{90}Yt secular equilibrium was achieved and the ^{90}Sr activity was half of the total activity.

Sequential extraction experiments were conducted to determine if ^{90}Sr retention by sediments changed with the addition of apatite solution containing 100 mM citrate, 40mM Ca, and 24 mM PO$_4$. The three different systems used for these experiments were: (1) untreated 100N sediment treated with an ^{90}Sr spike (described above), (2) apatite-laden sediment treated with a ^{90}Sr spike (described above), and (3) ^{90}Sr-laden sediment from the 13' depth (80 pCi/g) that was then treated with an apatite solution for 30 days. The first system is the control, and should show no change in ^{90}Sr ion exchange over time. The second and third systems are hypothesized to show a change from ^{90}Sr retention by ion exchange to slow incorporation into apatite. Because the ^{90}Sr in the third system has been in contact with sediment for decades, there may be some resistance to desorption from sediment onto the apatite. Therefore differences between the second and third system may reflect the influence of ^{90}Sr/sediment aging.

Sequential extractions consisted of selected chemical extraction to remove ion exchangeable ^{90}Sr, organic-bound ^{90}Sr, carbonate-bound ^{90}Sr, and remaining (residual) ^{90}Sr. Both Sr and ^{90}Sr were analyzed in extractions to determine if the Sr was retained differently from the ^{90}Sr. It was expected that Sr was geologically incorporated into many different sediment minerals, so should be more difficult to remove compared with ^{90}Sr, which was recently added to the systems. Sequential extractions were conducted on 5 g sediment per water samples removed from the 120 g per 70 mL systems at specified time intervals from hours to 4,000 h. The ion-exchangeable extraction consisted of the addition of 0.5 M KNO$_3$ to the sediment sample for 16 hours.[39] The organic-bound extraction conducted after the ion-exchangeable extraction consisted of 0.5 M NaOH for 16 hours.[40] The carbonate-bound extraction conducted after the organic-bound extraction consisted of the addition of 0.05 M Na$_3$EDTA for six hours.[41,42] The residual extraction conducted after the carbonate-bound extraction consisted of the addition of 4 M HNO$_3$ at 80°C for 16 hours.[41] Apatite dissolution rates are highest at low pH,[43] so this extraction is expected to remove ^{90}Sr that is incorporated into the apatite. For every extraction, the mobilized Sr and ^{90}Sr was measured. The amount of carryover from one extraction to another was accounted for in calculations.

^{90}Sr added to aqueous solutions in contact with Hanford sediments was rapidly adsorbed (Figure 9). The sediment with no additional treatment (i.e. no apatite) showed that adsorption equilibrium (Figure 9, square symbols) was reached witin two hours, giving a Sr K$_d$ value of 14.8 cm^3/g. For the sediment that contained precipitated apatite, there was initial rapid uptake of ^{90}Sr (more than one hour) that was likely adsorption. In this case, the solution ^{90}Sr continued to decrease even after 120 hours, likely due to

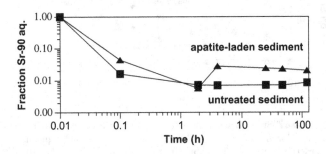

Figure 9.   Sr-90 adsorption rate to untreated and apatite solution-laden Hanford 100N sediment. The resulting retention by sediments, described by a distribution coefficient (K$_d$) was 24.7 cm3/g for the untreated sediment and 7.6 cm3/g for the apatite-laden sediment.

some uptake by the apatite. The apparent K_d at 120 hours was 7.6 cm^3/g, which was smaller than the untreated sediment due to the higher solution ionic strength (apatite solution ionic strength 97 mM versus groundwater ionic strength 11 mM). In a study of the influence of major ions on Sr retention by Hanford sediments, Routson and others[44] reported a Sr K_d of 49 cm^3/g (0.001 M NaNO$_3$) and K_d of 16 cm^3/g (0.1 M NaNO$_3$), so there was a 4 times decrease in the K_d value by increasing the ionic strength with Na$^+$, which is the predominant cation in the spent apatite solution.

Solid phase extractions showed that the addition of apatite to sediment resulted in stronger retention of ^{90}Sr, which is presumed to be caused by some incorporation into the apatite structure. For the untreated sediment, solid phase extractions showed that 83.6 ± 3.3% of the ^{90}Sr was bound by ion exchange (0.5 M KNO$_3$, Figure 10a), 0.0% organic bound; 5.2 ± 0.7% by a carbonate extraction (0.05 M EDTA); and 5.8 ± 0.6% by a mineral dissolution extraction (4 M HNO$_3$ at 80°C). The remaining 5.6% ^{90}Sr

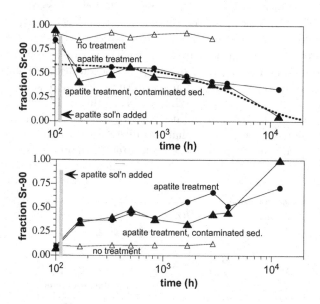

Figure 10.   Solid phase ^{90}Sr extractions on sediment without and with apatite treatment showing changes in ^{90}Sr mobility: (a) mobile fraction — combined aqueous and ion exchangeable extractions, and (b) immobile fraction — combined carbonate (EDTA) and mineral dissolution (HNO$_3$) extractions. Extractions are shown for the Hanford 100N sediment with no treatment (open triangle), apatite solution addition to sediment with ^{90}Sr added (circles), and apatite solution addition to ^{90}Sr field-contaminated sediment (solid triangles). The apatite solution was in the experimental system for the entire time.

was aqueous. Therefore, ^{90}Sr was retained on untreated Hanford sediments predominantly by ion exchange, as expected.

The addition of the apatite solution to sediment and subsequent apatite precipitation resulted in ^{90}Sr being incorporated into the apatite structure by one or more mechanisms: (i) during initial precipitation of apatite, some Sr-90 is incorporated into the apatite structure substituted for Ca; (ii) solid phase Ca-hydroxyapatite can incorporate Sr/Sr-90 by solid phase dissolution/reprecipitation; (iii) solid phase Ca-hydroxyapatite can incorporate Sr/Sr-90 by crystal lattice rearrangement of adsorbed Sr. The rate of incorporation of Sr-90 during the initial precipitation phase (mechanism (i)) should be fairly rapid (the precipitation rate, 10s to 100s of hours), whereas the solid phase incorporation rate is slow (1,000s of hours).

A significant amount of ^{90}Sr (40–50%) was no longer in the ion exchangeable fraction by 200 h (Figure 10a, both circles and solid triangles), likely indicating ^{90}Sr uptake during the initial apatite precipitation. There were corresponding increases in EDTA and HNO$_3$ extractable amounts (Figure 10b). This could not be due to sorption onto the apatite, which would show up in the ion exchangeable fraction), so is likely due to ^{90}Sr incorporation into apatite during the initial precipitation phase. The organic-bound extraction (0.5M NaOH) extracted no ^{90}Sr in all cases, as there was no natural organic matter in these subsurface sediments.

Over the course of 1,000s of hours (months) the Sr-90 became less mobile as shown by the mobile fraction decreasing (Figure 10a), and the immobile fraction increasing (Figure 10b). The rate of Sr-90 immobilization (defined by the extraction change) was well approximated by a simple first-order reaction (black dashed line in Figure 10a) with a half-life of 180.5 days (4,332 h). It is hypothesized that this ^{90}Sr uptake is caused by the slow dissolution of Ca-apatite and precipitation of ^{90}Sr-apatite. The actual rates that would be observed in the field would be dependent on the aquifer temperature, the mass of apatite in the sediment, and the presence of the high ionic strength apatite solution. For example, an aquifer temperature of 15°C is likely to have a 2× slower rate than observed in these 22°C laboratory studies. Greater mass of apatite loading in the sediment (by higher concentration or multiple apatite injections) will uptake ^{90}Sr more rapidly. Solid phase extractions shown (Figure 10) were conducted with the high ionic strength solution remaining in contact with the sediment for the entire experiment, whereas other experiments in which the solution was removed showed more rapid ^{90}Sr uptake. In the field, the spent apatite solution will migrate downgradient, which would result in higher adsorption of the ^{90}Sr to the apatite, and as a consequence more rapid uptake in the apatite structure.

Conclusions

A method has been described for the formation of nanosize apatite crystals in sediment that has applications for formation of reactive barriers and contaminated site stabilization. The method is based on complexing calcium in calcium phosphate solutions to prevent precipitation of calcium phosphate solids before the solution can be injected into sediment to be treated. As the indigenous soil microorganisms degrade the citrate, the calcium is made available to form calcium phosphate precipitates. These precipitates change to apatite over time. In batch sorption experiments, the apatite demonstrated strong sorption and retention of ^{90}Sr. Based on sorption data from literature, other radionuclides and heavy metals including U, Np, Pu and Pb should exhibit the same behavior. The method offers an alternative approach to conventional methods such as jet grouting, excavation and backfilling for construction of reactive barriers and contaminated site stabilization. It also may be applied to remediation of certain problematic sites where conventional methods are not feasible.

Acknowledgements

This work was supported by Sandia National Laboratories under contract with the U.S. Department of Energy under contract DE-AC04-94AL85000. Pacific Northwest National Laboratory operated by Battelle for the United States Department of Energy under Contract DE-AC06-76RL01830.

References

1. R. D. Spence and C. Shi (eds.), Solidification and Stabilization of Hazardous, Radioactive and Mixed Wastes, CRC Press, Boca Raton, FL (2005).
2. P. H. Ribbe (ed.), Reviews in Mineralogy & Geochemistry, Vol. 48: Phosphates, Geochemical, Geobiological and Material Importance. Mineralogical Society of America (2002).
3. United States Department of Energy, Passive Reactive Barrier, Subsurface Contaminant Focus Area Innovative Technology Summary Report. http://www.em.doe.gov/ost under "Publications" (2002).
4. J. S. Arey, J. C. Seaman and P. M. Bertsch, Immobilization of uranium in contaminated sediments by hydroxyapatite addition, Environ. Sci. Technol. **33**(2), 337–342 (1999).
5. C. C. Fuller, J. R. Bargar, J. A. Davis and M. J. Piana, Mechanisms of uranium interactions with hydroxyapatite: Implications for groundwater remediation, Environ. Sci. and Technol. **36**(2), 158–165 (2002).

6. J. Jeanjean, J. C. Rouchaud, L. Tran and M. Fedoroff, Sorption of uranium and other heavy metals on hydroxyapatite, *J. Radioanal. Nucl. Chem. Lett.* **201**(6), 529–539 (1995).

7. R. C. Moore, C. Sanchez, K. Holt, P. Zhang, H. Xu and G. Choppin, Formation of hydroxyapatite in soils using calcium citrate and sodium phosphate for control of strontium migration, *Radiochimica Acta* **92**(9–11), 719–723 (2004).

8. R. C. Moore, M. Gasser, N. Awwad, K. Holt, F. Salas, A. Hasan, M. Hasan, H. Zhao and C. Sanchez, Sorption of plutonium(VI) by hydroxyapatite, *J. Radioanaly. and Nuc. Chem.* **263**(1), 97–101 (2005).

9. S. Lazic and Z. Vukovic, Ion exchange of strontium on synthetic hydroxyapatite, *J. Radioanal. Nucl. Chem.* **149**(1), 161–168 (1991).

10. R. Z. LeGeros, G. Quirolgico and J. P. LeGeros, Incorporation of strontium in apatite: Effect of pH, *J. Dental Research* **58**, 169 (1979).

11. Z. Vukovic, S. Lazic, I. Tutunovic and S. Raicevic, On the mechanism of strontium incorporation into calcium phosphates, *J. Serbian Chem. Soc.* **63**(5), 387–393 (1998)

12. S. Bailliez, A. Nzihou, E. Beche and G. Flamant, Removal of lead (Pb) by hydroxyapatite sorbent, *Process Safety Environ. Prot.* **82**(2), 175–180 (2004).

13. E. Mavropoulos, A. M. Rossi, A. M. Costa, C. A. C. Perez, J. C. Moreira and M. Saldanha, Studies on the mechanisms of lead immobilization by hydroxyapatite, *Environ. Sci. and Technol* **36**(7), 1625–1629 (2002).

14. I. Smiciklas, A. Onjia, J. Markovic and S. Raicevic, Comparison of hydroxyapatite sorption properties towards lead, zinc and strontium ions, *Mat. Sci. Forum* **494**, 405–410 (2005).

15. E. Andronescu, E. Stefan, E. Dinu and C. Ghitulica, Hydroxyapatite synthesis, *Key Engineering Materials* **206**(2), 1595–1598 (2002).

16. A. D. Papargyris, A. I. Botis and S. A. Papargyri, Synthetic routes for hydroxyapatite powder production, *Key Engineering Materials* **206**(2), 83–86 (2002).

17. A. J. Tofe, Chemical decontamination using natural or artificial bone, U.S. Patent 5,711,015 (1998).

18. R. Z. LeGeros, *Calcium Phosphates in Oral Biology and Medicine.* Basel, Switzerland: S. Karger (1991).

19. R. C. Moore and R. Bontchev, *Apatite Formation in Sediment for In Situ Reactive Barrier Formation* (SAND 2005–6869 Sandia National Laboratories, in press, 2005).

20. R. P. Shellis, A. R. Lee and R. M, Observations on the solubility of carbonate-apatites, *J. Coll Interface Sci.* **218**(2), 351–358 (1999).

21. A. B. Barry, H. H. Zhuang, A. A. Baig and W. A. Higuchi, Effect of fluoride pretreatment on the solubility of synthetic carbonated apatite, *Cal. Tissue Int.* **72**(3), 236–242 (2003).

22. P. A. W. Van Hees, D. L. Jones and D. L. Godbold, Biodegradation of low molecular weight organic acids in coniferous forest podzolic soils, *Soil Biol and Biochem* **34**(9), 1261–1272 (2002).

23. L. Brynhildsen and T. Rosswall, Effects of metals on the microbial mineralization of organic acids, *Water Air Soil Poll.* **94**(1–2), 45–57 (1997).

24. M. J. Truex, Pacific Northwest National Laboratory, unpublished data (2005).

25. A. E. Martell and R. M. Smith, Other Organic Ligands, *Critical Stability Constants* (New York, NY: Plenum Press, 1979).

26. J. A. M. Van der Houwen and E. Valsami-Jones, Influence of organic ligands on the precipitation of calcium phosphate at neutral pH, *Proceedings of the 11th Annual V.M. Goldschmidt Conference* (2001).

27. C. M. Johnson, P. R. Stout, T. C. Boyer and A. B. Carlton, Comparative chlorine requirements of different plant species, *Plant and Soil* **8**, 337–353 (1957).

28. G. H Nancollas and M. S. Mohan, The growth of hydroxyapatite crystals, *Arch. Oral. Biol.* **15**, 731–745 (1970).

29. D. N. Misra, Interaction of citric acid with hyrdoxyapatite: Surface exchange of ions and precipitation of calcium citrate, *J. Dent. Res.* **75**(6), 1418–1425 (1996).

30. J. A. M. G. Van der Houwen, Cressy, B. A. Cressey and E. Valsami-Jones, The effect of organic ligands on the crystallinity of calcium phosphate, *J. Cry. Growth* **249**, 572–583 (2003).

31. J. E. Bailey and D. F. Ollis, *Biochemical Engineering Fundamentals*, 2nd edn. (McGraw-Hill, Inc., New York, 1986), pp. 246–250.

32. J. M. Hughes and J. Rakovan, The crystal structure of apatite, $Ca_5(PO_4)_3(F,OH,Cl)$, in *Phosphates: Geochemical, Geobiological and Materials Importance, Reviews in Mineralogy and Geochemistry*, Vol. 48 (Mineralogical Society of America, Washington, DC, 2002), pp. 1–12.

33. G. Waychunas, Luminescence, X-ray emission and new spectroscopies, *Rev Mineral* **18**, 638–698 (1989).

34. R. C. Ewing, The design and evaluation of nuclear waste forms: Clues from mineralogy, *Can Mineral* **39**, 697–715 (2001).

35. J. C. Elliot, P. E. Mackie and R. A. Young, Monoclinic hydroxyapatite, *Science* **180**, 1055–1057 (1973).

36. J. M. Hughes, M. Cameron and K. D. Crowley, Structural variations in natural F, OH and Cl apatites. *Amer Mineral* **74**, 870–876 (1989).

37. Szecsody Sorption of Sr-90 by apatite formed in sediment and sequential extraction, PNNL report, in press (2005).

38. R. J. Serne and V. L. LeGore, Strontium-90 adsorption-desorption properties and sediment characterization at the 100N-area. Pacific Northwest National Laboratories, PNL-10899/UC-702 (1996).

39. C. Amrhein and D. L. Suarez, Procedure for determining sodium-calcium selectivity in calcareous and gypsiferous soils, *Soil Sci. Soc. Am. J.* **54**, 999–1007 (1990).

40. G. Sposito, L. Lund and A. Chang, Trace metal chemistry in arid-zone field soils amended with sewage sludge: 1. Fractionation of Ni, Cu, Zn, Cd, and Pb in solid phases, *Soil Sci. Soc. Am. J.* **46**, 260–264 (1982).

41. G. Sposito, C. LeVesque, J. LeClaire and A. Chang, Trace metal chemistry in arid-zone field soils with sewage sludge: 3. effect of time on the extraction of trace metals, *Soil Sci. Soc. Am. J.* **47**, 898–902 (1983).

42. C. Steefel, Evaluation of the field-scale cation exchange capacity of Hanford sediments, Water-Rock Interaction, eds. E. Wanty and P. Seal (Taylor and Francis Group, London, 2004), pp. 999–1002.

43. C. Chairate, E. Oelkers, S. Kohler, N. Harouiya and J. Lartigue, An experimental study of the dissolution rates of apatite and britholite as a function of solution composition and pH from 1 to 12, Water-Rock Interaction, eds. E. Wanty and P. Seal (Taylor and Francis Group, London, 2004), pp. 671–674.

44. R. C. Routson, G. Barney, R. Smith, C. Delegard and L. Jensen, Fission product sorption parameters for Hanford 200 area sediment types, RHO-ST-35, Rockwell Hanford Operations, Richland, Washington (1981).

45. C. C. Fuller, J. R. Bargar and J. A. Davis, Molecular-scale characterization of uranium sorption by bone apatite materials for a permeable reactive barrier demonstration, *Environ. Sci. Technol.* **37**(20) 4642–4649 (2003).

46. R. C. Moore, K. C. Holt and K. B. Helean, Temperature dependence of citrate degradation by soil microorganisms, SAND 2005–6870. Sandia National Laboratories, in press (2005).

47. R. C. Moore, K. Holt, H. Zhao, A. Hasan, N. Awwad, M. Gasser and C. Sanchez, Sorption of Np(V) by synthetic hydroxyapatite, *Radiochim. Acta.* **91**, 721–727 (2003).

48. G. H. Nancollas and B. B. Tomazic, Growth of calcium phosphate on hydroxyapatite crystals, effect of supersaturation and ionic medium, *J. Phy. Chem.* **79**, 2218–2225 (1974)

49. K. Onuma, A. Oyane, K. Tsutsui, K. Tanaka, G. Treboux, N. Kanzaki and A. Ito "Precipitation kinetics of hydroxyapatite revealed by the continuous-angle laser light-scattering technique, *Journal of Physical Chemistry B* **104**(45), 10563–10568 (2000).

50. R. P Singh, Y. D. Yeboah, E. R Pambid and P. Debayle, Stability constant of the calcium-citrate(3-) ion pair complex, *J. Chem. Eng. Data* **36**, 52–54 (1991).

51. D. L. Sparks (ed.), Methods of soil analysis, Part 3, *Chemical Methods* (Soil Science Society of America Book Series 5, 1996).

Chapter 5

Functionalized Nanoporous Sorbents for Adsorption of Radioiodine from Groundwater and Waste Glass Leachates

S. V. Mattigod, G. E. Fryxell and K. E. Parker

Pacific Northwest National Laboratory
Richland, Washington 99352, USA

Introduction

Radioactive iodine is one of the fission products from nuclear industry that ends up encased in waste forms for ultimate disposal. Iodine in the environment is highly mobile because it does not either form solids of limited solubility or irreversibly adsorb on to soil mineral surfaces.[1] Radioiodine, in particular [129]I, is of significant environmental concern due to its very long half life (1.7×10^7 years). Therefore, considerable effort has been focused on finding suitable adsorptive materials that can immobilize or delay the transport of radioiodine that would be released from physically and chemically degrading waste packages. For instance, metallic Cu and its oxides,[2] sulfide minerals,[3-11] cementitious forms,[12] various types of clay and oxide minerals,[6,12-18] modified zeolites,[19] and organophilic clays[20-25] have been tested and evaluated for their adsorptive properties for radioiodine.

When exposed to groundwater, vitrified or cement waste forms used for immobilizing radioactive waste, will leach high concentrations of dissolved constituents containing moderate to high alkalinity. Under such conditions, several complex geochemical reactions are known to occur in waste forms, neighboring engineered structures, and the surrounding sediments, which include: dissolution of several carbonate and silicate minerals, precipitation of secondary and tertiary mineral phases, radionuclide adsorption onto minerals (primary, secondary, and tertiary), and sequestration of radionuclides into secondary and tertiary mineral phases.

Under Eh-pH conditions typically encountered in ground and surface waters, iodine exists mainly as iodide. In highly oxidizing and alkaline conditions however, iodine may exist as iodate. Therefore, significant fraction of iodine in highly oxidizing alkaline leachates from glass and cement waste forms may also exist in the form of iodate. A moderate pH value (< 8.0) and the absence of any strong oxidants in the glass leachate used in our experiments indicated that iodine in solution would exist principally as iodide.

According to the criteria enumerated by Viani,[26] successful radionuclide sorbent materials should: (1) have at least moderate adsorption/immobilization properties; (2) be stable for a long time in the post-closure environment; (3) not adversely affect water chemistry; and (4) not be prohibitively expensive. The objective of this study was to evaluate novel adsorbent materials we have developed as high affinity sorbents for radioiodine and to compare their adsorptive performance with selected natural minerals. In this study, we investigated the radioiodine sorptive performance of functionalized nanoporous materials.

Experimental

The iodide-specific sorbents consisted of a series of materials developed from nanoporous ceramics. These nanoporous ceramic substrates have very high surface areas ($\sim 1000 \, \mathrm{m^2/g}$) and specially tailored pore sizes that range from 20–100 Å. These high-surface area substrates are functionalized with monolayers of well-ordered functional groups that have a high affinity and specificity for specific types of free or complex cations or anions. These self-assembled monolayers on mesoporous silica (SAMMS) materials with high adsorption properties have been tested successfully on a series of heavy metal cations (Hg, Cu, Cd, and Pb) and oxyanions (As, Cr, Mo, and Se). Detailed descriptions of the synthesis, fabrication, and adsorptive properties of these novel materials have been published.[27–34] The nanoporous sorbents selected for these tests consisted of thiol-functionalized SAMMS material. One of these samples was prepared by saturating the sorption sites with mercury and the other was generated by site saturation with silver ions.

Certain sulfide minerals are known to adsorb radioactive iodide. Therefore, six sulfide minerals were selected for tests of their adsorptive performance with synthetic novel sorbents (Table 1). All minerals samples except argentite were obtained from Ward's Geology, Rochester, New York. The argentite was synthesized by adding 1 gram of reagent grade silver nitrate to 1.4 grams of reagent grade sodium sulfide monohydrate dissolved in 200 mL of DI water. The precipitate was separated and washed three times with deionized distilled water and dried overnight at 105°C. The mineral samples

Table 1. Experimental matrix used in adsorption experiments.

^{125}I Spike (Bq/mL)	Solution Matrix	Sorbents
~ 36,500	Groundwater	Hg-thiol SAMMS, Ag thiol SAMMS
~ 32,500	Glass leachate	Hg-thiol SAMMS, Ag thiol SAMMS, Argentite (Ag_2S), Chalcocite (Cu_2S), Chalcopyrite ($CuFeS_2$), Cinnabar (HgS), Galena (PbS), Stibnite (Sb_2S_3)

were prepared for adsorption experiments by crushing with an agate mortar and pestle and then sieving to separate particles between < 0.150 and > 0.075 mm.

The groundwater used in these experiments was collected from an uncontaminated well located in the Hanford Site. The groundwater was filtered through a 0.45 micron filter and then analyzed using standard techniques. Inductively coupled plasma-atomic emission spectroscopy and ion chromatography were used to determine dissolved cation and anion concentrations, respectively. The composition of the groundwater indicated that calcium, magnesium, sodium, and potassium are the principal cationic constituents with chloride, sulfate, and bicarbonate the dominant anions (Table 2).

Glass leachate was prepared by equilibrating, for more than seven days, a quantity of crushed simulated waste glass with groundwater from the Hanford Site. Following equilibration, the glass leachate was filtered through a 0.45 micron filter (Table 2).

Adsorption experiments were conducted by equilibrating each sorbent with aliquots of groundwater or simulated glass leachate spiked with ^{125}I (Table 1). Solution-to-solid ratios of ~100, ~500, ~1,000, ~5,000 and ~10,000 mL/g were used to evaluate the degree of loading radionuclide. A positive control containing the ^{125}I-spiked groundwater and sorbent was used to evaluate ^{125}I sorption to labware and filters. The mixture was gently agitated for ~20 hours at $25 \pm 3°C$ and portions of equilibrated solutions were removed, filtered and counted for residual ^{125}I activity. Analysis of ^{125}I in liquid samples was conducted by gamma-ray spectrometry, using a calibrated Wallac® 1480 Wizard™ 3-in NaI detector with built-in software.

Results and Discussion

The adsorption data indicated that the nanoporous sorbents (Hg-thiol SAMMS, and Ag-thiol SAMMS) had adsorbed as much as ~3 × 10^8 Bq/g of radioiodine (Table 3). Among the sulfide minerals, the

Table 2. Chemical composition of groundwater samples and waste glass leachate.

Constituent	Groundwater (mg/L)	Glass Leachate (mg/L)
pH (SU)	8.1	7.8
Cond. (mS/cm)	0.40	0.56
Alk (as CO_3^{2-})	54.1	67.5
Cl	7.8	22.0
Br	0.10	< 0.01
F	0.17	< 0.01
I	< 0.005	< 0.005
NO_2^-	0.68	< 0.01
NO_3^-	27.2	1.7
PO_4^{-3}	< 0.01	0.3
SO_4^{-2}	82.5	108.0
Al	0.01	0.02
B	< 0.05	0.11
Ba	< 0.03	0.03
Be	< 0.01	0.01
Ca	49.5	61.4
Fe	0.07	< 0.05
K	1.7	8.3
Mg	14.6	16.1
Mn	0.17	0.03
Na	13.2	46.0
Si	16.5	16.6

highest observed adsorption for cinnabar, argentite, chalcocite, and chalcopyrite were $\sim 7 \times 10^7$ Bq/g, $\sim 1.8 \times 10^8$ Bq/g, $\sim 3.5 \times 10^6$ Bq/g, and $\sim 6.9 \times 10^5$ Bq/g, respectively (Table 3). Positive control experiment indicated that the labware and filters did not adsorb any radioiodine. The other two sulfide minerals that were tested, galena and stibnite, did not adsorb iodide to any measurable extent. These data showed that the novel sorbents had the highest iodide adsorption capacities that were about two to four times more than cinnabar and argentite and about two- to three-orders of magnitude higher than the chalcocite and chalcopyrite (Tables 3).

The selectivity (affinity) of a sorbent for a contaminant is typically expressed as a distribution coefficient (K_d) which defines the partitioning of the contaminant between sorbent and solution phase at equilibrium. Distribution coefficient is the measure of an exchange substrate's selectivity or specificity for adsorbing a specific contaminant or a group of contaminants from matrix solutions, such as waste streams. The distribution coefficient (sometimes referred to as the partition coefficient at equilibrium) is defined as a ratio of the adsorption density to the final contaminant concentration

Table 3.   Iodide-125 adsorption by novel sorbents and minerals.

Soln/Sol Ratio (mL/g)	Eq. Act. (Bq/mL)	Ads (Bq/g)	Ads (%)	Soln/Sol Ratio (mL/g)	Eq. Act. (Bq/mL)	Ads (Bq/g)	Ads (%)
			Groundwater				
	Hg-thiol SAMMS				Ag-thiol SAMMS		
99	1093	3.51×10^6	97	98	108	3.58×10^6	100
489	611	1.76×10^7	98	495	409	1.79×10^7	99
1007	942	3.59×10^7	97	1021	848	3.64×10^7	98
5255	1911	1.82×10^8	95	5068	2299	1.73×10^8	94
9618	2801	3.24×10^8	92	10432	2961	3.50×10^8	92
			Glass Leachate				
	Hg-thiol SAMMS				Ag-thiol SAMMS		
480	163	1.55×10^7	100	96	7	3.13×10^6	100
968	165	3.13×10^7	100	485	64	1.57×10^7	100
4860	427	1.56×10^8	99	973	124	3.15×10^7	100
9745	647	3.10×10^8	98	4825	2188	1.46×10^8	94
				9741	2274	2.94×10^8	93
	Cinnabar				Argentite		
97	1078	3.03×10^6	97	96	67	3.12×10^6	100
469	2402	1.41×10^7	93	488	781	1.55×10^7	98
953	7471	2.38×10^7	77	832	2119	6.77×10^7	93
4253	21060	4.86×10^7	35	4057	7461	1.02×10^8	77
10950	26020	7.08×10^7	14	8026	9528	1.84×10^8	71
	Chalcocite				Chalcopyrite		
98	18614	1.35×10^6	43	96	29562	2.82×10^5	9
444	28206	1.90×10^6	13	461	31587	4.35×10^5	3
941	29380	2.92×10^6	10	953	31766	6.86×10^5	2
8920	32096	3.48×10^6	1	4046	33067	0	0
				8109	32559	0	0

in solution at equilibrium. This measure of selectivity is defined as

$$K_d = \frac{(x/m)_{eq}}{c_{eq}} \qquad (1)$$

where, K_d is the distribution coefficient (mL/g), $(x/m)_{eq}$ is the equilibrium adsorption density (Bq of iodide per gram of adsorbing substrate), and c_{eq} is the iodide concentration (Bq/mL) in contacting solution at equilibrium.

The novel sorbents exhibited very high distribution coefficients (K_d) for ^{125}I in both Hanford groundwater and the simulated glass leachate

Table 4.   Iodine-125 adsorption affinity parameters K_d (mL/g) for novel sorbent
materials and minerals.

Adsorbent	Groundwater[a]	Surface Water[b]	Glass Leachate[a]	Concrete Leachate[b]
Hg-thiol SAMMS	$3 \times 10^4 - 1 \times 10^5$	$2 \times 10^3 - 2 \times 10^6$	$8 \times 10^4 - 5 \times 10^5$	$8 \times 10^3 - 6 \times 10^5$
Ag-thiol SAMMS	$3 \times 10^4 - 1 \times 10^5$	$2 \times 10^4 - 3 \times 10^5$	$6 \times 10^4 - 5 \times 10^5$	$9 \times 10^3 - 6 \times 10^5$
Cinnabar (HgS)	—	$7 \times 10^2 - 3 \times 10^3$	$2 \times 10^3 - 7 \times 10^3$	$7 \times 10^1 - 1 \times 10^2$
Argentite (Ag$_2$S)	—	$3 \times 10^2 - 8 \times 10^4$	$1 \times 10^4 - 5 \times 10^4$	$2 \times 10^4 - 7 \times 10^5$

[a]This Study.
[b]Reference (Kaplan et al., 2000).

(Table 4). The iodide K_d value for Hg-thiol SAMMS substrate in groundwater (3×10^4 to 1×10^5 mL/g) was noticeably lower that the K_d value in glass leachate (8×10^4 to 5×10^5 mL/g). This difference in I adsorption affinity of Hg-SAMMS can be attributed to the probable surface sites that exist in these two solution matrices. Based on Hg hydrolysis and speciation data of Baes and Mesmer,[35] we estimated that the iodide binding sites on Hg-SAMMS equilibrated in groundwater matrix were likely to be R-Hg(OH)$_2$, whereas, due to lower pH and higher concentration of chloride, the binding sites on Hg-SAMMS reacting in glass leachate medium were probably in the form of R-HgOHCl. If this were the case, the I adsorption in these two cases can be represented by the following ligand exchange reactions [Equations (1) & (2)]

$$R\text{-Hg(OH)}_2^0 + 2I^-(aq) = 2OH^-(aq) + R\text{-HgI}_2^0 \qquad (2)$$

$$R\text{-HgOHCl}^0 + 2I^-(aq) = Cl^-(aq) + OH^+(aq) + R\text{-HgI}_2^0 \qquad (3)$$

Although association constants for surface species (R-Hg(OH)$_2^0$, R-HgOHCl^0, R-HgI$_2^0$) are unknown, using association constants for aqueous species,[36] the magnitude of I adsorption constants (ligand exchange) for these reactions were estimated to be $\sim 7 \times 10^2$ and $\sim 4 \times 10^6$ respectively. Using these constants, the free energies of iodide adsorption were calculated to be ~ -16 kJ/mol and ~ -38 kJ/mol respectively. These calculated adsorption free energy values indicated that although iodide adsorption on to both the sites (R-Hg(OH)$_2^0$ and R-HgOHCl^0) are energetically favored, adsorption through ligand exchange on to R-HgOHCl^0 is likely to occur with significantly higher free energy release than the ligand exchange reaction with R-Hg(OH)$_2^0$. These qualitative differences in the energetics of ligand exchange reaction appeared to be reflected in the higher distribution coefficient (K_d) observed in glass leachate medium.

The iodide K_d values for Ag-thiol SAMMS in both groundwater and glass leachate matrices are 3×10^4 to 1×10^5 mL/g and 6×10^4 to 5×10^5 mL/g respectively. The magnitude of these values is similar to that of Hg-thiol SAMMS. Therefore, the adsorption mechanisms of iodide onto Ag-thiol SAMMS sites is expected to be similar to the mechanism suggested for Hg-thiol SAMMS sites.

As compared to novel sorbents, the iodide K_d values for cinnabar and argentite were 2×10^3 to 7×10^3 mL/g and 1×10^4 to 5×10^4 mL/g, respectively. These data indicate that the novel sorbents performed significantly better in that they adsorbed two to four times more ^{125}I and exhibited significantly higher adsorption affinity (distribution coefficients one to two orders of magnitude higher) than the sulfide minerals, cinnabar and argentite. Such improved performance can be attributed to higher numbers of accessible metal sites on novel sorbents as compared to the accessible adsorption sites on metal sulfides. Previous measurements indicated that Hg-thiol and Ag-thiol SAMMS have about 2 and 2.8 mmol of adsorption sites per gram of substrate. By comparing the adsorption performance of novel sorbents and the minerals, we estimated that cinnabar has a maximum of $\sim$0.45 mmol of iodide adsorption sites per gram of material, whereas argentite has a maximum of $\sim$1.47 mmol of iodide adsorption sites per gram of material. Therefore the observed maximum ^{125}I loading in these experiments ($\sim$7 $\times 10^{-8}$ mmol/g for novel getters, and 2×10^{-8} to 4×10^{-8} mmol/g for mineral sorbents) indicate occupancy of a negligible fraction of potential I adsorption sites for both the novel and mineral sorbents.

Although the energetics of iodide adsorption through ligand exchange reactions are similar for Hg-SAMMS and cinnabar, the higher maximum adsorption and distribution coefficient exhibited by the novel sorbents can be attributed to its two times higher number of available sites per unit mass ($\sim$ 2 mmol/g). Similarly, the higher maximum adsorption and distribution coefficient exhibited by Ag-SAMMS in comparison to argentite can also be ascribed to its significantly higher number of potential I adsorption sites.

For comparison, iodide (^{129}I) data for these same sorbents obtained in a previous study[11] in surface water and cement leachate matrices are also included in Table 4. These data confirmed that the novel sorbents are capable of adsorbing radioiodine with high specificity. For instance, these experiments were conducted in different media with pH and solution conductivity ranging from a low of 5.65 SU and 0.021 mS/cm for surface water to 11.8 SU and 4.9 mS/cm cement leachate respectively. These data suggest that both Hg-thiol SAMMS and Ag-thiol SAMMS can adsorb radioiodine

Table 5. Iodide distribution coefficient data for natural and synthetic sorbent materials.

Sorbent	K_d (mL/g)	Reference
Hg-thiol SAMMS	3×10^4–5×10^5	This Study
	2×10^3–2×10^6	11
Ag-thiol SAMMS	3×10^4–5×10^5	This Study
	2×10^4–5×10^5	11
Cinnabar	2×10^3–7×10^3	This Study
	1×10^1–1×10^2	4
	0–5×10^1	7
	7×10^1–3×10^3	8, 9
	7×10^2–3×10^3	11
	4×10^3–2×10^4	15
Argentite	1×10^4–5×10^4	This Study
	3×10^2–8×10^4	11
Other Sulfides: bornite,	1×10^1–3×10^3	This study,
Chalcocite, chalcopyrite, galena,		3, 4, 5, 6, 8,
pyrite, enargite		9, 11, 15
Oxides: Al-, Cu-, Fe-, Pb-, Ti-	0–1×10^2	2, 4, 5, 6, 16
Clays: allophane, attapulgite,	0–3×10^1	4, 6, 15, 17
bentonite, clinoptilolite, kaolinite,		
illite, sepiolite, vermiculite		
Modified clays	2×10^0–5×10^3	21, 22, 24, 25

from diverse media with differing pH, chemical composition and ionic strength.

To gain some understanding of the iodide adsorption behavior of the two novel sorbents (Hg-thiol, and Ag-thiol SAMMS), the data obtained in our tests were compared with the published iodide performance data for natural and modified mineral sorbents (Table 5). These data show that most of the natural minerals, except sulfides, have K_d values of $< 10^3$ mL/g, whereas sulfide minerals such as cinnabar and argentite have K_d values that typically range from 10^2 to 10^4 mL/g. Modified clays were reported to exhibit enhanced affinity (2×10^0 to 5×10^3 mL/g) to adsorb iodide. However, these data also showed that the performance of these modified minerals was no better than the iodide adsorption performance of natural sulfide minerals such as cinnabar and argentite. Additionally, most of these sorbent tests were conducted in solution matrices in which the competing halide ions (Cl, Br, and F) were either absent or their concentrations were low.

The novel sorbents Hg-thiol and Ag-thiol SAMMS showed very high affinities for adsorption of radioiodine ($K_d \sim 10^4$–10^5 mL/g). The performance of these novel sorbents was remarkable because these sorbents were tested in matrices of groundwater, surface water, waste glass, and concrete leachates, which are typical environments in which these sorbents are expected to perform. Also, these novel sorbent tests were conducted in matrices of natural environmental solutions that contain other halide ions such as Cl, Br, and F in concentrations that far exceeded the concentrations of radioiodine. Calculated molar concentration ratios showed that Cl, Br, and F concentrations in these natural solution matrices exceeded radioiodine concentrations by eight to ten orders of magnitude. Therefore, the magnitude of the measured K_d values indicated that the novel sorbents were adsorbing iodide with very high selectivity.

Conclusions

In summary, performance tests were conducted using novel sorbent materials. These tests revealed that these sorbent materials can immobilize or delay the transport of radioiodine that would be released during physical and chemical degradation of solidified low-level waste packages. The results showed that metal-capped novel sorbents such as Hg-thiol and Ag-thiol Self-Assembled Monolayers on Mesoporous Silica (SAMMS), designed specifically to adsorb soft anions such as I^-, had very high affinities for adsorption of radioiodine ($K_d \sim 1 \times 10^4$ to 4×10^5 mL/g). The iodide adsorption performance of these novel sorbents was from one to two orders of magnitude better than many natural mineral and modified mineral sorbents. These data indicate that the novel nanoporous sorbent materials are capable of significantly retarding the mobility of radioiodine leaching from physically and chemically weathered low-level waste packages during various physical and chemical weathering reactions expected during long-term disposal.

Acknowledgements

Pacific Northwest National Laboratory is a multiprogram national laboratory operated for the U.S. Department of Energy by Battelle Memorial Institute under Contract DE-AC06-76RLO 1830.

References

1. J. C. Wren, J. M. Ball and G. A. Glowa, *Nucl. Technol.* **129**, 297–325 (2000).
2. Z. Haq, G. M. Bancroft, W. S. Fyfe, G. Bird and V. J. Lopata, *Env. Sci. Technol.* **14**, 1106–1110 (1980).
3. R. Strickert, A. M. Friedman and S. Fried, *Nucl. Technol.* **49**, 253–266 (1980).

4. G. Allard, B. Torstenfelt, K. Andersson and J. Rydberg, Possible retention of iodine in the ground, in *Material Research Society Symposium Proceedings*, Vol. 176 (Material Research Society, Pittsburgh, Pennsylvania, 1981), pp. 673–680.

5. Z. Huie, Z. Jishu and Z. Lanying, *Radiochimica Acta.* **44/45**, 143–145 (1988).

6. D. Rancon, *Radiochimica Acta.* **88**, 87–193 (1988).

7. Y. Ikeda, M. Sazarashi, M. Tsuji and R. Seki, *Radiochimica Acta.* **65**, 195–198 (1994).

8. S. D. Balsley, P. V. Brady, J. M. Krumhansl and H. L. Anderson, *Env. Sci. Technol.* **30**, 3025–3027 (1997).

9. S. D. Balsley, P. V. Brady, J. M. Krumhansl and H. L. Anderson, ^{129}I and ^{99}TcO$_4$ Scavengers for Low Level Radioactive Waste Backfills, SAND95-2978 (Sandia National Laboratories, Albuquerque, New Mexico, 1997).

10. S. D. Balsley, P. V. Brady, J. M. Krumhansl and H. L. Anderson, *J. Soil. Contamination* **7**, 125–141 (1998).

11. D. Kaplan, S. V. Mattigod, K. E. Parker and G. Iversen, *Experimental Work in Support of the ^{129}I-Disposal Special Analysis.* WSRC-TR-2000-00283, Rev 0. Westinghouse Savannah River Company, Savannah River Site, Aiken, South Carolina (2000).

12. M. Atkins and F. P. Glasser, Encapsulation of radioiodine in cementitious waste forms, in *Material Research Society Symposium Proceedings*, Vol. 176 (Material Research Society, Pittsburgh, Pennsylvania, 1990), pp. 15–22.

13. R. A. Couture and M. G. Seitz, *Nucl. Chem Waste Management* **4**, 301–306 (1983).

14. P. Taylor, *A Review of Methods for Immobilizing Iodine-129 Arising from a Nuclear Fuel Cycle Plant, with Emphasis on Waste-Form Chemistry*, AECL-10163, Atomic Energy Canada Ltd., Whiteshell Nuclear Research Establishment, Pinawa, Canada (1990).

15. M. Sazarashi, Y. Ikeda, R. Seki and H. Yoshikawa, *J. Nucl. Sci. Tech.* **31**, 620–622 (1994).

16. N. Hakem, B. Fourest, R. Guillaumont and N. Marmier, *Radiochimica Acta.* **74**, 225–230 (1996).

17. D. Kaplan, R. J. Serne, K. E. Parker and I. V. Kutnyakov, *Env. Sci. Technol.* **34**, 399–405 (2000).

18. M. J. Kang, K. S. Chun, S. W. Rhee and Y. Do, *Radiochimica Acta.* **85**, 57–63 (1999).

19. G. D. Gradev, *J. Radioanal. Nucl. Chem.* **116**, 341–346 (1987).

20. J, Bors, *Radiochimica Acta.* **51**, 139–143 (1990).

21. J. Bors, *Radiochimica Acta.* **58/59**, 235–238 (1992).

22. J. Bors, A. Gorny and S. Dultz, *Radiochimica Acta.* **66/67**, 309–313 (1994).

23. J. Bors, A. Gorny and S. Dultz, *Radiochimica Acta.* **74**, 231–234 (1996).

24. J. Bors, S. Dultz and A. Gorny, *Radiochimica Acta.* **74**, 231–234 (1998).

25. M. Sazarashi, Y. Ikeda, R. Seki, H. Yoshikawa and Y. Takashima, Adsorption behavior of I ions on minerals for geological disposal of ^{129}I Wastes, in *Material Research Society Symposium Proceedings*, Vol. 353 (Material Research Society, Pittsburgh, Pennsylvania, 1995), pp. 1037–1043.

26. B. Viani, *Assessing Materials ("Getters") to Immobilize or Retard the Transport of Techetium Through Engineered Barrier System at the Potential Yucca Mountain Nuclear Waste Repository*, UCRL-ID-133596. Lawrence Livermore National Laboratory, Livermore, California (1999).
27. G. E. Fryxell, J. Liu, A. A. Hauser, Z. Nie, K. F. Ferris, S. V. Mattigod, M. Gong and R. T. Hallen, *Chem. Mat.* **11**, 2148–2154 (1999).
28. G. E. Fryxell, J. Liu and S. V. Mattigod, *Mat. Tech. and Adv. Perf. Mat.* **14**, 188–191 (1999).
29. G. E. Fryxell, J. Liu, S. V. Mattigod, L. Q. Wang, M. Gong, T. A. Hauser, Y. Lin, K. F. Ferris and X. Feng, Environmental applications of interfacially-modified mesoporous ceramics, in *Proceedings of the 101st National Meetings of the American Ceramic Society* (1999).
30. K. M. Kemner, X. Feng, J. Liu, G. E. Fryxell, L. Q. Wang, A. Y. Kim, M. Gong and S. V. Mattigod, *J. Synchrotron Rad.* **6**, 633–635 (1999).
31. J. Liu, G. E. Fryxell, S. V. Mattigod, M. Gong, Z. Nie, X. Feng and K. N. Raymond, *Self-Assembled Monolayers on Mesoporous Support (SAMMS) Technology for Contaminant Removal and Stabilization.* PNL-12006. Pacific Northwest National Laboratory, Richland, Washington.
32. J. Liu, G. E. Fryxell, S. V. Mattigod, T. S. Zemanian, Y. Shin and L. Q. Wang, *Studies in Surface Science and Catalysis* **129**, 729–738 (2000).
33. S. V. Mattigod, X. Feng, G. E. Fryxell, J. Liu, M. Gong, C. Ghormley, S. Baskran, Z. Nie and K. T. Klasson, *Mercury Separation from Concentrated Potassium Iodide/Iodine Leachate Using Self-Assembled Mercaptan on Mesoporous Silica Technology*, PNNL-11714. Pacific Northwest National Laboratory, Richland, Washington (1997).
34. S. V. Mattigod, J. Liu, G. E. Fryxell, S. Baskaran, M. Gong, Z. Nie, X. Feng and K. T. Klasson *Fabrication and Testing of Engineered Forms of Self Assembled Monolayers on Mesoporous Silica (SAMMS) Material*, PNNL-12007. Pacific Northwest National Laboratory, Richland, Washington (1998).
35. C. F. Baes and R. E. Mesmer, *The Hydrolysis of Cations* (John Wiley and Sons, New York, 1976).
36. R. M. Smith and A. E. Martell, *Critical Stability Constants* (Plenum Press, New York, 1976).

Nanoporous Organic/Inorganic Hybrid Materials

Chapter 6

Nature's Nanoparticles: Group 4 Phosphonates

Abraham Clearfield

Texas A&M University, College Station, TX 77843-3255, USA
clearfield@mail.chem.tamu.edu

Introduction

Natural Nanoparticle Formation

From the time nanoscience and technology became fashionable, practitioners have devised a very large number of methods by which nano-sized particles or systems can be produced. In my own work it is nature that has been the producer of nanoparticles. We have just been the recipient. This has been true for zirconium phosphates, several types of zirconium phosphonates, aluminum phosphonates and our latest discovery tin phosphonates. These compounds have been utilized to develop layer by layer films, as ion exchangers, additives for proton conducting fuel cells, as catalysts and catalyst carriers and in many diverse uses as nanoparticles. We shall begin our odyssey with a description of the synthesis, structure and properties of α-zirconium phosphate, $Zr(O_3POH)_2 \cdot H_2O$.

α-Zirconium Phosphate and Nanoparticles

History and structure

The emergence of zirconium phosphate as a compound of interest arose from work done at Oak Ridge National Laboratories in the 1950s.[1] There was a need for ion exchangers to remove radioactive species from reactor cooling water. The organic ion exchange resins of that time were degraded by the radioactive species in hot reactor water. Therefore, a search was underway, world-wide for inorganic ion exchangers that would not be so affected. Hydrous oxides were considered and it was observed that hydrous zirconia sorbed large amounts of phosphate becoming a cation exchanger.[1] Subsequently, zirconium phosphate was prepared as a gelatinous amorphous precipitate on addition of phosphoric acid to a soluble zirconium salt. The dried gel possessed interesting cation exchange properties. However, in hot water a significant loss of phosphate ion resulted due to hydrolysis.

At that time, I was employed by a branch of NL Industries that manufactured zirconium chemicals. I suggested a study of a family of zirconium-based ion exchangers but met with only lukewarm enthusiasm. However, I was teaching in the evening school at Niagara University and proposed the project to a M.S. candidate James Stynes. The result was that we were able to convert the zirconium phosphate gel to crystals and establish the composition as $Zr(HPO_4)_2 \cdot H_2O$ and the layered nature of the compound.[2] The crystal structure was determined first by film methods[3a] and later by automated diffractometry.[3b] The unit cell dimensions are a $=$ 9.060(2), b $=$ 5.297(1), c $=$ 15.414(3) Å, ß $=$ 101.71(2)°C, monoclinic space group $P2_1/n$. A schematic drawing of the structure is shown in Figure 1. The Zr atoms are slightly above and below the mean plane of the layer and are six coordinate to oxygens from six phosphate groups. Each monohydrogen phosphate bonds to three Zr atoms arranged at the apices of a near equilateral triangle. The P-OH group points into the interlayer space and hydrogen bonds to the water molecule. The water in turn hydrogen bonds to a framework oxygen in the same layer. There are no interlayer hydrogen bonds but only van der Waals forces holding the layers together.

Crystal growth and ion exchange behavior

In order to grow single crystals for the first X-ray study[3a] the gel was held in a sealed quartz tube with 12 M H_3PO_4 at 170°C for four weeks. Subsequently, crystals could be grown in a day or two in an HF solution.[4] These differences in the rate of crystal growth were of interest because we had observed a rather pronounced difference in the ion exchange behavior of batches of crystals grown under different conditions. Examination of the dried gel particles showed that the particles had no crystalline type shape and were of the order of 10 to 40 nm in size. Even when refluxed in 0.35 M H_3PO_4 for several days the particles did not grow significantly. Figure 2 illustrates the condition of the crystals grown in increasing concentrations of phosphoric acid. The slow crystal growth is the result of the low solubility of zirconium phosphate in the solutions of low phosphoric acid concentration. The estimated solubility of crystalline zirconium phosphate in 1 M H_3PO_4 at 25°C is estimated to be of the order of 10^{-6} g/L. A graph of solubility in phosphoric acid concentration of 7 M and greater at several

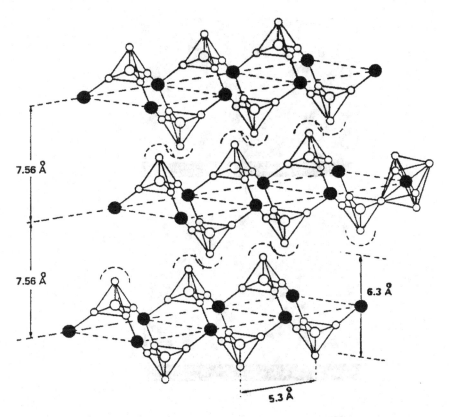

Figure 1. Schematic representation of α-zirconium phosphate layers as viewed down the b-axis direction: Zr, ●; P, ◯; O, ∘.

temperatures is given in Figure 3.[5] The crystallization mechanism is that of Ostwald ripening in accord with the solubility shown in Figure 3. This fact is illustrated by the X-ray patterns in Figure 4.[6]

The picture that arises from our studies is that of crystal perfection accompanying particle growth. The behavior of the particles is strongly tied to their crystallinity. We shall illustrate by their ion exchange behavior and intercalation and exfoliation properties. Because there are several additional phases of zirconium phosphate, we shall refer to the present one as the alpha phase or α-ZrP. The variation of the ion exchange curves as a function of crystallinity is shown in Figures 5a and 5b. To understand why the titration curves change in shape as the crystallinity increases we first consider the behavior of a strong acid polystyrene sulfonic acid ion exchange resin. These resins are influenced by the degree of cross-linking by divinyl benzene. For an eight percent cross-linked resin, the cross-links

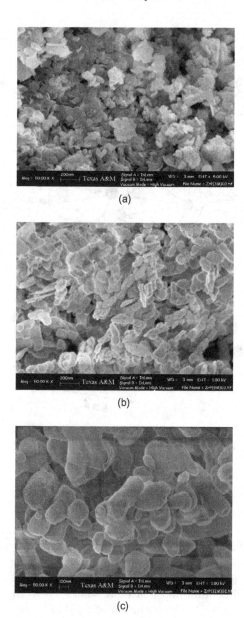

Figure 2. Scanning Electron micrographs of α-ZrP as grown by refluxing the gel in increasing concentrations of H_3PO_4. Conditions, 10 g $ZrOCl_2 \cdot 8H_2O$ in 100 ml H_3PO_4 of molarity (a) 3, (b) 9, and (c) 12.

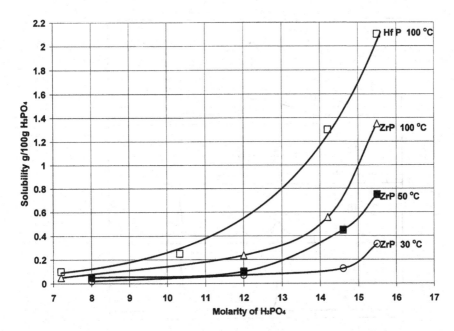

Figure 3. Solubility of α-ZrP and α-HfP in grams per 100 g of aqueous phosphoric acid.

between the linear polystyrene chains create cavities that are about 50 Å in diameter. The cavities are filled with water causing the resin beads to swell. Consider the exchange of Na^+ for protons of the sulfonic acid groups, which are present in the cavities as hydronium ions. As the protons are displaced to the outer solution, the Na^+ spreads uniformly throughout the resin bead so that only one solid phase is present. Using the terminology of the phase rule, the system has three components. One choice of components is the hydrogen ion displaced or the pH, the total sodium ion added and ion exchange capacity which gives us the amount of H^+ left in the solid phase. At constant temperature and pressure the phase rule equation, $f = c - p + 2$ becomes $f = c - p + 0$, where f = degrees of freedom, c = number of components, p = number of phases. There are two phases present, the solid exchanger and the solution phase so that the system has a degree of freedom. Thus, for each addition of NaOH, the pH increases slightly until the capacity is reached, whereupon the pH increases sharply. The titration curve is analogous to that of a strong acid-strong base titration.

For the ZrP gel and also for 0.5:48 (Figure 4) where the crystallinity is poorly developed, the added Na^+ spreads throughout the entire ZrP

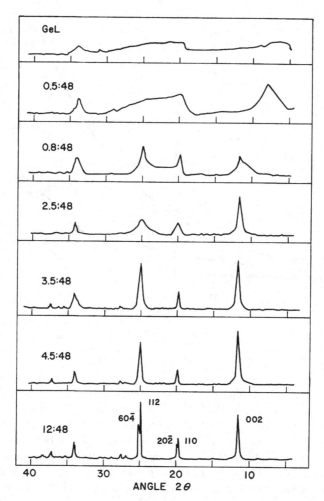

Figure 4. X-ray diffraction powder patterns of α-ZrP prepared by refluxing the gel in different concentrations of H_3PO_4 for 48 hours. The concentration of H_3PO_4 in molarity is the number preceding 48.

nanoparticle. Thus, the pH increases with each addition because of the one degree of freedom. The pH rises more steeply than for the sulfonic acid resin because the PO-H groups are weak acid groups and some sodium ion hydrolysis occurs leaving some NaOH in the solution phase. An interesting feature of the exchange is the fact that as the 0.5:48 sample becomes infused with water the interlayer spacing originally at ∼ 8 Å increases to 11.2 Å. The

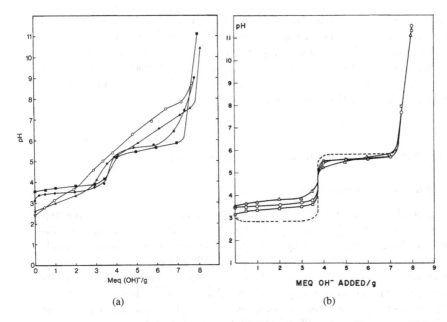

Figure 5. Potentiometric titration curves of α-ZrP for (a) samples of low crystallinity and (b) intermediate crystallinity. Dashed line represents the theoretical curve for a fully crystalline sample. Conditions: (a) amorphous gel refluxed in H_3PO_4 for 48 h in 0.8 M (o) 2.5 M ($\blacktriangle$), 4.5 M ($\bullet$), 12 M ($\blacksquare$); (b) amorphous gel in 12 M H_3PO_4 for 48 h ($\triangle$), 96 h (o), 190 h ($\square$); Tritrant 0.1 M (NaCl + NaOH).

sodium ion is then able to diffuse equally throughout the particle. In the case of the fully crystalline phase, which requires about three weeks of refluxing in 12 M H_3PO_4 to achieve this level of crystallinity,[7] a second phase of composition $Zr(NaPO_4)(HPO_4) \bullet 5H_2O$ with an 11.8 Å interlayer spacing forms at the first uptake of Na^+. This phase spreads inward from the edges until the crystallite is completely converted to the half-exchanged phase. Because two solid phases are always present, there are no degrees of freedom and the system is invariant until the endpoint is reached at 3.53 mequiv/g of $Zr(HPO_4)_2$. A second exchange process converts the half-exchanged phase to $Zr(NaPO_4)_2 \bullet 3H_2O$. The ideal titration curve would then resemble the dashed line in Figure 5(b). In between the gel-like phases and the fully crystalline phase there are two phases formed at low sodium ion uptake and one of them is a solid solution. Eventually the phase with a low level of Na^+ is converted to the solid solution phase which proceeds to completion of the sodium uptake. As the crystallinity of the exchanger increases, the solid solution ranges become narrower but the number of changes of phase increase. The situation is quite complex.[8] A good summary of these and other aspects of the ion exchange processes has been presented by Alberti.[9]

The individual particles in Figure 2(a) are as small as 20 to 50 nm parallel to the layers with some particles as large as 200 nm but of poor crystal shape. The thickness of the particles is much less and may range as low as 2 to 5 nm or three to seven layers.

Nanoparticle Formation and Its Consequences

Intercalation and exfoliation

By intercalation we mean the process by which molecules are inserted between the layers topotactically. This process should be reversible. In contrast, exfoliation is a process by which the layers are separated into individual colloidal particles. The thickness of the colloidal particles are generally of the order of 1 nm, i.e. one layer thick. Many layered compounds undergo intercalation reactions and may also be exfoliated. An excellent introduction to the subject is provided in the books by Whittingham and Jacobson[10] and Muller-Warmuth and Schollhorn.[11] α-ZrP is a solid acid and as such readily intercalates amines. The same is true for the other isostructural group IV phosphates and arsenates. Extensive chapters on amine intercalation by the group IV phosphates are provided in the aforementioned books[10,11] as well as in Comprehensive Supramolecular Chemistry.[12]

The maximum uptake of n-alkylamines by α-ZrP is two moles per formula weight or one amine molecule per P-OH group. For most amines, transfer of the phosphate proton to the amine group accompanies the intercalation reaction with formation of a bilayer. If the amine is added as a titrant, several phases are found to form before the saturation point.[13] Titration to the half-end point with propylamine results in swelling of the layers and addition of water with sonication results in exfoliation of the layers. Exfoliation may also be induced by intercalation of ammonium salts such as tetramethyl ammonium hydroxide or the tetrabutylammonium cation. Exfoliation takes place because the attraction of the layers for each other becomes very weak as the layers are spread apart by the intercalation reaction. There is also a very low van der Waals attraction for the propylamine chains for each other at the half-exchange point because the amines are 10.6 Å apart, twice the lateral distance of P-OH groups on the layers. However, we have found some very interesting features in the process.

One of the interesting aspects of the exfoliation of α-ZrP is that the sheets lose phosphate to the solution.[14] The phosphate is replaced by hydroxyl groups and the process proceeds from the edges inward. By lowering the temperature to near zero degrees, the rate of hydrolysis is reduced. Hydrolysis has been observed in other ways. Addition of even the highly crystalline α-ZrP to water results in slow hydrolysis with accompanying lowering of pH. In carrying out measurements of heats of exchange[15] or titrations, the amount of hydrolysis needs to be considered.[16] Hydrolysis can be largely prevented by addition of phosphate to the solution. Furthermore, the lower the degree of crystallinity the less exfoliant is required for complete exfoliation,[17] see below.

Polymer-Zirconium Phosphate Nanocomposites

One of our interests in exfoliation was as a means to prepare polymer-layered inorganic nanocomposites. In the conventional usage, smectite clays, generally montmorillonites, are incorporated into the polymers. To effect a good distribution of the clay in the polymer, amines are first intercalated into the clay so as to move the layers apart and increase the hydrophobicity of the clay-amine intercalate. The composites exhibit large changes in their physical and mechanical properties.[18] For example, improvements in modules, yield strength and elongation at break as well as resistance to oxygen diffusion have been observed. Generally three outcomes of the clay incorporation into the polymer have been noted (Figure 6); conventional, where stacks of layers are inserted into the polymer; intercalated, in which the layers are spread apart by an intercalant; and exfoliated, where a true nanocomposite is formed with the exfoliated layers.

The use of clays has several drawbacks. It is extremely difficult to achieve 100% purity of the clay or narrow particle size distribution and therefore a controlled aspect ratio of the clay nanofiller. In most cases, the clay is

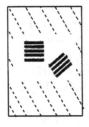 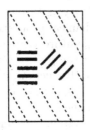 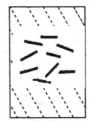

Conventional   Intercalated (nano)   Exfoliated (nano)
(No polymer intercalation)   (limited intercalation)   (extensive intercalation)

Figure 6. Schematic illustrations of the three possible types of polymer-clay composites (from Ref. 18a with permission).

not completely exfoliated and is rarely uniformly dispersed throughout the polymer.[19,20] It is therefore difficult to achieve reproducibility of physical properties and to determine the effect of changing the aspect ratio of the layered filler on the properties of the nanocomposites. Our idea was to use a synthetic layered material such as α-zirconium phosphate to prepare the composites. To make the layers compatible with an epoxy resin, we choose to intercalate polyether amines into α-ZrP. Jeffamines are commercially available polyether amines containing both ethylene and isopropylene groups. They are designated as for example, M300 where M stands for monoamine and 300 is the approximate formula weight. It was found that the ease of amine intercalation and subsequent exfoliation depended upon the crystallinity of the α-ZrP sample.

Samples of α-ZrP were prepared by refluxing 10 g of zirconyl chloride, $ZrOCl_2 \cdot 8H_2O$, in 100 ml of 3, 6, 9, 12, M H_3PO_4 for 24 hours.[17] Preparations were also carried out hydrothermally and refluxed with addition of HF. Hydrofluoric acid solubilizes the α-ZrP and speeds up the crystal growth process as does the hydrothermal procedure. M715 was chosen as the intercalant and added to the α-ZrP samples. The results for the refluxed samples are shown in Figure 7 and not surprisingly the lower the crystallinity, the less M715 is required for complete layer intercalation. As little as 0.13 mol of M715 per mol of Zr was required for full expansion of the layers, whereas 1.5 mol was required for the 12 M preparation.

The reaction was performed in methyl ethyl ketone (MEK) and the ratio of M715 to α-ZrP was 2:1.[21] A transparent yellowish gel was obtained that gave an X-ray pattern with nine orders of 00ℓ reflections with $d_{001} = 73$ Å. The epoxy monomer was also dissolved in MEK to which 1.9% by volume of M715, α-ZrP gel was added. The solvent was removed and polymerization effected at 130°C. A Transmission Electron Microscopy (TEM) micrograph of the resultant composite is shown in Figure 8. It shows a relatively uniform dispersion of exfoliated α-ZrP layers throughout the polymer.[22] The composite exhibited improved mechanical properties but suffered from the fact that the glass transition temperature (Tg) was reduced from 227°C for the neat epoxy to 90°C in the composite. Control experiments indicated the culprit was the Jeffamine. Reduction of the amount of M715 to 0.5 mol resulted in an epoxy nanocomposite with a Tg of 160°C. Further improvement in the Tg is being pursued.

As a general rule, we may find systems of nanocomposites for a range of polymers by properly substituting organic groups between the α-ZrP layers that are compatible with the organic structure of the polymer; for example, ethylphenyl group for polystyrene. Organic groups can be grafted onto the exfoliated zirconium phosphate[23] layers as a means of gaining compatibility with the polymer. Proton conducting properties of α-ZrP

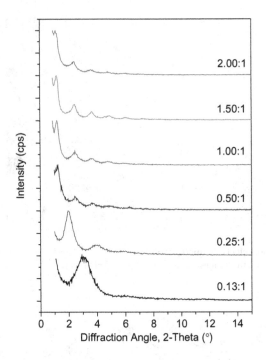

Figure 7. XRD powder patterns of α-ZrP prepared in three molar H_3PO_4 intercalated by Jeffamine M715 in the ratios indicated.

and their use in fuel cell membranes will be described in the section on sulfophosphonates.

Nanoparticle α-Zirconium Phenylphosphonate

Zirconium Phenylphosphonate (ZrPP) was first prepared by Alberti *et al.*[24] Although its composition is stoichiometric, $Zr(O_3PC_6H_5)_2$, its X-ray powder pattern contained only a handful of broad peaks. Refluxing in the presence of HF improves the crystallinity but not sufficiently to obtain single crystals. In fact, it was not until 1993 that the crystal structure was solved utilizing crystals that had been grown by Prof. Alberti hydrothermally in HF at 200°C for 30 days (Figure 9). The layers are like those of α-ZrP but with phenyl groups replacing the -OH pendent groups in the phosphate. The reason for the 30° tilt of the phenyl rings arises from the fact that the Zr atoms are in the plane; whereas in α-ZrP, the metal atoms are slightly above and below the mean plane. The particle size as determined by

Figure 8. Transmission electron micrograph of α-ZrP/epoxy polymer at high magnification showing the uniform dispersion of exfoliated α-ZrP layers.

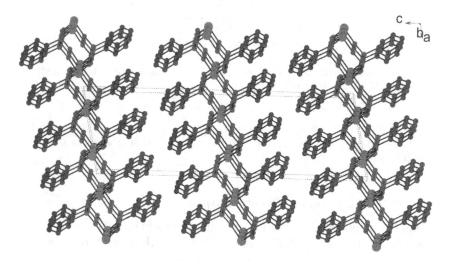

Figure 9. Structure of zirconium phenylphosphonate as viewed down the b-axis direction.

Scanning Electron Microscopy (SEM) was in the range of 20–60 nm when prepared hydrothermally at 140°C for a 3 day hold time. These particles may be as thin as 7 nm or about 9 layers thick (Figure 10). Smaller particles are produced by milder treatment.

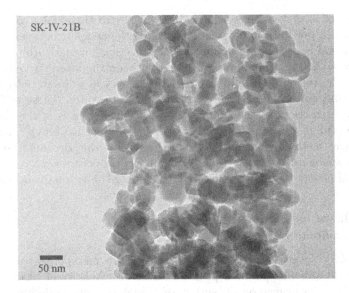

Figure 10. Scanning electron micrograph of zirconium phenylphosphonate particles grown hydrothermallly at 140°C, 3 days, HF/Zr = 3.

Sulfonated Zirconium Phenylphosphonate

ZrPP can be sulfonated in fuming sulfuric acid to give $Zr(O_3PC_6H_4SO_3H)_{2-x}(O_3PC_6H_5)_x$.[26] However, the distance between phenyl groups along the layers is 5.3 Å. Because of the ring thickness (3.2 Å) there is barely enough room to insert an $-SO_3H$ group in the meta-position relative to P–C bond. The sulfonation reaction is more facile if a mixed derivative $Zr(O_3POH)(O_3PC_6H_5)$ is used. As a sulfonate the ZrPP becomes a polyelectrolyte with complete ionization of the sulfonic acid groups. This property results in the complete exfoliation of the sulfonate in aqueous media.[26] We carried out light scattering examination of one of these colloids with the result that about 95% of the particles could be assigned a layered shape with the in-plane dimensions of 4.5 nm × 4.5 nm and 1.5 nm thick.[27] This size is considerably smaller in length and breadth than the particles pictured in Figure 10 which may result from the aggressive sulfonation treatment. Further investigation is in progress. Several applications of the sulfonates are under examination. Their strong acid character imparts ion exchange capabilities to these compounds.[28] The ion exchange selectivity was found to increase as the cation size and charge increases. Free energies, $\Delta G°$ in kJ/mol, for the reactions $H^+ - M^+$ (M = Na^+, Cs^+) and M^{2+} (M = Mg^{2+}, Ba^{2+}) were determined (28b) as 0.27 (Na^+), −9.7 (Cs^+), −3.2 (Mg^{2+}) and −11.2 (Ba^{2+}).

Addition of large, positively charged complexes to the exfoliated sulfonate results in precipitation by sequestration of the complex between the sulfonate layers. For example, charge transfer reactions, using $Ru(bipy)_3^{2+}$ (where bipy = 2,2'-bipyridine) encapsulated between the layers, were carried out.[29] The ru(bipy)$_3$ exhibited pronounced spectral shifts in both its absorption and emission spectra relative to those in water. These shifts result from interactions of the complex with the phenyl rings of the host and the bipyridine. The excited state reactions between $Ru(bipy)_3^{2+}$ and the quencher, methylviologen (MV^{2+}), held within the zirconium sulfophenylphosphonate were measured.[30] The quenching occurs via a combination of diffusional (dynamic/collisional) and sphere of action quenching.

Sensors Based on Zirconium Phosphate and Phenylphosphonate

Alberti et al.,[31,32] were able to devise hydrogen sensors based on zirconium phosphate and sulfophosphonate proton conductors. A layer of titanium hydride (TiH_x) was formed on the surface of a small disk of titanium metal by heating at 600°C in the presence of H_2. The hydride was then covered by a thin layer of exfoliated α-ZrP or $Zr(O_3PC_6H_4SO_3H)_x(HPO_4)_{2-x}$. A second electrode was added by deposition of a thin layer of platinum over the protonic conductor. The electromotive force (E) of this cell is given by:

$$E = -0.029 \log(P_{H_2}) + C \tag{1}$$

where P_{H_2} is the pressure exerted by H_2 and C is a constant.
In the presence of oxygen, a mixed potential

$$E = (RT/aF)\ln(P_{O_2}{}^{1/4})(P_{H_2})^{-1/2} + C \tag{2}$$

is measured (P_{O_2} is the pressure of oxygen and F is Faraday's constant). If the pressure of H_2 is held constant by use of a zirconium hydride electrode, the cell may be used as an oxygen sensor. Sensors for NO_x and SO_x are also under development.

Layer-by-Layer Constructs

Thin Films

Thin films have great practical value in many applications such as optical coatings and microelectronics. One technique for growth of thin films that allows for exquisite control of the process is to grow the film layer by layer. A general summary of the techniques and results has been provided in a new book on Nanochemistry.[33] We will limit ourselves to the use of phosphonates to construct multilayer thin films by a Langmuir-Blodgett films (LB) type procedure. Mallouk and coworkers[34,35] pointed out the similarity between LB films and those formed from zirconium alkylphosphonates and devised a scheme to prepare thin films based upon this similarity. A substrate such as a gold film or silica surface is primed by binding one end of a long-chain molecule to the surface (Figure 11). The other end contains

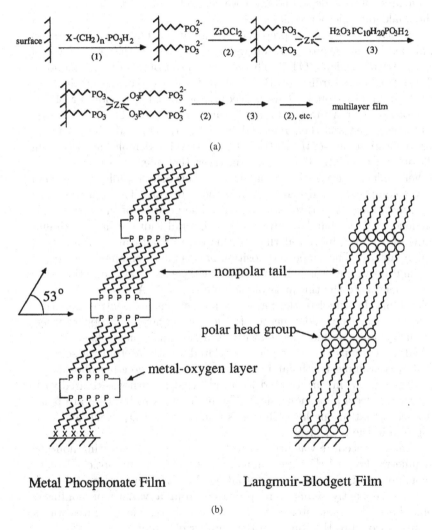

(a)

(b)

Metal Phosphonate Film   **Langmuir-Blodgett Film**

Figure 11.   (a) Procedure for step-by-step assembly of zirconium diphosphonate thin films and (b) structural analogy between Langmuir-Blodgett films and the layer by layer films (from Ref. 35 with permission).

a phosphonic acid group. This surface is then dipped into a Zr(IV) solution, which then binds to the free phosphonic acid end of the monolayer. That the metal is actually bonded is evidenced by its not being removed on washing and subsequent adsorption of an α, ω-bis(phosphonic acid). The process is repeated as many times as desired. Characterization by ellipsometry and angle resolved X-ray Photoelectron Spectroscopy (XPS) demonstrated that the thickness of each deposited layer was close to that of the layer spacing in the analogous microcrystalline solid.

Talham and co-workers[36] prepared zirconium phosphonate films by a LB technique. The surface of a silicon wafer was coated with a monolayer of octadecyltrichlorosilane (OTS). Then a single layer of octadecylphosphonic acid was transferred from an LB trough to the surface of the OTS-silicon wafer held in the trough. This procedure produced a monolayer of the phosphonic acid in a tail-to-tail arrangement with the OTS on the silicon wafer. The wafer was then removed from the trough and dipped into a beaker containing a 5-mM solution of Zr^{4+} to self-assemble the zirconium at the organic template. To complete the layer, the wafer was rinsed in water and then replaced into the LB trough where a new octadecylphosphonic acid film was compressed and transferred to the substrate. The film was built up by repetition of this three step process. The position of the asymmetric methylene (v_a CH_2) band at 2918 cm^{-1} with a full width at half-maximum (fwhm) of 20 cm^{-1} indicated that in this case an all-trans, close-packed template formed. The shape and position of this band remained unchanged as the multilayer film was built up. Ellipsometry and X-ray diffraction were in agreement as to the thickness of the bilayer (51–52 Å).

The films described in the foregoing paragraphs are centrosymmetric. Putvinski *et al.*,[37] devised a method for depositing layers with polar order. 11-Hydroxyundecanethiol was deposited as a monolayer on a gold substrate. The hydroxyl groups were then phosphorylated with an acetonitrile solution of POCl$_3$ containing triethylamine to yield the corresponding phosphonic acid. The acid was in turn treated with Zr^{4+} that in turn was treated with 11-hydroxyundecyl phosphonic acid. The multilayers so built up are polar in the sense that in one layer the Zr is bonded to a PO$_4$ group and to a PO$_3$ group in the next layer.

A similar procedure was utilized to fix azo-dyes into repeating noncentrosymmetric layers.[38,39] These films exhibited nonlinear optical behavior comparable in magnitude to that of LiNbO$_3$ and were stable to 150°C. Subsequently, films were prepared from a variety of nonlinear chromophores.[39,40] The intensities of the second harmonic generated waves were proportional to the square of the number of layers.

The zirconium-phosphonate bonding in the layer-by-layer thin films is thought to yield the same arrangement as in the layers of α-zirconium phosphate. The area available to the phosphonate is 24 Å^2.[3]

This requirement is met for most alkyl and linear aryl chains. However, the AT&T group has prepared films from phosphonated porphyrins and other ligands[41] whose cross-sectional areas exceeded 24 Å^2. The XPS analysis of such films shows that the amount of zirconium in the film is larger than expected for an α-ZrP-type structure. In final section, we shall show that in the preparation of bulk zirconium phosphates with ligands exceeding the required 24 Å^2, new structure-types form.

Recent work on thin-film growth has involved adsorption of oppositely charged polyelectrolytes.[42] Mallouk and co-workers[43] extended this technique to layering of structurally well-defined, two-dimensional colloidal polyanions and polymeric cations. The α-ZrP layers were completely exfoliated by intercalating a small amine [MeNH$_2$, CH$_3$CH$_2$CH$_2$NH$_2$, (Me)$_4$NH$^+$, etc.] between the layers, diluting and sonicating the suspended solid.[44] When such a suspension is placed in contact with an amine-modified gold or silicon surface, the surface-NH$_3^+$ groups displace the loosely held amine associated with one side of the ZrP layer. This procedure electrostatically attaches the α-ZrP layer to the gold or silicon surface. A number of polymeric or oligomeric cations were then attached to the exposed side of the α-ZrP layer by an ion exchange reaction. Among the ions utilized were the aluminum Keggin ion [Al$_{13}$O$_4$(OH)$_{24}$(H$_2$O)$_{12}$]$_7^+$, polyallylamine hydrochloride (PAH) and cytochrome c.

As an example, we consider the PAH film in which the PAH was tagged with dye molecules, fluorescein and rhodamine B. Sequential adsorption of these polymers and α-ZrP was carried out on glass slides to form the films.[43] Energy transfer between the two dyes proceeds by a Förster mechanism and provides a quantitative probe of intermolecular distances within the film. By varying the number of α-ZrP/PAH spacer layers between the two dyes it was possible to measure the energy transfer efficiencies from steady state emission spectra of the thin films. Energy transfer efficiencies near unity were achieved with properly designed systems. Based on these results, creation of more complex systems which combine both energy and electron transfer were contemplated. A valuable review has been presented by Mallouk *et al.*,[45] in which many methodologies of preparing thin films and their potential applications are described. We present only a few in what follows.

Applications of Thin Films

Nonlinear Optical (NLO) thin films

We have already mentioned the zirconium phosphonate films containing azo-dyes as exhibiting NLO behavior. The three-step adsorption used to prepare these films ensures the polar orientation of these molecules in the film. Because the bonding is covalent they are stable to 150°C, a relatively high temperature for NLO media.[40]

Sensors

Mallouk et al.,[46] prepared a sensor for ammonia using a copper phosphonate LBL thin film anchored to a gold substrate. The external layer was terminated by a Cu^{2+} layer. The sensing device was a quartz crystal microbalance (QCM) to which an oscillating current is applied, causing a shear oscillation of the crystal. Mass changes on either electrode of the balance results in a dampening of the QCM frequency. Ammonia is sorbed onto the Cu atoms and the change in frequency recorded which is calibrated to the mass uptake.

A similar technique was used to design a carbon dioxide sensor.[47] Three molecules that react with CO_2 to form carbonates and carbamates, 3-aminopropanol (3-aminopropyl) methyldihydroxysilane and p-xylylemediamine were individually incorporated into the $Cu_2[O_3P(CH_2)_8PO_3]$ thin films and found to respond to the presence of CO_2.

We shall conclude this section by discussing a recent paper where many of these thin film techniques have been utilized to address the question of "Light to Chemical Energy Conversion".[48] The stumbling block as with all photochemical systems is to prevent the back electron transfer of the promoted electron. "It follows that successful photosystems must maintain sensitizers, electron relays, and catalysts in the proper spacial arrangement and chemical environment to prevent the recombination reactions".[48] The Mallouk group developed "onion type" composites rather than linearly stacked thin films.[49,50] Fumed silica particles of ~ 50 nm average diameter were derivatized with (3-aminopropyl) trimethoxysilane. The particles were then coated with exfoliated sheets of α-ZrP about $15 \times 15 \times 0.8$ nm in size. The next layer consisted of methylviologen-functionalized polystyrene followed sequentially by another layer of α-ZrP sheets and a polystyrene layer functionalized with a Ru(II) poly(pyridyl) photosensitizer p-[Ru](p-bpy)(bpy)_2]^2, (P-Ru^{2+}) where bpy is 2,2'-bipyridine and p-bpy is the 2,2'-bipyridine ligand attached to the polymeric backbone.

Visible light absorption by p-Ru^{2+} produced a metal to ligand charge transfer (MLCT) excited state, p-[Ru^{III}(p-bpy)(bpy)(bpy•$^-$)]$^{2+}$ that is a potent reductant capable of transferring an electron to p-MV^{2+}. However, by adding methoxy-N,N-bis (ethylsulfonate), ($MDESA^{2-}$) to the system a more satisfactory charge separated state was achieved with a quantum yield of $\sim 30\%$. However, it is necessary to have good contact between the P-Ru layer and the $MDESA^{2-}$.

Another interesting layer by layer system was developed by Mark Thompson *et al.*[51] They used thin films prepared from viologen derivatives and porphyrins. The arrangement was dictated by the redox potentials and optical energy gaps of the two types of layers. These films were photoactive when exposed to visible light producing photocurrent quantum yields of ~4% and a fill factor of 50%.

Pillared Porous Zirconium Diphosphonates

Nanoparticle Synthesis

Martin Dines *et al.*,[52] conceived the idea of producing porous materials by cross-linking diphosphonic acids. The stoichiometric compound, $Zr(O_3PC_{12}H_8PO_3)$, was depicted ideally as in Figure 12. Because the pillars are 5.3 Å apart (as in α-ZrP) such materials should have no internal pores. However, by introducing spacer molecules such as the phosphorous ion O_3PH^{2-}, the Dines group was able to obtain porous "house of cards"

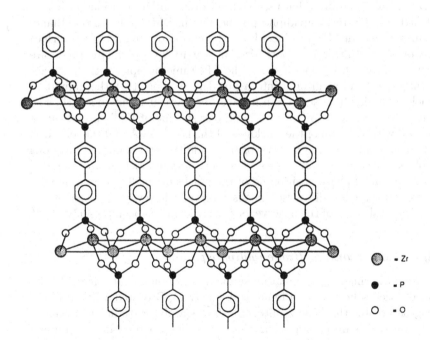

Figure 12. Idealized depiction of $Zr(O_3PC_{12}H_8PO_3)$ based on the structure of the α-ZrP layer. The biphenyl groups may be tilted if the layer is similar to that in zirconium phenylphosphonate.

type products with high surface areas. Unfortunately, they did not measure the complete N_2 sorption isotherms to obtain information about the pores. One of their interesting results was the fact that the fully pillared compound, in which no spacer group was added also exhibited a very high surface area. Subsequent work by us has shown that if the reactions are carried out in aqueous media the pore size distribution is very broad and mainly of the mesoporous type,[53] indicative of external interparticle pores.

Alberti et al.,[54] prepared a number of zirconium diphosphonates in aqueous media that exhibited low surface areas ($< 10\,m^2/g$). However, with the addition of excess H_3PO_3 and DMSO as solvent, high surface areas were obtained ($300–500\,m^2/g$) with the bulk of the pores in the 20–40 Å range. Such large pores must result from stacking of very small particles to produce interparticle mesoporosity. Subsequent work by us showed that considerable microporosity can develop in the particles if prepared in DMSO without addition of H_3PO_3.[55] This microporosity was accompanied by a considerable percentage of mesoporosity. We also found, as will be detailed below, that the micron sized particles were aggregates of much smaller nano-sized particles.

Our group discovered that use of excess zirconium in reactions with 4,4'-biphenyl and 4,4'-terphenyldiphosphonic acid in DMSO or DMSO-ethanol mixtures with added HF produces layered microporous hybrids.[56,57] Surface areas of $350–420\,m^2/g$ and pore sizes in the 10–20 Å diameter range are routinely obtained. Figure 13 shows the TEM micrographs of the particles as synthesized and as ground into a powder. We note that the individual particles contain about seven to nine layers and are 60–80 nm in length. The X-ray powder patterns provide an interlayer spacing of 13.65 Å; and this is in accord with the sum of the thickness of the layer, 6.6 Å and the length of the biphenyl pillar ~ 7 Å. Only two or three additional broadened reflections are observed at higher 2θ values. The very few number of layers and a somewhat out of phase positioning of the stacks of small particles may be the reason for the few diffraction peaks observed. However, a mechanism by which internal pores of the observed size may form has been presented.[56,57]

Sulfonated Zirconium Biphenyldiphosphonate

The aromatic rings may be readily sulfonated in fuming sulfuric acid. The washed, dried solid has been found to be a strong Bronsted acid with acid strengths greater than the zeolite ZSM-5.[57] Catalytic studies in terms of alkylation and hydrocarbon isomerization are in progress. In the sulfonic acid-form, the cavities fill with water and behave as strong acid ion exchangers. The difference between such exchangers and the commercially available sulfonic acid resins is that the phosphonates have much smaller pores, 15 Å

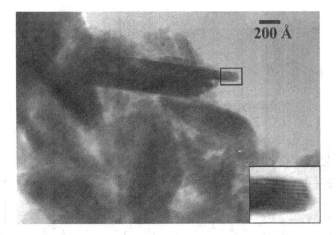

Figure 13.   (a) Scanning electron micrograph of zirconium biphenyldisphosphonate prepared solvothermally in DMSO at 80°C for three days and (b) a transmission electron micrograph of the same sample as a fine powder. The inset shows a portion of an individual particle containing eight layers.

diameter versus about 50 Å for 8% cross-linked sulfonic acid resins. Furthermore they are non-swelling resins, both characteristics conducive to higher selectivities. The possibility that the non-sulfonated materials can be used as molecular sieves with control of the hydrophobic-hydrophilic character of the pores by judicious functionalization of the aromatic rings is also suggested. The formation of nanoparticles in these and other metal systems appears to be a common occurrence for a range of ligands as detailed below.

Crown Ether Zirconium Phosphonates

Previously we had prepared a series of aza-crown ethers essentially attached to zirconium phosphate type layers. The compounds were of two types, those to which a $CH_2-PO_3H_2$ group was attached to a single aza group or those to which the attachment was to a 1,4-diazacrown. These phosphonic acids were then reacted with Zr(IV) in the presence of varying amounts of phosphoric acid.[58] A schematic of the results for the cross-linked diazacrown compounds is presented in Figure 14. The X-ray patterns are similar to those obtained for the zirconium biphenyldiphosphonic acid materials. For the mono-aza-compound with 1:1 phosphate to phosphonate composition the interlayer spacing is 20 Å indicative of a bilayer formation. The compound prepared with excess phosphate has a 13 Å interlayer spacing

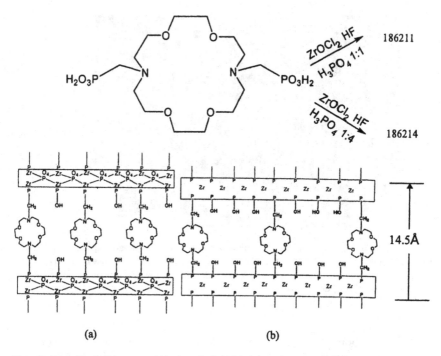

Figure 14. Zirconium complexes of 1,10-bis(phosphonomethyl)-18-crown-6. Top: schematic of preparation procedure; (a) at low HPO_4^{2-} content (186211) with PO_4 groups within the layers and (b) at high HPO_4^{2-} content with an α-ZrP type layer. Oxygen atoms within the layers have been omitted.

which arises from interdigitation of the crown ethers. In the case of the cross-linked products, 186211 and 186214, the interlayer spacing is the same irrespective of the phosphate-phosphonate ratio. A TEM micrograph of 186214 is shown in Figure 15. The particles are about 50–200 nm long and 20–40 nm wide. However, they may be only 10–20 nm or 8–16 layers thick. It should be mentioned that these compounds were prepared at 60–70°C and the biphenyl derivatives at 80–90°C in sealed vessels. Higher temperatures result in mixtures.

Non-Porous Diphosphonates

It was also discovered that if the zirconium biphenyl-bis(phosphonic acid) is prepared in ethanol or ethanol-water mixtures with no HF added, a high surface area product is also obtained but the bulk of this surface area is external. The same is the case if aluminum is substituted for zirconium but

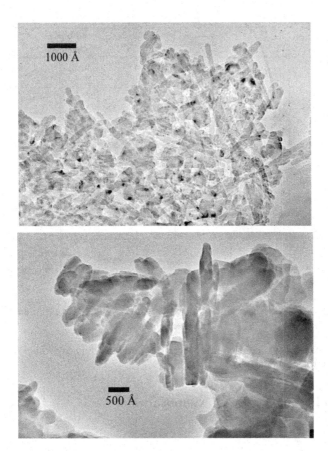

Figure 15. Transmission electron micrographs of sample 186211 (top) and 186214 (bottom) showing the crystal morphologies of the cross-linked zirconium azacrown phosphonates.

in an 8:3 mol ratio of metal to phosphonic acid. A typical N_2 sorption-desorption isotherm is shown in Figure 16 and has all the characteristics of a type IV isotherm with almost no internal porosity. An electron micrograph of an aluminum biphenylene diphosphonate is shown in Figure 17 demonstrating that the particles contain about 5–8 layers. Surface areas of 150–400 m^2/g are obtained based on the preparative method. These particles must have a very high level of phosphonic acid groups on the surface. If the groups are capped by metal atoms, then it should be possible to remove the metal by acid treatment leaving phosphonic acid groups on the surface free to react with cations, charged complexes, bases, etc.

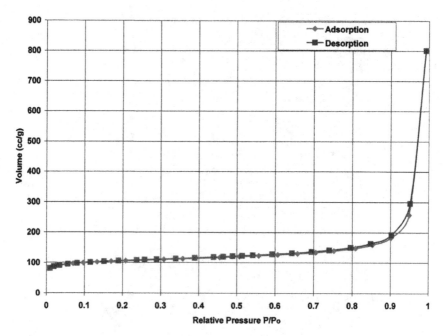

Figure 16. An N_2 sorption-desorption isotherm of aluminum biphenyldiphosphonate. S.A. = $417\,m^2/g$ that is mainly external.

Proton Conducting Fuel Cell Membranes

Background

There is currently a great effort directed towards development of workable fuel cells because of their energy efficiency. Efficiencies as high as 70–80% are possible for fixed site units and 40–50% for transportation applications versus the current 20–35% with internal combustion engines.[59] For transportation purposes, proton-exchange membrane fuel cells (PEMFC) using H_2 as the fuel are preferred. A schematic drawing of such a fuel cell is given in Figure 18. The key ingredient in the PEMFC is the solid polymer membrane which conducts protons from the anode to the cathode. The most preferred membranes are fluorocarbon in nature to which are affixed sulfonic acid groups. These membranes are excellent electronic insulators with high proton conductivities of the order of $10^{-2}\,S\,cm^{-1}$ (Siemans per cm or $(ohm\ cm)^{-1}$). The membranes require water as the main proton conduction medium and so work best at temperatures below 100°C. The problem arises from the fact that the H_2 must be generated by an on-site fuel processor from methane or methanol. In the process, CO is also generated which

Figure 17.   A transmission electron micrograph of an aluminum biphenyldiphosphonate showing particles with eight or less cross-linked layers.

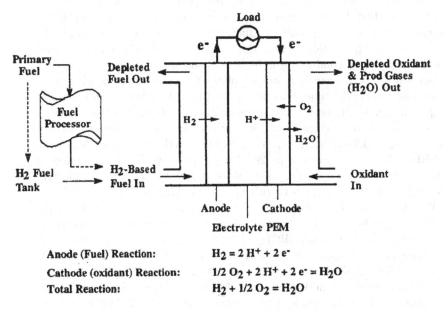

Anode (Fuel) Reaction: $H_2 = 2 H^+ + 2 e^-$

Cathode (oxidant) Reaction: $1/2 O_2 + 2 H^+ + 2 e^- = H_2O$

Total Reaction: $H_2 + 1/2 O_2 = H_2O$

Figure 18.   A schematic diagram of a proton-exchange membrane fuel cell (PEMFC) using an on board fuel cell processor (from Ref. 59 with permission).

poisons the anode catalyst, platinum.[60] However, operation of the fuel cell at 120–130°C would eliminate this problem but the membrane requires a considerable external pressure to avoid water loss.[59,60] At temperatures of 140–160°C direct methanol fuel cells (DMFC), where methanol is used directly as the fuel are possible. Suitable membranes are not yet available for use at these temperatures. In the most recent work, polymer composites containing zirconium phosphates or sulfonated zirconium phenylphosphonate show considerable promise.

With the advent of nanochemistry, research has developed with the aim of creating polymer composites in which inorganic nanoparticles are dispersed within a polymer.[61] We have already described polymer-clay composites and our work with exfoliated α-ZrP and an epoxy resin. Similar work has been done with Nafion and other proton conducting polymers. We briefly review the proton conduction behavior of α-ZrP and the several sulfonated phenylphosphonates as a preliminary to describing the membranes.

Proton Conductivity of α-Zirconium Phosphate and Sulfophosphonates

The proton conductivity of α-ZrP depends upon its crystallinity and the relative humidity. Table 1 summarizes the conductivity of α-ZrP of different crystallinities. The specific conductance decreases as the crystallinity increases. The explanation is that the surface area decreases in the same order[63] and proton conduction is higher on the surface because of higher surface water content and greater number of surface protons. The effect of humidity on conductivity was determined for a highly crystalline α-ZrP sample that was exfoliated with propylamine, deintercalated to condense the sheets and then compressed into discs. The water content of the discs averaged 0.7 mol for a relative humidity (RH) of 5% to 1.3 mol for 90% RH. The dc conductives at 20°C were found to lie in the range of 3–$0.9 \times 10^{-4}\,\mathrm{S\,cm^{-1}}$ when the layers were oriented parallel to the electric field and seven times lower perpendicular to the layers.[64]

We have already described the synthesis of $Zr(O_3C_6H_4SO_3H)$ (O_3POH).[26] Alberti et al.,[65] prepared a similar compound of composition $Zr(O_3C_6H_4SO_3H)_{0.73}(O_3PCH_2OH)_{1.27}$. This compound was shown to be a pure proton conductor. Arrhenius plots at several RH are shown in Figure 19. The highest conductivity at 295 K was $1.65 \times 10^{-2}\,\mathrm{S\,cm^{-1}}$. Somewhat higher conductivities were obtained at 278 K for more highly sulfonated Zr and Ti compounds as shown in Table 2. The zirconium and titanium phenylphosphonates were sulfonated in fuming sulfuric acid[66] and recovered by addition of methanol to the diluted acid followed by centrifugation.

Table 1. Specific conductance of α-ZrP samples obtained by different methods of preparation and ordered according to their degree of crystallinity (T = 25°C).

Sample	Preparation Method	Specific Conductance $(\Omega^{-1}\ cm^{-1})$
1	Precipitation at room temperature, amorphous	8.4×10^{-3}
2	Precipitation at room temperature, amorphous	3.5×10^{-3}
3	Refluxing method $(7{:}48)^a$, semicrystalline	6.6×10^{-4}
4	Refluxing method $(10{:}100)$, crystalline	9.4×10^{-5}
5	Refluxing method $(12{:}500)$, crystalline	3.7×10^{-5}
6	Slow precipitation from HF solutions, crystalline	3.0×10^{-5}

[a]Numbers in parentheses indicate the concentration of H_3PO_4 in molarity and the number of hours refluxed, respectively.

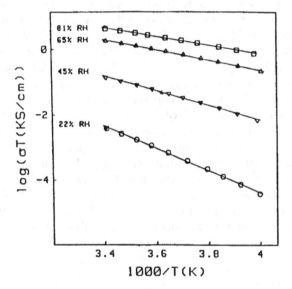

Figure 19. Arrhenius plots of log (σT) as a function of 1000/T for $Zr(O_3PC_6H_4SO_3H)_{0.73}(O_3PCH_2OH)_{1.27}$ at different relative humidities (reprinted from Ref. 65 with permission).

The titanium compound experiences a certain amount of cleavage of the P-C bond as the formula derived from elemental analysis and thermogravimetric analysis was $Ti(O_3POH)_{0.25}(O_3PC_6H_5)_{0.12}$-$(O_3PC_6H_4SO_3H)_{1.63}\cdot 3.64H_2O$ as opposed to the fully sulfonated zirconium phosphonate, $Zr(O_3PC_6H_4SO_3)_2\cdot 3.6H_2O$. These sulfonated derivatives are among the best known proton conductors.

Table 2.   Conductivity in reciprocal Ω cm at 5°C as a function of relative humidity for zirconium and titanium sulfophenylphosphonates.

Sample	Relative humidity (%)				
	20(3)[a]	30(3)	50(4)	65(4)	85(5)
EWS-3-89 (Zr)	5×10^{-6}	2×10^{-4}	1.1×10^{-3}	7.8×10^{-3}	2.1×10^{-2}
EWS-4-1 (Ti)	4×10^{-5}	3×10^{-3}	1.2×10^{-2}	7.2×10^{-2}	1.3×10^{-1}

[a]Number in parenthesis is the estimated error in the humidity measurement.

Membrane Preparation

Nafion membranes are among the most used as proton conductors. However, they suffer from a decrease in proton conductivity due to water loss as the temperature of operation increases.[67] This limits the use of nafion membranes to about 80°C[68] in fuel cells. Thus, either new membrane types need to be developed or the nafion membranes must be improved to operate above 80°C particularly above 120°C. This has been done by dispersing insoluble inorganic solids such as metal oxides, phosphates and phosphonates within the nafion. The use of nanoparticles such as 50 nm silica spheres or α-ZrP lamellae or sulfonated zirconium and titanium phenylphosphonates ensures good distribution and strong interactions between the particles and the membrane. Much improved nafion membranes have been produced in terms of water retention and conductivity by forming such composites especially with the phenylsulfonates. A similar technique to form composites with other polymers has been similarly successful.[69] Still, membranes that operate in actual fuel cells above 120°C are not yet commercial, but continued research using lamellar nanoparticles is being actively pursued.

Tin(IV) Phosphonates

Synthesis and Nanoparticle Formation

A recent paper that caught our attention described the synthesis of tin(IV) phenylphosphonate prepared in the presence of sodium decylsulfonate (SDS) as template.[70] It was found that this compound had a composition $Sn(O_3PC_6H_5)_2$, a surface area of 255 m^2/g and an average pore size of 3.3 nm. What was strange was that preparations without template also had high surface areas (197 and 135 m^2/g) but the pore distribution range was 20–30 nm. We repeated this preparation and found a quite regular micropore distribution ($\sim$ 10 Å diameter) and a surface area of $\sim$ 300 m^2/g

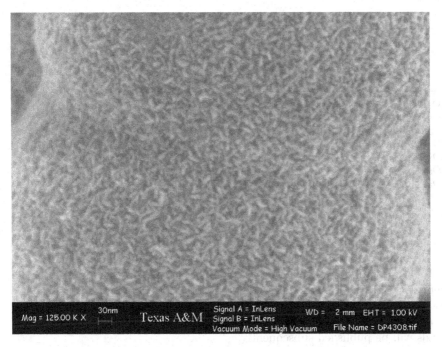

Figure 20. Transmission electron micrograph of $Sn(O_3PC_6H_5)_2$ at high magnification showing the aggregates of small particles and the porous nature of the spherical aggregate.

with a near type I isotherm.[71] The question arises as to how such a structure can form from a presumably layered compound similar to zirconium phenylphosphonate (based upon X-ray diffraction patterns). Scanning electron micrographs at high magnification revealed that the solid consists of nanoparticles that aggregate into spheres that are micron-sized (Figure 20). The pores arise from the "house of cards" face to edge arrangement of the particles. The particles are very small and must have a narrow size distribution to yield such regular pores. Small angle neutron scattering confirmed the nano-size of the particles. The phenomenon appears to be general as we have now made about twenty derivatives and all of them show similar characteristics as shown in Table 3. We note that different average pore sizes result from different ligands. The only compound that was different was the methylphosphonate. It crystallized as platelets aggregating into rosettes. By longer reaction times and higher temperatures, the tin methylphosphonate yielded an X-ray powder pattern amenable to solution. The compound is eight coordinate with bonding similar to that in rare earth phenyl phosphonates.[72] Continued hydrothermal treatment at 180°C

Table 3.   Ideal formulas and properties of selected tin(IV) phosphonates.

Ideal Formula	Surface Area (m^2/g)			Max in Pore Size Distribution Curve (Å)	TGA wt. Loss (%)
	d_{001} (Å)	Total	Micro		
$Sn(O_3PC_6H_5)_{1.95}F_{0.05} \cdot H_2O$	15.35	367	350	9	34.09
$Sn(O_3PC_6H_5)_{0.95}(O_3PCH_3)_{0.95}$ $F_{0.10} \cdot 0.4H_2O$	15.22	514	493	11	24.05
$Sn(O_3PCH_3) \cdot 0.4H_2O$	9.74	68	0		6.8
$Sn(O_3PC_{12}H_8PO_3) \cdot 0.8H_2O$	13.8	416	388	17	36.0
$Sn(O_3PC_{10}H_{21})_2$	28.07	264	220	9	47.6
$Sn(O_3PCH_2CH_2COOH)_2 \cdot 0.8H_2O$	11.07	224	102	27	33.09

for eight days produced larger platelets of average dimensions 50 nm long and 10–15 nm thick or 6–10 layers thick. Thus, we conclude that the nano-sized feature is due to their very low solubility in aqueous media even in the presence of HF. It should also be noted from Table 3 that the pore size varies with the type of ligand used to prepare the tin phosphonate. It also should vary with preparative conditions and this has been observed by us. Details will be published subsequently.

Conclusions

We have only mentioned γ-zirconium phosphate in a peripheral way for reasons given in what follows. It was discovered shortly after the alpha phase[73] and its structure determined much later.[74] The reason is that no single crystals were grown and only when X-ray powder methods were developed was a solution of the structure obtained. The formula for γ-ZrP is $Zr(PO_4)(H_2PO_4) \cdot 2H_2O$. It is layered with a 12.2 Å interlayer spacing. The orthophosphate group resides within the layers bonding to four Zr atoms forming linear chains. The chains are bridged into layers utilizing the two unprotonated oxygens of the dihydrogen phosphate groups. The -OH groups project away from the layers into the interlayer space forming hydrogen bonds with the two water molecules. It is almost impossible to form phosphonate compounds of γ-ZrP by direct synthesis procedures. Instead, they are obtained by ester interchange reactions of the type

$$Zr(PO_4)(O_2P(OH)_2) \cdot 2H_2O + RPO_3H_2 \rightleftharpoons Zr(PO_4)\left(O_2P{<}^{R}_{OH}\right) + H_3PO_4$$

$$(3)$$

As a result, the particle size is generally the particle size of the starting γ-ZrP sheets. A good introduction to the chemistry of γ-ZrP is given in a review article by Giulio Alberti.[75] Undoubtedly, a rich nanochemistry exists in this system.

The recognition that the group 4 phosphates and phosphonates form nanoparticles because of their great insolubility in aqueous media is a key fact in understanding and controlling their chemistries. For example, the ease of exfoliation of the layers depends upon the total force between layers and thus on the size of the layers. By controlling the size, we are able to exfoliate α-ZrP even in non-aqueous systems for use in dispersion in polymers. The size of the particles may also play a significant role in proton conducting polymer composites for fuel cells. It will surely play a role in catalysis where control of surface area, pore size and active groups per unit of surface area are important variables. We are just beginning to explore the Sn(IV) phosphonates and many surprises may still be in store if a similar silicon chemistry is possible. It could hardly be imagined that the discovery of the crystalline zirconium phosphates in the early 1960s and the subsequent synthesis of the first zirconium phosphonates in 1978 that such a vast new field of chemistry would result. But it is not finished, many new developments will certainly arise from continued research in this field.

Acknowledgements

We gratefully acknowledge the financial support of the National Science Foundation, Grant No. DMR-0332453 and the Welch Foundation, Grant No. A-0673.

References

1. (a) K. A. Kraus and H. O. Phillips, *J. Amer. Chem. Soc.* **78**, 694 (1956);
   (b) K. A. Kraus, H. O. Phillips, T. A. Carlson and J. S. Johnson, *Proc. 2nd Int. Conf. Peaceful Uses of Atomic Energy*, Geneva 1958 paper No. 15/P/1832. United Nations.
2. A. Clearfield and J. A. Stynes, *J. Inorg. Nucl. Chem.* **26**, 117–129 (1964).
3. (a) A. Clearfield and G. D. Smith, *Inorg. Chem.* **8**, 431–436 (1969);
   (b) J. M. Troup, A. Clearfield, *Inorg. Chem.* **16**, 3311–3314 (1977).
4. G. Alberti and J. Torracca, *J. Inorg. Nucl. Chem.* **30**, 317–318 (1968).
5. A. Clearfield and J. R. Thomas, *Inorg. Nucl. Chem. Lett.* **5**, 775–779 (1969).
6. A. Clearfield, *Ann. Rev. Mater. Sci.* **14**, 205–229 (1984).
7. A. Clearfield, L. Kullberg and A. Oskarsson, *J. Phys. Chem.* **78**, 1150–1153 (1974).

8. A. Clearfield, A. Oskarsson and C. Oskarsson, *Ion Exchange and Membranes.* **1**, 91–107 (1972) (Journal out of print).
9. G. Alberti, *Acc. Chem. Res.* **9**, 163–170 (1976).
10. M. S. Whittingham and A. J. Jacobson (eds.), *Intercalation Chemistry* (Academic Press: New York, NY, 1982).
11. W. Müller-Warmuth and R. Schollhorn (eds.), *Progress in Intercalation Research* (Kluwer Academic Publ.: Dordrect, 1994).
12. A. Clearfield and U. Costantino, *Comprehensive Supramolecular Chemistry*, eds. G. Alberti and T. Bein, Vol. 7 (Pergamon Press: New York, NY, 1996), p. 107.
13. R. M. Tindwa, D. K. Ellis, G.-Z. Peng and A. Clearfield, *J. Chem. Soc., Faraday Trans.* **81**, 545–552 (1985).
14. D. M. Kaschak, S. A. Johnson, R. E. Hooks, H. N. Kim, M. D. Ward and T. E. Mallouk, *J. Am. Chem. Soc.* **120**, 10887–10894 (1998).
15. A. Clearfield and L. H. Kullberg, *J. Phys. Chem.* **78**, 1812–1817 (1974).
16. A. Clearfield and A. Oskarsson, *Ion Exch. Memb.* **1**, 205–213 (1974).
17. L. Sun, W. J. Boo, R. L. Browning, H.-J. Sue and A. Clearfield, *Chem. Mater.* **17**, 5606–5609 (2005).
18. (a) T. Lan, P. D. Kavirnatna and T. J. Pinnavaia, *Chem. Mater.* **7**, 2144–2150 (1995); (b) P. B. Messersmith and E. P. Giannelis, *Chem. Mater.* **6**, 1719–1725 (1994).
19. T. Lan and T. J. Pinnavaia, *Chem. Mater.* **6**, 2216–2219 (1994).
20. Y. Kurokawa, H. Yasuda, M. Kashiwagi and A. Oyo, *J. Mater. Sci. Lett.* **16**, 1670–1672 (1997).
21. N. Bestaoui, N. A. Spurr and A. Clearfield, *J. Mater. Chem.*, DOI:10.1039/b511351b.
22. H.-J. Sue, K. T. Gam, N. Bestaoui, N. A. Spurr and A. Clearfield, *Chem. Mater.* **16**, 242–249 (2004).
23. S. Yamanaka, *Inorg. Chem.* **15**, 2811–2817 (1976).
24. G. Alberti, U. Costantino, S. Allulli and N. Tomassini, *J. Inorg. Nucl. Chem.* **40**, 1113–1117 (1978).
25. D. M. Poojary, H.-L. Hu, F. L. Campbell, III and A. Clearfield, *Acta Crystallogr.* **B49**, 996–1001 (1993).
26. C.-Y. Yang and A. Clearfield, *Reactive Polymers.* **5**, 13–21 (1987).
27. A. Clearfield, (unpublished data).
28. (a) L. H. Kullberg and A. Clearfield, *Solv, Extr. Ion Exch.* **7**, 527–540 (1989); (b) L. H. Kullberg and A. Clearfield, *Solv. Extr. Ion Exch.* **8**(1), 187–197 (1990).
29. J. L. Colon, C.-Y. Yang, A. Clearfield and C. R. Martin, *J. Phys. Chem.* **92**, 5777–5781 (1988).
30. J. L. Colon, D. S. Thakur, C.-Y. Yang, A. Clearfield and C. R. Martin, *J. Catal.* **124**, 148–159 (1990).
31. G. Alberti and R. Polambari, *Solid State Ionics.* **35**, 153–156 (1989).
32. G. Alberti, M. Casciola and R. Polambari, *Solid State Ionics.* **52**, 291–295 (1992).

33. G. A. Ozin and A. C. Arsenault (eds.), *Nanochemistry, A Chemical Approach to Nanomaterials* (RSC Publishing, Cambridge, UK, 2005), Ch. 3, pp. 95–125.

34. H. Lee, L. J. Kepley, T. E. Mallouk, H. G. Hong and S. Akhter, *J. Phys. Chem.* **92**, 2597–2601 (1988).

35. H. Lee, L. J. Kepley, H. G. Hong and T. E. Mallouk, *J. Am. Chem. Soc.* **110**, 618–620 (1988).

36. H. Byrd, J. K. Pike and D. R. Talham, *Chem. Mater.* **5**, 709–715 (1993).

37. T. M. Putvinski, M. L. Shilling, H. E. Katz, C. E. D. Chidsey, A. M. Mujsce and A. B. Emerson, *Langmuir.* **6**, 1567–1571 (1990).

38. H. E. Katz, G. Scheller, T. M. Putvinski, M. L. Schilling, W. L. Wilson and C. E. D. Chidsey, *Science.* **254**, 1485–1487 (1991).

39. H. E. Katz, M. L. Schilling, S. B. Ungashe, T. M. Putvinski and C. E. D. Chidsey, *Supramolecular Chemistry*, ed. T. Bein (American Chemical Society Symposium Series 499: Washington, DC, 1992), p. 24.

40. H. E. Katz, W. L. Wilson and G. Scheller, *J. Am. Chem. Soc.* **116**, 6636–6640 (1994).

41. S. B. Ungashe, W. L. Wilson, H. E. Katz, G. R. Scheller and T. M. Putvinski, *J. Am. Chem. Soc.* **114**, 8717–8719 (1992).

42. S. W. Keller, H.-N. Kim and T. E. Mallouk, *J. Am. Chem. Soc.* **116**, 8817–8818 (1994).

43. D. M. Kaschak and T. E. Mallouk, *J. Am. Chem Soc.* **118**, 4222–4223 (1996).

44. M. Fang, D. M. Kaschak, A. C. Sutorik and T. E. Mallouk, *J. Am. Chem. Soc.* **119**, 12184–12191 (1997).

45. T. E. Mallouk, H.-N. Kim, P. J. Olliver and S. W. Keller, *Comprehensive Supramolecular Chemistry*, ed. G. Alberti, Vol. 7 (Pergamon Press: New York, NY, 1996), pp. 189–217.

46. L. C. Brousseau and T. E. Mallouk, *Anal. Chem.* **69**, 679–687 (1997).

47. L. C. Brousseau, D. J. Aurentz, A. J. Benesi and T. E. Mallouk, *Anal. Chem.* **69**, 688–694 (1997).

48. P. G. Hoertz and T. E. Mallouk, *Inorg. Chem.* **44**, 6828–6840 (2005).

49. S. E. Keller, S. A. Johnson, E. S. Brigham, E. H. Yonemato and T. E. Mallouk, *J. Am. Chem. Soc.* **117**, 12879–12880 (1995).

50. D. M. Kaschak, S. A. Johnson, C. C. Waraksa, J. Pogue and T. E. Mallouk, *Coord. Chem. Rev.* **186**, 403–416 (1999).

51. F. B. Abddrazzaq, R. C. Kwong and M. E. Thompson, *J. Am. Chem. Soc.* **124**, 4796–4803 (2002).

52. M. B. Dines, P. M. Di Giacomo, K. P. Callahan, P. C. Griffith, R. H. Lane and R. E. Cooksey, *Chemically Modified Surfaces in Catalysis and Electrocatalysis*, ed. J. S. Miller, Amer. Chem. Soc. Symposium Series 192, Wash. D. C. Amer. Chem. Soc. (1982), pp. 223–240.

53. A. Clearfield, *Design of New Materials*, eds. A. Clearfield and D. A. Cocke (Plenum Press: New York, NY, 1986), pp. 121–134.

54. G. Alberti, U. Costantino, R. Vivani and P. Zappelli, *Synthesis/ Characterization and Novel Applications of Molecular Sieve Materials*, eds. R. L. Bedard *et al.*, Vol. 233 (Materials Research Society Symp. Proc., 1991), pp. 101–106.

55. A. Clearfield, *Chem. Mater.* **10**, 2801–2810 (1998).
56. A. Clearfield and Z. Wang, *J. Chem. Soc. Dalton Trans.* 2937–2947 (2002).
57. Z. Wang, J. M. Heising and A. Clearfield, *J. Am. Chem. Soc.* **125**, 10375–10383 (2003).
58. B. Zhang and A. Clearfield, *J. Am. Chem. Soc.* **119**, 2751–2752 (1997).
59. C. Song, *Catal. Today* **77**, 17–49 (2002).
60. G. Alberti, M. Casciola, L. Massinelli and B. Bauer, *J. Membrane Sci.* **185**, 73–81 (2001).
61. R. Dagni, *Chem. Eng. News.* **77**, 25–37 (1999).
62. G. Alberti, M. Casciola, U. Costantino, G. Levi and G. Ricciardi, *J. Inorg. Nucl. Chem.* **40**, 533–537 (1978).
63. A. Clearfield and J. Berman, *J. Inorg. Nucl. Chem.* **43**, 2141–2142 (1981).
64. M. Casciola and U. Costantino, *Solid State Ionics* **20**, 69–73 (1986).
65. G. Alberti, M. Casciola, U. Costantino, E. Peraio and E. Montonero, *Solid State Ionics.* **50**, 315–322 (1992).
66. E. W. Stein, A. Clearfield and M. A. Subramanian, *Solid State Ionics.* **83**, 113–124 (1996).
67. Y. Sone, P. Ekdunge and D. Simonsson, *J. Electrochem. Soc.* **143**, 1254–9 (1996).
68. G. Alberti and M. Casciola, *Ann. Rev. Mater. Res.* **33**, 129–154 (2003).
69. B. Bonnet, D. J. Jones, J. Roziére, L. Tchicaya and G. Alberti, *J. New Mater. Electrochem. Systems* **3**, 87–92 (2000).
70. N. K. Mal, M. Fujiwara,Y. Yamada and M. Masahiko, *Chem. Lett.* **32**, 292–293 (2003).
71. A. Subbiah, D. Pyle, A. Rowland, J. Huang, R. A. Narayanan, P. Thiyagarajan, J. Zon and A. Clearfield, *J. Am. Chem. Soc.* **127**, 10826–10827 (2005).
72. R. C. Wang, Y.-P. Zhang, H. Hu, R. R. Frausto and A. Clearfield, *Chem. Mater.* **4**, 864–871 (1992).
73. A. Clearfield, R. H. Blessing and J. A. Stynes, *J. Inorg. Nucl. Chem.* **30**, 2249–2258 (1968).
74. (a) A. N. Christensen, E. K. Anderson, I. G. Anderson, G. Alberti, M. Nielsen and M. S. Lehmann, *Acta*; (b) D. M. Poojary, B. G. Shpeizer and A. Clearfield, *J. Chem. Soc. Dalton Trans.* 111–113 (1995).
75. G. Alberti, *Comprehensive Supramolecular Chemistry*, eds. G. Alberti and T. Bein (Pergamon, Elsevier: Tarrytown, NY, 1996), pp. 151–188.

Chapter 7

Synthesis of Nanostructured Hybrid Sorbent Materials Using Organosilane Self-Assembly on Mesoporous Ceramic Oxides

Glen E. Fryxell

Pacific Northwest National Laboratory
Mailstop K2-44, PO Box 999
Richland, WA, 99352, USA
glen.fryxell@pnl.gov

Introduction

The single most important factor in determining quality of life in human society is the availability of pure, clean drinking water. Wars have been fought, and will continue to be fought, over access and control of clean water. Drinking water has two major classes of contamination, biological contamination and chemical contamination. Bacterial contamination can be dealt with by a number of well-established technologies (e.g. chlorination, ozone, UV, etc.), but chemical contamination is a somewhat more challenging target. Common organic contaminants, such as pesticides, agricultural chemicals, industrial solvents, and fuels can be removed by treatment with UV/ozone, activated-carbon or plasma technologies. Toxic heavy metals like mercury, lead and cadmium can be partially addressed by using traditional sorbent materials like alumina, but these materials bind metal ions non-specifically and can easily be saturated with harmless, ubiquitous species like calcium, magnesium and zinc (which are actually nutrients, and do not need to be removed). Another weakness of these traditional sorbent materials is that metal ion sorption to a ceramic oxide surface is a reversible process, meaning they can easily desorb back into the drinking water supply.

A chemically specific sorbent material, capable of permanently sequestering these toxic metal ions from groundwater, or sequestering them from industrial waste, thereby preventing their entry into the groundwater, is needed. Ideally, the kinetics of heavy metal sorption should be fast, allowing for high throughput in the process stream, and a high binding capacity for the target heavy metal is clearly desirable. In addition, as acceptable drinking water contamination limits get lower and lower, more sensitive analytical methods are needed in order to detect such contamination, so if these sorbent materials could enhance the sensing and detection of these toxic analytes, that would be an added benefit.

The surfactant templated synthesis of mesoporous ceramics offers the synthetic chemist the ability to create highly uniform pore structures where the pores are only a few nanometers in diameter. This versatile and powerful methodology creates morphologies that condense a huge amount of surface area into a very small volume. For example, a typical procedure for making MCM-41 (an Exxon/Mobil catalyst support) provides a highly ordered hexagonal honeycomb of parallel 20–25 Å pores, with wall thicknesses of about 12–14 Å, and a surface area of about $1000\,m^2/g$.[1,2] Thus, a 10 g sample of this material (about 2 rounded tablespoons) has more surface area than an entire football field, endzones included.

While high surface area provides the promise of high binding capacity for a sorbent material, this backbone is still a simple ceramic oxide, and while ceramic oxides commonly physisorb heavy metals, they tend to do so non-selectively and reversibly, making them inefficient for waste remediation or water purification. What is needed is a way of enhancing the chemical selectivity of the interface, as well as making the binding chemistry more robust. This is done by chemically modifying these mesoporous materials with self-assembled monolayers that are terminated with chemically selective ligands. Molecular self-assembly allows the promise of high loading capacity to be realized by coating the mesoporous interface with a dense, covalently anchored functional coating. With this functional coating in place, the tethered ligand field provides an avenue for covalent chelation, delivering a significant driving force for sequestering a specific metal ion, and the close proximity of these ligands to one another allows for multiple metal-ligand interactions, further enhancing the ligand's affinity for a specific metal ion. By tailoring the nature of this interfacial ligand field, it is possible to selectively remove different classes of metal ions. The rigid, open pore structure provided by the mesoporous ceramic backbone prevents solvent swelling (which can close off pores in polymeric resins). Since all of the binding sites are available all of the time, sorption kinetics are rapid. Although the mesoporous supports are more expensive than typical sorbent materials on a per-unit-weight basis (e.g. activated-carbon, polymer-based ion-exchange resins, etc.), the very high functional density of the self-assembled monolayers on mesoporous supports (SAMMS) makes them cheaper to use in the final analysis as a result of their superior performance

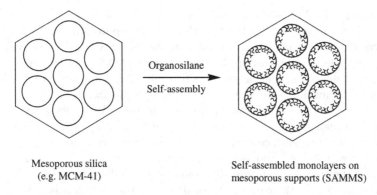

Mesoporous silica
(e.g. MCM-41)

Self-assembled monolayers on
mesoporous supports (SAMMS)

Figure 1. Construction of a terminally functionalized self-assembled monolayer inside of a mesoporous ceramic creates a powerful and versatile, foundation upon which to build chemically selective sorbent materials. By varying the nature of the interfacial ligand field the chemical selectivity of the sorbent can be tuned for specific targets.

and much reduced waste disposal costs (which result from the fact that far less sorbent material is needed to effect the separation). Since the chemical selectivity of the SAMMS materials is dictated by the interfacial ligand field, the following discussion will be organized by ligand type.

Sulfur-Based Ligands

When designing ligands to selectively bind specific metal species it is important to capture the appropriate stereochemical, electronic and steric attributes in the ligand field. A key concept here is that of "hard-soft acid-base theory", introduced by Pearson in the 1960s.[3,4] In short, this theory focuses on the polarizability of the acid-base pair, and the impact that it has on their affinity for one another. Highly polarizable Lewis acids (i.e. "soft" — large atom, electron rich species in which the electron cloud is readily distorted by an external influence) "prefer" to react with polarizable (i.e. "soft") Lewis bases. For example, Hg species have a very high affinity for thiol (aka "mercaptan" for its ability to capture mercury) ligands. Non-polarizable Lewis acids (i.e. "hard" — a highly electropositive species in which the electron cloud is either non-existent or not easily distorted) "prefer" to react with non-polarizable (i.e. "hard") Lewis bases. An example of a "hard" acid-base pair reaction would be that of a bare proton reacting with fluoride ion. Thus, when designing a sorbent material for a "soft" target ion (e.g. Hg), it is important to build it around a "soft" ligand field. When aiming towards a "hard" target ion, one should start with a "hard" ligand field. We will start this discussion on the early work done with Hg, using a ligand field of thiols within the mesoporous architecture.

The typical recipe that we have used to make thiol-SAMMS[5,6] involves pre-hydrating the mesoporous silica (65 Å pores and $900\,m^2/g$) interface by stirring a toluene suspension of the silica with 2 monolayers worth of water (based on available surface area, assuming a monolayer to be composed of 5×10^{18} molecules of water per square meter). Water plays a very important in organosilane self-assembly.[7] While chloro- and alkoxysilanes can indeed react with surface silanols, Maciel determined many years ago that this is slow process,[8] and that the relevant process generally involved hydrolysis of the silane by interfacially bound water, followed by condensation (acid catalyzed in the case of chlorosilanes) of the hydroxysilane with a surface silanol (or siloxane). The intermediate tris(hydroxy)silane is commonly overlooked in discussions of organosilane self-assembly, but to do so is to ignore the very species that makes the self-assembly process possible. After hydrolysis, the tris(hydroxy)silane is hydrogen bound to the silica surface. Through a process of making and breaking H-bonds, this species can "crab-walk" across the silica surface until it encounters another such molecule. The van der Waal's attractive forces (as well as any other attractive forces, such as H-bonds, dipole-dipole interactions, etc.) between these molecules drive the self-assembly process, forming a critical nucleus in the aggregation process. As this agglomerated assembly grows, it becomes less and less mobile. With the extended contact time and close proximity[9] of these tris(hydroxyl)silanes and the silica surface, these molecules begin to undergo condensation processes, both with the surface and between themselves, ultimately forming a covalently anchored and crosslinked monolayer structure. For chlorosilanes, this process is catalyzed the HCl formed by hydrolysis of the silane, while the alkoxysilanes generally require added heat for these condensation processes to take place. Without interfacial water, none of this would be possible, as the only reaction pathway available to the silane is simple capping of whatever silanols may be available.

This process results in a defective monolayer structure. There are a number of dangling hydroxyls left in a monolayer made by this route, as well as pinhole defects (vacancies). These defects can be preserved by any excess water, or by the alcohol resulting from silane hydrolysis. For example, a typical case of hydrated MCM-41 being treated with an excess of 3-mercaptopropyltrimethoxysilane (MPTMS), results in a monolayer population density of about 4 silanes/nm^2.[10] If the reaction mixture is subjected to distillation (to remove the methanol by-product and water/toluene azeotrope) after the four to six hour reflux period however, it is possible to increase the population density to about 5 silanes/nm^2, (this equates to about 3.3 mmole/g) as well as to significantly increase the degree of cross-linking.[10] By removing the water from the reaction mixture it is possible to drive the condensation equilibria farther towards the ideal cross-linked structure, and by removing the H-bonding molecules (i.e. water and methanol), the pinhole defects can be filled in. The post-deposition distillation drives both of these processes.

The materials made by this route offer high surface area that is densely covered by the thiol ligand field, thereby creating a very high population of Hg binding sites, while simultaneously keeping all of these binding sites at the sorbent interface, and therefore accessible. The rigid open pore structure of the ceramic backbone prevents any solvent swelling phenomenona, which can limit access to binding sites by closing off pores. This helps to maintain free access to all of the ligands and allows the sorption kinetics to be rapid. When thiol-SAMMS are used to remove Hg from a 10,000 ppb $Hg(NO_3)_2$ solution, over 98% of the Hg is sequestered in less than five minutes.[11] When a similar experiment was carried out using a thiolated resin (Rohm-Haas GT-73), it took eight hours to remove only 90% of the Hg. Both materials are using the same functionality to bind the Hg (a thiol), but the rigid open architecture of the thiol-SAMMS provides for a 500-fold rate difference by keeping all of the binding sites available all the time. Thiol SAMMS was found to be able to sequester over 600 mg of Hg per gram of sorbent.[11]

The nanoporous honeycombed morphology of SAMMS not only provide high binding capacity and rapid sorption kinetics, it also blocks microbial access to the vast majority of the sorbent's active surface. This is a concern because certain bacteria can actually metabolize bound mercury, methylating it, making it not only more mobile, but also more toxic. By binding the Hg inside a nanoscale pore, a bacterium cannot gain access to it, thereby preventing these deadly processes.

Competition experiments revealed that thiol-SAMMS do not bind any of the ubiquitous cations commonly found in groundwater, like Na, K, Mg, Ca, Fe, Ni or Zn.[12] Thiol-SAMMS do bind other "soft" cations, like Cd, Ag and Au. Extended X-ray Absorbance Fine Structure (EXAFS) studies of the Hg-thiol-SAMMS adduct revealed a 1:1 Hg/S stoichiometry, and that each Hg was interacting with 2 S atoms, with a third O ligand.[13] This is explained by a divalent Hg cation binding to one thiolate and one oxygenated ligand (i.e. either a nitrate anion or a water molecule), and coordinating to the S atom of a neighboring thiol (see Figure 2).

Pinnavaia and co-workers have developed a similar, but decidedly unique synthetic strategy for making some similar heavy metal sorbent materials.[14] In the Pinnavaia approach, neutral amines are used as the surfactants to form the micelles. Instead of calcining the as-synthesized greenbody to remove the surfactant, the template is removed via Soxhlet extraction (typically using ethanol). This is an expeditious method, and has the advantage

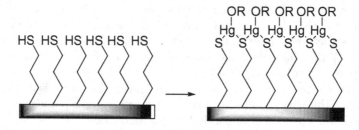

Figure 2.   Hg bonding in thiol-SAMMS.[13]

of leaving the pore walls fully hydroxylated (and hydrated), and therefore receptive to silanation. However, by avoiding calcination, this method also leaves the pre-ceramic pore walls incompletely cross-linked, and therefore of uncertain long-term stability. When these disordered assemblies of wormlike channels are subsequently decorated with a 3-mercaptopropylsilane coating, they found that these materials had a surface area of approximately $800 \, m^2/g$ and pore diameters on the order of $30 \, \text{Å}$.[14] This material was found to contain approximately 1.5 mmole of thiol per gram of sorbent and had a Hg(II) capacity of 310 mg/g (these numbers are roughly half those of the PNNL material described above). This procedure was found to be both fast and easy, but led to incomplete functionalization.

A follow-on study compared the performance of these materials with other related sorbent materials.[15] In this study, the MCM-41 derived materials were prepared after being vacuum dried. In the absence of water there can be no possibility of a fully dense monolayer being formed. This procedure only allows the remaining surface silanols to be capped, ultimately producing a product with only $0.28 \, silanes/nm^2$ (or 0.57 mmole/g). The 1:1 binding stoichiometry of Hg/S was confirmed once again. No competition was observed from a series of representative transition metal cations.[16]

Pinnavaia subsequently streamlined this synthetic strategy even further by making it a "1-pot" co-condensation procedure using octylamine or dodecylamine as template.[17] The neutral amine template was once again removed via a separate Soxhlet extraction step (not calcination). Pores of the mercaptan derivatives were found to be in the range of 1.48 to 1.75 nm. Up to 8 mole-percent of the functional organosilane was successfully installed in these materials. This methodology was further optimized by moving to longer chain alkyl amines and performing the assembly process at moderately elevated temperatures (e.g. 65°C).[18] Very high degrees of functional loading (i.e. 50 mole-percent organosilane) and framework cross-linking were obtained using this optimized procedure. Excellent hydrothermal stability was reported (material composed of 50 mole-percent organosilane was reported to be stable in boiling water for at least 10 hours).[18]

Materials made using these methods revealed that Hg can be bound in multiple ways.[19,20] One X-ray Absorbance Spectroscopy (XAS) study[19] found that the Hg-S coordination number never exceeded 1, and that at low Hg loading that Hg was bound as a monodentate S complex, with an oxygenated ligand. No evidence was found for any bidentate S-Hg-S complexes in this study. At higher Hg loadings, they found that condensation to form Hg-O-Hg-OH bridges became a significant pathway. A follow-up study,[20] using atomic pair distribution function (PDF) analysis of synchrotron X-ray powder diffraction data found no evidence of any Hg-O bonds. At low Hg loadings the authors suggest that the Hg is tetrahedrally bound to 4 different S atoms. This interesting bonding scheme suggests either a great deal of flexibility of the mesoporous ceramic support, or that opposing faces of the functional pores structure are very close to one another. At higher Hg loadings, they see a 1:1 Hg/S stoichiometry, with the Hg bound to two bridging thiolate ligands,[20] somewhat similar to previous measurements.[13]

Metal-ligand binding is an equilibrium process. Solid-phase sorbent binding chemistry reflects not only the equilibrium between the metal and ligand, but is also complicated by the impact that the binding of the first few cations has on the binding chemistry of the final cations (i.e. the nature of the ligand field evolves as the binding process proceeds). After Pearson reported his concepts of hard-soft acid-base theory,[3,4] a number of efforts were made to quantify this model. None of these attempts were completely successful (for a discussion of these efforts, see Ref. 21). In 1967, Misono presented a model summarizing an extensive set of binding studies in which he fit the binding constants to a 2 parameter model.[22] One of these parameters was called the metal cation's "softness parameter". It was found that a useful correlation was obtained when the saturation binding capacity for a given cation was plotted against the Misono softness parameter.[23] One of the things that makes this correlation particularly intriguing is that some of the metal cations involved have notably different binding geometries from one another (square planar, octahedral, etc.), meaning that these cations interact with the thiolated interface in very different ways, and with different numbers of thiols. It is noteworthy that the correlation is found with the binding capacity expressed in terms of milliequivalents per gram rather than millimoles per gram, indicating that charge accumulation at the interface is a significant consideration in determining the position of equilibrium. It seems that the accumulation of Lewis acids on the monolayer interface decreases the effective basicity of the remaining thiol ligand field by shifting the equilibrium to alleviate the repulsive electropositive interactions between the cations. The impact of this effect is felt sooner for those Lewis acids with weaker binding affinities, therefore saturation occurs at a lower loading density. While the lower end of the binding curve is dominated by equilibrium effects, the upper end of the curve is undoubtedly limited by stoichiometry, as it is unrealistic to suggest that the ligand field will have any significant effect after the ligand field is saturated via inner sphere coordination. For divalent cations, this would suggest an asymptotic limit of approximately 6.6 meq/g (the population of thiols in these monolayers is approximately 3.3 mmole/g).

Similar thiol functionalized mesoporous sorbents have also been used to bind Pt and Pd.[24,25] These materials clearly have promise in catalyst recovery.

Supercritical fluids have been found to be a particularly powerful, and "green", reaction medium for organosilane self-assembly.[10,26,27] Supercritical carbon dioxide ($SCCO_2$) effectively solvates the siloxane monomers, but it also doesn't inhibit the attractive van der Waal's forces between the hydrocarbon chains since CO_2 is a small linear molecule, thus allowing self-assembly to proceed smoothly. In addition, since self-assembly is an associative process, it can be accelerated by carrying it out under conditions of high pressure. This significantly speeds up the production of SAMMS, and at the same time virtually eliminates the wastestream resulting from production (a critical component of "green manufacturing"). In addition, the monolayers so formed have a lower defect density as a result of some novel annealing mechanisms that come into play under these reaction conditions. The low viscosity of $SCCO_2$ also facilitates mass transport of silane throughout the nanoporous ceramic matrix. When the monolayer deposition is complete, the pores of the SAMMS are clear and dry, and not filled with residual solvent that must be removed before the materials can be used.

Jaroniec and co-workers have also done some elegant work incorporating novel S-based ligands into mesoporous scaffolds, and the reader is referred to Prof. Jaroniec's chapter in this book for further details.

Periodic mesostructured organosilicas (PMOs)[28] is a related class of functional mesoporous materials that has received a great deal of attention in recent years.[29] A remarkable example of Hg affinity was reported for a functional PMO containing a tetrasulfide linkage.[30] This material was made using P123 as the template and a disilane containing a tetrasulfide linkage was included up to 15 mole-percent. These materials were found to have surface areas in the range of 242 to 654 m^2/g, and surface area was found to decrease with increased tetrasulfide loading. The Hg loading capacity was found to range from 627 mg/g (for the 2 mole-percent material), up to an incredible 2710 mg/g for the 15 mole-percent material. The Hg laden materials were found have Hg/S ratios of 2–3, suggesting that some sort of unusual binding scheme is taking place in these materials. It was possible to strip the Hg using concentrated acid. Subsequent Hg binding capacity was found to be approximately 40% of the original activity, suggesting that perhaps some ligand decomposition took place during re-generation.

Amines

Aminoalkylsiloxanes have been used as coupling agents for composite materials for several decades now. Historically, there have been some questions (and controversy) as to how these molecules interact with a silica surface, whether through acid/base chemistry between the amine and a surface silanol, through hydrogen bonding of the hydrolyzed siloxane to the silica surface, or through condensation chemistry between the hydrolyzed siloxane and surface silanol.[31] Extensive investigation, by many groups around the world, seems to have led to the consensus that each of these interactions are important, but each at a different stage of the overall process. The first thing that happens, as the aminoalkylsiloxane is sorbed to the surface, is a proton transfer between the surface silanol and the amine to form the ammonium silanolate salt. The surface silanol is more acidic[32] than is the ammonium ion, and this reaction is exothermic. If there is water adsorbed to the silica surface, then the siloxane anchor can undergo hydrolysis on the surface, generating the corresponding tris(hydroxy)silane that is so key to the self-assembly process. The mobility of the tris(hydroxy)silane may very well be limited due to the tethered ammonium silanolate (in a fully hydroxylated surface, this species should be able to move around easily as a result of proton hopping, but in more sparsely hydroxylated surfaces such mobility is likely to be more limited). As more and more of these tris(hydroxyl)silanes accumulate on the surface they will begin to aggregate as a result of H-bonding and van der Waal's interactions. It is important to recognize that many of these molecules will still be "bent over", with both the tris(hydroxyl)silane and the ammonium ion headgroup interacting with the silica surface. As a result, when the condensation processes start, giving rise to the anchored, cross-linked monolayer structure, there will be a significant number of defects, both in terms of pinhole defects (vacancies) and in terms of cross-linking defects ("dangling hydroxyls"), both resulting from the blocking action of the "bent over" molecules. As a result, these monolayers tend to be somewhat more fragile than other systems (and as will be discussed later, these can be thermally cured post-complexation to enhance their stability).

Mesoporous silica has been functionalized with organosilanes terminated with a simple primary amine. For example, SBA-15 has been coated with 3-aminopropyltriethoxysilane and characterized.[33] Similar materials were prepared and evaluated as base catalysts for condensation chemistry.[34] Site isolated behavior has been demonstrated for amine functionalized mesoporous silicas.[35] These simple amine terminated materials are versatile building blocks for creating more complex molecular architectures, but are not generally used directly for environmental applications.

Similar amine modified mesoporous materials have also been used to graft various metal species to surfaces. For example, 2,2'-bipyridine and 1,10-phenanthroline complexes of Cu and Mn have been anchored to mesoporous silicas through an aminopropylsilane tether, and these complexes were studied spectroscopically and electrochemically.[36–41] The general focus of this work seems to be primarily catalyst oriented, but there is also potential for environmental application.

3-(2-aminoethyl)-aminopropyl trimethoxysilane is commercially available and provides ready access to the chelating ethylenediamine (EDA) ligand. Use of this organosilane to construct EDA-SAMMS for the reversible capture of CO_2[42] is discussed in detail in the chapter by Dr. Zheng, and the reader is referred to that chapter for details.

An interesting aminated material was made in which macrocyclic "cyclam" ligands were incorporated into a PMO.[43] In this case the ligand loading density was approximately 10 mole-percent of the total siloxane content. The surface area of these macrocycle containing PMOs ranged from 420–703 m^2/g and pore size varied from 51 to 87 Å. These materials were shown to be a versatile foundation upon which to build more complex ligand structures (macrocycles, phosphonates, etc.). There is clearly great promise in this area for the design and construction of sophisticated molecular architectures, perhaps with discriminating metal-binding properties.

Transition Metal Complexes

While the amine-terminated organosilanes tend to give rise to monolayers that have a fair number of defects in them, these amines can still be coordinated to a metal center, especially so for the chelating diamines as they form strong complexes with many transition metals. Once these amines are tied up with a metal center, then these defects are simply dangling hydroxyl groups and pinhole vacancies (i.e. the amine is no longer serving as a blocking group, preventing access to these defects). Therefore, it is possible to remove some of these defects by thermally curing the monolayer after complexation, promoting the condensation of the some of the dangling hydroxyls, resulting in a more stable monolayer structure. This is conveniently done by refluxing the sample in toluene and removing the water using a Dean-Stark trap.

Since these chelating diamine ligands are neutral, they form cationic transition metal complexes. These are of interest since they can serve as anion exchange materials. For example, EDA-SAMMS have been used to form complexes with a variety of transition metal cations. The Cu(II) complex is particularly interesting.[44] Being a d^9 species, an octahedral coordination environment would result in orbital degeneracy. Thus the complex "wants" to undergo some sort of Jahn-Teller distortion to alleviate this degeneracy. The bond angles and distances imposed by the chelated EDA ligand make this difficult. As a result, portions of the ligand field are more weakly held than the rest. Detailed EXAFS studies have shown that when a tetrahedral oxometallate anion comes in and associates with the Cu(II)-EDA complex that the primary amines from two adjacent EDA ligands are lost and the anion forms a direct, monodentate bond to the metal center[45] (see Figure 3). This results in a trigonal bipyramidal coordination geometry and alleviates the orbital degeneracy. This bond formation is potentially reversible, and offers a level of chemical selectivity for these sorbent materials. For tetrahedral anions that form soluble Cu(II) salts (e.g. sulfate), this process is readily reversed and the sulfate anion easily displaced from the complex; for tetrahedral anions that form insoluble Cu(II) salts (e.g. arsenate), this adduct does not readily dissociate and the anion is firmly held. This strategy has been successfully employed to sequester arsenate and chromate,[44] as well as pertechnetate.[46] This work has been built upon by the elegant research of Yoshitake and co-workers,[47–49] and the reader is referred to the chapter that Prof. Yoshitake has contributed to this book for details.

This approach can also be used as a functionalization strategy for subsequent elaboration of these metal-complex SAMMS. For example, treatment

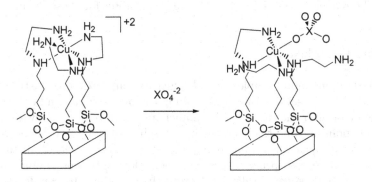

Figure 3.   Tetrahedral anion binding by the Cu(EDA)$_3$ complex.

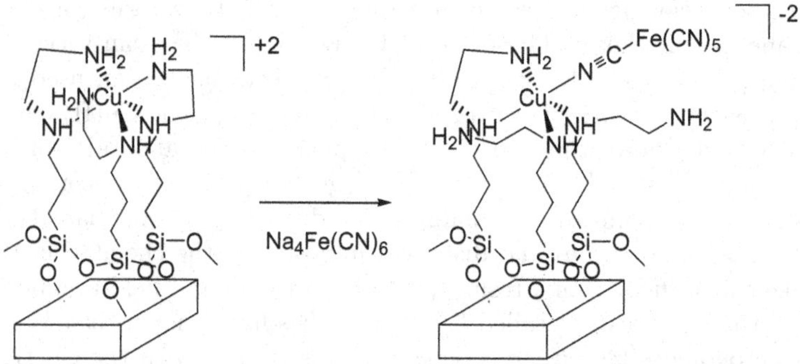

Figure 4. Preparation of Ferrocyanide SAMMS.

of Cu(EDA) SAMMS with sodium ferrocyanide results in the lavender colored ferrocyanide adduct.[50] This bimetallic complex is of interest because it has been long recognized that transition metal ferrocyanide complexes tend to form insoluble cesium salts. Cs-137 has a halflife of about 30 years and is one of the primary radioactive isotopes of concern in nuclear waste and radioactive fallout.[51] Removing Cs in the presence of large excesses of Na (or K) has been a challenge in the past, although some success has been had with hydrophilic crown ether-based approaches[52] and the crystalline silicotitanates.[53] Ferrocyanide-SAMMS has been shown to have very high affinity for Cs, even in the presence of huge excesses of Na or K, and sorption kinetics are reasonably fast, with equilibrium being reached in less than an hour.[50]

Other metal complexes have also been used to remove other classes of anions. For example, the Ag and Hg adducts of thiol-SAMMS, being "soft" Lewis acids have been demonstrated to be effective sorbents for removing iodide. The reader is referred to Dr. Mattigod's chapter for those details.

Chelating Complexants

When constructing more complex ligand structures, amidation chemistry is a convenient and versatile strategy for joining the pieces. This can be done either before or after the silane has been anchored to the surface. However, it is worth pointing out that performing this linkage after the silane is anchored to the surface is an inherently limiting process, especially if the ligand has a large "footprint", as it can then block adjacent amidation sites.[54] Thus, as a general rule, it is probably preferable to couple the fragments first, then perform the deposition onto the silica surface. This also allows the favorable H-bond interaction between the N-H and an adjacent C=O to help drive the self-assembly process.[55]

As a general rule, the rare earth cations tend to be fairly "hard" Lewis acids as a result of their high oxidation states. Thus, when designing a ligand field to selectively bind the lanthanide and actinide cations, it behooves the chemist to incorporate "hard" Lewis bases into the ligand structure. In addition, there is an interesting phenomenon that comes into play with these cations, in which the binding of one class of ligand to the cation enhances the binding affinity of another class of ligand. For the rare earth cations, the most important types of synergistic ligands are the amide carbonyl group and those ligands containing the phosphoryl group (P=O double bond, commonly loosely grouped under the term "phosphine oxide"). Given the versatility of the amide coupling process described in the previous paragraph, and the availability of the requisite precursor organosilanes, it is no surprise that the amide linkage figures strongly in the ligand design strategy for the selective sequestration of rare earth cations.

Perhaps the most commonly encountered chelating complexants are those composed of one of more glycine subunit (e.g. EDTA), and indeed this strategy has been employed in these functional mesoporous sorbent materials. For example, treating the commercially available 3-isocyanatopropyltrimethoxysilane with glycine (in the presence of triethylamine as buffer), very cleanly gives rise to the corresponding glycinyl-urea silane, as shown in Figure 5.[56] Deposition of this silane in an MCM-41 matrix resulted in a monolayer population density of over 5 silanes/nm^2. This high population density seems to be due to the additional driving forces for self-assembly in this molecule (2 NH's to interact with an adjacent C=O, dipole-dipole interactions, etc.). This ligand field was found to be an effective sorbent phase for both lanthanide and actinide cations, but only at pHs above about 4.5 (for the lanthanides, which is presumably the approximate pKa of the carboxylic acid),[56] and above about 2 for the actinides (the actinides are stronger Lewis acids than are the lanthanides, and actinide coordination enhances the dissociation, thereby increasing the acidity of the carboxylate).[57] Sorption kinetics were found to be rapid, with equilibrium being achieved in less than a minute.[56] There was no significant competition observed in the presence of excess transition metal cations, and

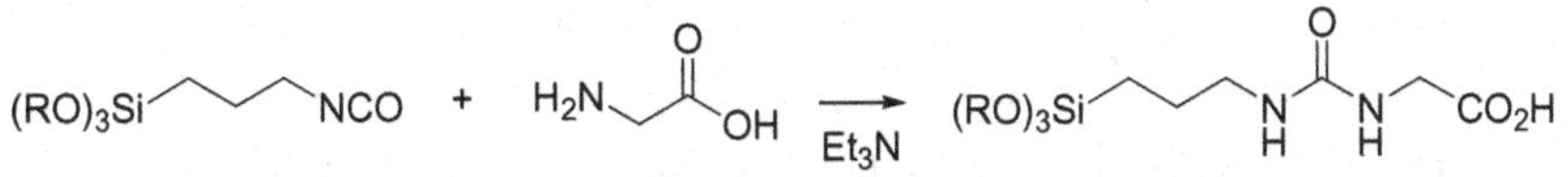

Figure 5. Preparation of the Glycinyl-urea (Gly-UR) silane.

it was found that this ligand field could easily be stripped and re-generated by a simple acid wash.[56]

Given that the rare earth cations are "hard" Lewis acids, then it comes as no surprise that they form strong complexes with "hard" Lewis bases like alkoxides. However, typical alcohols are not very acidic, and given that most lanthanide and actinide manipulations are done under acidic conditions (for solubility reasons), these rather basic ligands would be quickly protonated and cleaved at the pHs typically used to handle these materials. Thus it was necessary to design a chelating alkoxide that was sufficiently acidic to form stable complexes on the acid side. This was attempted by using a phenol instead of an alcohol, and by placing a carbonyl group in the ortho position, where it could both chelate the metal ion and enhance the phenol group's acidity (see Figure 6). Salicylamide (Sal) SAMMS was found to be an effective sorbent at near neutral conditions (e.g. pH = 6.5), where it was particularly effective. Under more acidic conditions, its affinity for both lanthanide and actinide cations was quite limited.[56,57]

Detailed EXAFS analysis of the Eu(III) adduct of Sal-SAMMS revealed an average Eu-O coordination number of 8 and a Eu-O radial distance of 2.40 ± 0.015 Å.[56] These results are consistent with the Eu(III) cation being 8-coordinate (typical for the lanthanides), in either a cubic or distorted square antiprism geometry (see Figure 7). These EXAFS results support

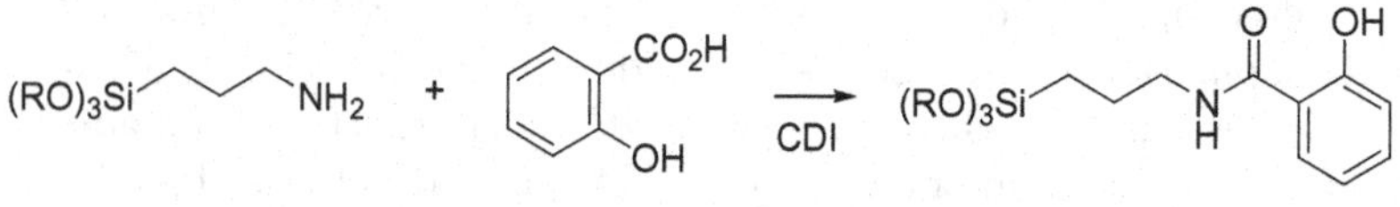

Figure 6.   Preparation of Salicylamide (Sal) silane.

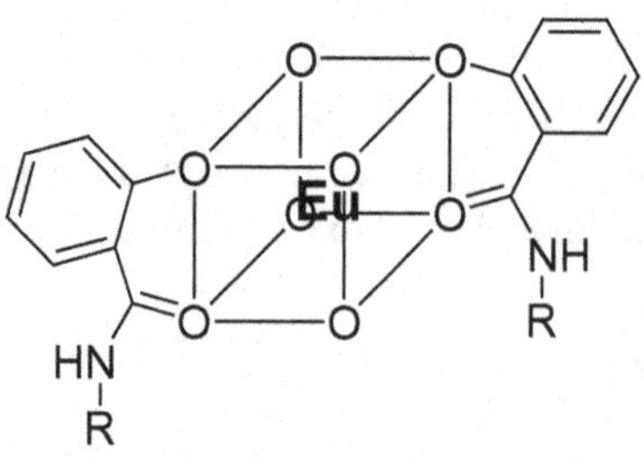

Figure 7.   Cartoon showing one possible bonding geometry of the 4:1 Salicylamide/Eu(III) complex (only two ligands are shown for clarity purposes). The square antiprism is also a possible.

the conclusion that the close proximity of the ligands in the monolayer interface allow for multiple ligands to interact with a single metal cation. In this case we observe a 4 to 1 ligand:metal interaction, which clearly contributes to enhancing the binding affinity between a given ligand and metal cation.

Phosphates and phosphonates are known to have high affinity for rare earth cations, making these functionalities logical targets for incorporation into the SAMMS motif. Once again, the amide linkage was chosen for its desirable synergistic properties. The commercially available diethylphosphonoacetic acid was activated with carbonyl diimidazole (CDI), then subsequently treated with 3-aminopropyltrimethoxysilane (APS) to afford the acetamide phosphonate (Ac-Phos) silane cleanly and in high yield (see Figure 8).[58] Both the ester and the acid form of the Ac-Phos SAMMS were found to be effective sorbent materials for both the lanthanide and actinide species. In the case of the ester, it appears that the central enolizable methylene is sufficiently acidic to lose a proton upon complexation, even under fairly acidic conditions (e.g. pH = 2, or even lower in some cases). This is undoubtedly driven by the strength of the complex formed. The acid form of the Ac-Phos ligand very good affinity for lanthanide and actinide cations over a wide range of pHs, with no competition from common transition metal and alkaline earth metal cations.[56,57]

In an effort to probe the effect of chelate cavity size on the efficacy of these SAMMS materials, the homolog of Ac-Phos SAMMS was also prepared (see Figure 9). Michael addition of diethylphophonous acid to trifluoroethylacrylate proceeded smoothly and in high yield.[58] Displacement of trifluoroethanol by APS was found to be clean and to afford the corresponding propionamide phosphonate (Prop-Phos) in virtually quantitative yield. Deposition of the Prop-Phos silane in MCM-41 resulted in a monolayer with a population density of 2.1 silanes/nm^2.[56] In contrast to the Ac-Phos ester SAMMS described above, the Prop-Phos ester SAMMS had no affinity at all for the lanthanide cations.[56] In general, the Prop-Phos ester SAMMS had little affinity for the actinide cations, the exception being Pu (IV), for which this material demonstrated an excellent affinity. Like the Ac-Phos acid SAMMS, the Prop-Phos acid SAMMS were found to have very rapid sorption kinetics and good affinity for the actinides in general (i.e. U, Pu,

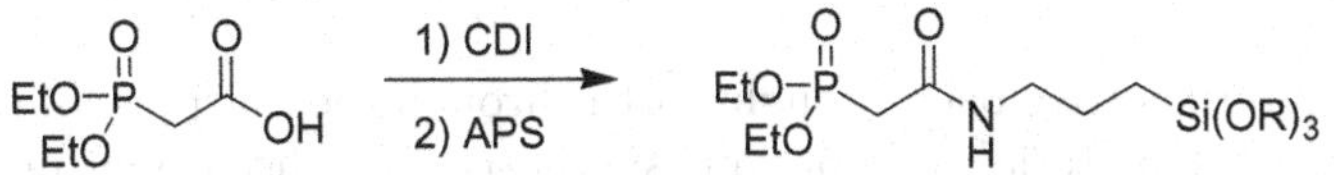

Figure 8. Preparation of the Acetamide phosphonate (Ac-Phos) silane.

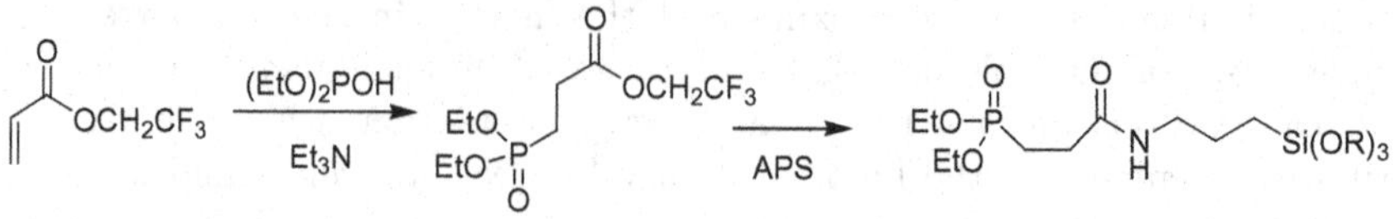

Figure 9. Preparation of the Propionamide phosphonate (Prop-Phos) silane.

Am, etc.). The larger chelation cavity of the Prop-Phos ligand was found to provide no significant advantage over the Ac-Phos ligand. Curiously, neither the Ac-Phos nor Prop-Phos acid SAMMS were effective sorbents for Np(V).

The Raymond group at Berkeley has spent over 20 years designing actinide specific ligands for the selective chelation and *in vivo* extraction of these species.[59] One of the most powerful classes of ligands to result from these studies are the hydroxypyridinone (HOPO) class of ligands.[60] Due to the hydroxyl groups on these ligands, they must be installed in protected form to prevent competing reaction with the siloxane anchor. The benzyl protected HOPO acids were activated with CDI, and subsequently coupled with APS, to afford the benzyl protected silane.[61,62] This was deposited in the usual manner in MCM-41, then the benzyl protecting group was removed using HBr in glacial acetic acid. Solid state NMR studies suggested a population density of about 0.5 to 1 silanes/nm^2 for these systems (presumably due to the steric bulk of the benzyl protecting groups during deposition).[62] Of the various HOPO SAMMS tested, the 3,2-HOPO ligand (shown in Figure 10) was clearly superior, offering excellent binding affinity for all of the actinides studied, including Np(V).

Aside from the clearly superior electronic interactions of this ligand with the actinide species, one possible contributing factor to 3,2-HOPO SAMMS affinity for Np(V) is the little bit of extra "wiggle room" left in the monolayer as a result of doing the deposition with the benzyl protected form. Once this protecting group is removed there is some additional conformational freedom that allows these ligands to adopt a "horizontal" attitude, allowing 3 of them to chelate the Np(V) around its equatorial plane, in a hexagonal bipyramidal fashion, as shown in Figure 11.[62]

Conclusions

These new materials have been brought forth through the marriage of templated nanoporous materials and organic synthesis. New discoveries in this field are limited only by the creative application of ligand design and the construction of more complex organic architectures within the mesoporous

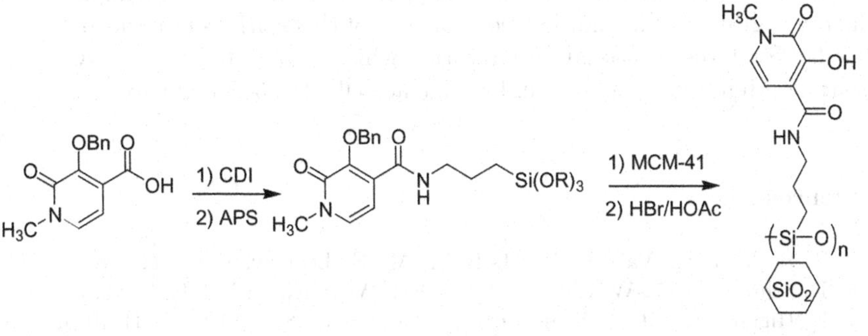

Figure 10.   Preparation of 3,2-HOPO SAMMS.

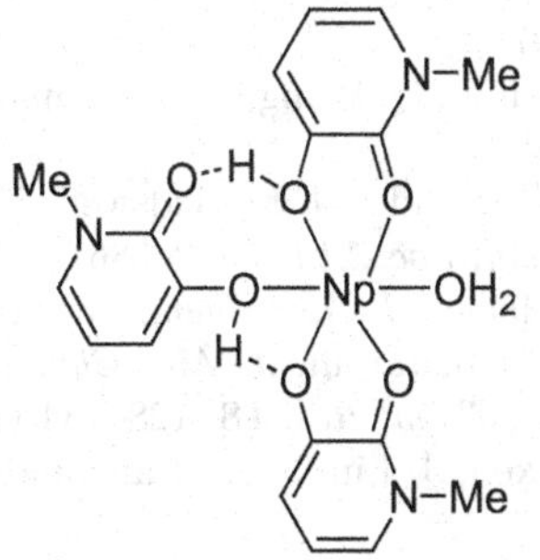

Figure 11.   Chelation geometry of the 3,2-HOPO ligand binding the neptonyl (NpO_2^+) cation (looking down the O-Np-O axis, with the 3 HOPO ligands and a water molecule in the equatorial plane; the hydrogen bonds between the HOPO ligands contribute to the driving force for complexation).

framework. It is the self-assembly and molecular recognition provided by the organic ligand field that provide these materials with much of their remarkable performance. As new concepts and molecular architectures are dreamed up and reduced to practice we will see many exciting new discoveries in this area, for years to come.

Acknowledgements

I would like to thank my co-authors for all their dedicated efforts and creative thinking while contributing to the various projects that gave rise to many of these nanomaterials. Research support from the U. S. Department of Energy is gratefully acknowledged. Portions of this work were performed at Pacific Northwest National Laboratories, which is operated for the DOE by Battelle Memorial Institute under contract DE AC06-76RLO 1830.

References

1. J. S. Beck, J. C. Vartuli, W. J. Roth, M. E. Leonowicz, C. T. Kresge, K. D. Schmitt, C. T.-W. Chu, D. H. Olson, E. W. Sheppard, S. B. McCullen, J. B. Higgins and J. L. Schlenker, *J. Am. Chem. Soc.* **114**, 10834–10842 (1992).
2. C. T. Kresge, M. E. Leonowicz, W. J. Roth, J. C. Vartuli and J. S. Beck, *Nature* **359**, 710–712 (1992).
3. R. G. Pearson, *J. Am. Chem. Soc.* **85**, 3533–3539 (1963).
4. R. G. Pearson, *Science* **151**, 172–177 (1966).
5. X. D. Feng, G. E. Fryxell, L. Q. Wang, A. Y. Kim, J. Liu and K. Kemner, *Science* **276**, 923–926 (1997).
6. J. Liu, X. Feng, G. E. Fryxell, L. Q. Wang, A. Y. Kim and M. Gong, *Advanced Materials* **10**, 161–165 (1998).
7. S. R. Wasserman, G. M. Whitesides, I. M. Tidswell, M. Ocko, P. S. Pershan and J. D. Axe, *J. Am. Chem. Soc.* **111**, 5852–5861 (1989).
8. D. W. Sindorf and G. E. Maciel, *J. Am. Chem. Soc.* **105**, 3767–3776 (1983).
9. F. M. Menger and U. V. Venkataram, *J. Am. Chem. Soc.* **107**, 4706–4709 (1985); F. M. Menger, *Acc. Chem. Res.* **18**, 128–134 (1985).
10. T. S. Zemanian, G. E. Fryxell, J. Liu, J. A. Franz and Z. Nie, *Langmuir* **17**, 8172–8177 (2001).
11. S. V. Mattigod, X. Feng, G. E. Fryxell, J. Liu and M. Gong, *Sep. Sci. Technol.* **34**, 2329–2345 (1999).
12. J. Liu, G. E. Fryxell, S. V. Mattigod, M. Gong, Z. Nie, X. Feng and K. Raymond, Self-Assembled Mercaptan on Mesoporous Supports (SAMMS) technology for contaminant removal and stabilization, PNNL-12006; September (1998).
13. K. Kemner, X. Feng, J. Liu, G. E. Fryxell, L.-Q. Wang, A. Y. Kim, M. Gong and S. V. Mattigod, *J. Synchrotron Rad.* **6**, 633–635 (1999).
14. L. Mercier and T. J. Pinnavaia, *Adv. Mat.* **9**, 500–503 (1997).
15. L. Mercier and T. J. Pinnavaia, *Env. Sci. Technol.* **32**, 2749–2754 (1998).
16. J. Brown, L. Mercier and T. J. Pinnavaia, *Chem. Comm.* 69–70 (1999).
17. L. Mercier and T. J. Pinnavaia, *Chem. Mater.* **12**, 188–196 (2000).
18. Y. Mori and T. J. Pinnavaia, *Chem. Mater.* **13**, 2173–2178 (2001).
19. C. C. Chen, E. J. McKimmy, T. J. Pinnavaia and K. F. Hayes, *Env. Sci. Technol.* **38**, 4758–4762 (2004).
20. S. J. L. Billinge, E. J. McKimmey, M. Shatnawi, H. J. Kim, V. Petkov, D. Wermeillle and T. J. Pinnavaia, *J. Am. Chem. Soc.* **127**, 8492–8498 (2005).

21. R. G. Parr and R. G. Pearson, *J. Am. Chem. Soc.* **105**, 7512–7516 (1983).
22. M. Misono, E. Ochiai, Y. Saito and Y. Yoneda, *J. Inorg. Nucl. Chem.* **29**, 2685–2691 (1967).
23. S. V. Mattigod, K. Parker and G. E. Fryxell, *Inorg. Chem. Comm.* **9**, 96–98 (2006).
24. T. Kang, Y. Park and J. Yi, *Ind. Eng. Chem. Res.* **43**, 1478–1484 (2004).
25. T. Kang, Y. Park and J. C. Park, *Studies in Surf. Sci. Catalysis* **146**, 527–530 (2003).
26. Y. Shin, T. S. Zemanian, G. E. Fryxell, L. Q. Wang and J. Liu, *Micropor. Mesopor. Mater.* **37**, 49–56 (2000).
27. T. S. Zemanian, G. E. Fryxell, O. Ustyugov, J. C. Birnbaum and Y. Lin, Synthesis of nanostructured sorbent materials using supercritical fluids, in *Supercritical Fluids and Nanomaterials*, eds. Gopalan, Wai and Jacobs, Chap. 24 (American Chemical Society, Washington DC, 2003), pp. 370–386.
28. T. Asefa, M. J. Maclachlan, N. Coombs and G. A. Ozin, *Nature* **402**, 867–871 (1999).
29. B. Hatton, K. Landskron, W. Whitnall, D. Perovic and G. A. Ozin, *Acc. Chem. Res.* **38**, 305–312 (2005).
30. L. Zhang, W. Zhang, J. Shi, Z. Hua, Y. Li and J. Yan, *Chem. Comm.* 210–211 (2003).
31. G. S. Caravajal, D. E. Leyden, G. R. Quinting, and G. E. Maciel, For an excellent discussion of the complexity of how this molecule binds to a silica surface, *Anal. Chem.* **60**, 1776–1786 (1988).
32. R. K. Iler, *Chemistry of Silica* (John Wiley and Sons, New York, 1979).
33. A. S. M. Chong and X. S. Zhao, *J. Phys. Chem. B* **107**, 12650–12657 (2003).
34. C. Yang, X. P. Jia, Y. D. Cao *et al.*, *Studies in Surf. Sci. Catalysis* **146**, 485-488 (2003).
35. M. W. McKittrick and C. W. Jones, *Chem. Mater.* **15**, 1132–1139 (2003).
36. S. Zheng, L. Gao and J. K. Guo, *J. Inorg. Mater.* **15**, 844–848 (2000).
37. S. Zheng, L. Gao and J. K. Guo, *J. Inorg. Mater.* **15**, 1015–1020 (2000).
38. S. Zheng, L. Gao and J. K. Guo, *J. Solid State Chem.* **152**, 447–452 (2000).
39. L. Gao and S. Zheng, *Mater. Transactions* **42**, 1688–1690 (2001).
40. S. Zheng, L. Gao and J. K. Guo, *Mater. Chem. Phys.* **71**, 174–178 (2001).
41. S. Zheng, L. Gao and J. K. Guo, *J. Inorg. Mater.* **16**, 459–464 (2001).
42. F. Zheng, D. N. Tran, B. Busche, G. E. Fryxell, R. S. Addleman, T. S. Zemanian and C. L. Aardahl, *Ind. Eng. Chem. Res.* **44**, 3099–3105 (2005).
43. R. J. P. Corriu, A. Mehdi, C. Reyé and C. Thieuleux, *Chem. Comm.* 1382–1383 (2002).
44. G. E. Fryxell, J. Liu, M. Gong, T. A. Hauser, Z. Nie, R. T. Hallen, M. Qian and K. F. Ferris, *Chem. Mater.* **11**, 2148–2154 (1999).
45. S. Kelly, K. Kemner, G. E. Fryxell, J. Liu, S. V. Mattigod and K. F. Ferris, *J. Phys. Chem. B* **105**, 6337–6346 (2001).
46. S. V. Mattigod, G. E. Fryxell, K. Alford, T. Gilmore, K. Parker and J. Serne, *Env. Sci. Technol.* **39**, 7306–7310 (2005).
47. H. Yoshitake, T. Yokoi and T. Tatsumi, *Chem. Mater.* **14**, 4603–4610 (2002).

48. H. Yoshitake, T. Yokoi and T. Tatsumi, *Chem. Mater.* **15**, 1713–1721 (2003).
49. H. Yoshitake, E. Koiso, T. Tatsumi, H. Horie and H. Yoshimura, *Chem. Lett.* **33**, 872–873 (2004).
50. Y. Lin, G. E. Fryxell, H. Wu and M. Englehard, *Env. Sci. Technol.* **35**, 3962–3966 (2001).
51. "Radiation in the Environment" in "Cleaning Our Environment: A Chemical Perspective" by the ACS, 2nd edn. (1978), pp. 378–452.
52. F. Chitry, S. Pellet-Rostaing, L. Nicod, J.-L. Gass, J. Foos, A. Guy and M. Lemaire, *J. Phys. Chem. A.* **104**, 4121–4128 (2000).
53. D. M. Poojary, R. A. Cahill and A. Clearfield, *Chem. Mater.* **6**, 2364–2368 (1994).
54. G. E. Fryxell, P. C. Rieke, L. L. Woods, M. H. Engelhard, R. E. Williford, G. L. Graff, A. A. Campbell, R. J. Wiacek, L. Lee and A. Halverson, *Langmuir.* **12**, 5064–5075 (1996).
55. S. W. Tam-Chang, H. A. Biebuyck, G. M. Whitesides, N. Jeon and R. G. Nuzzo, *Langmuir.* **11**, 4371–4382 (1995).
56. G. E. Fryxell, H. Wu, Y. Lin, W. J. Shaw, J. C. Birnbaum, J. C. Linehan, Z. Nie, K. Kemner and S. J. Kelly, *Mater. Chem.* **14**, 3356–3363 (2004).
57. G. E. Fryxell, Y. Lin, S. Fiskum, J. C. Birnbaum, H. Wu, K. Kemner and S. Kelly, *Env. Sci. Tech.* **39**, 1324–1331 (2005).
58. J. C. Birnbaum, B. Busche, W. Shaw and G. E. Fryxell, *Chem. Comm.* 1374–1375 (2002).
59. A. E. V. Gorden, J. Xu, K. N. Raymond and P. W. Durbin, *Chem. Rev.* **103**, 4207–4282 (2003); J. Xu, P. W. Durbin, B. Kullgren, S. N. Ebbe, L. C. Uhlir and K. N. Raymond, *J. Med. Chem.* **45**, 3963–3971 (2002); J. Xu, B. Kullgren, P. W. Durbin and K. N. Raymond, *J. Med. Chem.* **38**, 2606 (1995).
60. J. Xu, D. W. Whisenhunt, Jr., A. C. Veeck, L. C. Uhlir and K. N. Raymond, *Inorg. Chem.* **42**, 2665–2674 (2003); J. Xu, E. Radkov, M. Ziegler and K. N. Raymond, *Inorg. Chem.* **39**, 4156–4164 (2000); P. W. Durbin, B. Kullgren, J. Xu, K. N. Raymond, M. H. Hengè-Napoli, T. Bailly and R. Burgada, *Rad. Prot. Dosim.* **105**, 503–508 (2003); D. L. White, P. W. Durbin, N. Jeung and K. N. Raymond, *J. Med. Chem.* **31**, 11 (1988).
61. W. Yantasee, G. E. Fryxell, Y. Lin, H. Wu, K. N. Raymond and J. Xu, *J. Nanosci. Nanotechnol.* **5**, 1537–1540 (2005).
62. Y. Lin, S. K. Fiskum, W. Yantasee, H. Wu, S. V. Mattigod, G. E. Fryxell, K. N. Raymond and J. Xu, *Env. Sci. Tech.* **39**, 1332–1337 (2005).

Chapter 8

Chemically-Modified Mesoporous Silicas and Organosilicas for Adsorption and Detection of Heavy Metal Ions

Oksana Olkhovyk and Mietek Jaroniec*

*Department of Chemistry, Kent State University
Kent, Ohio 44242, USA
jaroniec@kent.edu

Foreword

The aim of this brief review is to give the current standpoint and trends in the development of nanomaterials for environmental applications, specifically for removal and detection of heavy metal ions from water. Recent advances in nanoscience and nanotechnology have been possible due to a tremendous development in the design of nanomaterials, i.e. materials consisting of nano-sized chemical domains and/or materials having nano-sized morphology and/or pores. Numerous research activities in this area are focused on the design and synthesis of nanomaterials with tunable structure and chemistry to achieve superior properties in comparison to conventional porous materials. An eminent example of this approach is the discovery of self-assembly (soft templating) synthesis, which afforded a new family of materials with ordered structures of nanopores[1] and allowed their post-synthesis decoration (functionalization) with various surface groups (see Scheme 1). A growing interest in utilizing unique properties of these materials for environmental applications has a stimulating effect on the development of novel nanomaterials, especially silica-based nanomaterials. Numerous interesting materials have been synthesized

---

*Corresponding author.

surfactant molecules   block-copolymer

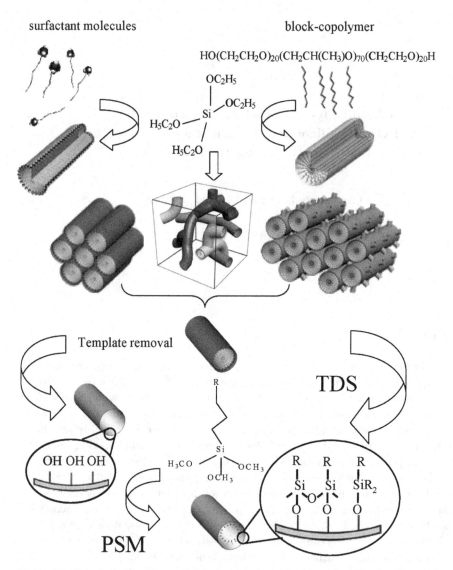

Scheme 1.   Schematic illustration of the synthesis of ordered mesoporous silicas (OMS) via surfactant (left) or polymeric (right) templating routes. After formation of siliceous framework (showed as representative hexagonal or cubic arrangement of pores), post-synthesis modification (PSM) of the template-free OMS or template displacement synthesis (TDS) can be carried out to decorate pore walls with desired surface ligands (R).

on the basis of silica framework by introducing structural order, uniform nanopores and desired surface chemistry (Scheme 1). With discovery of ordered siliceous nanostructures,[1] the well-established modification methods for chemical immobilization of reactive ligands onto the silica surface got a "second life." Nanomaterials designed by the aforementioned chemical modification possess interactive surface groups on the walls of ordered mesopores. These novel nanomaterials, in addition to the stable and rigid silica framework, feature high and accessible specific surface area due to the presence of ordered and uniform nanopores of tailorable size, shape and surface chemistry, which make them much more attractive for many environmental applications in comparison to the conventional materials.

The first part of this review deals with the synthesis, characterization and comparison of properties and performance of mesoporous silicas with attached various organic ligands as well as their application for adsorption of heavy metal ions. There are some disadvantages of the post-synthesis surface modification of nanoporous silicas related to the reduction of accessible porosity for adsorption and limited selection of ligands to be introduced by this method, which narrow the range of possible applications of these materials. The first important disadvantage of the aforementioned modification is its multi-step nature, which involves the synthesis of nanoporous silicas and their subsequent derivatization with desired ligands. Since among existing silanols not all of them are available and suitable for chemical reaction, their complete replacement with other functional groups is a challenging task, which often requires a multi-step process. Additionally, the geometrical factors related to the porous structure as well as size and shape of pores prevent achieving a full replacement of silanols by organic ligands. Therefore, one step co-condensation synthesis and template-displacement synthesis with properly selected hydrolysable silanes seem to be the methods of choice to reduce the cost and time of synthesis of novel mesoporous materials and simultaneously to retain the possibility of controlling their structural and surface properties.

The framework modification approach was considered as one that would permit overcoming the aforementioned disadvantages of the post-synthesis modification of porous silicas. The framework-modified porous silicas represent a successful combination of organic and inorganic building components and retain the template-predefined structural properties, while the incorporated organic fragments into the framework alter its chemical and mechanical properties (see Scheme 2). The availability of organosilane precursors bearing desired functional groups and capable for self-assembly in the presence of templating agents is the only limitation in this rapidly growing area

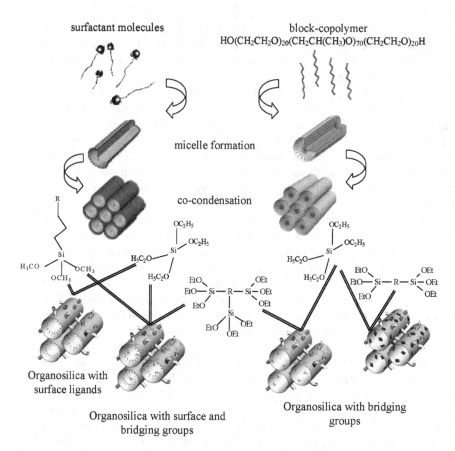

surfactant molecules

block-copolymer
$HO(CH_2CH_2O)_{20}(CH_2CH(CH_3)O)_{70}(CH_2CH_2O)_{20}H$

micelle formation

co-condensation

Organosilica with
surface ligands

Organosilica with surface and
bridging groups

Organosilica with bridging
groups

Scheme 2. Schematic illustration of the co-condensation synthesis of periodic meso-porous organosilicas (PMO) via surfactant (left) or polymeric (right) templating routes. Co-condensation of silica precursors bearing organic functionality in the presence of organic template affords mesoporous organosilica with incorporated surface, bridging, or surface and bridging groups. The PMO structures are often formed via cooperative mechanism, which does not involves prior formation of micelles by template species.

of research. The second part of this review is focused on the design and synthesis of framework-modified mesostructures, known as periodic meso-porous organosilicas (PMO), for sequestering metal ions. The PMO-related materials are extremely attractive for environmental applications because of almost unlimited opportunities to design their chemical composition as well as interfacial and structural properties.

Environmental Significance and Main Concepts of Designing Adsorbents Based on Ordered Mesoporous Materials

Scientific and social significance of novel adsorbents for heavy metal ions stems from the extremely high toxicity of heavy metals. These toxic persistent pollutants are considered by the U.S. Environmental Protection Agency as major hazardous contaminants of aqueous ecosystem. A part per billion is the maximum allowed concentration of such major pollutants as mercury, cadmium and lead in water. The toxic effect of heavy metal ions on living cells occurs through their selective binding to important biomolecules such as glutathione, cysteine and homocysteine, which consequently affect the protein structure,[2] disturb the enzyme activity and intoxicate cell metabolic functions. The knowledge of the affinity and binding mechanism of particular heavy metal ions to functional groups, which are involved in the intoxication process, is helpful to design proper functional materials that mimic naturally occurring chelating mechanism and utilize it for the detection and specific adsorption of heavy metal ions. Such metal-sequestering ligands usually contain oxygen (phenolic, carbonyl, carboxylic, hydroxyl, ether, phosphoryl), nitrogen (amine, nitro, azo, diazo, nitrile), sulfur (thiol, thioether, thiocarbamate) and other functional groups that exhibit high affinity to heavy metal ions.[3] Various types of materials were employed as supports for immobilization of functional groups capable for heavy metal ion adsorption. Polymers,[4−6] resins,[7−9] polystyrene,[10] fibers,[11] clays,[12] activated carbons,[13,14] natural zeolites, such as clinoptilolite, mordenite, chabbazite, erionite, and phillipsite,[15] chitosan and other low-cost materials[13] along with currently available commercial materials such as TMT (trimercaptotriazine), STC (potassium/sodium thiocarbonate), and HMP-2000 sodium dimethylthiocarbonate (SDTC)[16] compose an enormous library of nanoporous adsorbents used for sequestering of heavy metal ions. Among aforementioned supports, various silicas (aerosils, aerosilogels, silica gels, porous glasses, etc.)[17−20] have been extensively used for the immobilization of various groups of high selectivity and affinity towards heavy metal ions.[21,22]

Impregnation, covalent grafting, sol-gel synthesis and other methods for obtaining hybrid silica-based materials for selective adsorption, extraction and pre-concentration of heavy metal ions have been extensively studied (see review[23] and references therein) because these materials possess high chemical specificity and selectivity due to the attached ligands as well as stable and accessible porosity due to the silica support. Thus, the key issue in the development of silica-based adsorbents is in the selection of proper surface ligands as well as silica-supports of high surface area, accessible and well-developed porosity and high stability. Therefore, novel siliceous nanos-tructures are of high demand in the design of highly effective adsorbents for heavy metal ions that meet the EPA requirements and perform better than the existing materials.

Heavy Metal Ions Adsorbents Based on Ordered Mesoporous Silicas

A significant enhancement of adsorption properties of porous solids for environmental applications was accomplished by the discovery of a new family of ordered nanoporous materials[1] synthesized via "bottom-up" self-assembly approach. According to the IUPAC classification of porous systems,[24] the achievable pore sizes in these novel materials lie in the range of meso-pores (widths from 2 to 50 nm), thus they are known as ordered meso-porous materials (OMM). Among them, materials having purely siliceous framework are known as ordered mesoporous silicas (OMS). Since the pioneering Mobil publications in 1992,[1,25] a novel family of materials have grown significantly with respect to the type of porous structure (network of pores of definite size and shapes and its crystallographic symmetry) and chemical composition formed during self-assembly of properly selected organic templates and inorganic and/or hybrid organic-inorganic precursors, e.g. tetraethyl orthosilicate (TEOS) and organosilanes. Surfactant- and polymer-templated OMS with hexagonal arrangement of mesopores are MCM-41,[25–27] SBA-2,[28] SBA-3,[29,30] HMS,[31] SBA-15,[32,33] and FSM-16.[34] SBA-1 and MCM-48 possess cubic bi-continuous arrangement of inter-connected and uniform channels.[28,35,36] There are also popular cage-like mesostructures (e.g. FDU-1 and SBA-16) prepared by using block copoly-mers as templates.[37] The availability of various precursors that are able to self-assemble in the presence of surfactants and block copolymers creates almost unlimited opportunities for the synthesis of ordered mesoporous materials (OMM) with tailorable properties.[38–40] A proper selection of synthesis conditions and the structure-directing agents allows one to control the pore size ranging from ∼2 to 8 nm when surfactants are employed as templates and from ∼8 to 30 nm when block copolymers are used. Especially, the use of block copolymers was beneficial for the development of inexpensive and reproducible large-pore materials with channel-like and cage-like structures.[41–48]

Ordered mesoporous organosilicas due to the tunable surface chemistry and nanoporous structure represent excellent substitutes for ordered meso-porous silicas (OMS) in numerous applications.[49–51] The main advantage of these materials is their well-defined surface chemistry achieved by various surface modifications. Particular advantages arising from modification of OMS instead of conventional silicas are related to the structural organization of the former, high surface area (*ca.* $1000\,m^2/g$), greatly reactive surface,[52] and large mesopores, which could be tuned by judicious choice of templates. The accessibility of binding sites,[53,54] effects of confined geometry on the ligand layer formation, limited ligand bonding density[55] have been extensively studied to advance the design of OMM and improve their applicability for commercial uses[56] as well as to improve our understanding about formation of these materials.[57−59]

Post-Synthesis Modification of Template-Free OMS

OMS with monofunctional groups

An attractive approach to obtain high capacity adsorbents on the basis of OMS considers their functionalization with organic groups of specific properties for desired environmental applications. The resulting materials should possess functional groups over the entire surface capable to interact selectively with heavy metal ions. A high selectivity could be achieved by proper choice of functional group containing metal-chelating center of high affinity to a particular heavy metal ion. The Pearson's classification of soft acids and bases[60] suggests preferential binding of "Lewis soft acids" (e.g. mercury, cadmium, lead) to "soft bases" (sulfur), which is not interfered significantly by the presence of "hard acids" such as magnesium, calcium, and so on. The silane-type coupling agents possessing thiol groups were initially employed to obtain the OMS-based adsorbents for mercury ions; see the pioneering work from PNNL,[61] which initiated an extensive research on the design and synthesis of functionalized OMS for adsorption of heavy metal ions and other pollutants.

A conventional approach to decorate the pore walls of OMS with various functional groups, such as metal-chelating ligands, involves chemical reaction of the template-free OMS with organosilanes analogous to that used widely for the preparation of silica-based chemically bonded phases for chromatographic separations. Scheme 1 shows a simplified chart of the OMS synthesis followed by the template removal (usually calcination or extraction) and its post-synthesis modification involving the reaction of the template-free OMS with proper organosilanes. Namely, in this modification silanols present on the pore walls of OMS are replaced by desired surface ligands via single- or multi-step ligand attachment. Often a solvent (e.g. toluene), and/or catalyst (e.g. pyridine), or both are used to carry out the reaction of organosilane with silanol groups. Subsequently, the resulting material is extensively washed with polar and nonpolar solvents to eliminate the silane excess and reaction by-products. The main issue in this modification is to prevent the formation of silane multilayer on the mesopore walls instead of desired dense monolayer. The self-assembled monolayers of thiol functionality onto the mesopore walls of OMS afforded adsorbents with high specific surface area, in which ca. 75% of available silanols was replaced by thiol groups. An indicator of high adsorption affinity of the thiol-modified OMS towards mercury ions is a large value of the distribution coefficient, $K_d = 3.4 \times 10^5$. This coefficient is defined as the amount of the adsorbed metal in micrograms per gram of adsorbent divided by the concentration of the metal ion (in micrograms per milliliter) remaining in the eluent after adsorption experiment. The well-defined and open mesoporosity assured a fast diffusion and mass transfer of metal ions. Up to 0.6 g of Hg^{2+} per gram of thiol-functionalized OMS was reported.[62] The self-assembled monolayers on mesoporous supports (SAMMS) consisting of thiol functionality were shown to be efficient adsorbents for the removal of mercury ions from water and reduction of their concentration to that allowed by national pollution discharge emission (~ 12 ppt). Also, these materials passed the Toxicity Characterization Leaching Procedure tests of the U.S. Environmental Protection Agency, which manifest their potential usage in environmental clean-up.[63]

The MCM-41 and MCM-48 silicas exhibiting hexagonal or cubic arrangement of mesopores as well as SBA-15 with hexagonally arranged large mesopores interconnected by small micropores were extensively used to prepare highly effective adsorbents for heavy metal ions.[64,65] The study of mercaptopropyl-modified adsorbents prepared by using silica supports of different structures suggested that three-dimensional networks give better adsorbents for heavy metal ions (Hg^{2+} and Pb^{2+}).[66] It was possible to reduce Hg^{2+} content to the ppb level (with initial ppm Hg^{2+} concentration) when passing the model water through column packed with thiol-grafted mesoporous molecular sieve.[67] The factors affecting the desired characteristics of ideal heavy metal ion adsorbents, such as silanol density, channel dimensions, charge density and particle morphology, were studied to find the optimal properties of siliceous-supports that would assure their high performance in remediation processes.[68,69] At present, it is clear that inorganic-organic hybrid materials with functional monolayers have a great potential for environmental clean-up.[70,71] For instance, thiolated mesoporous solids showed a great selectivity for Hg^{2+} ions in presence of other metal ions (e.g. Cd^{2+}, Pb^{2+}, Zn^{2+}, Co^{3+}, Fe^{3+}, Cu^{2+} and Ni^{2+}) in various aqueous environments[72,73] and were reported to adsorb tetrahedral oxyanions of As, Cr and other relevant toxic metals; various radionuclides[74,75]; electroplating metals (e.g. Cu, Ni, Ag)[76]; actinides[77]; and noble metal ions (Pt^{2+}, Pd^{2+}).[78] Also, thiolated HMS materials showed prospective applicability for mining industry to recover rhodium ions from typical mine effluents containing also Cu, Ni and Zn elements; their high selectivity for rhodium ions in the presence of some reducing agents was observed, which permitted the recovery of these ions at ultra-low concentrations.[79] Also, incorporation of dithiocarbamate into the mesopores of MCM-41 afforded material with high adsorption capacity towards mercury ions too.[80] However, aminopropyl-modified mesoporous adsorbents were found to be capable of binding Cu^{2+}, Zn^{2+}, Cr^{3+} and Ni^{2+} ions,[65] while imidazole-functionalized nanostructures possessed great capability to preconcentrate noble metal ions.[81] The self-assembled mono-layers of carbamoylphosphonic acids, such as acetamide phosphonic acid and propionamide phosphonic acid, grafted onto the MCM-41 type of materials, were reported for adsorption of heavy and transition metal ions (Cd^{2+}, Co^{2+}, Cu^{2+}, Cr^{3+}, Pb^{2+}, Ni^{2+}, Zn^{2+}, and Mn^{2+}); adsorption equilibration in these systems was achieved within a very short period of time.[82]

OMS with multifunctional groups

Despite the fact that an unprecedented ligand coverage[61] allowed to achieve the maximum 1:1 heavy metal ion : ligand ratio[68] and high adsorption capacity, a main limitation of adsorbents with monofunctional groups is the ability to attract only one metal ion per ligand. In addition, the surface concentration of reactive silanols in ordered mesoporous silicas is usually smaller (often $\sim 30\%$) than that in conventional silicas.[55] Thus, the concentration of ligands in the monolayer formed on the mesopore walls of OMS is not high due to the smaller concentration of silanols, pore curvature and related geometrical limitations. Therefore, a fully accessible porosity (3D structure) is a key factor in the design of adsorbents with enhanced transport properties. An interesting approach, the so-called pore expansion method,[83] utilizing expanders such as *N*,*N*-dimethylalkylamine afforded the MCM-41 analogues with controlled pore sizes from 3.5 up to 25 nm and pore volumes from 0.8 up to 3.6 cm^3/g suitable for adsorption of Co^{2+}, Ni^{2+}, and Cu^{2+} ions and organic pollutants such as 4-chloroguaiacol or 2,6-dinitrophenol from water.[84] Also, their regeneration with alkylamines was possible.[84]

One of the possible alternatives to increase the adsorption capacity of the OMS-based adsorbents is to attach multifunctional ligands because they are able to bind more than one metal ion per ligand. Several multifunctional ligands such as 1-allyl-3-propylthiourea,[85] 1-benzoyl-3-propylthiourea,[86−88] and 2,5-dimercapto-1,3,4-thiadiazole ligands[89] were grafted on the meso-pores walls of MCM-41,[85,86] MCM-48[88] and SBA-15.[89] It was shown that those ligands possess ability to coordinate more than one metal ion per ligand; consequently, the adsorption capacity of OMS with those ligands may exceed one gram of mercury(II) per gram of the adsorbent or 5.0 mmol Hg^{2+}/g. While previously reported adsorbents obeyed Langmuir-type of adsorption,[90] the adsorption isotherm for mercury(II) ions on benzoyl-thiourea-modified OMS showed a more complex adsorption isotherm with two Langmuir-type constants of 1.41×10^5 L/mol and 1.08×10^8 L/mol indicating a two-step adsorption process. In other words, the aforementioned adsorbent showed two groups of sites of different adsorption energies. Over 70% of the initial adsorption capacity was retained after mild regeneration of this adsorbent.[86] A comparative study of 2D (MCM-41) and 3D (MCM-48) mesostructures with bonded benzoyl-thiourea ligands showed that the performance of 3D-type adsorbents is better than those having 2D structures.[88]

The attachment of more complex ligands to the pore walls of silica, especially those that are not commercially available in the form of coupling agents, can be performed via two or more reactions. Initially, the silica surface is modified by bonding organosilane that contains a group inert towards silica but capable to attach some specific molecular segments. This approach, which involves successive reactions, is especially convenient to build up large groups on the pore walls. In the case of large pores and good pore connectivity, a prior chemical modification of the silica surface may not be necessary because the entire surface ligand can be attached in a single step by using proper organosilane coupling agent, which is known as the homogeneous route of surface modification.[90] An example of this modification was attachment of 2-mercaptopyridine into the mesopore walls of MCM-41 and SBA-15, which afforded adsorbents for mercury ions that showed higher capacity than analogous materials prepared by multi-step modification (heterogeneous route).[92]

The multi-step surface modification provides much more flexibility for the design of surface functionality and avoids some undesired geometrical constrains. Initially, in this procedure some simple reactive groups such as chloroalkyl or aminoalkyl are attached to the mesopore walls via conventional chemical reaction of OMS with proper organosilanes, which is carried out in solvent and/or pyridine at elevated temperature under reflux conditions.[93] This initial modification of OMS with reactive groups is followed by another modification step, which involves reaction of these groups with proper chemical compounds in order to achieve the desired surface functionality. If the latter reaction proceeds to completion, the resulting material contains one type of surface groups only. In the case of an incomplete conversion of reactive groups, the resulting material contains some residual reactive groups in addition to the created multifunctional ligands. For instance, reaction of aminopropyl-OMS with benzoyl isothiocyanate resulted in $\sim 70\%$ conversion of aminopropyl groups into 1-benzoyl-3-propylthiourea ligands giving bifunctional OMS with benzoyl-propylthiourea ligands and residual aminopropyl groups.[86]

Surface Modification of OMS via Co-Condensation Synthesis

Incorporation of surface groups into OMS via direct co-condensation synthesis possesses numerous advantages over post-synthesis surface modification because it permits control of the hydrolysis and subsequent condensation of silica precursors such as TEOS or TMOS[94,95] and to achieve a high loading of surface groups. Either the surfactant-templated co-condensation synthesis[91,96−102] or block copolymer- directed structure formation[103−105] were used to obtain high capacity adsorbents for heavy metal ions.[59,106−110] The co-condensation synthesis is much simpler in comparison to the post-synthesis grafting because it reduces the number of synthesis steps. It permits the introduction of various reactive groups without pore blocking, which can be further derivatized to achieve the desired surface functionality. The co-condensation of primary (TEOS) and secondary (silane bearing reactive group) silica sources in the presence of structure-directing agents is also advantageous because it affords uniform distribution of these groups and simultaneously permits control of the structure and porosity of the resulting adsorbents.[111−114] For instance, a direct incorporation of chloropropyl or aminopropyl reactive groups via co-condensation synthesis affords OMS with those groups, a further modification of which is feasible. This approach was used to create cyclam-type functionality and manganese(III) Schiff-base complexes in OMS.[115,116] Some recent reports on the incorporation of acidic groups (phosphonic, humic) into mesoporous materials afforded effective adsorbents for chromium(III)[117] and cadmium(II) ions[118] from aqueous solutions. In the latter case the hybrid adsorbent was prepared by using a microwave-assisted synthesis, which is much faster than the conventional procedure and especially suitable for high-throughput screening of a wide range of precursors and experimental conditions for OMS-based adsorbents.[119−122]

To increase adsorption capacity of modified OMS, two or more surface groups could be introduced. Bifunctional OMS with aminopropyl and thiol groups synthesized via co-condensation were shown to have high adsorption capacity for several heavy metal ions.[123] The co-condensation synthesis afforded ordered mesostructures up to 60% of organic in the synthesis gel and worm-like mesostructures were obtained in the remaining concentration range up to 100% of organosilane.[124] Since the aforementioned synthesis provided adsorbents with high ligand loading and tailorable porosity, their adsorption performance was shown to be superior as evidenced by high values of the distribution coefficient ($ca.$ 10^7) for several heavy metal ions.

In conclusion, the co-condensation synthesis is a very simple and feasible approach to form a dense organic layer of functional groups and simultaneously create an ordered siliceous mesostructure of tailorable and accessible porosity. These mesostructures were found to be attractive not only in such traditional areas as adsorption, catalysis, separations and environmental clean up but also in nanotechnology and biotechnology (see articles[125–128] and references therein).

Functionalization of Template-Containing OMS

Simultaneous template displacement and attachment of functional groups into the pore walls of OMS is possible via so-called template displacement synthesis (TDS).[104,129–131] This method involves the removal (displacement) of organic templates from mesopores of the template-containing OMS accompanied by simultaneous attachment of desired surface ligands into the pore walls. This one-step process involving template removal and surface modification utilizes a high affinity of the template-containing OMS towards reactive organosilanes, which results in the replacement of self-assembled template-silica interface by a covalently bonded layer of organic groups on the silica surface. In contrast to the post-synthesis grafting of the template-free OMS, the TDS process has been found to be attractive for a direct incorporation of simple ligands into OMS that resulted in high affinity adsorbents for mercury ions[132] because it does not require prior template removal by extraction and/or calcination (see Scheme 1).

A combination of the co-condensation synthesis with template displacement process allows one to form more complex ligands inside mesopores of OMS. This approach was used to prepare high capacity adsorbents for mercury; and yielded adsorbents that contain multifunctional 2,5-dimercapto-1,3,4-thiadiazole ligand, which is capable of interacting with mercury ions via at least three active sites.[89] This work shows that a two-step post-synthesis modification (PSM) of the template-free OMS afforded materials with low surface area and small ligand coverage. In contrast, the one-pot (co-condensation) synthesis (OPS) of chloropropyl-modified OMS followed by either template displacement synthesis (TDS) or post-synthesis modification (PSM) afforded materials with large surface area, open porosity and high concentration of multifunctional ligands, which were able to adsorb up to 8.5 mmol of mercury(II) per gram of the adsorbent having ligand concentration of ~ 2.7 mmol/g. This high adsorption capacity suggests the presence of at least three binding sites for mercury(II) ions per attached ligand. The aforementioned adsorption capacity for the 2,5-dimercapto-1,3,4-thiadiazole-modified OMS studied is much higher than that for a conventional silica with analogous surface ligand.[133]

Periodic Mesoporous Organosilicas (PMO) as Adsorbents for Heavy Metal Ions

First successful attempts to incorporate some heteroatoms into silica framework were elaborated in early nineties.[134,135] This area of research is still growing because metal-containing silica mesostructures are of great importance for heterogeneous catalysis. However, a tremendous potential of framework-modified silicas has been demonstrated by the discovery of the so-called periodic mesoporous organosilicas (PMO) in 1999.[106,108,109,137] This "perfect marriage" of organic and inorganic chemistry had greatly accelerated the development of self-assembled mesostructured materials. In these materials not only a hybrid interface was created, but also a hybrid pore framework was achieved by co-condensation of organosilane precursors having the following general formula: $[(R'O)_3Si]_nR$, where R' refers to methyl or ethyl in hydrolysable alkoxy group and R is a functional group that constitutes an organic bridging unit within the inorganic framework of OMS. Thus, this organic group is homogeneously dispersed within the material framework. The co-condensation of $[(R'O)_3Si]_nR$ in the presence of surfactant or polymeric templates affords materials with high concentration of bridging groups, accessible porosity and tailored surface and structural properties. The chemical and structural properties of $[(R'O)_3Si]_nR$ precursor, especially its bridging group, predefine the location of this group within the material's framework. Numerous $[(R'O)_3Si]_nR$ precursors, mainly those with relatively simple bridging groups, are able to self-assemble into ordered mesostructures in the presence of proper surfactant or polymeric templates. However, the self-assembly of more complex precursors, especially those with large bridging groups, requires the addition of easily hydrolysable TEOS or TMOS in order to obtain organosilicas with ordered mesoporosity. Scheme 2 shows a general synthesis route to obtain periodic mesoporous organosilicas, including those with surface, surface/framework or framework functionalities. Depending on the nature of silica precursors, PMO materials with ultra low dielectric constant,[138] unique electron acceptor ability, photo- and thermochromic responsiveness,[139] chiral activity,[140] electrogenerated chemiluminescence,[141] photoactivity and many other unique properties[142] can be obtained. The easiness of their synthesis was a major driving force in the development of novel heavy metal ions adsorbents containing ligands that could not be introduced via conventional post-synthesis modification due to their bulky structure. For instance, cyclam units and dibenzo-18-crown-6 ether moieties that are capable of forming complexes with Cu(II) and alkali cations (Na$^+$ and/or K$^+$), respectively, were incorporated into mesoporous silicas.[143-145] Organosilanes with trialkoxysilyl groups containing amine functionality connected via linear, square, or tetrahedral bridges of coordination compounds of nickel, cadmium and zinc were used to synthesize hybrid mesoporous materials of various symmetry.[146] A novel method for the synthesis of lamellar silicas with high loading of amine groups was prepared by taking advantage of the reversible covalent binding of CO_2 to amines, where CO_2 gas acted as a structure directing agent. The reported aminosilicas possess high adsorption capacity towards transition metals and lanthanide ions such as Cu(II), Eu(III), and Gd(III) ions.[147] Another study showed that mono-, di-, and triaminofunctionalized MCM-41 and SBA-1 silicas adsorb oxyanions (chromate and arsenate) depending on the ligand density and type of OMS structure.[148] Hybrid materials that have capability to effectively bind heavy metal, inorganic, organic, charged, and neutral mercury compounds from not only aqueous solutions but also from other media such as oil or gas were reported too.[149]

A challenging aspect of introducing multifunctional organics into a small volume of mesopores was accomplished by co-condensation synthesis of PMO. Multifunctional PMO with various concentrations of organic bridging groups, morphology of the mesoporous network, and structural symmetry of the pores were reported in literature.[150-153] The co-condensation synthesis of organosilane mixtures[154,155] or co-condensation of precursors bearing multiple functionality[156] was investigated to develop PMO having multifunctional bridges within the silica framework. The main concern, arising from bulky nature of bridging groups employed in the synthesis, was a possible structure deterioration.[52,157-159] Recent efforts to design PMO with large bridging groups afforded nanomaterials with high surface area and ordered large pores.[160,161] Some large bridging groups such as 1,4,8,11-tetraazacyclotetradecane,[162] capable to adsorb heavy metal ions, were reported to provide materials with a small percentage of these groups and relatively poor structural ordering.[163] Nevertheless, the attempt to graft a metal-N-triethoxysilylpropylcyclam complex inside the mesopores with subsequent incorporation of another metal salt into the framework afforded bifunctional material and showed the possibility for designing the surface and framework properties of mesostructures.[164]

The design of framework composition of nanomaterials is an interesting and challenging area of research. Recently a PMO, obtained via one-pot synthesis of tris[3-(trimethoxysilyl)-propyl] isocyanurate as secondary silica precursor in the presence of TEOS and poly(ethylene oxide)-poly(propylene oxide)-poly(ethylene oxide) block copolymer, was reported.[165] A successful incorporation of this heterocyclic bridging group afforded highly ordered mesoporous materials with hexagonal symmetry, large surface area of $\sim 600\,m^2/g$ and large pores of ~ 8–$10\,nm$. The maximum adsorption of mercury(II) on these materials was $\sim 1.8\,g$ of Hg^{2+}. The corresponding distribution coefficients were about 10^8, which indicates high affinity of this bridging group towards mercury ions. Thioether functionalized organic-inorganic mesoporous materials,[166] synthesized by co-condensation of (1,4)-bis(triethoxysilyl)propane tetrasufide and tetraethoxysilane in the presence of block copolymer, poly(ethylene oxide)-poly(propylene oxide)-poly(ethylene oxide), showed also high selectively towards mercury ions in the presence of copper, cadmium, lead and zinc ions; their capacity was about $2.7\,g$ of Hg^{2+} per gram of adsorbent. The latter result shows a superior performance of PMO-based adsorbents; for instance, poly-thioether chelating resin would adsorb one-order lower amount of mercury ions.[167]

A great potential of multifunctional versus monofunctional adsorbents was demonstrated recently for bifunctional OMS with anchored ethylenediamine and thiol groups synthesized by combining direct and post-synthesis chemical modification methods, where supercritical fluid was chosen as reactive media for chemical grafting. Introduction of thiol functionality into ethylenediamine-modified MCM-41 materials generated a high capacity adsorbent for Cd^{2+} ions, which could be reused several times.[168]

Another study[169,170] shows the possibility of tailoring the interfacial and framework chemistry of PMO adsorbents by incorporation of two types of mercury-specific ligands. Introduction of thiol and isocyanurate ligands into PMO via co-condensation synthesis afforded high capacity adsorbent for mercury ions from aqueous solutions.[169] While the adsorption capacity of $1.8\,g\ Hg^{2+}/g$ was obtained for isocyanurate-containing organosilica with disordered mesoporous structure having high concentration of bridging groups, the bifunctional adsorbent with relatively low concentration of thiol and isocyanurate groups was structurally ordered and adsorbed $1.13\,g$ of Hg^{2+}/g.[171] However, introduction of an additional surface ligand into the PMO structure induced a structural change. Namely, the hexagonal ordering of channel-like mesopores in isocyanurate-containing PMO[170] changed to the cubic $I4_1 32$ symmetry group when additional mercaptopropyl surface groups were introduced during co-condensation.

Scheme 3 illustrates the concept of binding mechanism between heavy metal ions and multifunctional surface and bridging groups in PMO adsorbents. In this scheme, triangles and rectangles represent organic bridging groups in PMO that possess triple or quadruple binding sites for heavy

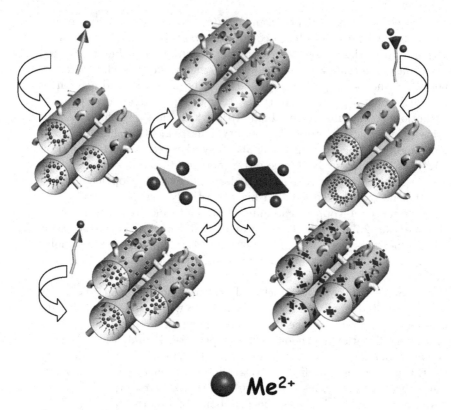

Scheme 3. Hypothetical models for binding of heavy metal ions (Me^{2+}) by various types of functional groups present in mesoporous organosilicas.

metal ions. The adsorption capacity of PMO adsorbents can be further enlarged by introduction of additional surface groups that are able to bind one or more heavy metal ions. Table 1 provides the list of metal-chelating ligands used to modify the siliceous mesopores, the methods used for their modification as well as adsorption capacities for many OMS-based adsorbents for heavy metal ions. The chemical structures of these ligands are shown in Schemes 4 and 5. Scheme 4 presents structures for the most common surface ligands, whereas Scheme 5 shows structures of the bridging groups used to design PMO adsorbents. The latter belong to a large group of hybrid organic-inorganic ordered nanostructures, which continue to gain popularity because of their potential applications in many fields of science and technology (see reviews[142,172-175] and references therein).

Table 1. List of functional ligands incorporated into ordered mesoporous materials via various synthesis methods and the corresponding metal ions adsorption capacities reported in literature.

Ligand	Modification Method	Reference	Metal Ion	Adsorption (mmol/g)
(3-Mercaptopropyl)-trimethoxysilane	PSM	61	Hg^{2+}	1.04
		62	Hg^{2+}	2.99
		63	Hg^{2+}	3.04
		65	Hg^{2+}	1.46
		67	Hg^{2+}	1.54
		70	Hg^{2+}	2.99
		68	Hg^{2+}	1.5
Dithiocarbamate	PSM	80	Hg^{2+}	0.2
N-(3-triethoxysilylpropyl)-4,5-dihydroimidazole	PSM	81	Pt^{2+}	0.09
			Pd^{2+}	0.09
Acetamide phosphonic acid	PSM	82	Cd^{2+}	0.32
Propionamide phosphonic acid	PSM	82	Cd^{2+}	0.32
N,N-dimethyldecylamine	PSM	84	Co^{2+}	1.05
			Ni^{2+}	0.93
			Cu^{2+}	1.67
1-Allyl-3-propylthiourea	PSM	85	Hg^{2+}	1.49
1-Benzoyl-3-propylthiourea	PSM	86	Hg^{2+}	4.98
		88	Hg^{2+}	6.7
3-Trimethoxypropyl-thioethylamine	PSM	90	Co^{2+}	1.08(0.72)
			Ni^{2+}	1.20(1.74)
			Cu^{2+}	1.70(1.91)
			Pb^{2+}	1.34(2.19)
			Hg^{2+}	4.02(2.89) homogeneous (heterogeneous synthesis route)
2-Mercaptopyridine	PSM	92	Hg^{2+}	0.16
1,4,8,11-Tetraazacyclo-tetradecane	PSM	115	Cu^{2+}	0.77
			Co^{2+}	0.80
(3-Mercaptopropyl)-trimethoxysilane	OPS	91	Hg^{2+}	1.26
		94	Hg^{2+}	2.1
		97	Hg^{2+}	2.3

Table 1. (*Continued*)

Ligand	Modification Method	Reference	Metal Ion	Adsorption (mmol/g)
		98	Hg^{2+}	1.49
		99	Cd^{2+}	0.20
		100	Hg^{2+}	1.37
		112	Hg^{2+}	0.59
Aminopropyl, [amino-ethylamino]propyl, [(2-aminoethylamino)-ethylamino]propyl	OPS	99	Cu^{2+} Ni^{2+} Co^{2+} Cd^{2+}	0.55 0.15 0.10 0.01
3-(2-Aminoethylamino)-propyltrimethoxysilane	OPS	114	Cd^{2+} Zn^{2+}	0.51 0.57
3-Aminopropyltrimethoxysilane	OPS	148	Chromate Arsenate	1.19 0.25
[1-(2-Aminoethyl)-3-aminopropyl]trimethoxysilane	OPS	148	Chromate Arsenate	0.27 0.28
(Trimethoxysilyl)-propyldiethylenetriamine	OPS	148	Chromate Arsenate	0.54 0.36
1-Allyl-3-propylthiourea	TDS	132	Hg^{2+}	0.80
2,5-Dimercapto-1,3,4-thiadiazole	TDS	89	Hg^{2+}	8.47
Trimethoxysilylpropyl diethylphosphonate	PMO	117	Cr^{2+}	1.58
Humic acid	PMO	118	Cd^{2+}	0.002
3-Aminopropyltriethoxysilane and 3-mercaptopropyltri-methoxysilane	PMO	123	Hg^{2+}	1.51
1,4,8,11-Tetrakis[3-(triethoxysilyl)-propyl]-1,4,8,11-tetraaza-cyclotetradecane	PMO	144	Cu^{2+}	1.31
N-(2-aminoethyl)-3-aminopropyltrimethoxysilane and N-(6-aminohexyl)-3-aminopropyl-trimethoxysilane	PMO	147	Cu^{2+} Eu^{3+}	6.5 4.8
(1,4)-Bis(triethoxysilyl)propane tetrasufide	PMO	166	Hg^{2+}	13.5
Tris[3-(trimethoxysilyl)propyl] isocyanurate	PMO	170	Hg^{2+}	8.9
Tris[3-(trimethoxysilyl)propyl] isocyanurate and (3-mercaptopropyl)-trimethoxysilane	PMO	169 171	Hg^{2+}	5.6

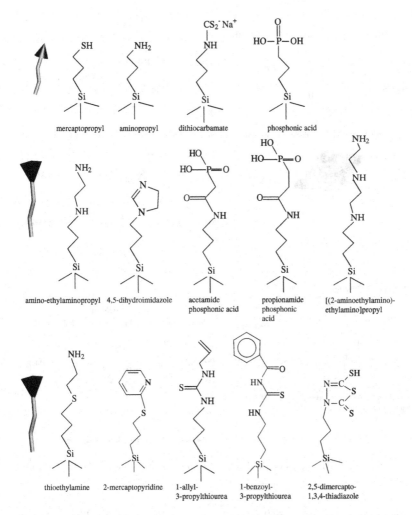

Scheme 4. Chemical structures of the metal-chelating ligands used for surface modification of OMS silicas and co-condensation synthesis of PMO.

A Brief Overview of Characterization of Mesostructured Adsorbents

Several techniques are used for characterization of adsorption, surface and structural properties of mesoporous organosilicas. One of the key issues in this characterization is the identification of functional groups present on the silica surface and/or in its framework. It is important to know the chemical

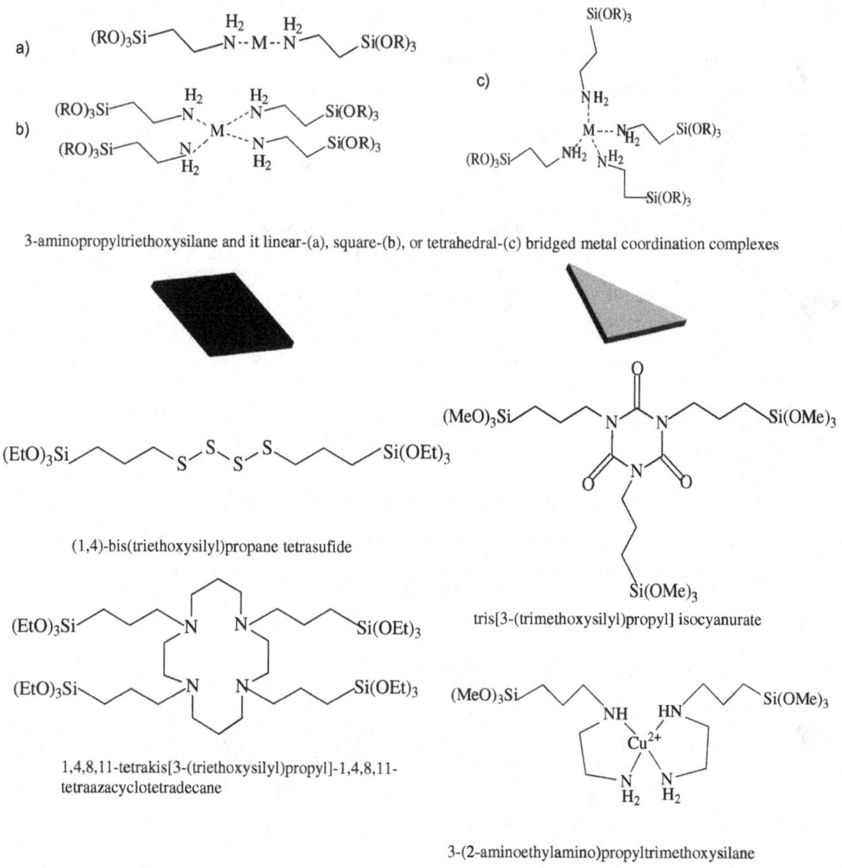

3-aminopropyltriethoxysilane and it linear-(a), square-(b), or tetrahedral-(c) bridged metal coordination complexes

(1,4)-bis(triethoxysilyl)propane tetrasufide

tris[3-(trimethoxysilyl)propyl] isocyanurate

1,4,8,11-tetrakis[3-(triethoxysilyl)propyl]-1,4,8,11-tetraazacyclotetradecane

3-(2-aminoethylamino)propyltrimethoxysilane

Scheme 5. Chemical structures of the bridging groups used in the synthesis of PMO-based adsorbents.

structure of these groups as well as the type of bonding involved, their location (pore walls, external surface, framework) and concentration. Depending on the type of adsorbent, its surface functionality is usually characterized by a combination of techniques such as elemental analysis, Fourier-Transform Infrared (FTIR) spectroscopy, solid state Nuclear Magnetic Resonance (NMR), high resolution thermogravimetry (TGA) and adsorption. Finally, physicochemical properties of surface groups present and their accessibility determine the adsorbent's selectivity, adsorption capacity and its ability for regeneration and reuse.

The structural ordering of pores (hexagonal, cubic, lamellar), pore size and shape (cylindrical, cage-like, slit-like) as well as material's morphology are usually accessed by Transmission Electron Microscopy (TEM), Scanning Electron Microscopy (SEM), Atomic Force Microscopy (AFM), X-ray diffraction (XRD), Small Angle X-ray Scattering (SAXS) and related techniques. Although all aforementioned techniques provide extremely valuable information about structural ordering of porous networks and surfaces, they are less convenient for assessment of the overall quality of nanoporous adsorbents. For instance, the alteration of porosity upon chemical modification, which is manifested by changes in the pore size, pore accessibility, pore size distribution as well as pore chemistry, can be effectively monitored by adsorption. This technique was shown to be an indispensable tool for the evaluation of the specific surface area, total pore volume, volume of micropores, pore size distribution, pore connectivity, surface heterogeneity and adsorption energy. Ordered mesoporous solids represent a family of nanoporous materials that exhibit characteristic types of nitrogen adsorption isotherms and hysteresis loops depending on their structural and surface properties. The relative pressure, at which monolayer of an adsorbate is formed, is commonly used for evaluation of the specific surface area.[176] The standard procedure, so-called BET method,[177] employs nitrogen adsorption data at 77 K at the relative pressure range from ~ 0.05 to ~ 0.3 to evaluate the monolayer capacity according to the BET equation, which after multiplication by the molecular area of nitrogen is converted to the specific surface area of the adsorbent studied. The total pore volume is calculated by the conversion of the amount adsorbed at the relative pressure about 0.99 to the volume of the liquid adsorbate. The presence and shape of hysteresis loop provides valuable information about adsorbent's porosity including size and shape of pores as well as pore connectivity.

Argon and nitrogen adsorption is a technique of choice to monitor the changes in the porous structure upon surface modification. For instance, the reduction in the pore size upon silanization indicates a successful attachment of ligands into the mesopore walls. A chemically bonded layer inside mesopores reduces their size, which is manifested by a shift of the capillary condensation step in direction of lower relative pressures; also, its height is reduced indicating a decrease in the volume of primary mesopores. Thus, the pore width can be easily tuned by selecting proper length of the attached ligand. The steepness of the capillary condensation step provides indication about pore size uniformity; steep capillary condensation step reflects narrow pore size distribution. This distribution can be evaluated from adsorption isotherms by using different methods, for example the KJS method,[178] which is based on the BJH algorithm[179] and was developed for hexagonally ordered mesostructures (MCM-41). This method is especially suited for the evaluation of pore size distribution in the range of small mesopores.

Figures 1 to 3 show typical adsorption isotherms and the corresponding pore size distributions for various types of mesoporous organosilicas. Nitrogen adsorption isotherms for various types of template-free (calcined) OMS are shown in Figure 1. The corresponding adsorption isotherms for those silicas subjected to the post-synthesis modification with various organosilanes are presented in Figure 2. Namely, this figure shows isotherms for mesoporous organosilicas with 3-mercaptopropyl (SH), 1-allyl-3-propylthiourea (ATU) and 1-benzoyl-3-propylthiourea (BTU) ligands. For 2,5-dimercapto-1,3,4-thiadiazole (DMT)-modified material, the template displacement synthesis was used as illustrated in Scheme 2. The second set of adsorption isotherms corresponds to the materials obtained via co-condensation synthesis (Figure 3), which feature either one type of framework groups, tris[3-(trimethoxysilyl)propyl] isocyanurate (ICS), or two types of groups, 3-mercaptopropyl surface ligands and tris[3-(trimethoxysilyl)propyl] isocyanurate framework groups. As can be seen from Figures 1 and 2, the post-synthesis surface modification of OMS can be easily monitored by the shift of the capillary condensation step in direction of lower relative pressures, which indicates a successful immobilization of surface liagnds inside mesopores. In contrast to the post-synthesis modification, for the materials obtained by co-condensation synthesis the pore size is mainly determined by the type of template and its dependence on the size and concentration of incorporated groups is much less pronounced. Therefore, organosilicas obtained by co-condensation synthesis feature high

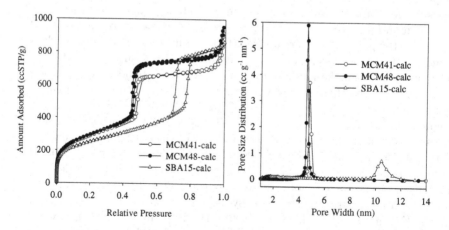

Figure 1. Nitrogen adsorption isotherms and the corresponding pore size distributions for calcined ordered mesoporous silicas of different structures before surface functionalization. Adsorption data from Refs. 86 and 88.

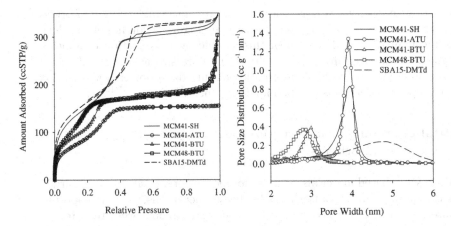

Figure 2.   Nitrogen adsorption isotherms and the corresponding pore size distributions for ordered mesoporous silicas after surface functionalization with multifunctional ligands: 3-mercaptopropyl (SH), 1-allyl-3-propylthiourea (ATU), 1-benzoyl-3-propylthiourea (BTU), and 2,5-dimercapto-1,3,4-thiadiazole (DMT). Adsorption data from Refs. 85, 86, 88 and 89.

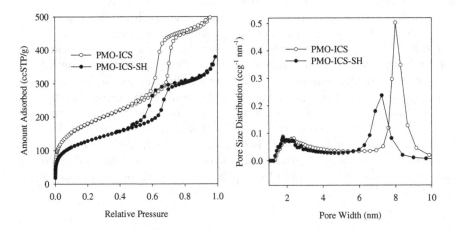

Figure 3.   Nitrogen adsorption isotherms and the corresponding pore size distributions for periodic mesoporous organosilicas with multifunctional bridging and surface groups: 3-mercaptopropyl (SH) and tris[3-(trimethoxysilyl)propyl] isocyanurate (ICS). Adsorption data from Refs. 165 and 169.

adsorption capacity and large pore size despite of the high content of bridging groups; e.g. see adsorption isotherm in Figure 2 for modified silica. However, further modification of this sample by introduction of another segment to chloropropyl ligand reduces the pore size and pore volume of the sample. A more detailed information about adsorption characterization of mesoporous materials can be found elsewhere.[86−89,118,165,169−171]

Metal-Containing Nanostructures as Heavy Metal Ions Adsorbents

While silica-based materials such as mesoporous organosilicas are regarded as the most promising adsorbents for heavy metal ions, other nanoporous materials such as surface-modified titania and zirconia were reported as promising adsorbents too.[180] Another example is MCM-41 with Fe^{3+}-coordinated chelating ligands, which possesses a high adsorption capacity for toxic oxyanions such as arsenate, chromate, selenate, and molybdate.[181]

Nanostructured metals or metal-containing materials were also reported as potential heavy metal ions adsorbents. Poly(1-vinylimidazole)-grafted magnetic (maghemite, γ-Fe_2O_3) nanoparticles[182] showed selective binding of divalent metal ions in the following order: $Cu^{2+} \gg Ni^{2+} > Co^{2+}$. Hexagonally ordered zirconia[183] was shown to have high affinity towards Co(II), Ni(II), Cu(II), and Zn(II) ions. Titania-incorporated mesoporous composites[184,185] featured a remarkable adsorption affinity towards heavy metal ions such as Pb(II) and mercury, with particular potential to control a vapor of elemental mercury[186,187] and other divalent metals.[188] Some advantages of employing mesostructured materials over conventionally-used La(III)-impregnated alumina and silica gels for arsenate removal were shown for SBA-15 material with incorporated various amounts of lanthanide oxide, for which about 10–14 times larger adsorption capacity was obtained.[189] Also, porous mixed-oxides with tailored properties derived from substituted layered double hydroxides were tested for adsorption of this pollutant.[190]

Mesostructures for Detection of Heavy Metal Ions

Since co-condensation synthesis of ordered mesoporous organosilicas (OMO) permits the introduction of a high amount of functional groups without blockage of porous structure and significant reduction of the surface area and porosity in the resulting materials, their use in sensing devices allow one to lower detection limits and enhance response for the targeted substances. Therefore, there is a great interest in the development of OMS-based materials for detection of heavy metal ions. For instance, high surface area and large pore size of chemically modified SBA-15 materials with 5,10,15,20-tetraphenylporphinetetrasulfonic acid, anchored by N-trimethoxysilylpropyl-N,N,N-trimethylammonium chloride,[191] were capable to detect Hg^{2+} at concentrations as low as 2.5×10^{-8} M. Also, a high detection limit (3.3×10^{-7} mol L^{-1} in water) and great selectivity for mercury ions was achieved in the presence of interfering cations such as Na^+, K^+, Ca^{2+}, Cu^{2+}, Cd^{2+}, Pb^{2+} on chemically modified SBA-15 material, in which calixarene bearing two dansyl fluorophores was an optically-sensing moiety.[192] Mesoporous materials were also found to be very efficient in the determination of heavy metal ions via electrochemical methods. For example, a MCM-41-modified graphite paste electrode gave high selectivity for hydroxylated mercury(II) species.[193] A good voltammetric response for Hg(II), Pb(II), Ag(I) and Cu(II) cations was observed for mesoporous silica possessing thick walls.[194] The thiol monolayers grafted in silica mesopores, known for their exceptional affinity and selectivity for heavy metal ions, were used to prepare the electrode sensing layer for detection of lead(II) in aqueous solutions. These electrodes exhibited fast responses in the parts per billion (ppb) range of concentration when square wave adsorptive stripping voltammetry technique was employed.[195] However, a simultaneous detection of lead(II) and mercury(II) in aqueous solutions was possible with detection limits of 0.5 ppb for Pb^{2+} and 3 ppb for Hg^{2+} when a self-assembled monolayer of thiol groups on mesoporous silica combined with carbon paste was used as the electrode.[196] Glycinylurea and carbamoylphosphonic self-assembled monolayers on mesoporous silicas anchored to the carbon paste electrodes were shown to have detection limits of 1 ppb and fast detection time of 2–3 minutes for lead, mercury, copper and cadmium ions using adsorptive stripping voltammetry.[197,198] The aforementioned studies distinctly show a great benefit of using functionalized ordered mesoporous materials for the development of sensing devices to detect heavy metal ions in water streams. Their high surface area, non-swelling and hydrothermally stable matrix, accessible mesopores with highly-specific ligands assure high selectivity and capacity towards heavy metal ions; moreover tunable width of mesopores prevents adsorption of bulk organics from waste waters.

Mesoporous Materials for Capturing Toxic Pollutants other then Heavy Metals

The preceding sections described the great potential of properly-designed ordered mesoporous organosilicas for adsorption and detection of heavy metal ions. It is noteworthy that these materials are also attractive for adsorption of other pollutants such as dioxins,[198] sulfides[200] and various volatile organic compounds (VOC) including carbon tetrachloride, *n*-hexane, benzene, toluene, and *o*-xylene[201,202]; 4-n-heptylaniline and 4-nonylphenol, endocrine disrupters,[203,204] alkylphenols and alkylanilines,[205] chlorinated hydrocarbons,[206] cancerogenic nitrosamines[207,208] and other VOC.[209]

Prospective

This review shows that both surface modification of ordered silica mesostructures as well as co-condensation synthesis of organosilanes in the presence of surfactant and polymeric templates can be used to design highly selective and high capacity adsorbents for heavy metal ions. Especially, the latter method is feasible for the design and synthesis of adsorbents with high loading of surface and bridging groups without sacrificing their surface area, pore size and pore volume. Also, the co-condensation synthesis in the presence of properly selected templates is the method of choice to generate mesostructures with highly accessible porosity.

Novel sources of non-toxic and biodegradable precursors and low cost synthesis would play a key role in the development of methods for large scale production of future materials, including those for environmental applications. Since nanocomposite and hybrid materials are mainly built-up via "bottom-up" approach, there is a great deal of attention to assure that the resulting nanomaterials do not present a health risk and they are environmentally friendly, i.e. their synthesis and use fulfill requirements of green chemistry principles and safer technological processes. Since a major feature of nanomaterials is the possibility to introduce a variety of unique properties into confined nanospace, this research area is expected to grow rapidly in the coming years, especially in relation to advanced applications of nanomaterials.

Acknowledgment

The NSF support is acknowledged under grant CTS-0086512.

References

1. J. S. Beck, J. C. Vartuli, W. J. Roth, M. E. Leonowicz, C. T. Kresge, K. D. Schmitt, C. T.-W. Chu, D. H. Olson, E. W. Sheppard, S. B. McCullen, J. B. Higgins and J. L. Schlenker, *J. Am. Chem. Soc.* **114**, 10834–10843 (1992).

2. B. Hultberg, A. Andersson and A. Isaksson, *Toxicology* **126**, 203–212 (1998).

3. E. S. Raper, *Coordination Chemistry Reviews* **61**, 115–184 (1985).

4. R. V. Davies, J. Kennedy, E. S. Lane and J. L. Willans, *J. Appl. Chem.* **9**, 368–371 (1959).

5. C. Kantipuly, *Talanta* **37**, 491–517 (1990).

6. S. D. Alexandratos and D. L. Wilson, *Macromolecules* **19**, 280–287 (1986).

7. A. R. Tuker and A. Tunceli, *F. J. Anal. Chem.* **345**, 755–758 (1993).

8. A. Deratani and B. Sebille, *Anal. Chem.* **53**, 1742–1746 (1981).

9. R. J. Philips and J. S. Fritz, *Anal. Chem.* **50**, 1504–1508 (1978).

10. M. Griesbach and K. H. Lieser, *Angew. Makromol. Chem.* **90**, 143–153 (1980).

11. R. Liu, H. Tang and B. Zhang, *Chemosphere* **38**, 3169–3179 (1999).

12. R. Celis, M. Carmen Hermosin and J. Cornejo, *Environ. Sci. Technol.* **34**, 4593–4599 (2000).

13. S. Babel and T. A. Kurniawan, *J. Hazard. Mater. B* **97**, 219–243 (2003).

14. D. Mohan, V. K. Gupta, S. K. Srivastava and S. Chander, *Coll. Surfaces A: Physicochem. Eng. Aspects* **177**, 169–181 (2001).

15. M. J. Zamzow, B. R. Eichbaum, K. R. Sandgren and D. E. Shanks, *Separation Sci. Technology* **25**, 1555–1569 (1990).

16. M. M. Matlock, K. R. Henke and D. A. Atwood, *J. Hazard. Mater.* **92**, 129–142 (2002).

17. T. I. Tikhomirova, V. I. Fadeeva, G. V. Kudryavtsev, P. N. Nesterenko, V. M. Ivanov, A. T. Savitchev and N. S. Smirnova, *Talanta* **38**, 267–274 (1991).

18. H. Watanabe, K. Goto, S. Taguchi, J. W. McLaren, S. S. Berman and D. S. Russell, *Anal. Chem.* **53**, 738–739 (1981).

19. A. G. S. Prado, L. N. H. Arakaki and C. Airoldi, *Dalton Trans.* 2206–2209 (2001).

20. G.-Z. Fang, J. Tan and X.-P. Yan, *Anal. Chem.* **77**, 1734–1739 (2005).

21. E. F. Vansant, P. Van Der Voort and K. C. Vrancken (eds.), *Characterization and Chemical Modification of the Silica Surface*, Studies in Surface Science and Catalysis, Vol. 93 (1995), pp. 572.

22. G. V. Lisichkin, G. V. Kudryavtsev, A. A. Serdan, S. M. Staroverov and A. Ya. Yuffa, *Modified Silicas in Sorption, Catalysis, and Chromatography* (1986), pp. 248.

23. P. K. Jal, S. Patel and B. K. Mishra, *Talanta* **62**, 1005–1028 (2004).

24. J. Rouquerol, D. Avnir, C. W. Fairbridge, D. H. Everett, J. H. Haynes, N. Pernicone, J. D. F. Ramsay, K. S. W. Sing and K. K. Unger, *Pure Appl. Chem.* **66**, 1739–1758 (1994).

25. C. T. Kresge, M. E. Leonowicz, W. J. Roth, J. C. Vartuli and J. Beck, *Nature* **359**, 710–712 (1992).

26. J. S. Beck, J. C. Vartuli, W. J. Roth, M. E. Leonowicz, C. T. Kresge, K. D. Schmitt, C. T. W. Chu, D. H. Olson, E. W. Sheppard *et al.*, *J. Am. Chem. Soc.* **114**, 10834–10843 (1992).

27. C. Chen Yan, H. X. Li and M. E. Davis, *Microporous Materials* **2**, 17–26 (1993).

28. Q. Huo, R. Leon, P. M. Petroff and G. D. Stucky, *Science* **268**, 1324–1327 (1995).
29. Q. Huo, D. I. Margolese, U. Ciesla, D. G. Demuth, P. Feng, T. E. Gier, P. Sieger, A. Firouzi, B. F. Chmelka *et al.*, *Chem. Mater.* **6**, 1176–1191 (1994).
30. Q. Huo, D. I. Margolese, U. Ciesla, P. Feng, T. E. Gier, P. Sieger, R. Leon, P. M. Petroff, F. Schueth and G. D. Stucky *Nature* **368**, 317–321 (1994).
31. P. T. Tanev and T. J. Pinnavaia, *Science* **267**, 865–867 (1995).
32. D. Zhao, J. Feng, Q. Huo, N. Melosh, G. H. Frederickson, B. F. Chmelka and G. D. Stucky, *Science* **79**, 548–552 (1998).
33. D. Zhao, Q. Huo, J. Feng, B. F. Chmelka and G. D. Stucky, *J. Am. Chem. Soc.* **120**, 6024–6036 (1998).
34. S. Inagaki, Y. Fukushima and K. Kuroda, *Chem. Commun.* 680–682 (1993).
35. A. Monnier, F. Schuth, Q. Huo, D. Kumar, D. Margolese, R. S. Maxwell, G. D. Stucky, M. Krishnamurty, P. Petroff *et al.*, *Science* **261**, 1299–303 (1993).
36. K. Schumacher, M. Grun and K. K. Unger, *Microporous Mesoporous Mater.* **27**, 201–206 (1999).
37. C. Yu, Y. Yu and D. Zhao, *Chem. Commun.* 575–576 (2000).
38. Q. Huo, D. I. Margolese and D. Stucky Galen, *Chem. Mater.* **8**, 1147–60 (1996).
39. J. S. Beck and J. C. Vartuli, *Cur. Opinion Solid State Mater. Sci.* **1**, 76–87 (1996).
40. C. G. Goltner and M. Antonietti, *Adv. Mater.* **9**, 431–436 (1997).
41. C. Yu, Y. Yu, L. Miao and D. Zhao, *Microporous Mesoporous Mater.* **44/45**, 65–72 (2001).
42. D. Y. Zhao, P. D. Dong, N. Melosh, F. Y. Lin, B. F. Chmelka and G. Stucky, *Adv. Mater.* **10**, 1380–1385 (1998).
43. P. Yang, D. Zhao, B. F. Chmelka and G. D. Stucky, *Chem. Mater.* **10**, 2033–2036 (1998).
44. C. Yu, B. Tian, J. Fan, G. D. Stucky and D. Zhao, *J. Am. Chem. Soc.* **124**, 4556–4557 (2002).
45. J. R. Matos, L. P. Mercuri, M. Kruk and M. Jaroniec, *Langmuir* **18**, 884–890 (2002).
46. P. Van Der Voort, M. Benjelloun and E. F. Vansant, *J. Phys. Chem. B* **106**, 9027–9032 (2002).
47. C. E. Tattershall, S. J. Aslam and P. M. Budd, *J. Mater. Chem.* **12**, 2286–2291 (2002).
48. J. R. Matos, M. Kruk, L. P. Mercuri, M. Jaroniec, L. Zhao, T. Kamiyama, O. Terasaki, T. J. Pinnavaia and Y. Liu, *J. Am. Chem. Soc.* **125**, 821–829 (2003).
49. G. S. Attard, J. C. Glyde and C. G. Goltner, *Nature* **378**, 366–368 (1995).
50. T. Linssen, K. Cassiers, P. Cool and E. F. Vansant, *Adv. Colloid Interface Sci.* **103**, 121–147 (2003).
51. A. Sayari, *Chem. Mater.* **8**, 1840–1852 (1996).

52. A. Vinu, K. Z. Hossain, and K. Ariga, *J. Nanosci. Nanotech.* **5**, 347–371 (2005).
53. A. Walcarius, M. Etienne and J. Bessiere, *J. Chem. Mater.* **14**, 2757–2766 (2002).
54. A. Walcarius, M. Etienne and B. Lebeau, *Chem. Mater.* **15**, 2161–2173 (2003).
55. B. P. Feuston and J. B. Higgins, *J. Phys. Chem.* **98**, 4459–4462 (1994).
56. S. Polarz and B. Smarsly, *J. Nanosci. Nanotech.* **2**, 581–612 (2002).
57. U. Ciesla and F. Schuth, *Microporous Mesoporous Mater.* **27**, 131–149 (1999).
58. J. Y. Ying, C. P. Mehnert and M. S. Wong, *Angew. Chemie Int. Ed.* **38**, 56–77 (1999).
59. A. Sayari and S. Hamoudi, *Chem. Mater.* **13**, 3151–3168 (2001).
60. R. G. Pearson, *J. Am. Chem. Soc.* **85**, 3533–3539 (1963).
61. X. Feng, G. E. Fryxell, L.-Q. Wang, A. Y. Kim, J. Liu and K. M. Kemner, *Science* **276**, 923–926 (1997).
62. J. Liu, X. D. Feng, G. E. Fryxell, L. Q. Wang, A. Y. Kim and M. L. Gong, *Adv. Mater.* **10**, 161–165 (1998).
63. X. Chen, X. Feng, J. Liu, G. E. Fryxell and M. Gong, *Separ. Sci. Tech.* **34**, 1121–1132 (1999).
64. K. Moller and T. Bein, *Chem. Mater.* **10**, 2950–2963 (1998).
65. A. M. Liu, K. Hidajat, S. Kawi and D. Y. Zhao, *Chem. Commun.* 1145–1146 (2000).
66. Y. Kim, B. Lee and J. Yi, *Separation Sci. Technol.* **39**, 1427–1442 (2004).
67. L. Mercier and T. J. Pinnavaia, *Adv. Mater.* **9**, 500–503 (1997).
68. L. Mercier and T. J. Pinnavaia, *Envir. Sci. Technol.* **32**, 2749–2754 (1998).
69. K. M. Kemner, X. Feng, J. Liu, G. E. Fryxell, L.-Q. Wang, A. Y. Kim, M. Gong and S. Mattigod, *J. Synchrotron Radiation* **6**, 633–635 (1999).
70. J. Liu, X. Feng, G. E. Fryxell, L. Q. Wang, A. Y. Kim and M. Gong, *Chem. Eng. Technol.* **21**, 97–100 (1998).
71. Z. Luan, J. A. Fournier, J. B. Wooten, D. E. Miser and M. J. Chang, *Stud. Surf. Sci. Catal.* **156**, 897–906 (2005).
72. S. V. Mattigod, X. Feng, G. E. Fryxell, J. Liu and M. Gong, *Separ. Sci. Technol.* **34**, 2329–2345 (1999).
73. C. Lesaint, F. Fré bault, C. Delacôte, B. Lebeau, C. Marichal, A. Walcarius and J. Patarin, *Stud. Surf. Sci. Catal.* **156**, 925–932 (2005).
74. G. E. Fryxell, J. Liu, T. A. Hauser, Z. Nie, K. F. Ferris, S. Mattigod, M. Gong and R. T. Hallen, *Chem. Mater.* **11**, 2148–2154 (1999).
75. G. E. Fryxell, J. Liu and S. Mattigod, *Mater. Technol.* **14**, 188–191 (1999).
76. G. E. Fryxell, J. Liu, S. V. Mattigod, L. Q. Wang, M. Gong, T. A. Hauser, Y. Lin, K. F. Ferris and X. Feng, *Ceramic Transactions* **107**, 29–37 (2000).
77. G. Fryxell, Y. S. Fiscum, Y. C. Birnbaum and H. Wu, *Environ. Sci. Technol.* **39**, 1324–1331 (2005).
78. T. Kang, Y. Park and J. Yi, *Ind. Engineer. Chem. Research* **43**, 1478–1484 (2004).

79. A. Abughusa, L. Amaratunga and L. Mercier, *Stud. Surf. Sci. Catal.* **156**, 957–962 (2005).

80. K. A. Venkatesan, T. G. Srinivasan and P. R. Vasudeva Rao, *J. Radioanal. Nuclear Chem.* **256**, 213–218 (2003).

81. T. Kang, Y. Park, K. Choi, J. S. Lee and J. Yi, *J. Mater. Chem.* **14**, 1043–1049 (2004).

82. W. Yantasee, Y. Lin, G. E. Fryxell, B. J. Busche and J. C. Birnbaum, *Separation Sci. Technol.* **38**, 3809–3825 (2003).

83. A. Sayari, *Angew. Chemie Int. Ed.* **39**, 2920–2922 (2000).

84. A. Sayari, S. Hamoudi and Y. Yang, *Chem. Mater.* **17**, 212–216 (2005).

85. V. Antochshuk and M. Jaroniec, *Chem. Commun.* 258–259 (2002).

86. V. Antochshuk, O. Olkhovyk, M. Jaroniec, I.-S. Park and R. Ryoo, *Langmuir* **19**, 3031–3034 (2003).

87. O. Olkhovyk, V. Antochshuk and M. Jaroniec, *Analyst* **130**, 104–108 (2005).

88. O. Olkhovyk, V. Antochshuk and M. Jaroniec, *Coll. Surfaces A.* **236**, 69–72 (2004).

89. O. Olkhovyk and M. Jaroniec, *Adsorption* **11**, 205–214 (2005).

90. A. G. S. Prado, L. N. H. Arakaki and C. Airodi, *Green Chemistry* **4**, 42–46 (2002).

91. R. I. Nooney, M. Kalyanaraman, G. Kennedy and E. J. Maginn, *Langmuir* **17**, 528–533 (2001).

92. D. Pérez-Quintanilla, I. del Hierro, M. Fajardo and I. Sierra, *Microporous Mesoporous Mater.* 2005, in press.

93. C. P. Jaroniec, M. Kruk, M. Jaroniec and A. J. Sayari, *Phys. Chem. B* **102**, 5503–5510 (1998).

94. M. H. Lim, C. F. Blanford and A. Steinm, *Chem. Mater.* **10**, 467–470 (1998).

95. W. M. Van Rhijn, D. E. De Vos, B. F. Sels, W. D. Bossaert and P. A. Jacobs, *Chem. Commun.* **1998**, 317–318.

96. Y. Mori and T. J. Pinnavaia, *Chem. Mater.* **13**, 2173–2178 (2001).

97. J. Brown, R. Richer and L. Mercier, *Microporous Mesoporous Mater.* **37**, 41–48 (2000).

98. A. Bibby and L. Mercier, *Chem. Mater.* **14**, 1591–1597 (2002).

99. L. Bois, A. Bonhomme, A. Ribes, B. Pais, G. Raffin and F. Tessier, *Coll. Surfaces A* **221**, 221–230 (2003).

100. H.-J. Im, C. E. Barnes, S. Dai and Z. Xue, *Microporous Mesoporous Mater.* **70**, 57–62 (2004).

101. S. Huh, J. W. Wiench, J.-C. Yoo, M. Pruski and V. S.-Y. Lin, *Chem. Mater.* **15**, 4247–4256 (2003).

102. K. Kosuge, T. Murakami, N. Kikukawa and M. Takemori, *Chem. Mater.* **15**, 3184–3189 (2003).

103. D. Margolese, J. A. Melero, S. C. Christiansen, B. F. Chmelka and G. D. Stucky, *Chem. Mater.* **12**, 2448–2459 (2000).

104. H.-P. Lin, L.-Y. Yang, C.-Y. Mou, S.-B. Liu and H.-K. Lee, *New Journal Chem.* **24**, 253–255 (2000).

105. Y.-H. Liu, H.-P. Lin and C.-Y. Mou, *Langmuir* **20**, 3231–3239 (2004).

106. S. Inagaki, S. Guan, Y. Fukushima, T. Ohsuna and O. Terasaki, *J. Am. Chem. Soc.* **121**, 9611–9614 (1999).
107. S. Guan, S. Inagaki, T. Ohsuna and O. Terasaki, *J. Am. Chem. Soc.* **122**, 5660–5661 (2000).
108. T. Asefa, M. J. MacLachlan, N. Coombs and G. A. Ozin, *Nature* **402**, 867–871 (1999).
109. B. J. Melde, B. T. Holland, C. F. Blanford and A. Stein, *Chem. Mater.* **11**, 3302–3308 (1999).
110. S. Inagaki, S. Guan, T. Ohsuna and O. Terasaki, *Nature* **416**, 304–307 (2002).
111. S. Dai, M. C. Burleigh, Y. H. Ju, H. J. Gao, J. S. Lin, S. J. Pennycook, C. E. Barnes and Z. L. Xue, *J. Am. Chem. Soc.* **122**, 992–993 (2000).
112. J. Brown, L. Mercier and T. J. Pinnavaia, *Chem. Commun.* 69–70 (1999).
113. D. J. Macquarrie, *Chem. Commun.* 1961–1962 (1996).
114. Y.-K. Lu and X.-P. Yan, *Anal. Chem.* **76**, 453–457 (2004).
115. R. J. P. Corriu, A. Mehdi, C. Reye and C. Thieuleux, *Chem. Mater.* **16**, 159–166 (2004).
116. P. Sutra and D. Brunel, *Chem. Commun.* 2485–2486 (1996).
117. K. H. Nam and L. L. Tavlarides, *Chem. Mater.* **17**, 1597–1604 (2005).
118. L. C. Cides da Silva, G. Abate, N. Andrea, M. C. A. Fantini, J. C. Masini, L. P. Mercuri, O. Olkhovyk, M. Jaroniec and J. R. Matos, *Stud. Surface Sci. Catal.* **155**, 941–950 (2005).
119. B. L. Newalkar, S. Komarneni and H. Katsuki, *Chem. Commun.* 2389–2390 (2005).
120. B. L. Newalkar, J. Olanrewaju and S. Komarneni, *Chem. Mater.* **13**, 552–557 (2001).
121. S. E. Park, J. S. Chang, Y. K. Hwang, D. S. Kim, S. H. Jhung and J. S. Hwang, *Catalysis Surveys from Asia* **8**, 91–110 (2004).
122. H. Katsuki and S. Komarneni, *J. Am. Ceramic Soc.* **84**, 2313–2317 (2001).
123. B. Lee, Y. Kim, H. Lee and J. Yi, *Microporous Mesoporous Mater.* **50**, 77–90 (2001).
124. A. Walcarius and C. Delacote, *Chem. Mater.* **15**, 4181–4192 (2003).
125. S. Mann, S. L. Burkett, S. A. Davis, C. E. Fowler, N. H. Mendelson, S. D. Sims, D. Walsh and T. W. Nicola, *Chem. Mater.* **9**, 2300–2310 (1997).
126. M. H. Valkenberg and W. F. Holderich, *Catalysis Rev. Sci. Eng.* **44**, 321–374 (2002).
127. A. Stein, B. J. Melde and R. C. Schroden, *Adv. Mater.* **12**, 1403–1419 (2000).
128. H. Yoshitake, *New Journal Chem.* **29**, 1107–1117 (2005).
129. V. Antochshuk and M. Jaroniec, *Chem. Commun.* 2373–2374 (1999).
130. S. Dai, Y. Shin, Y. Ju, M. C. Burleigh, J.-S. Lin, C. E. Barnes and Z. Xue, *Adv. Mater.* **11**, 1226–1230 (1999).
131. V. Antochshuk, A. S. Araujo and M. Jaroniec, *J. Phys. Chem. B* **104**, 9713–9719 (2000).
132. V. Antochshuk, M. Jaroniec, S. H. Joo and R. Ryoo, *Stud. Surf. Sci. Catal.* **141**, 607–614 (2002).

133. P. M. de Padilha, L. A. De Melo Gomes, C. C. F. Padilha, J. C. Moreira and N. L. Dias Filho, *Analytical Letters* **32**, 1807–1820 (1999).
134. P. T. Tanev, M. Chibwe and T. J. Pinnavaia, *Nature* **368**, 321–323 (1994).
135. K. M. Reddy, I. Moudrakovski and A. Sayari, *Chem. Commun.* 1059–1060 (1994).
136. D. Antonelli and J. Y. Ying, *Angew. Chem. Int. Ed.* **35**, 426–430 (1996).
137. C. Yoshina-Ishii, T. Asefa, N. Coombs, M. J. MacLachlan and G. Ozin, *Chem. Commun.* 2539–2540 (1999).
138. S. Yang, P. A. Mirau, C.-S. Pai, O. Nalamasu, E. Reichmanis, J. C. Pai, Y. S. Obeng, J. Seputro, E. K. Lin, H.-J. Lee, J. Sun and D. W. Gidley, *Chem. Mater.* **14**, 369–374 (2002).
139. M. Alvaro, B. Ferrer, V. Fornes and H. Garcia, *Chem. Commun.* 2546–2547 (2001).
140. C. Baleizao, B. Gigante, D. Das, M. Alvaro, H. Garcia and A. Corma, *Chem. Commun.* 1860–1861 (2003).
141. J.-K. Lee, S.-H. Lee, M. Kim, H. Kim and W.-Y. Lee, *Chem. Commun.* 1602–1603 (2003).
142. G. Wirnsberger and G. D. Stucky, *Chem. Phys. Chem.* **1**, 90–92 (2000).
143. G. Dubois, R. J. P. Corriu, C. Reye, S. Brandes, F. Denat and R. Guilard, *Chem. Commun.* 2283–2284 (1999).
144. G. Dubois, C. Reye, R. J. P. Corriu, S. Brandes, F. Denat and R. Guilard, *Angew. Chem. Int. Ed.* **40**, 1087–1090 (2001).
145. C. Chuit, R. J. P. Corriu, G. Dubois and C. Reye, *Chem. Commun.* 723–724 (1999).
146. Z. Zhang and S. Dai, *J. Am. Chem. Soc.* **123**, 9204–9205 (2001).
147. J. Alauzun, A. Mehdi, C. Reye and R. J. P. Corriu, *J. Am. Chem. Soc.* **127**, 11204–11205 (2005).
148. H. Yoshitake, T. Yokoi and T. Tatsumi, *Chem. Mater.* **14**, 4603–4610 (2002).
149. C. Liu and J. Economy, *PMSE Preprints* **91**, 1037–1038 (2004).
150. S. R. Hall, C. E. Fowler, B. Lebeau and S. Mann, *Chem. Commun.* 201–202 (1999).
151. T. Asefa, M. Kruk, M. J. MacLachlan, N. Coombs, H. Grondey, M. Jaroniec and G. Ozin, *J. Am. Chem. Soc.* **123**, 8520–8530 (2001).
152. M. C. Burleigh, M. A. Markowitz, M. S. Spector and B. P. Gaber, *J. Phys. Chem. B.* **105**, 9935–9942 (2001).
153. Q. Yang, M. P. Kapoor and S. Inagaki, *J. Am. Chem. Soc.* **124**, 9694–9695 (2002).
154. W. J. Hunks and G. A. Ozin, *J. Mater. Chem.* **15**, 764–771 (2005).
155. S. Huh, J. W. Wiench, B. G. Trewyn, S. Song, M. Pruski and V. S.-Y. Lin, *Chem. Commun.* 2364–2365 (2003).
156. W. J. Hunks and G. A. Ozin, *Adv. Funct. Mater.* 259–266 (2005).
157. S. Huh, H.-T. Chen, J. W. Wiench, M. Pruski and V. S.-Y. Lin, *J. Am. Chem. Soc.* **126**, 1010–1011 (2004).
158. M. A. Wahab, I. Imae, Y. Kawakami and C.-S. Ha, *Chem. Mater.* **17**, 2165–2174 (2005).
159. R. P. Hodgkins, A. E. Garcia-Bennett and P. A. Wright, *Microporous Mesoporous Mater.* **79**, 241 (2005).

160. M. Kuroki, T. Asefa, W. Whitnal, M. Kruk, C. Yoshina-Ishii, M. Jaroniec and G. A. Ozin, *J. Am. Chem. Soc.* **124**, 13886–13895 (2002).

161. K. Landskron, B. D. Hatton, D. D. Perovic and G. A. Ozin, *Science* **302**, 266–269 (2003).

162. R. J. P. Corriu, A. Mehdi, C. Reye and C. Thieuleux, *Chem. Commun.* 1382–1383 (2002).

163. R. J. P. Corriu, A. Mehdi, C. Reye and C. Thieuleux, *New Journal Chem.* **27**, 905–908 (2003).

164. R. Corriu, A. Mehdi and C. Reye, *J. Organometal. Chem.* **689**, 4437–4450 (2004).

165. O. Olkhovyk and M. Jaroniec, *J. Am. Chem. Soc.* **127**, 60–61 (2005).

166. L. Zhang, W. Zhang, J. Shi, Z. Hua, Y. Li and J. Yan, *Chem. Commun.* 210–211 (2003).

167. Yu. A. Zolotov, O. M. Petrukhin, G. I. Malofeeva, E. V. Marcheva, O. A. Shiryaeva, V. A. Shestakov, V. G. Miskar'yants, V. I. Nefedov, Yu. I. Murinov and Yu. E. Nikitin, *Analytica Chimica Acta* **148**, 135–57 (1983).

168. W.-H. Zhang, X.-B. Lu, J.-H. Xiu, Z.-L. Hua, L.-X. Zhang, M. Robertson, J.-L. Shi, D.-S. Yan and J. D. Holmes, *Adv. Func. Mater.* **14**, 544–552 (2004).

169. O. Olkhovyk, S. Pikus and M. Jaroniec, *J. Mater. Chem.* **15**, 1517–1519 (2005).

170. O. Olkhovyk and M. Jaroniec, *Stud. Surface Sci. Catal.* **155**, 197–204 (2005).

171. O. Olkhovyk and M. Jaroniec, *SPIE Proceedings* **5929**, 176–183 (2005).

172. T. Asefa, C. Yoshina-Ishii, M. J. MacLachlan and G. A. Ozin, *J. Mater. Chem.* **10**, 1751–1755 (2000).

173. M. J. MacLachlan, T. Asefa and G. A. Ozin, *Chem. Eur. Journal* **6**, 2507–2511 (2000).

174. W. Hunks and G. A. Ozin, *J. Mater. Chem.* **15**, 3716–3724 (2005).

175. B. Hatton, Landskron, K. W. Whitnall, D. Perovic and G. A. Ozin, *Acc. Chem. Research* **38**, 305–312 (2005).

176. S. J. Gregg and K. S. W. Sing, *Adsorption, Surface Area and Porosity* (Academic Press, London, 1982).

177. S. Brunauer, P. H. Emmett and E. Teller, *J. Am. Chem. Soc.* **60**, 309–319 (1938).

178. M. Kruk, M. Jaroniec and A. Sayari, *Langmuir* **13**, 6267–6273 (1997).

179. E. P. Barrett, L. G. Joyner and P. H. Halenda, *J. Am. Chem. Soc.* **73**, 373–380 (1951).

180. R. C. Schroden, M. Al-Daous, S. Sokolov, B. J. Melde, J. C. Lytle, A. Stein, M. C. Carbajo, Fernandez, J. Torralvo and E. E. Rodriguez, *J. Mater. Chem.* **12**, 3261–3267 (2002).

181. T. Yokoi, T. Tatsumi and H. Yoshitake, *J. Coll. Interface Sci.* **274**, 451–457 (2004).

182. M. Takafuji, S. Ide, H. Ihara and Z. Xu, *Chem. Mater.* **16**, 1977–1983 (2004).

183. R. F. de Farias, A. A. S. do Nascimento and C. W. B. Bezerra, *J. Coll. Interface Sci.* **277**, 19–22 (2004).

184. Y.-M. Xu, R.-S. Wang and F. Wu, *J. Coll. Interface. Sci.* **209**, 380–385 (1999).
185. G. X. S. Zhao, J. L. Lee and P. A. Chia, *Langmuir* **19**, 1977–1979 (2003).
186. E. Pitoniak, C.-Y. Wu, D. Londeree, D. Mazyck, J.-C. Bonzongo, K. Powers and W. Sigmund, *J. Nanoparticle Research* **5**, 281–292 (2003).
187. E. Pitoniak, C.-Y. Wu, D. W. Mazyck, K. W. Powers and W. Sigmund, *Environ. Sci. Technol.* **39**, 1269–1274 (2005).
188. L. Lv, F. Su and G. X. S. Zhao, *Stud. Surf. Sci. Catal.* **156**, 933–940 (2005).
189. M. Jang, J. K. Park and E. W. Shin, *Microporous Mesoporous Mater.* **75**, 159–168 (2004).
190. G. Carja, R. Nakamura and H. Niiyama, *Microporous Mesoporous Mater.* **83**, 94–100 (2005).
191. T. Balaji, M. Sasidharan and H. Matsunaga, *Analyst* **130**, 1162–1167 (2005).
192. R. Metivier, I. Leray, B. Lebeau and B. Valeur, *J. Mater. Chem.* **15**, 2965–2973 (2005).
193. A. Walcarius and J. Bessiere, *Chem. Mater.* **11**, 3009–3011 (1999).
194. A. M. Bond, W. Miao, T. D. Smith and J. Jamis, *Analytica Chimica Acta* **396**, 203–213 (1999).
195. W. Yantasee, Y. Lin, X. Li, G. E. Fryxell, T. S. Zemanian and V. V. Viswanathan, *Analyst* **128**, 899–904 (2003).
196. W. Yantasee, Y. Lin, T. S. Zemanian and G. E. Fryxell, *Analyst* **128**, 467–472 (2003).
197. W. Yantasee, G. E. Fryxell, M. M. Conner and Y. Lin, *J. Nanosci. Nanotechnology* **5**, 1537–1540 (2005).
198. W. Yantasee, Y. Lin, G. E. Fryxell and B. J. Busche, *Analytica Chimica Acta.* **502**, 207–212 (2004).
199. R. T. Yang, R. Q. Long, J. Padin, A. Takahashi and T. Takahashi, *Ind. Eng. Chem. Res.* **38**, 2726–2731 (1999).
200. W. Z. Shen, J. T. Zheng and Q. J. Guo, *Stud. Surf. Sci. Catal.* **156**, 951–956 (2005).
201. X. S. Zhao, Q. Ma and G. Q. M. Lu, *Energy Fuels* **12**, 1051–1054 (1998).
202. Y. Ueno, T. Horiuchi, M. Tomita, O. Niwa, H-S. Zhou, T. Yamada and I. Honma, *Anal. Chem.* **74**, 5257–5262 (2002).
203. K. Inumaru, Y. Inoue, S. Kakii, T. Nakano and S. Yamanaka, *Chem. Lett.* **32**, 1110–1111 (2003).
204. K. Inumaru, J. Kiyoto and S. Yamanaka, *Chem. Commun.* 903–904 (2000).
205. K. Inumaru, Y. Inoue, S. Kakii, T. Nakano and S. Yamanaka, *Phys. Chem. Chem. Phys.* **6**, 3133–3139 (2004).
206. J. Lee, Y. Park, P. Kim, H. Kim and J. Yi, *J. Mater. Chem.* **14**, 1050–1056 (2004).
207. C. F. Zhou, Y. M. Wang, J. H. Xu, T. T. Zhuang, Y. Wang, Z. Y. Wu and J. H. Zhu, *Stud. Surf. Sci. Catal.* **156**, 907–916 (2005).
208. A. Ji, L. Y. Shi, Y. Cao and Y. Wang, *Stud. Surf. Sci. Catal.* **156**, 917–924 (2005).
209. D. P. Serrano, G. Calleja, J. A. Botas and F. J. Gutierrez, *Ind. Eng. Chem. Res.* **43**, 7010–7018 (2004).

Chapter 9

Hierarchically Imprinted Adsorbents

Hyunjung Kim, Chengdu Liang and Sheng Dai

Oak Ridge National Laboratory
Oak Ridge, Tennessee, USA

Introduction

The production of selective adsorbents to detect and separate various entities in environmental waste is a growing field with broad applications and critical importance. One of the important entities in environmental waste is metal ions. Industry produces a vast amount of metals and metal wastes. Nuclear energy and past weapons production facilities have also created challenges in the area of metal ion separations. Another significant component of environmental waste is biological molecules (e.g. proteins and viruses from wastewater). Recognition of these entities may be also integrated into sensors to detect bio-warfare agents or various environmental contaminants. This chapter describes the synthesis of these selective adsorbents to detect these various compounds in environmental applications using the hierarchical imprinting approach.

Molecular Imprinting

The concept of molecular imprinting is based on an early "lock and key" concept, used to explain enzyme function and antibody formation.[1–3] Although this idea has since been proven incorrect, it was used by Frank Dickey in the 1940s to develop artificial counterparts to enzymes and antibodies originally in silical gels to develop selective silica adsorbents for different types of dye molecules.[4] By acidifying a silicate solution in the presence of methyl-orange dye, drying the resultant gel, and then washing the dye from the gel, Dickey obtained silica gels capable of specific adsorption of the methyl-orange 1.4 times greater than than for ethyl-orange.

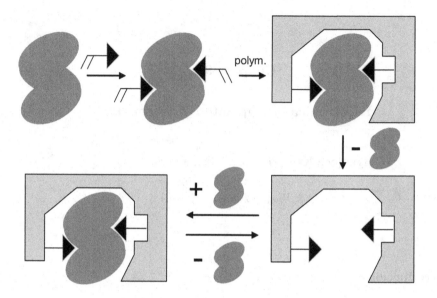

Figure 1.   Outline of the molecular imprinting process in organic polymers.

Building upon Dickey's findings, Wulff and Sarhan, in the 1970s, developed the current strategy to synthesize selective adsorbents by molecular imprinting in organic polymers.[5] This current molecular imprinting method to synthesize selective adsorbents in organic polymers is shown schematically in Figure 1. This molecular imprinting process utilizes solution interactions between a target molecule (template) and appropriate functional monomers, by either covalent[6] or noncovalent interactions[7,8] such as ionic, coordinative, hydrophobic, or hydrogen-bonding interactions. These pre-organized solution complexes are then immobilized into a polymer matrix by co-polymerization with a cross-linking monomer. Removal of the template leaves a cavity in the polymer matrix complementary to the shape, size, and functionalities of the template. This method of molecular imprinting in organic polymers has been used to imprint various templates from small metal ions to large biological molecules for applications in the areas of separation, catalysis, chemosensors, biomimic enzymes, drug delivery, and so on.[9]

Hierarchical Imprinting

One major drawback associated with the current bulk molecular imprinting technique is that the pre-organized solution complexes are immobilized into heterogeneous organic polymer matrices without any independent control over structural parameters, such as pore sizes and surface areas. In the current bulk molecular imprinting technique, the pore sizes and surface areas are controlled by a porogenic solvent, where these structural parameters in polymers are determined by the rate and the extent of phase separation between growing polymer chains and the porogenic solvent. However, the reported "porogenic effect"[10–13] on these bulk imprinted polymers provides evidence that the imprinting effects on the polymers are affected by the different degrees of swelling of the polymers in the porogenic solvent, in addition to the desired effects from the immobilized solution complexes in polymer matrices. Also, the process of optimizing porogenic solvents to produce desired pore sizes and surface areas is countered by the strong interactions required between a template and functional monomers.[14] This lack of independent control on the structural parameters of bulk imprinted polymers has caused difficulties in optimizing imprinted materials for easy accessibility and fast mass transfer of substrates on the imprinted binding sites without affecting their binding properties.[15–19]

The hierarchical imprinting approach provides a solution to control the structural parameters independent of binding parameters, which are controlled by the immobilized solution complexes in solid matrices. The original concept of the hierarchical imprinting approach involves the imprinting synthesis using multiple templates over several discrete dimension scales for metal ions.[20] This concept has been extended from its original applications for metal ions to various target compounds.[21,22] Hierarchically imprinted structures can be assembled by two methodologies: (1) co-assembly and (2) step-wise assembly. The following sections address synthesis, characterization, and performance of the hierarchically imprinted adsorbents generated by these two methods.

Co-Assembly Approach

The co-assembly approach to generate hierarchically imprinted adsorbents is based on the self-assembly of various structural components via simultaneous uses of multiple templates. Surfactants have been used to control structural parameters of the hierarchically imprinted silica gels and polymers.

Mesoporous Silica Adsorbents

Two templates (metal ions and surfactant micelles) on different length scales were used to generate hierarchically Cu^{2+}-imprinted silica adsorbents as shown in Figure 2.

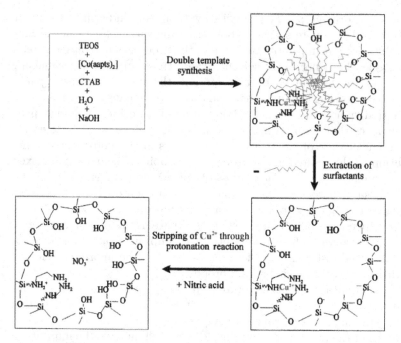

Figure 2. Schematic diagram of the synthesis of hierarchically imprinted adsorbents using Cu^{2+} and CTAB simultaneously as templates.[20]

On the microporous level, the removal of the metal ion from the resultant silica adsorbent generates recognition sites for the metal ion. On the mesoporous level, removal of the surfactant templates results in the formation of relatively large cylindrical pores (diameters of 25 to 60 Å) that give the gel an overall porosity.

The synthetic process to produce these hierarchically imprinted adsorbents using Cu^{2+} and cetryltrimethylammonium bromide (CTAB) is as follows. The bifunctional ligand — 3-(2-amino-ethylamino) proplytrimethoxysilane, $H_2NCH_2CH_2NH-CH_2CH_2CH_2-Si(OMe)_3$ (appt) — was used to complex with the Cu^{2+} template. The ethylenediamine group in this ligand forms strong bidentative interactions with the Cu^{2+} template. The imprinting complex precursor used in this study was $[Cu(appt)_2]^{2+}$, synthesized according to standard literature procedures.[23] The $[Cu(appt)_2]^{2+}$ complexes, CTAB, tetrathylorothosilicate (TEOS), water, and base (NaOH) were mixed and heated to imprint the two templates (Cu^{2+} and CTAB) into a silica matrix via the base-catalyzed hydrolytic condensation of TEOS. The blue solid products were recovered from filtration. Ethanol/HCl was used to reflux these products, and to extract the surfactant and the Cu^{2+} templates from the silica matrix. The final material was washed with large volumes of 1 N HNO_3 to ensure complete removal of the Cu^{2+} template. Control adsorbents, using the same procedure without the Cu^{2+} template, were also prepared. In this process, other surfactants — the anionic sodium dodecylsulfate (SDS) and the neutral dodecylamine (DDA) — were used to prepare Cu^{2+}-imprinted hierarchical structures and their corresponding control adsorbents.

These hierarchically imprinted silica gels and the corresponding control adsorbents have large surface areas (in the range of 200 to $600\,m^2/g$) in comparison with those of all other adsorbents imprinted only with $[Cu(appt)_2]^{2+}$ ($< 30\,m^2/g$). The UV-vis spectrum of $[Cu(appt)_2]^{2+}$ in methanol solution and that of $[Cu(appt)_2]^{2+}$ covalently immobilized in a mesoporous silica-gel prepared with CTAB show close agreement (Figure 3).

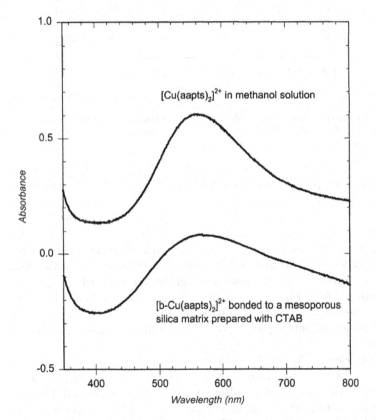

Figure 3. UV-Vis spectra of $[Cu(appt)_2]^{2+}$ in methanol solution and $[Cu(appt)_2]^{2+}$ covalently immobilized in a mesoporous silica gel prepared with CTAB.[20]

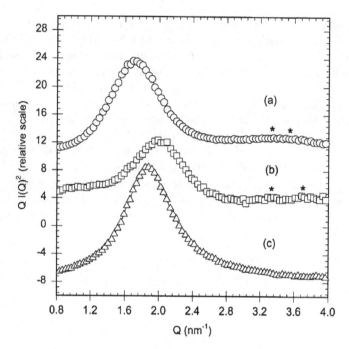

Figure 4. Small-angle scattering of hierarchically imprinted adsorbents prepared using (a) CTAB, (b) SDS, and (c) DDA, where $Q = (4\pi/\lambda) \sin(\theta/2)$, λ = X-ray wavelength (0.154 nm), and θ = scattering angle.[20]

This close match indicates that the environment of the copper ion is similar and that the $[Cu(appt)_2]^{2+}$ complexes are immobilized in the silica adsorbents.[24] Small-angle X-ray Scatterings (Figure 4) from the hierarchically Cu^{2+}-imprinted silica adsorbents with CTAB, SDS, and DDA show the common peak around $Q = 1.2$–2.4 nm, in agreement with the pore size independently measured in nitrogen adsorption experiments. Additional peaks present from the scattering on the silica adsorbents prepared either with CTAB (Figure 4a) or SDS (Figure 4b) indicate hexagonal structures, and the absence of this additional peak from the scattering on the silica adsorbents prepared by DDA (Figure 4c) indicates worm-like structures.

These hierarchically imprinted silica adsorbents and their corresponding control adsorbent are neutralized to a pH of 7 and dried using a vacuum oven at 50°C for 6 hours before adsorption tests were conducted. Competitive adsorption studies were conducted with Cu^{2+} and Zn^{2+} ions in order to measure the selectivity of these hierarchically Cu^{2+}-imprinted adsorbents. The selectivity coefficient, k, for the binding of a specific ion in the presence of competitor species is obtained from Equation (2):

$$M_1(solution) + M_2(adsorbent) \leftrightarrow M_2(solution) + M_1(adsorbent) \quad (1)$$

$$k = \frac{[M_2(solution)][M_1(adsorbent)]}{[M_1(solution)][M_2(adsorbent)]} = \frac{k_d[Cu^{2+}]}{k_d[Zn^{2+}]} \quad (2)$$

where M_1 and M_2 correspond to Cu^{2+} and Zn^{2+}, respectively.

The distribution coefficient, k_d, is calculated from Equation (3):

$$k_d = (C_i - C_f)/C_f \times volume \ of \ solution \ (mL)/mass \ of \ gel \ (g) \quad (3)$$

where C_i is the initial solution concentration and C_f is the final solution concentration.

A measure of the increase in selectivity due to the imprinting process of the Cu^{2+} template (k') is calculated by the ratio of the selectivity coefficient (k) of the imprinted adsorbents to that of the control adsorbents:

$$k' = k \ (imprinted)/k \ (control) \quad (4)$$

The k and k' values for these hierarchically imprinted silica adsorbents prepared with CTAB, SDS, or DDA range between 180 and 33,000, and between 5 and 280, respectively. These values are much higher than those obtained for imprinted silica adsorbents prepared without the surfactants ($k = 20$ and $k' = 1.3$). This finding can be attributed to the low surface areas obtained on the imprinted silica adsorbents without the surfactant ($26 \, m^2/g$) as compared with those obtained on the hierarchically imprinted silica adsorbents (240–$520 \, m^2/g$). These results demonstrate the importance of using surfactant templates to generate large surface areas in these hierarchically Cu^{2+}-imprinted silica adsorbents.

Further optimization to increase the selectivity was explored by adding various ratios of methyltrimthyoxysilane ($CH_3Si(OCH_3)_3$; MTMOS) or phenyltrimethoxysilane ($C_6H_5Si(OCH_3)_3$; PTMOS) to the mixtures of $[Cu(appt)_2]^{2+}$ and tetramethylorothosilicate ($Si(OCH_3)_4$; TMSO),[25] using adequate co-solvents,[24] and with bridged polysilsesquioxane matrices.[26] These results demonstrated that an enhanced imprinting effect and selectivity of these hierarchically Cu^{2+}-imprinted silica adsorbents toward the template ion can be obtained by enhancing the hydrophobicity of the sol-gel materials. This double-imprinting technique was also successfully used in creating Cd^{2+}-selective recognition sites on silica mesoporous adsorbents.[27]

Surface Imprinted Polymeric Adsorbents

The co-assembly approach to generate hierarchically imprinted adsorbents has been also explored in polymer matrices by creating microspheres with imprinted external cavities using emulsion polymerization. This method was

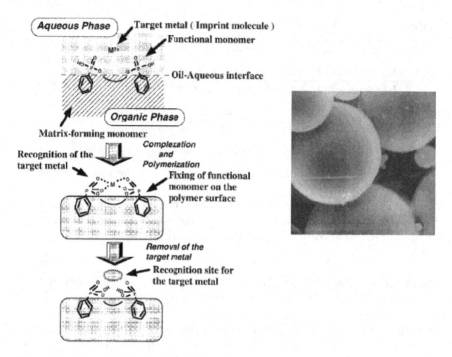

Figure 5. Schematic illustration of the surface imprinting technique and a scanning electron microscopy micrograph of the resulting polymers using W/O/W emulsion polymerization.[34]

originally pioneered by Uezu and coworkers to generate surface-imprinted polymers for Zn^{2+} templates in water-in-oil (W/O) emulsions.[28] After thorough optimization studies,[29-34] successful hierarchically imprinted microspheres for the Zn^{2+} template were synthesized in water-in-oil-in-water (W/O/W) emulsions (Figure 5).[34]

These microspheres have hierarchical structures composed of spherical surfaces and cavities scattered on the surfaces. The morphology of the microspheres was controlled by the following parameters: (1) Magnesium ion was used to control ionic strength to prevent shrinking of W/O emulsions during the polymerization. (2) The negatively-charged surfactant SDS was used at optimum concentration to prevent aggregation of the spherical W/O emulsions. The position of the cavities on the surfaces of the microspheres was achieved using the bifunctional surfactant 1,12-dodecanediol-O-O′-diphenyl phosphonic acid (DDDPA). The bifunctional surfactant, which is amphiphilic in nature, complexes with the Zn^{2+} template and positions the complexes at the interface between the water and oil surfaces. This method also provides an additional advantage in imprinting water-soluble templates such as metal ions and biological components such as amino acids and proteins.

The steps in the synthesis of these surface-imprinted polymers are as follows: the bifunctional surfactant (DDDPA), the emulsion stabilize (L-glutamic acid dioleylester; $2C_{18}\Delta^9GE$), and the cross-linking monomer (trimethoylpropane trimethacrylate; TRIM) were dissolved in organic solvents (i.e. toluene with 5 vol % 2-ethyhexyl alcohol). This oil mixture was added to an aqueous solution of $Zn(NO_3)_2$ at pH 3.5, and the oil and water mixture was sonicated to produce a W/O emulsion. The W/O emulsion was then placed in an aqueous buffer solution (pH 3.5) that contained an ionic strength adjuster [$Mg(NO_3)_2$] and the surfactant SDS to form the W/O/W emulsion. The W/O/W emulsion was polymerized with the initiator 2,2'-azobis(2,4'-dimethylvaleronitrile), AIBN at 55°C for four hours. After the polymerization, the resulting microspheres were washed with 1 M hydrochloric acid to remove the Zn^{2+} template. These surface-imprinted polymers have been successfully used for the selective uptake of Zn^{2+} over Cu^{2+} in buffered aqueous solutions. These studies identified several critical parameters to optimize the selectivity of these surface-imprinted polymers. It was found that increasing rigidity of the phosphonic acid-functionalized molecule significantly increases the selectivity of the surface-imprinted polymers toward metal ions.[30] In the procedure using the bifunctional surfactant DDDPA, it was also found that the length of the spacer connecting the bifunctional phosphonic groups is an important factor in ensuring optimized recognition sites and rigidity of these sites on the surfaces of the polymer.[31]

This method of generating hierarchically imprinted polymers has been also extended to organic molecules, such as enantiomers of derivatized amino acids[35,36] and bifunctional amino acids,[37] as well as to nucleotides.[38] This idea of using a functionalized surfactant to generate hierarchically surface-imprinted polymers has been also applied with a core-shell morphology of the polymers for cholesterol,[39] caffeine, and tripeptides containing the Gly-Gly sequence.[40,41]

Step-Wise Assembly Approach

The step-wise assembly approach to generate hierarchically imprinted adsorbents involves sequential generation of hierarchical structures. In this approach, various forms of inorganic hosts have been used as modules to control the structural parameters on the hierarchically imprinted adsorbents. The step-wise assembly approach has been used to generate surface-imprinted inorganic, polymeric, and inorganic-organic hybrid adsorbents.

Surface Imprinted Sol-Gel Adsorbents

The step-wise assembly approach to generate hierarchically imprinted adsorbents was originally demonstrated by Dai and co-workers.[42] The key to this method is to coat the pore surface of an ordered mesoporous silica host (MCM-41) with the complex precursor $[Cu(appt)_2]^{2+}$ via the hydrolysis and condensation reactions of silicon alkoxide groups in appt (Figure 6).

A mesoporous ordered silica host was synthesized with optimum pore diameter to match the stereochemical requirements for the surface imprinting of the complexes. The prepared mesorporous silica samples have surface areas of over $1000\,m^2/g$, mesopore volumes of $0.98\,cm^3/g$, and average pore sizes of $\sim$25 Å, as measured by powder X-ray Diffraction (XRD) and Small-angle X-ray Scattering. This pore diameter of the mesoporous silica samples is ideal to fit the complexes $[Cu(appt)_2]^{2+}$ ($\sim$16 to 25 Å). These hierarchically imprinted adsorbents showed a k' value of 40 for the Cu^{2+} template over Zn^{2+}. When the $[Cu(appt)_2]^{2+}$ was coated on the pore surfaces of commercial amorphous silica gel (Aldrich; $\bar{d}$ = 60 Å; surface area = $600\,m^2/g$), the resulting imprinted materials did not show any significant selectivity toward the Cu^{2+} template (k' = 1.54). This demonstrates the importance of the optimized pore diameter and ordered structure of the inorganic hosts for the optimized selectivity toward the metal ions.

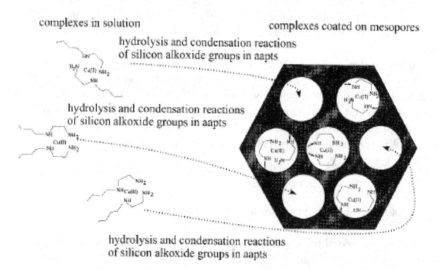

Figure 6. Schematic diagram of the imprint-coating process. First, the complexes are introduced between target metal ions and bifunctional ligands. Then the siloxane groups in the bifunctional ligands are hydrolyzed, and finally the complexes are covalently coated on the mesopore surfaces.[42]

Liu and co-workers[43] have successfully explored the step-wise assembly strategy to imprint organic molecules inside mesopores, allowing the control over both binding properties and pore structures. The ordered mesoporous silica hosts (MCM-41) were also successfully used to imprint dopamine and glucosamine by creating a multi-functionalized mesopore that can interact with the template by covalent and non-covalent interactions.[44] The importance of the ordered mesopores in creating cavities was also demonstrated in this study by the lack of selectivity towards the template when an amorphous silica was used as the host.[44] Another successful application of the mesorporous silica hosts was demonstrated by creating the surface-imprinted sol-gel adsorbents for estrone in the inorganic hosts.[45]

Even though the mesoporous silica hosts used in the previous study have a perfect nano-scale order, the wall structure of the mesoporous materials is amorphous, which could collapse under hydrothermal conditions. Thus, the use of porous zeolitic crystalline hosts to imprint Cu^{2+} were tested for the possible improved imprinting effect, which may come from the generation of uniform imprints and the limited choices of coordination environments.[46] Due to the small pore opening of the L-zeolite used in this study (7.1–7.8 Å), a step-wise "ship-in-bottle" imprinted methodology was used (Figure 7). The Cu^{2+} template was first used to replace K^+ or Na^+ in the L-zeolite through the ion-exchange interaction. This step was then followed by the complexations of the surface Cu^{2+} template with aapt ligands and the surface coupling of aapts with neighboring surface SiO^- or $-SiOH$ groups. By extracting the Cu^{2+} template via protonation of the amine groups on the complexes, a zeolite imprinted on the channel surface with the Cu^{2+} ion was prepared. Using the same procedure without the Cu^{2+}, control zeolite adsorbents were also prepared.

The exchange of the Cu^{2+} ion with the precursor complex $[Cu(aapt)_x]^{2+}$ in the zeolite-imprinted adsorbent was confirmed by the UV-Vis and Fourier-Transform Infrared (FTIR) spectra. In the UV-Vis spectrum, the maximum band position moves from 681 nm to 585 nm in agreement with the change of the coordination environment of the Cu^{2+} ion from water to

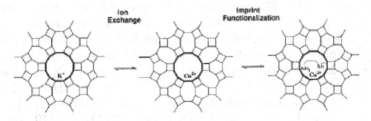

Figure 7.   Schematic diagram of the Cu^{2+} imprinting in L-zeolite hosts.[46]

amine. In the FTIR spectrum, the C-H stretching around $2900\,cm^{-1}$ can be attributed to the C-H groups of the aapt ligands. The quantification of the C-H stretching peak also showed that a similar amount of the aapt ligands was present in the imprinted and the control zeolite adsorbents. Powder XRD was used to observe the crystalinity of the zeolite structure during the various stages of the preparation. The XRD patterns of the imprinted and the control zeolite adsorbents show that the crystallinity of the zeolite adsorbents was maintained during the preparation process and that the space distribution of the functional ligands in the imprinted zeolite adsorbent was different from that in the control zeolite adsorbents, as would be expected in the imprinting process. The selectivity of the imprinted zeolite adsorbents towards the Cu^{2+} template in aqueous Cu^{2+}/Zn^{2+} mixtures was tested using a typical batch procedure. The imprinted zeolite adsorbents showed k and k' values of 29 and 12, respectively, for the imprinted Cu^{2+} in aqueous Cu^{2+}/Zn^{2+} mixtures. These values for the imprinted zeolite adsorbents are higher than these obtained for the control adsorbents. However, the values for the former are much lower than these obtained from the imprinted mesoporous silica hosts ($k = 91$ and $k' = 40$).[42]

Several other forms of inorganic hosts, such as layered nanomaterials[47] and nanomembranes,[48] have also been successfully used to produce surface-imprinted sol-gel materials via the step-wise assembly approach. For example, Zhang *et al.*, used polysilicate magadiite as a layered nanomaterial to imprint Cu^{2+} ions at the interface of the layered magadiite (Figure 8).[47]

The layered magadiite is composed of one or more negatively-charged sheets of SiO_4 tetrahedra with abundant silanol-terminated surfaces where these negative charges are balanced by either Na^+ or H^+ in the interlayer spacing. The layered magadiite provides unique features of tunable gallery spacing and stable crystalline for hierarchically imprinted structures. The synthesis steps for the imprinting of Cu^{2+} ions in the layered magadiite are as follows: First, the Na^+-magadiite host was synthesized according to previously published methods.[49] Next, via ion-exchange, the Na^+ in the galleries of the layered magadiite was replaced by a much larger CTA^+ cation (cetyltrimthylammonium), which was then exchanged with the precursor complex $[Cu(aapt)_2]^{2+}$. The complexes exchanged into the galleries were then covalently attached to the surfaces through condensation reactions of silicon alkoxide groups in the bifunctional ligands with neighboring surface SiO^- or $SiOH$ groups. The product was recovered by filtration and washed with large volumes of 1 M HNO_3 to remove the Cu^{2+} template. The successful exchange process from Na^+, to CTA^+ cation, and then to the complex $[Cu(aapt)_2]^{2+}$ was confirmed by the XRD patterns and the UV-Vis spectra. The ^{29}Si Cross-Polarization Magic-angle Spinning (CP/MAS) NMR spectra of the imprinted magadiite adsorbents confirmed that the crystallinity

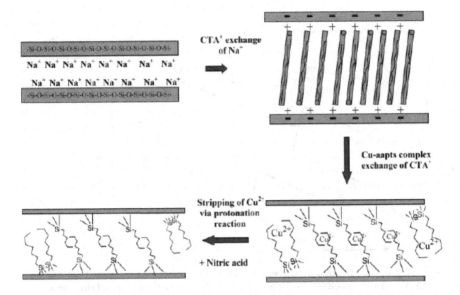

Figure 8. Schematic diagram of imprinting Cu^{2+} ions in magadiite silicate.[47]

of the host was maintained during the preparation process. The amount of the appts ligands incorporated in the galleries of the magadiite hosts was higher in the imprinted magadiite than in the control magadiite, as indicated by the basal spacing calculated from the corresponding XRD patterns and the resonance peaks from the corresponding ^{29}Si CP/MAS NMR spectra. The results from the adsorption studies on the magadiite adsorbents toward the Cu^{2+} template in aqueous Cu^{2+}/Zn^{2+} mixtures showed the preferential adsorption of the Cu^{2+} template on the imprinted magadiite adsorbents ($k = 140.03$) over that on the control magaditte adsorbents ($k = 45.47$), resulting in the imprinting factor (k') of 3.24.

Recently, Yang *et al.*, applied a molecular imprinting technique to imprint estrone into the wall of the sol-gel nanotubes (Figure 9).[48] The silica nanotubes were synthesized within the pores of nanopore alumina template membranes using the sol-gel method.

First, silica monomer- template complexes were synthesized by reacting estrone with 3-(triethoylsilyl)propyl isocyanate. The silica monomer-estrone complexes and TEOS were then added to a mixture of acetate buffer at pH 5.1 with ~ 10 vol % ethanol. After the mixture was stirred for five minutes, the alumina membrane was immersed. After 0.5 hour at low pressure, the alumina membrane was removed by phosphoric acid, rinsed with ethanol, and cured at 150°C for one hour. To extract the estrone template, the imprinted silica nanotube membrane was heated at 180°C in a

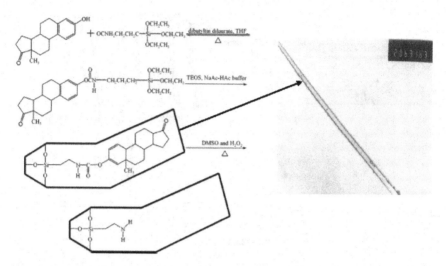

Figure 9. Schematic diagram of the estrone imprinting using the alumina template membrane. The transmission electron microscopy (TEM) shows the imprinted silica nanotubes (with 100 nm diameter) after the removal of the alumina template membranes by phosphoric acid.[48]

mixture of dimethyl sulfoxide (DMSO) and water and then washed with the mixture at room temperature. The amount of estrone that could not be removed via this process was ~1.2%, as estimated from the carbonyl peak of the urethane group of the silica monomer-estrone complexes at $1736\,cm^{-1}$ in the Infrared (IR) spectroscopy. After the alumina template had been removed by phosphoric acid, Transmission Electron Microscopy (TEM) of the imprinted silica nanotubes showed pore diameters of 100 nm (Figure 9). Calculations from a steady-state binding method showed that the ratio of the selectivity of the imprinted silica nanotubes towards the estrone template to that of the corresponding control silica nanotubes was 6.94 in chloroform.

Surface Imprinted Polymeric Adsorbents

The surface-imprinted polymeric adsorbents using silica hosts were first demonstrated by Yilmetz et al., for theophylline templates (Figure 10).[21] The major objective in employing this method is to covalently immobilize the template in the silica hosts by reacting 8-carboxypropy-derivatized theophylline with aminopropy-derivatized silica hosts to form amide bonds using

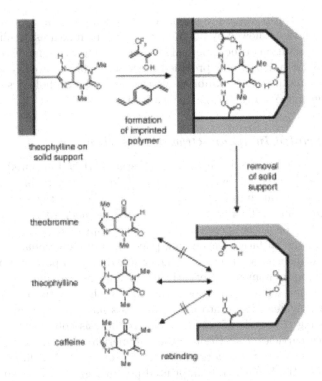

Figure 10.   Schematic diagram of surface imprinting method using immobilized template on the silica hosts.[21]

carbodiimide chemistry. Subsequently, the functional monomers (trifluoromethylacrylic acid; TFMAA) and the cross-linking monomers (divinlybenzene; DVB) are added to the silica hosts with the immobilized template, and these mixtures are then polymerized. The removal of the silica hosts by aqueous hydrofluoric acid generates the surface-imprinted polymeric adsorbents. The successful imprinting effect on these surface-imprinted polymeric adsorbents for the theophylline templates was demonstrated from a radioligand binding studies in toluene. This concept of immobilizing templates on silica hosts to independently control the structural parameters of the imprinted adsorbents was also demonstrated for 9-ethyladenine or triaminopyrimidine,[22] peptides,[50] and hemoglobin.[51] Yang et al.,[52] used this concept to prepare imprinted nanowires for glutamic acid template using alumina membranes as inorganic hosts instead of the spherical silica hosts. Rather than immobilizing templates in the pores of the spherical silica hosts, Yilmaz et al.,[53] generated (-)-isoproterenol-imprinted polymers by adding the mixtures of the pre-polymer components within the pores of spherical silica hosts, followed by polymerization of the mixtures within the pores and the resolution of the spherical silica hosts. The Scanning Electron Microscopy (SEM) micrographs provided in these studies showed that the resulting imprinted polymers exhibited a structure and morphology similar to the "mirror image" of the original silica hosts.

Surface Imprinted Inorganic-Organic Adsorbents

Instead of removing the inorganic hosts to generate surface-imprinted polymers, surface-imprinted inorganic-organic hybrid adsorbents have been developed, in which an imprinted polymer is attached to the surface of porous inorganic hosts. There are essentially two methods to generate an imprinted surface on these inorganic hosts. The imprinted polymers can be grafted onto the surface by attaching polymerizable functional groups, which are then incorporated into a polymer matrix by co-polymerization with cross-linking monomers. Alternatively, an imprinted polymer can be grown onto the surface, which has been modified with an initiator.

Imprinted polymer grafts onto inorganic hosts have included methods involving coating the walls of capillary electrophoresis columns[54,55] and the use of silica particles.[56–59] Prior to the polymerization, a vinyl group is attached to the surface of the inorganic hosts. For example, Arnold et al.,[56] employed this method to generate imprinted polymer grafts on the surface of inorganic hosts for a series of mono- and bis-imidazole templates. The anchoring sites for the imprinted polymers were provided by 10 μm porous silica particles (1000 Å) modified with propylmethacrylate using a silanizing reagent, 3-(trimethosylsilyl)-propylmethacrylate (Figure 11).

One of the templates, the functional monomer (copper(II) [N-(4-vinylbenzyl)imino]diacetate • 2H$_2$O) and the cross-linking monomer (ethylene glycol dimethacrylate; EGDMA), were mixed with the modified silica suspended in 80% aqueous methanol. These mixtures were then polymerized at 40°C for 48 hours with ammonium peroxylsulfate (APS) as an initiator. The resulting adsorbents were washed extensively with solutions of EDTA and 1,4,7-triazacyclononane to remove the template and incorporated copper ions. The SEM morphology of the resulting adsorbents

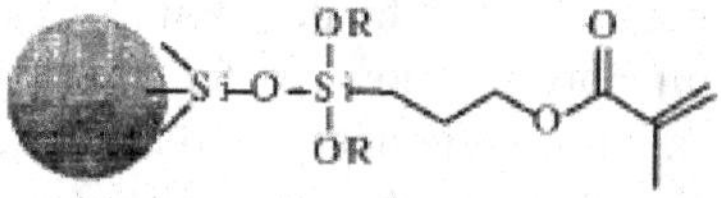

Figure 11. Modification of the silica hosts using chemically attached polymerizable group.

showed the continuous polymer nanospheres (in the 50 to 100 nm range) coated in a thin layer (on the order of 1 μm or less) on the silica hosts. The selectivity of these hybrid adsorbents towards the template, tested using High-Performance Liquid Chromatography (HPLC), showed a strong imprinting effect even when the copper ion was replaced with zinc ion. However, this approach to graft the imprinted polymers on the surfaces of the hosts requires that the monomer mixture be applied as a thin liquid film on the surface prior to the polymerization because the initiator is in solution. Thus, control of the thickness of the polymer layer is difficult, and the maximum density of grafted polymer chains is limited due to kinetic and steric factors, as demonstrated by Bruggemann *et al.*[54]

A better control of the morphology of inorganic-organic hybrid adsorbents was obtained when the imprinted polymers were grown on the surface of the inorganic hosts modified with an initiator.[60,61] A free-radical initiator was attached to a surface either chemically (Figures 12a and 12b) or by physical adsorption (Figure 12c).[60]

The covalent attachment of the initiator on the surface of the silica hosts was comprised of the two successive surface reactions. The free silanol groups ($\sim$8 μmol/m^2) on the silica surface, re-hydroxylated

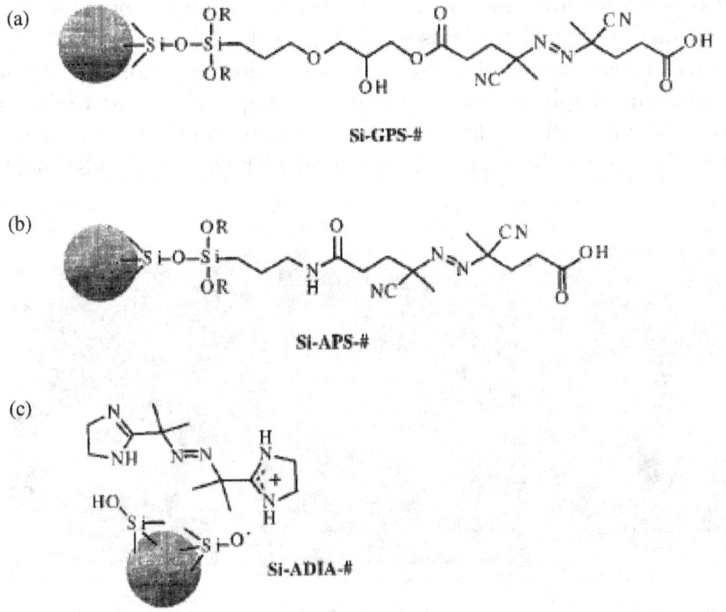

Figure 12. Modification of the silica hosts either (a and b) using covalently attached initiators or (c) non-covalently attached initiators.

according to standard procedures,[62] were reacted with trimethoxylgly-cidoxypropylsilane (GPA) or (3-aminopropyl)triethoxylsilane (APS), fol-lowed by reaction of the epoxy or amino groups with an azo-initiator such as azobis(cyanopentanoic acid) (ACPA), which led to the formation of an ester or amide bond between the surface and the azo-initiator. The results from the elemental analysis after the sequential coupling to the silica hosts showed an overall conversion of the silanol groups of $\sim 10\%$ (i.e. 0.7 to $0.8\,\mu mol/m^2$). In the noncovalent approach to incorporating the initiator on the surface of the silica hosts, a strongly basic amidine-containing initia-tor (2,2'-azobis-(N,N'-dimethyleneisobutyramidine; ADIA) was used. This basic initiator was adsorbed to bare silica particles from chloroform, result-ing in a reproducible and slightly higher surface coverage ($1\,\mu mol/m^2$) than those generated by the covalent attachments. These initiator-modified par-ticles were then immersed in the monomer mixture — the template (L-phenylalanine anilide; L-PA), the functional monomer (methacrylic acid; MAA), the cross-linking monomer (EGDMA), and the porogen (toluene or dichloromethane) — followed by photopolymerization. The SEM micro-graphs of the resulting particles (Figure 13) showed morphology changes by the imprinting process on the surface of the silica hosts (Figure 13b) in comparison to the unmodified silica hosts. At higher grafting levels with increasing polymerization time, aggregated silica hosts held together by a web-like polymer structure were observed (Figure 13c).

Systematic investigation of these imprinted inorganic-organic hybrid adsorbents was accomplished by measuring the separation and the reso-lution parameters in HPLC. The noncovalent approach resulted in a much lower separation and resolution factor compared with that of the adsorbents

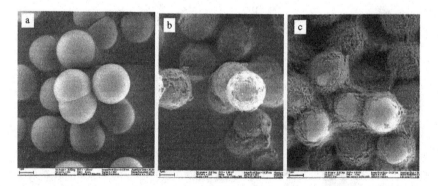

Figure 13. SEM micrographs of (a) unmodified silica hosts, (b) modified silica hosts with covalently attached APS initiator containing 3.6% carbon, and (c) modified silica hosts with covalently attached APS initiator containing 16% carbon.[60]

prepared by covalently attached initiators. Chromatographic performance of the adsorbents prepared by covalently attached initiators showed that the column efficiency depends on the thickness of the imprinted polymers on the silica hosts. Imprinted polymers grafted as thin film (~ 0.8 nm) showed a higher column efficiency, which decreased with increasing thickness of the polymer films. On the other hand, the sample loading capacity and separation factor were observed at a higher thickness of the imprinted polymers (~ 7.0 nm). These results show that these inorganic-organic hybrid adsorbents can be rendered tunable, by controlling the thickness of the imprinted polymers (e.g. via polymerization time), for different applications such as in HPLC stationary-phases (i.e. analytical and preparative chromatography).

To gain further control of the morphology on these inorganic-organic hybrid adsorbents, Sellergren *et al.*, used a so-called "iniferter" (Figure 14)[63] to prevent undesired polymerization in solution, which invariably accompanies the use of the covalently attached azo-initiators.

An "iniferter" is an initiator for free-radical polymerization, which undergoes primary radical termination and ordinarily avoids irreversible biomolecular termination. Iniferters decompose to give two different radicals: one able to initiate polymerization and the other being stable but capable of terminating the growing polymer chains by recombination. Recently, Ruckert *et al.*, used Scanning mode Transmission Electron Microscopy (STEM) and Energy Dispersive X-ray spectroscopy (EDX) to monitor the distribution of the imprinted polymers grafted on the surface of silica hosts using iniferters.[64] This living nature of the iniferters was also applied to consecutively graft two polymer layers imprinted with two different templates, or one imprinted and one non-imprinted layer in any order.[65] Recent comparative studies of these different formats of hierarchically imprinted adsorbents showed better accessibility to substrates when thin films of imprinted polymers were grafted on the silica hosts than with other formats, generated either by filling the pores of the silica hosts with polymerization mixtures and subsequent polymerization (leading to silica-imprinted polymer composites) or spherical imprinted adsorbents after the dissolution of the silica hosts.[66] These different morphologies of the imprinted adsorbents were shown to affect the mass transfer kinetics of the substrates without any significant influences on the molecular recognition properties of these imprinted adsorbents.[67]

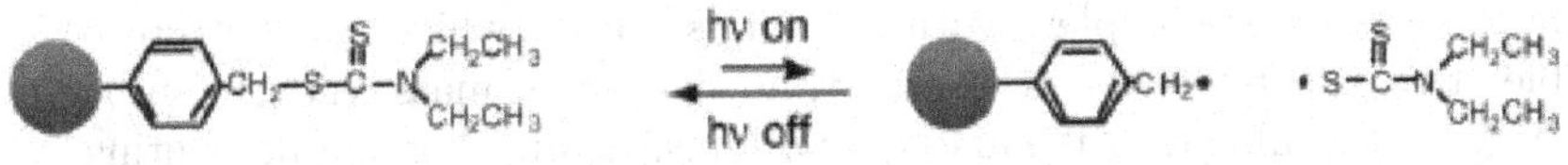

Figure 14.   Modifications of the silica hosts with a dithiocarbamate iniferter.

Surface Imprinted Titania Gel Film

Even though the above methods to prepare hierarchically imprinted adsorbents by the surface modifications provide satisfactory substrate selectivity, the molecular details of these modified surfaces are not clear. Forming ultrathin films on the surfaces provides an opportunity to study the molecular details of these modified surfaces and the practical application in sensor areas. In principle, the use of organized molecular films such as Langmuir-Blodgett multilayers and surface-bound monolayers is not compatible with the imprinting process, which requires the adaptable structural modification. Well-characterized imprinted ultrathin films can be formed on metal oxides by chemical vapor deposition, as shown by Kodakari et al., using preadsorbed benzoate anion as a template.[68] However, this chemical vapor deposition method requires a high temperature, at which thermal-labile organic compounds cannot be used as templates.

Recently, Kunitake et al.,[69,70] developed a method to prepare imprinted ultrathin titania gel films by sequential chemisorption and activation, where individual metal oxide layers are formed with nanometer precision at mild preparatory conditions appropriate for thermal-labile organic templates (Figure 15).

First the titanium alkoxide [Ti(O-nBu$_4$] was interacted with the azobenzene carboxylica acid template (C$_3$AzoCO$_2$H) to form the complexes. These complexes were added with water and aged for several hours to produce Ti$_4$O$_4$(OH)$_4$ (O-nBu)$_4$C$_3$AzoCO$_2$H. A gold-coated quartz crystal microbalance (QCM) electrode modified with mercaptoethanol was allowed to interact with the template solutions to form covalently bound surface monolayers of the titanium oxide-template complexes. The physisorbed complexes were removed by washing with toluene for one minute. The chemisorbed surface layers of the template complexes were then hydrolyzed in air to give a new hydroxylated surface. This process comprises one cycle, and multilayers on the surface can be formed by repeating this cycle several times with the reproducible control of the thickness of each layer (0.8–2.7 nm). The templates in the ultrathin titania films were completely removed with 1% aqueous ammonia for 30 minutes as estimated by the UV absorbance and the QCM frequency measurements.

The imprinted ultrathin titania films (ten cycles) were tested for their adsorption kinetics and selectivity toward the template using in situ QCM experiments. Figure 16a shows that the mass increase due to the rebinding of the template was complete within 40–60 seconds, which can be reproducible at least three times. In contrast, no significant mass change can be observed on the ultrathin titania control films, confirming the imprinting effect. Figure 16b shows the highest binding of the C$_3$AzoCO$_2$H template

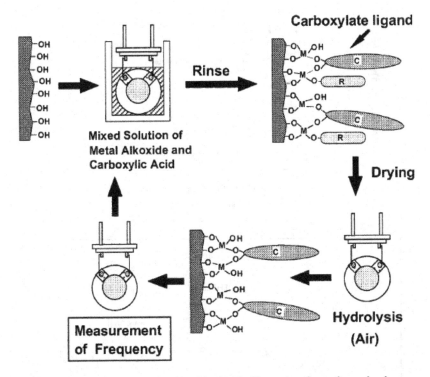

Figure 15.   Imprinting process in ultrathin titania films using the surface sol-gel process. "C" and "R" denote the template molecule and the unhydrolyzed butoxide group of Ti(O-nBu)$_4$.[69]

on the imprinted ultrathin titania films compared with other structural analogues. This method of generating well-controlled imprinted ultrathin titania films was extended for derivatized amino acids,[71] Mg^{2+} ions,[72] di- and tri-peptides,[73] and (+)-glucose.[74] Further refinements of this approach were also shown in the development of self-supporting imprinted ultrathin films[75] and liquid-phase deposition to make the titania ultrathin films.[76]

Conclusions

This chapter reviews the known available methods to generate hierarchically imprinted adsorbents for separation purposes. These hierarchically imprinted adsorbents allow us to independently control structural parameters (to facilitate accessibility and mass transfer kinetics) in addition to the desired adsorption parameters (i.e. selectivity toward imprinted molecules)

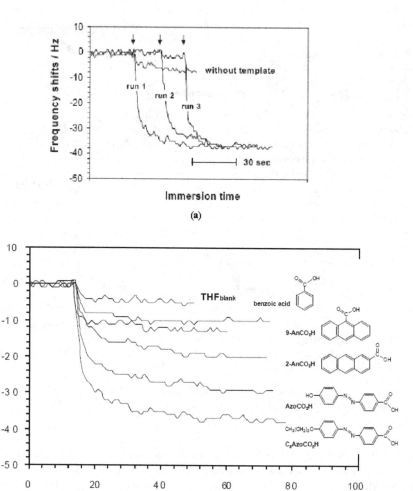

Figure 16. *In situ* QCM frequency decrease due to (a) rebinding of the template molecules and (b) rebinding of a series of carboxylic acids.[70]

using co-assembly and step-wise assembly synthesis. Thus, these materials should allow us to independently optimize these parameters for efficient separation and detection of various metal ions and biological molecules in environmental applications. Further systematic studies on these hierarchically imprinted adsorbents for their mass transfer kinetics and adsorption properties are also desired to facilitate the development of high-performance imprinted adsorbents for the practical applications in future.

References

1. E. Fisher, *Ber. Dtsch. Chem. Ges.* **27**, 2985 (1894).
2. S. Mudd, *J. Immunol.* **23**, 423 (1932).
3. L. Pauling, *J. Am. Chem. Soc.* **62**, 2643 (1940).
4. F. H. Dickey, *Proc. Natl. Acad. Sci. U.S.A.* **35**, 227 (1949).
5. G. Wulff and A. Sarhan, *Angew. Chem. Int. Ed. Engl.* **11**, 341 (1972).
6. G. Wulff, Enzyme-like catalysis by molecular imprinted polymers, *Chem. Rev.* **102**(1), 1–27 (2002).
7. K. Mosbach, Molecular imprinting, *Trends. Biochem. Sci.* **19**(1), 9–14 (1994).
8. B. Sellergren, Noncovalent molecular imprinting: Antibody-like molecular recognition in polymeric network materials, *Trends. Anal. Chem.* **16**(6), 310–320 (1997).
9. *Molecularly Imprinted Polymers. Man-Made Mimics of Antibodies and their Applications in Analytical Chemistry*, ed. B. Sellergren (Elsevier, 2003).
10. D. Spivak, M. A. Gilmore and K.J. Shea, Evaluation of binding and origins of specificity of 9-ethyladenine imprinted polymers, *J. Am. Chem. Soc.* **119**(19), 4388–4393 (1997).
11. H. Kim and G. Guiochon, Thermodynamic studies on the solvent effects in chromatography on molecularly imprinted polymers. 1. Nature of the organic modifiers, *Anal. Chem.* **77**, 1708–1717 (2005).
12. H. Kim and G. Guiochon, Thermodynamic studies on solvent effects in molecularly imprinted polymers. 2. Concentration of the organic modifier, *Anal. Chem.* **77**, 1718–1726 (2005).
13. H. Kim and G. Guiochon, Thermodynamic studies of the solvent effects in chromatography on molecularly imprinted polymers. 3. Nature of the organic mobile phase, *Anal. Chem.* **77**, 2496–2504 (2005).
14. B. Sellergren and K. J. Shea, Influence of polymer morphology on the ability of imprinted network polymers to resolve enantiomers, *J. Chromatogr.* **115**, 3368–3369 (1993).
15. K. Miyabe and G. Guiochon, A study of mass transfer kinetics in an enantiomeric separation system using a polymeric imprinted stationary phase, *Biotechnol. Prog.* **16**, 617 (2000).
16. H. Kim, K. Kaczmarski and G. Guiochon, Mass transfer kinetics on the heterogeneous binding sites of molecularly imprinted polymers, *Chem. Eng. Sci.* **60**, 5425–5444 (2005).
17. H. Kim, K. Kaczmarski and G. Guiochon, Intraparticle mass transfer kinetics on molecularly imprinted polymers of structural analogues of a template, *Chem. Eng. Sci.* **61**, 1122 (2006).
18. H. Kim and G. Guiochon, Thermodynamic functions and intraparticle mass transfer kinetics of structural analogues of a template on molecularly imprinted polymers in liquid chromatography, *J. Chromatogr.* **1097**, 84 (2005).

19. H. Kim, K. Kaczmarski and G. Guiochon, Isotherm parameters and intra-particle mass transfer kinetics on molecularly imprinted polymers in acetonitrile/buffer mobile phases, *Chem. Eng. Sci.* **61**, 5249 (2006).

20. S. Dai, M. C. Burleigh, Y. H. Ju, H. J. Gao, J. S. Lin, S. J. Pennycook, C. E. Barnes and Z. L. Xue, Hierarchically imprinted sorbents for the separation of metal ions, *J. Am. Chem. Soc.* **122**, 992 (2000).

21. E. Yilmaz, K. Haupt and K. Mosbach, The use of immobilized templates — A new approach in molecular imprinting, *Angew. Chem. Int. Ed.* **39**(12), 2115 (2000).

22. M. M. Titirici, A. J. Hall and B. Sellergren, Hierarchically imprinted stationary phases: Mesoporous polymer beads containing surface-confined binding sites for adenine, *Chem. Mater.* **14** (2002).

23. G. De, M. Epifani and A. Licciulli, Copper-ruby monoliths by the sol-gel process, *J. Non-Cryst. Solids.* **201**, 250–255 (1996).

24. A. T. Baker, The ligand field spectra of cooper(II) complexes, *J. Chem. Educ.* **75**, 98 (1998).

25. R. D. Makote and S. Dai, Matrix-induced modifications of imprinting effect for Cu^{+2} adsorption in hybrid silica matrices, *Anal. Chim. Acta* **435**, 169–175 (2001).

26. M. C. Burleigh, S. Dai, E. W. Hagaman and J. S. Lin, Imprinted polysilssquioxanes for the enhanced recognition of metal ions, *Chem. Mater.* **13**, 2537–2546 (2001).

27. Y. K. Lu and X. P. Yan, An imprinted organic-inorganic hybrid sorbent for selective separation of cadmium from aqueous solution, *Anal. Chem.* **76**, 453–457 (2004).

28. K. Uezu, R. Nakamura, M. Goto, M. Murata, M. Maeda, M. Takagi and F. Nakashio, Metal ion-imprinted resins prepared by surface template polymerization with W/O emulsion, *J. Chem. Eng. Jpn.* **27**, 436–438 (1994).

29. M. Yoshida, K. Uezu, M. Goto and F. Nakashio, Metal ion-imprinted resins with novel bifunctional monomer by surface template polymerization, *J. Chem. Eng. Jpn.* **29**, 174–176 (1996).

30. M. Yoshida, K. Uezu, M. Goto and S. Furusaki, Required properties for functional monomers to produce a metal template effect by a surface molecular imprinting technique, *Macromolecules* **32**, 1237–1243 (1999).

31. M. Yoshida, K. Uezu, F. Nakashio and M. Goto, Spacer effect of novel bifunctional organophosphorus monomers in metal-imprinted polymers prepared by surface template polymerization, *J. Polym. Sci. Part A: Polym. Chem.* **36**, 2727–2734 (1998).

32. K. Uezu, H. Nakamura, J. Kanno, T. Sugo, M. Goto and F. Nakashio, Metal ion-imprinted polymer prepared by the combination of surface template polymerization with postirradition by gamma-rays, *Macromolecules* **13**, 3888–3891 (1997).

33. K. Uezu, H. Nakamura, M. Goto, F. Nakashio and S. Furusaki, Metal-imprinted microsphere prepared by surface template polymerization with W/O/W emulsions, *J. Chem. Eng. Jpn.* **32**, 262–267 (1999).

34. M. Yoshida, K. Uezu, M. Goto and S. Furusaki, Metal ion imprinted microsphere prepared by surface molecular imprinting technique using water-in-oil-in-water emulsions, *J. Appl. Polym. Sci.* **73**, 223–1230 (1999).

35. M. Yoshida, K. Uezu, M. Goto and S. Furusaki, Surface imprinted polymers recognizing amino acid chirality, *J. Appl. Polym. Sci.* **78**, 659–703 (2000).

36. M. Yoshida, Y. Hatate, K. Uezu, M. Goto and S. Furusaki, Chiral-recognition polymer prepared by surface molecular imprinting technique, *Colloids and Surfaces A: Physicochem. Eng. Aspects* **169**, 259–269 (2000).

37. K. Araki, M. Goto and S. Gurusaki, Enantioselective polymer prepared by surface imprinting technique using a bifunctional molecule, *Anal. Chim. Acta* **469**, 173–181 (2002).

38. H. Tsunemori, K. Araki, K. Uezu, M. Goto and S. Furusaki, Surface imprinting polymers for the recognition of nucleotides, *Bioseparation* **10**, 315–321 (2001).

39. N. Perez, M. J. Whitcombe and E. N. Vulfson, Surface imprinting of cholesterol on submicrometer core-shell emulsion particles, *Macromolecules* **34**, 830–836 (2001).

40. S. Carter, L. Shui-Yu and S. Rimmer, Core-shell molecular imprinted polymer colloids, *Supramolecular Chemistry* **15**, 213–220 (2003).

41. S. Carter and S. Rimmer, Surface imprinted polymer core-shell particles, *Advanced Functional Materials.* **14**, 553–561 (2004).

42. S. Dai, M. C. Burleigh, Y. Shin, C. C. Morrow, C. E. Barnes and Z. Xue, Imprint coating: A novel synthesis of selective functionalized ordered mesoporous sorbents, *Angew. Chem. Int. Ed.* **38**, 1235–1239 (1999).

43. Y. S. Shin, J. Liu, L. Q. Wang, Z. M. Nie, W. D. Samuels, G. E. Fryxell and G. J. Exarhos, Ordered hierarchical porous materials: Toward tunable size- and shape-selective microcavities in nanoporous channels, *Angew. Chem. Int. Ed. Eng.* **39**, 2702–2707 (2000).

44. V. S. Lin, C. Lai, J. Huang, S. Song and S. Xu, Molecular recognition inside of multifunctionalized mesoporous silicas: Toward selective fluorescence detection of dopamine and glucosamine, *J. Am. Chem. Soc.* **123**, 11510–11511 (2001).

45. C. D. Ki, C. Oh, S. Oh and J. Y. Chang, The use of a thermally reversible bond for molecular imprinting of silica spheres. *J. Am. Chem. Soc.* **124**, 14838–14839 (2002).

46. Z. Zhang, S. Dai, R. D. Hunt, Y. Wei and S. Qiu, Ion-imprinted zeolite: A surface functionalization methodology based on "ship-in-bottle" technique, *Adv. Mater.* **13**, 493–496 (2001).

47. Z. Zhang, S. Saengkerdsub and S. Dai, Interface ion-imprinting synthesis on layered magadiite hosts, *Chem. Mater.* **15**, 2921–2925 (2003).

48. H. Yang, S. Zhang, W. Yang, X. Chen, Z. Zhuang, J. Xu and X. Wang, Molecularly imprinted sol-gel nanotubes membranes for biochemical separations, *J. Am. Chem. Soc.* **126**, 4054–4055 (2004).

49. R. A. Fletcher and D. M. Bibby, Synthesis of kenyaite and magadiite in the presence of various anions, *Clays and Clay Minerals* **35**, 318–320 (1987).

50. M. M. Titirici and B. Sellergren, Peptide recognition via hierarchicl imprinting, *Anal. Bioanal. Chem.* **378**, 1913–1921 (2004).
51. T. Shiomi, M. Matsui, F. Mizukami and K. Sakaguchi, A method for the molecular imprinting of hemoglobin on silica surfaces using silanes. *Biomaterials.* **26**, 5564–5571 (2005).
52. H. Yang, S. Zhang, F. Tan, Z. Zhuang and X. Wang, Surface molecularly imprinted nanowires for biorecognition, *J. Am. Chem. Soc.* **127**, 1378–1379 (2005).
53. E. Yilmaz, O. Ramstrom, P. Moller, D. Sanchez and K. Mosbach, A facile method for preparing molecularly imprinted polymer spheres using spherical silica templates, *J. Mater. Chem.* **12**, 1577–1581 (2002).
54. O. Bruggermann, R. Freitag, M. J. Whitcombe and E. N. Vulfson, Comparison of polymer coatings of capillaries for capillary electrophoresis with respect to their applicability to molecular imprinting and electrochromatography, *J. Chromatogr. A* **781**, 43–53 (1997).
55. J. M. Lin, T. Nakagama, K. Uchiyama and T. Hobo, Capillary electrochromatographic separation of amino acid enantiomers using on-column prepared molecularly imprinted polymers, *J. Pharmaceut. Biomed. Anal.* **15**, 1351–1358 (1997).
56. S. D. Plunkett and F. H. Arnold, Molecularly imprinted polymers on silica: Selective supports for high-performance ligand-exchange chromatography, *J. Chromatogr. A* **708**, 19–29 (1995).
57. S. Vidyasankar, M. Ru and F. H. Arnold, Molecularly imprinted ligand-exchange adsorbents for the chiral separation of underivatized amino acids, *J. Chromatogr. A* **775**, 51–63 (1997).
58. T. H. Kim, K. C. Do, H. Cho, T. Y. Chang and J. Y. Chang, Facile preparation of core-shell type molecularly imprinted particles: Molecular imprinting into aromatic polyimide coated on silica spheres, *Macromolecules* **38**, 6423–6428 (2005).
59. E. J. Acosta, S. O. Gonzalez and E. E. Simanek, Synthesis, characterization, and application of melamine-based dendrimers supported on silica gel, *J. Polym. Sci. Part A. Polym. Chem.* **43**, 168–177 (2005).
60. C. Sulitzky, B. Ruckert, A. J. Hall, F. Lanza, K. Unger and B. Sellergren, Grafting of molecularly imprinted polymer films on silica supports containing surface-bound free radical initiators, *Macromolecules* **35**, 79–91 (2002).
61. L. Schweiz, Molecularly imprinted polymer coatings for open-tubular capillary electrochromatography prepared by surface initiation, *Anal. Chem.* **74**, 1192–1196 (2002).
62. K. K. Unger, *Adsorbents in Column Liquid Chromatography*, ed. K. K. Unger, Vol. 47 (Marcel Dekker Inc. New York, 1990), pp. 331–470.
63. B. Ruckert, A. J. Hall and B. Sellergren, Molecularly imprinted composite materials via iniferter-modified supports, *J. Mater. Chem.* **12**, 2275–2280 (2002).
64. B. Ruckert and U. Kolb, Distribution of molecularly imprinted layers on macroporous silica gel particles by STEM and EDX, *Micron* **36**, 247–260 (2005).

65. B. Sellergren, B. Ruckert and A. J. Hall, Layer-by-Layer grafting of molecularly imprinted polymers via iniferter modified supports, *Adv. Mater.* **14**, 1204–1208 (2002).

66. F. G. Tamayo, M. M. Titirici, A. B. Martin-Esteban and B. Sellergren, Synthesis and evaluation of new propazine-imprinted polymer formats for use as stationary phases in liquid chromatography, *Anal. Chim. Acta* **542**, 38–46 (2005).

67. C. Baggiani, P. Baravalle, L. Anfossi and C. Tozzi, Comparison of pyrimethanil-imprinted beads and bulk polymer as stationary phase by nonlinear chromatography, *Anal. Chim. Acta* (2005) 125–134.

68. N. Kodakari, N. Katada and M. Niwa, Molcular-sieving silica overlayer on tin oxide prepared using an organic template, *J. Chem. Soc. Chem. Comm.* **6**, 623–624 (1995).

69. S. Lee, I. Ichinose and T. Kunitake, Molecular imprinting of azobenzene carboxylic acid on a TiO_2 ultrathin film by the surface sol-gel process, *Langmuir.* **14**, 2857–2863 (1998).

70. T. Kunitake and S. W. Lee, Molcular imprinting in ultrathin titania gel films via surface sol-gel process, *Anal. Chim. Acta* **504**, 1–6 (2004).

71. S. Lee, I. Ichinose and T. Kunitake, Molecular imprinting of protected amino acids in ultrathin multilayers of TiO_2 gel, *Chem. Lett.* **12**, 1193–1194 (1998).

72. J. H. He, I. Ichinose, S. Fujikawa, T. Kunitake and A. Nakao, A general, efficient method of incorporation of metal ions into ultrathin TiO_2 films, *Chem. Mater.* **14**, 3493–3500 (2002).

73. I. Ichinose, T. Kikuchi, S. W. Lee and T. Kunitake, Imprinting and selective binding of di- and tri-peptides in ultrathin TiO_2-gel films in aqueous solutions, *Chem. Lett.* **1**, 104–105 (2002).

74. S. W. Lee and T. Kunitake, Adsorption of TiO_2 nanoparticles imprinted with D-glucose on a gold surface, *Molecular Crystal and Liquid Crystal* **371**, 111–114 (2001).

75. M. Hashizume and T. Kunitake, Preparation of self-supported ultrathin films of titania by spin coating, *Langmuir* **19**, 10172–10178 (2003).

76. L. Feng, Y. Liu and J. Hu, Molecular imprinted TiO_2 thin film by liquid phase deposition for the determination of L-glutamic acid, *Langmuir* **20**, 1786–1790 (2004).

Chapter 10

Functionalization of Periodic Mesoporous Silica and Its Application to the Adsorption of Toxic Anions

Hideaki Yoshitake

Graduate School of Environment and Information Sciences
Yokohama National University
79-7 Tokiwadai, Hodogaya-ku
Yokohama 240-8501, Japan
yos@ynu.ac.jp

Introduction

Why Mesoporous Materials?

While the practical application of porous solids have been extensively studied, the pores in solid are classified into three categories. According to the IUPAC compendium, a mesopore is the name given to pores with widths between those of micropores (pore size < 2 nm) and macropores (pore size > 50 nm). The difference in size does not simply lead to various molecular sieving effects, but also to a wide range of characteristic chemical phenomena occurring within the pores. To discover chemically interesting phenomena specific to a particular pore size, we need ordered porous material, which has a defined pore size and a well-developed porous structure.

Mesoporous silica with a periodic structure was first reported in the early 1990s.[1,2] Attentions has been drawn to this material because of the unique way in which the structure is formed, the narrow pore size distribution on the meso-scale and the periodicity of mesostructure. Various kinds of surfactant micelle have been used as templates for directing the structure of mesoporous silicas, in which the size of the pores is almost the same as that of the micelle and the arrangement of the pore channels traces self-organized pattern of the micelle.[3-5] Typical mesoporous silicas appearing most frequently in the literature are illustrated in Figure 1. The procedure

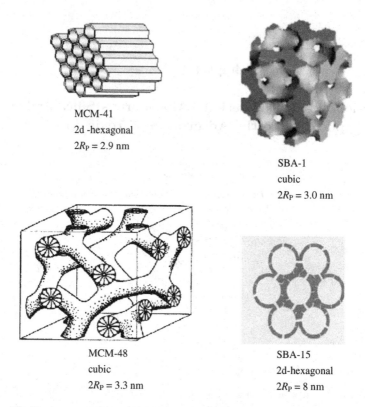

Figure 1. Porous structures of MCM-41, MCM-48, SBA-1 and SBA-15. Typical pore sizes are shown. The pore size distribution is narrow in these mesoporous silicas though the diameter $(2R_P)$ is usually controllable. Although the structure of the mesopore arrays of MCM-41 and SBA-15 (a cross sectional view is shown) is similar, these are connected to each other with micropores in the latter silica.

for synthesis is as follows; a silica precursor (e.g. tetraethyl orthosilicate, TEOS) and a surfactant are mixed in a basic or acidic solution. The mixture is maintained at 273 to 373 K for several days during which extensive hydrolysis occurs giving rise to the formation of siloxane (Si-O-Si) bonds. This polymerisation leads to precipitation and the solid is filtered out, washed and collected. Finally the product is calcined at 600 to 900 K or washed with a suitable solvent, such as hydrochloric acid or ethanol, to remove the surfactant. The space occupied by the surfactant micelle is converted into mesopores at lattice points in a certain space group. The structural characteristics of mesoporous silica, therefore, arise from micelle structure. From time to time, we call this synthesis "inverse replication." The pore size is distributed according to the micelle structure and the framework is usually amorphous. This is much different to the structure of crystalline zeolite prepared with a molecular ion template, which gives a much narrower pore size distribution than a surfactant micelle. The choice of surfactant is quite important in obtaining particular mesostructures. CTMAB (cetyltrimethylammonium bromide), CTEAB (cetyltriethylammonium bromide) and Pluronic® P123 are typically used for the synthesis of MCM-41 (and MCM-48, too), SBA-1 and SBA-15, respectively.

The application of this type of silica has been explored in most of the fields concerned with interfacial phenomena, such as molecular sieving, catalysis, sensor, and so on,[6–9] and a considerable number of studies have been contributed to the functionalization with reactive organic compounds, such as organic silanes. The adsorption of environmentally toxic cations and anions on such functionalized mesoporous silicas has been intensively investigated. This trend is not due to a simple coincidence of the rising importance of the chemistry of mesostructured materials and the increasing worldwide concern with environmental problems. If not, why have mesoporous silica frequently been employed as platforms for the synthesis of functionalized solids in the development of environmental remediation tools?

Large surface areas are often required in the application of solid materials. This "surface" should be, needless to say, accessible to the molecules and ions in the particular application, while the value evaluated by the BET surface area measurement, using nitrogen adsorption, does not always characterize the surface properly. Thus, when we conceive or plan a certain environmental application of a solid, we should further consider the physical and chemical properties of the surface, whether it is hydrophilic or hydrophobic, porous or nonporous, pore size distributions, etc. Since the surface energy increases with the specific surface area, fine particles tend to coagulate, resulting in a decrease in surface area and a deterioration in performance while they function. The porous structure is an important factor in extending the surface area while at the same time preventing such coagulation. This illustration is much simplified, but explains why the attention of researchers concerned with applications requiring large surface areas has been drawn towards porous solids.

The uniform pore size is considered to be one of the characteristics of practical importance. Among porous solids, zeolite[10] and carbon[11] have been most frequently studied. In the zeolites, which are found naturally or synthesized artificially, the pore size is at most 1.3 nm or less. This is not large enough for aqueous ions to diffuse rapidly into the pores. Moreover, the chemical nature of the zeolite surface is not always the most favourable for aqueous ions for the following reasons. The pores surface of pure silica zeolite is hydrophobic rather than hydrophilic due to its crystalline nature.

Higher hydrophilicity is obtained by choosing zeolites containing Al, which gives rise to defects with acidity. Hydrophobicity in zeolites is sometimes a disadvantage for the adsorption of toxic materials in an aqueous environment. Water is the most widely used medium also in industry. Furthermore, it is difficult to decorate the pore surface with organic functionalities due to the small pore size. This means that there are not many ways by which the structure of the adsorption sites for a particular target adsorbate can be optimized.

Active carbons have been widely used for the removal of harmful or unpleasant substances especially in closed environments. Although the BET specific surface area often exceeds $2000 \, m^2/g$, they have a wide pore size distribution from micro- to mesopores. In active carbons, the micropores usually contribute a significant fraction of the total pore volume and the BET surface area. This pattern of pore size distribution is not suitable for the solid materials to be used in solution. In fact, considerable effort has been devoted to diminishing the number of small micropores in order to increase the capacitance of carbon electrodes.[12] On the other hand, the surface hydrophilicity can be controlled, more or less, by a chemical treatment, because the organic functional groups terminate the lattice at the solid surface. However, for the development of active carbons as adsorbents or sensors in an aqueous environment, it should be noted that several kinds of surface groups, which have quite different reactivities, exist at the same time and the density of these functional groups is usually much lower than that of silanols on the surface of amorphous silica. These properties make the formation of a uniform and dense layer of functional groups difficult. The reaction of silane at the surface of amorphous silica mainly occurs between the surface silanols (Si-OH) and the alkoxy (Si-OR) bond in the silane molecule. In addition, the density of silanol in mesoporous silica is high, say 2 to $3 \, nm^{-2}$. These properties seem advantageous for the purpose of functionalization for synthesizing the adsorption sites of target ions.

In spite of the advantages of mesoporous silica for the adsorption of toxic substances, there are several defects *a priori*. Shape selectivity, such as that found in the molecular sieving effect of zeolite, is not to be expected for two reasons. One is that the pore sizes are distributed more than in zeolites. The zeolites are crystalline solid while mesoporous silicas are amorphous. The templates are a molecular ion and a micelle of surfactant, respectively. The second reason is that, even if we could prepare silica with well-defined pore size as in crystalline solids, the molecular sieving effect is less clear-cut in a mesopore window than in a microporous one. The variation of conformational isomers increases exponentially with molecular size, and this effect makes the dimensions of the molecules ambiguous. Therefore, it is not always reasonable to expect to have precise control in molecular sieving with mesoporous silica and the materials on a mesoporous silica platform. Another disadvantage of mesoporous silica is the instability of the mesoporous structure under hydrothermal conditions, such as in boiling water.[13] The fragility in water provokes a doubt about its durability in practical application. Once these disadvantages can be overcomed, mesoporous silica will provide an excellent platform for synthesized interfaces for remediation in aqueous environments.

Environmentally Toxic Anions

Drinking water is regulated as shown in Table 1; however excessive contaminant anions, such as arsenate, chromate, selenate and molybdate have often been found in the environment. It has been claimed in several different countries that chronic human health disorders caused by these toxins are discovered. The arsenic crisis in Bangladesh[14,15] is considered to be due to drinking water polluted with natural arsenate and arsenite in groundwater. An examination into the quality of groundwater in Chikugo Plain,

Table 1. Guidelines for the maximum amount of inorganic pollutants in drinking water (mg/L).

	WHO Guideline	US EPA Regulation	EU Direction	Japan Regulation
Antimony	0.005[a]	0.006	0.005	0.002[a]
Arsenic	0.01[a]	0.05	0.01	0.01
Barium	0.7	2		
Beryllium		0.004		
Boron	0.5[a]		1	1
Cadmium	0.003	0.005	0.005	0.01
Chloride			250	
Chromium	0.05[a]	0.1	0.05	0.05[b]
Copper	2[a]		2	
Cyanide	0.07	0.2[c]	0.05	0.01
Fluoride	1.5	4.0	1.5	0.8
Iron			0.2	
Lead	0.01		0.01	0.01
Manganese	0.5[a]		0.05	
Mercury	0.001	0.002[d]	0.001	0.0005
Molybdenum	0.07			0.07
Nickel	0.02[a]		0.02	0.01
Selenium	0.01	0.05	0.01	0.01
Sodium			200	
Sulfate			250	
Uranium	0.002[a]	0.03		0.002[a]
Thallium		0.002		

N.B. a: temporary value, b: as hexavalent Cr, c: as free cyanide, d: as inorganic mercury.

Japan demonstrated that about 25% of the 11,673 wells are contaminated with arsenic above the regulation limit for drinking water.[16] The chronic toxicity is also remembered as triggering the emergence of blackfoot disease on the south-western coast of Taiwan.[17,18] Exposure to arsenic from drinking water has also been reported from the American continents.[19] A major source of chromium pollution is industrial, such as from leather tanning and electroplating factories, where proper waste treatment facilities may be lacking.[20,21] These observations show that, although access to safe water is fundamental issue for living, it is often threatened in societies that depend on wells.

The exposure to anionic pollutants in groundwater is a universal phenomenon and the development of remediation methods is an urgent issue for environmental technology and materials chemistry. Building a water purification plant is normally an inappropriate measure for societies dependent on wells; and more cost-effective and environmentally friendly methods are needed. Chemical separation of harmful anions, such as chemisorption, can be a good solution; however, the difficulties will be encountered in the selectivity of such anions.

In the toxic oxyanions referred to above, the central element is penta- or hexavalent in the oxygen tetrahedron. These anions have a quite similar structure; the central atom — oxygen distances in AsO_4^{3-}, CrO_4^{2-}, SeO_4^{2-} and MoO_4^{2-} are 0.169, 0.165, 0.166 and 0.176 nm, respectively. This similarity in structure implies that the chemical behaviour is the same for adsorption on solid surfaces. From the environmental point of view, the selectivity of chloride, sulfate, nitrate, carbonate and phosphate will be more important in the removal of toxic oxyanions, because their typical concentration in fresh water is from several ppm to several tenths of ppm. Considering the environmental regulation, the adsorption of toxic oxyanions needs to be 100 to ~1000 times more selective than that of those abundant natural anions. Sulfate and phosphate are also tetrahedral oxyanions, even though the bond length between the central element and oxygen is smaller than in the pollutants: $r(\text{S-O}) = 0.148$ nm (in SO_4^{2-}) and $r(\text{P-O}) = 0.154$ nm (in PO_4^{3-}). The structural nature of such aqueous anions will cause difficulties in the separation of toxins. This problem would emerge as a severe inhibition of adsorption, because the number of adsorption sites would be limited. Fine control of the structure of the adsorption site is thus usually necessary for preparing a selective adsorbent.

Hierarchy of the Solid Structure and Adsorption

It has been demonstrated that solids synthesized by conventional methods, such as graining, impregnation, etc, have unsatisfactory adsorption strengths and capacities for toxic oxyanions.[22] For synthesizing effective adsorbents for a particular target ion, the structural factors of the solid should be considered both at the micro- and meso-scale. These views on different scale cannot always be separated but sometimes interfere with each other. Silica with a large surface area on which organic functional groups are grafted has been used as a synthetic adsorbate. In such a solid, the strength of the adsorption is controlled mainly by the microstructure of the adsorption site. This is critically important for oxyanion adsorbents, because a nearly complete removal (= a huge distribution coefficient) is necessary in order to meet the environmental regulations and the high selectivity to other bulk anions with a similar structure is also needed. The rate of diffusion during adsorption depends on the porous structure, which is to be evaluated at the micro- and meso-scale. The preparation of the surface, which is also called functionalization or decoration of the surface, also depends considerably on the micro- and mesostructure. It has been maintained for a long time that one of the examples of the study carried out on microstructures is the reactivity of silanol groups on a silica surface, which has, in fact, been argued for a long time.[23] On the other hand, it has been pointed out that uniform dispersion of organic functional groups is difficult in the functionalization of a mesoporous silica.[24] The structure of the surface organic groups and the results of adsorption experiments can vary considerably, even when the same kind of functional group is grafted on silica with a similar pore size and surface area, since they depend on the preparation method and the porous structure.

In functionalized mesoporous silicas, the microstructure revealed by conventional physicochemical techniques, such as Infrared Spectroscopy (IR), Nuclear Magnetic Resonance Spectroscopy (NMR), UV-Vis Spectroscopy, etc, is less like that predicted by the adsorption characteristics than in functionalized nonporous silicas. This is due to the interference by mesopores on the microscopic structure of the organic groups during preparation as well as in the working state. The adsorption is sometimes influenced more by the distribution of adsorption sites in the mesopores and the interactions between organic groups and the framework silica surface than the microstructure of adsorption sites. This cannot only be true for the case of functionalized mesoporous silica but also for functionalized fumed silica, polymers and natural substances. It is proper to mention that the adsorption properties sensitive to the mesostructure will appear explicitly only due to the usage of a defined and rigid mesoporous structure for the adsorbates. In this sense, choosing mesoporous silica with periodically defined structure for a platform for synthetic adsorbate is a good way to obtain an understanding of mesostructural factors in the disturbance of adsorption. The features of oxyanion adsorption under mesostructural influence are described in the following sections of this chapter.

Functionalization of Mesoporous Silicas

Decoration of the Pore Surface with an Organic Layer

The three methods described in this section are illustrated in Scheme 1.

One of the most simple preparation methods of organic layers in mesopores is grafting, in which reactive silane with a desired functionality is "grafted" via the reaction mainly with isolated silanols on the mesoporous silica. The loading of the organic groups and the uniformity of the organic

(a) graft

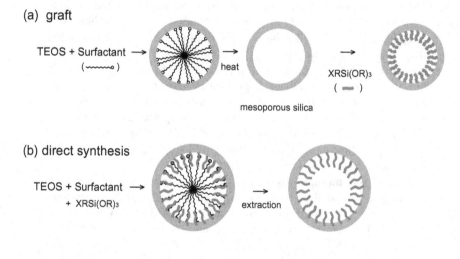

(b) direct synthesis

(c) functionalization in association with templating

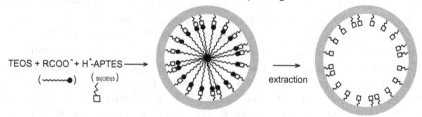

Scheme 1.   Preparation of functionalized mesoporous silica, illustrated with the formation of the silica framework: grafting, direct synthesis and functionalization in association with templating.

layer depend on water on the silica surface. Although this preparation method and its characteristics can be applied to all kinds of amorphous silicas, a specific problem arising from the mesoporous structure has been recognized. This is the disturbance of the uniform distribution of the functional groups.

Figure 2 shows the adsorption capacity of arsenate on amino-functionalized MCM-41 and SBA-1 prepared by grafting 3-aminopropyltrimethoxylsilane ($NH_2CH_2CH_2CH_2Si(OCH_3)_3$), [1-(2-aminoethyl)-3-aminopropyl]trimethoxysilane ($NH_2CH_2CH_2NHCH_2CH_2CH_2Si(OCH_3)_3$, which can be called EDA-propylsilane), and 1-[3-(trimethoxysilyl) propyl] diethylenetriamine ($NH_2CH_2CH_2NHCH_2CH_2NHCH_2CH_2CH_2 Si(OCH_3)_3$, which can be called DETA-propylsilane). The adsorption increases with the number of amino groups in the silane molecule, though, with careful observation, the lack of linearity can be found on the plot of MCM-41. In addition, in spite of the similar surface area and pore size, the adsorption is considerably different between these two mesoporous frameworks. This is certainly because the amino groups in a large organic group or in highly loaded organic groups adsorb less arsenate than those in small or dispersed organic groups.[25] The difference between the frameworks is more significant in a plot of the As/N ratio versus the amount of amino groups loaded on the

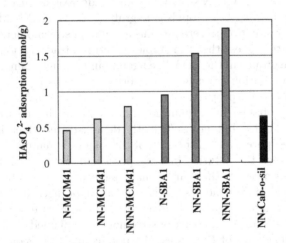

Figure 2. Adsorption capacity of arsenate on 3-aminopropyltrimethoxylsilane (N-), [1-(2-aminoethyl)-3-aminopropyl]trimethoxysilane (NN-), 1-[3-(trimethoxysilyl) propyl]diethylenetriamine (NNN-) grafted on MCM-41 and SBA-1. The BET specific surface area of these mesoporous silica frameworks is 1280 and 1220 m^2g^{-1}, respectively, and the pore size is 2.9 and 3.0 nm, respectively. The result on [1-(2-aminoethyl)-3-aminopropyl]trimethoxysilane grafted on non-porous silica (Cab-O-sil M-7D, BET surface area: 200 m^2g^{-1}) is shown for comparison.

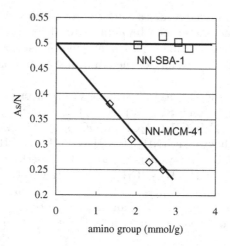

Figure 3. Arsenate adsorption capacity of [1-(2-aminoethyl)-3-aminopropyl] trimethoxysilane-grafted MCM-41 and SBA-1. The ratio of arsenic to nitrogen is plotted against the amount of amino groups.

solid (Figure 3). This apparent stoichiometry is constantly found in EDA-propylsilane-grafted SBA-1, NN-SBA-1, while it decreases with the loading of amino groups in EDA-propylsilane-grafted MCM-41, NN-MCM-41. However, the extrapolation to zero gives us a common stoichiometry: As/N = 0.5. These plots suggest that the stable structure at adsorption saturation is two amine groups in the EDA-like ligand binding one arsenate anion, unless there is inhibition by neighboring organic groups or by the pore wall, and that a certain proportion of the organic groups in NN-MCM-41 does not work, probably due to mutual interference in the adsorption. It has often been claimed but not sufficiently demonstrated that, in grafting silane on mesopore walls, dispersion of the silanes is hampered by the competition between the diffusion into the pores and the reaction with the pore surface.[24] As a result, the grafted silane molecules tend to be densely populated near the pore windows. Although this hypothetical mechanism might be closely related with the decrease of As/N in Figure 3, the difference between the framework structures cannot be explained. It could be caused by the curvature of the pores, the pore-connecting geometry, the higher surface reactivity, and other properties of MCM-41.

Since uniform dispersion is often unlikely when grafting silanes onto mesoporous silica, the co-condensation of TEOS and silane with a desired function is proposed for the preparation of functionalized mesoporous silica. This method is also called direct synthesis.[24] Although several problems have been pointed out in direct synthesis, the organic groups are much more

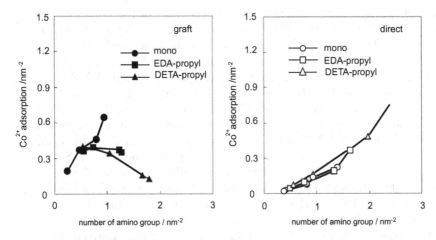

Figure 4. Adsorption of Co^{2+} on (a) 3-aminopropyltrimethoxylsilane (mono-), [1-(2-aminoethyl)-3-aminopropyl]trimethoxysilane (EDA-), 1-[3-(trimethoxysilyl) propyl]diethylenetriamine (DETA-) grafted MCM-41; and (b) those equivalent mesoporous functionalized silicas prepared by the direct co-condensation method. These are plotted against the amount of amino groups.

uniformly dispersed than in grafting when TEOS and silane are well mixed before hydrolysis. A distinct difference between amino-functionalized mesoporous silicas prepared by grafting and co-condensation has been found in cobalt(II) adsorptions (Figure 4).[26] The adsorption capacity is generally increased with the number of adsorption sites grafted, if the functional groups are grafted uniformly on the surface. On the contrary, the variation of adsorption depends on the type and the surface "density" of silanes. In grafted MCM-41 ($2R_p = 3.6\,nm$), the amount of Co^{2+} adsorption is not proportional to the amount of amino groups. The adsorption is even unchanged or decreased with the loading, when EDA-propylsilane or DETA-propylsilane is applied, respectively. These nonlinearities can be explained by the following mechanism. A large silane molecule can suffer from interference with its diffusion into the pores, while the reactivity of methoxy groups is almost the same as in short silanes, and, consequently, the silane may settle near the pore window. This settling can inactivate the amino groups by the "congestion" of organic chains as well as inhibit the diffusion of Co^{2+}.[26] In contrast, all the adsorption data for the functionalized MCM-41-motif prepared by direct synthesis appear almost on the same line, implying a homogeneous dispersion of amino groups in the pores, regardless of the surface density of the amino groups and the structure of the organic chains in the silane molecule.

Direct synthesis can inactivate a part of the functional groups through strong interaction with the surface or it can bury the organic groups in the framework.[24] This problem will be critical in amino-functionalization, considering the significant disturbance in the formation of ordered mesopore walls (suggesting a strong interaction with the silica framework).[27] A solution for avoiding the disturbance of amino groups during the siloxane bond formation is functionalization in association with templating (FAT).[24,28,29] In this modified direct synthesis, the template molecule and the functional group are strongly bonded during hydrothermal synthesis. The functional group is a part of the silane molecule and the other silane bond, Si-OR so far in the literature, is hydrolysed with a silica precursor to form a siloxane network. After the mesostructure is formed, the solid is treated with a suitable solvent to dissociate the bond in the surfactant-functional group complex. To prepare 3-aminopropyl-functionalized mesoporous silica, 3-aminopropyltriethoxysilane (APTES) and TEOS is hydrolysed in the presence of a carboxylate with a long alkyl chain. Under suitable pH conditions, where the amino groups are protonated and the rate of hydrolysis of $Si\text{-}OC_2H_5$ bonds is still high enough, amino-groups and carboxylate are bound by ionic interaction during hydrothermal synthesis. The hydrolysis of $Si\text{-}OC_2H_5$ in APTES and TEOS results in polymerisation and a solid formation around the micelles. The carboxylates are removed by washing with acidified CH_3CN to liberate the amino groups on the pore surface. Considering the lack of reports on the successful synthesis of mesoporous silica using TEOS and an anionic surfactant, the interaction between amino and carboxyl groups is essential for the mesostructure formation. It may also be claimed that all amino groups exist, theoretically, only on the pore surface and that their dispersion is uniform.

Figure 5 shows a comparison between argentometric titration and elemental analysis of nitrogen in functionalized mesoporous silica. The former measurement is based on the reaction: $Ag^+ + Cl^- \rightarrow AgCl$ (precipitated), where the Cl^- neutralizes the $-NH_3^+$ of the aminopropyl groups; and, hence, is a quantification of the $-NH_3^+$ exposed on the surface. The ratio of nitrogen measured by titration to that by elemental analysis, therefore, indicates the surface exposure of the amino groups. This is independent of the loading of amino groups in 3-aminopropyl-functionalized mesoporous silica prepared by the FAT method and the titration records 95% of nitrogens detected by the bulk elemental analysis. On the other hand, the equivalent mesoporous silica (MCM-41-motif) prepared by direct synthesis using CTMAB and the same silane (3-aminopropyltrimethoxysilane) provides a linearly decreasing function against loading. This distinct difference clearly demonstrates the advantage of FAT method in exposing the functional groups to the surface. The fact that almost all the amino groups are exposed to the surface

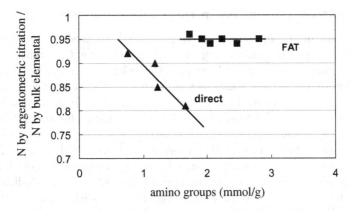

Figure 5. Ratio of nitrogen measured by argentometric titration to that measured by bulk elemental analysis in 3-aminopropyl-functionalized mesoporous silicas. The solid samples are prepared by functionalization in association with templating (FAT) and by direct synthesis (direct). The template surfactants are sodium laurate (LAS) and CTMAB, respectively.

over all loadings reveals that the attractive interaction between the amino and carboxyl groups really works during the mesostructure formation. The head group of CTAB that is used for direct synthesis can be repulsive to the amino groups during the mesostructure formation. This mode of interaction promotes the interaction between amines with silanol groups or expels the organic chain into the micropores in the wall.

Another advantage of the FAT method is an easy achievement of high loading of the functional group. More than 5.4 mmol/g of 3-aminopropyl groups can be found in functionalized mesoporous silica without loss of periodicity, when it is prepared by the FAT method.[24] On the other hand, in direct synthesis, 3-aminopropyltrimethoxysilane and CTAB provides a loading of 1.7 mmol/g.[26] Although the density of amino groups depends considerably on the pre-treatment of the MCM-41, the grafting provides around 1.5 to 2 mmol/g after the removal of physisorbed water to activate Si-OH groups on MCM-41.[25,26] The comparison with simple direct synthesis and grafting reveals the outstanding loading of functional groups by FAT.

Other valuable techniques — such as on-surface synthesis, graft on as-synthesized mesoporous silica, multiple-functionalization of specific sites (in the pores or on the exterior surface), and so on — in the preparation of organic functionalities on mesoporous silicas are described in a recent review.[24]

Transition Metal Cation-Incorporated Adsorption Sites

The adsorption of oxyanions on amino-functionalized mesoporous silica is an ion-exchange process. During the adsorption, the Cl^- that neutralizes the amino group is released into the solution. Thus the amino group is protonated (by the treatment with an HCl solution.) The amino groups, especially chelating ligands such as EDA, coordinate transition metal cations. Many of this kind of complex are usually stable over a wide pH range. Considering, roughly, the interaction between transition metal ions and oxyanions found in insoluble minerals, such as scorodite (iron(III) arsenate dihydrate, $FeAsO_4 \bullet 2H_2O$), much stronger chemical interactions with oxyanions can be obtained at the transition metal cation coordinate sites than the protonated amino group. Iron(III)-, cobalt(II)-, nickel(II)- and copper(II)-coordinated EDA-propyls have been prepared with [1-(2-aminoethyl)-3-aminopropyl]trimethoxysilyl mesoporous silicas.[30,31]

This is a typical on-surface synthesis (post-modification), where the coordination of the cation is carried out at the grafted functional groups, and, hence, attention should be paid to the uniformity. This is because the conversion of surface organic groups by substitution, oxidation, reduction, and so on, has been usually small in the studies on on-surface synthesis of mesoporous silicas. The examples are presented in Ref. 24. The cation coordination also provides unsatisfactory results for a uniform conversion as shown in Table 2. The nitrogen/metal cation ratios in the cation-coordinated amino-functionalized mesoporous silicas quite often have an unreasonable stoichiometry for the coordination complexes expected.[31,32]

Table 2. Uptake of transition metal cations by amino-functionalized mesoporous silicas. N/M^{n+} ratios measured by elemental analyses.

	N Density* (mmol/g)	Fe^{3+}	Co^{2+}	Ni^{2+}	Cu^{2+}
N-MCM-41	1.5	3.2	2.5	0.48	2.6
N-MCM-48	1.5	2.0	1.8	0.43	2.0
NN-MCM-41	2.8	3.8	3.3	0.55	3.6
NN-MCM-48	2.7	2.7	2.3	0.40	2.4
NNN-MCM-41	3.5	2.8	7.9	1.1	9.4
NNN-MCM-48	4.2	2.8	7.0	1.0	9.0

N.B. MCM-41 and MCM-48 functionalized by grafting 3-aminopropyltrimethoxylsilane (N-), [1-(2-aminoethyl)-3-aminopropyl]trimethoxysilane (NN-) and 1-[3-(trimethoxysilyl)propyl]diethylenetriamine (NNN-) are stirred in chlorides of these cations at room temperature.
*The specific amount of nitrogen in the solid before coordinations of metal cations. A part of them are lost after coordination, accompanied with a decrease of the BET surface area.

The ratio less than unity found in Ni^{2+}-coordinated mesoporous silicas suggests the adsorption of nickel(II) on the silica surface. In contrast, $N/M^{n+} = 7 \sim 9$ in NNN-functionalized mesoporous silicas indicates a large number of uncoordinated amino groups in the pores. Table 2 shows only the average compositions of the bulk solids and this may be the results from mixtures of several sites with different structures. However, the N/M^{n+} ratios for $Fe^{3+}/$ and Cu^{2+}/NN-MCM-41 suggest a structure in tetra-coordination of the amino groups.

The N/M^{n+} ratio depends on the loading of the EDA-propyl groups. It has been claimed from the results of EDS-SEM that, at an extremely high loading of the EDA-propyl groups, a $[Cu(EDA)_3]^{2+}$-type complex is formed in NN-MCM-41.[30,33]

Since the adsorption sites are not usually uniform and interference between the sites occurs, Langmuir type adsorption isotherms are not obtained in these adsorbents. Furthermore, high quality plots are hardly obtained due to the low concentration of oxyanions in the experimental solution. Instead, the strength of the adsorption should be evaluated by the distribution coefficient, K_d, between the solid and the solution. A typical K_d — adsorption plot is shown in Figure 6. The detection limit of the chemical analysis gives the maximum distribution coefficient, which is equal to 2×10^5 in Figure 6. In all the adsorbents, K_d decreases with adsorption, implying the distribution of the adsorption strength or interference between

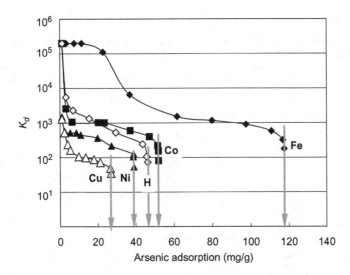

Figure 6. Distribution coefficient versus adsorption of arsenate on Fe^{3+}, Co^{2+}, Ni^{2+}, Cu^{2+} and H^+-coordinated NN-MCM-41.

the adsorption sites. The vertical drops that are observed with relatively low K_d indicate adsorption saturation. The adsorption capacity is given by these drops (with arrows in Figure 6). Both the strength and capacity of the adsorption are highest in Fe^{3+}/NN-MCM-41. The value of K_d for this adsorbent reached more than 10^5 in the region where that for Cu^{2+}/NN-MCM-41 was 10^2; the strength of arsenate adsorption differs by three orders of magnitude. The adsorption capacity of this iron-activated adsorbent is four times higher than the copper-activated one. The adsorption strength and capacity depend considerably on the cations in the adsorption site.

The adsorption capacity of these transition metal cation-activated amino-functionalized mesoporous silicas is summarized in Table 3.[25,31,32] Although in practice the uptakes of arsenate per unit amount of adsorbent have a large impact, and are, in fact, superior to those found in natural or nearly natural solids, such as geothite $(20\,mg/g)$,[34] natural iron ore $(0.4\,mg/g)$,[35] activated alumina $(15.9\,mg/g)$,[36] anatase with a BET surface area of $251\,m^2/g$ $(41.4\,mg/g)$,[37] and so on, the molar ratio of arsenic to cation is more important in exploring the mechanism and the structure of the adsorption site. Many of the As/M^{n+} values are irrational for a stoichiometric reaction, just as found in Table 2; the adsorption structures are likely to be mixtures. The As/M^{n+} ratio will contain structural information of the adsorbent only when the molar ratio of the nitrogen to the cation is nearly stoichiometric, Fe^{3+}/NN-MCM-41 and Cu^{2+}/NN-MCM-41 (Table 2). The values for these adsorbents are 2.8 and 0.7, respectively, suggesting that the

Table 3. The adsorption capacity of arsenate on cationized amino-functionalized mesoporous silicas.

		H^+	Fe^{3+}	Co^{2+}	Ni^{2+}	Cu^{2+}
N-MCM-41	C_{max}	64	122	84		
	As/M^{n+}	0.35	2.1	1.1		
N-MCM-48	C_{max}	71	146	124		
	As/M^{n+}	0.37	1.6	1.2		
NN-MCM-41	C_{max}	86	226	97	73	51
	As/M^{n+}	0.22	2.8	1.2	0.25	0.69
NN-MCM-48	C_{max}	122	353	141	121	71
	As/M^{n+}	0.38	2.7	1.2	0.32	0.65
NNN-MCM-41	C_{max}	110	226	74		
	As/M^{n+}	0.25	1.6	1.3		
NNN-MCM-48	C_{max}	182	251	88		
	As/M^{n+}	0.37	1.4	1.1		

N.B. C_{max} in $mg(g\text{-adsorbent})^{-1}$. The blanks indicate not measured.

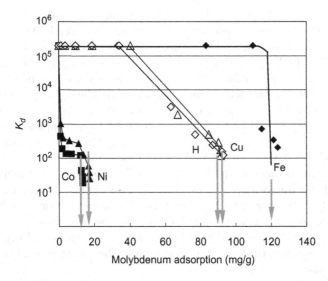

Figure 7. Distribution coefficient versus adsorption of molybdate on Fe^{3+}, Co^{2+}, Ni^{2+}, Cu^{2+} and H^+-coordinated NN-MCM-41.

stoichiometric ratio at saturation is $As:M^{n+} = 3:1$ and $1:1$ for iron and copper, respectively. Spectroscopic analysis can determine the structure (*vide infra*).

The K_d-adsorption plot is useful for an appraisal of the adsorption properties of environmentally toxic oxyanions.[31b] We can easily tell the behaviour of an ideal adsorbent; K_d is immeasurably high (meaning below the detection limit in the analysis of the concentration of the anions remaining in solution) till adsorption saturation is reached and the capacity is large enough. This is observed in the adsorption of molybdate on $Fe^{3+}/$ NN-MCM-41 as shown in Figure 7. The plots for the other cations are completely different from those in arsenate adsorption. The strength and capacity of Cu^{2+} and $H^+/$NN-MCM-41 are larger than Ni^{2+} and $Co^{2+}/$NN-MCM-41, whereas Co^{2+}, H^+ and Ni^{2+} show larger K_ds and capacities than Cu^{2+} in arsenate adsorption.

The Structure of the Adsorption Sites

In solution, tris(ethylenediamine)iron(III) sulphate has been known as a $Fe(EDA)_3$ complex.[38] The complexation constants, $\log K_1$, $\log K_2$ and $\log K_3$, for the EDA ligand with Co^{2+} are 5.97, 4.91 and 3.18 (in an ionic strength of $1.4\,mol/dm^3$), respectively, while those with Cu^{2+} are 10.72, 9.31 and -1.0. The third constant for Cu^{2+} suggests that $[Cu(EDA)_3]^{2+}$

is possible, only if the concentration is high enough.[39,40] These equilibrium constants in the homogenous chemistry imply that the three-fold coordination by the EDA ligand can be stable for these transition metal cations. It is well known that, in d^9 octahedral complexes, the number of electrons in the e_g orbital is odd and trans-$[Cu(EDA)_2(H_2O)_2]^{2+}$ is very stable. The fixation of the silane on the surface will considerably decrease the degree of freedom in the motion of EDA and, consequently, a high surface density (more than 3 EDAs per nm^2) is necessary to obtain an $M(EDA)_3$ complex. On the contrary, the surface EDA density ($= $ (N content)$/2\ S_{BET}$) is 0.9 and 1.4 per nm^2 for NN-MCM-41 and NN-MCM-48, respectively. [30] These values support the contention that $M(EDA)_2Cl_n^-$ and $M(EDA)Cl_n^-$ are the major species on these amino-functionalized mesoporous silicas. X-ray Absorption Near-Edge Structure (XANES) and Extend X-ray Absorption Fine-Structure (EXAFS) spectroscopies are suitable for the determination of the structure of the adsorption sites, both before and after oxyanion adsorption, when a transition metal cation is incorporated in the amino-functionalized mesoporous silica.

A structural determination of sulphate and arsenate adsorption on Cu^{2+}/NN-mesoporous silica ($2R_p = 6$ nm) has been carried out in detail.[33] It is claimed that parts of the EDA ligands in the $Cu(EDA)_3$ site dissociate when the oxyanion is coordinated. The local structure around Cu and As indicates direct Cu-O bond formation, a monodentate linkage between the anion and Cu^{2+} and a trigonal bipyramidal geometry of the Cu centre (Figure 8). With this adsorbent, however, the specific adsorption reached 142 mg/g.[30] This is significantly larger than the data in Table 3: 51 and

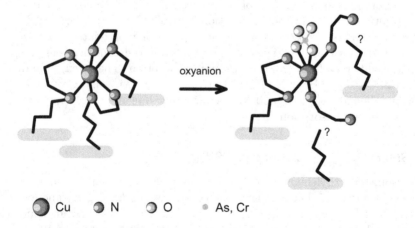

Figure 8. $[Cu(EDA)_3]$ centre for arsenate adsorption. A possible adsorption model.

71 mg/g for NN-MCM-41 and NN-MCM-48, respectively. This difference is likely to be due to a larger number of silanes being grafted and, hence, Cu^{2+} coordinated: 1.0–1.4 mmol/g in the study in Ref. 30, which is almost twice as large as the density in Cu^{2+}/NN-MCM-41 (Table 2). Interestingly, the As/Cu ratio in Table 3, 0.69, agrees remarkably with the adsorption on highly dense Cu^{2+}/NN-mesoporous silica, 0.71. These comparisons suggest that the capacity is explained simply by that of the loading of Cu^{2+} on amino-functionalized mesoporous silica and the mode of interaction of arsenate is not very sensitive to the coordination environment of Cu^{2+}.

The oxyanion adsorbed on Fe^{3+}/NN-MCM-41 has been investigated by X-ray Absorption Fine-Structure (XAFS) spectroscopies.[31,32] Figure 9 shows the Fe K-edge XANES of Fe^{3+}/NN-MCM-41 and those adsorbing arsenate, chromate, selenate and molybdate. Two unresolved peaks appear in the pre-edge region of the spectra for arsenate and selenate adsorbed Fe^{3+}/NN-MCM-41. The first component appears at about 7110.5 eV, which has a smaller area than the second one at about 7110.8 eV. These combined peaks are a typical feature of an O_h-like Fe(III) centre. Theoretical and simulative studies of the Fe K-edge 1s–3d transition of iron complexes revealed that the absorption at the lower energy is related to electric quadrupole transitions, whereas that at the higher energy is due to both quadrupole and dipole transitions.[41] In molybdate-adsorbed Fe^{3+}/NN-MCM-41, the intensity is almost the same as in the selenate-adsorbed Fe, but the peaks are not distinguished. The enhanced pre-edge peak is caused by the dipole-allowed transition by p-d hybridization. This feature is observed in the spectrum of Fe^{3+}/NN-MCM-41 and in that binding with chromate. After the adsorption of arsenate, selenate and probably molybdate, the structure of Fe^{3+} becomes more symmetric than the initial Fe^{3+}/NN-MCM-41, while no such structural change is observed in chromate adsorption. The difference in the pre-edge peak among the oxyanions suggests a variation in the adsorption strengths. The distribution coefficient in chromium adsorption is smaller than the other three oxyanions (e.g. comparing the adsorption of 20 mg/g, K_d for CrO_4^{2-} is 3000 while those for $HAsO_4^{2-}$, SeO_4^{2-} and MoO_4^{2-} are 110000, > 200000 and > 200000, respectively). This is not a simple coincidence but suggests a mechanism where the relatively weak interaction does not significantly disturb the coordination environment of Fe^{3+}.

The Fe K-edge EXAFS spectra measured before and after the adsorption of 1, 2 and 2.8 equivalent arsenates on Fe^{3+}/NN-MCM-41 follows the growth of the adsorption structure.[31] $k^3\chi(k)$ EXAFS oscillations with curve-fitting results and their Fourier transforms (FT) are shown in Figure 10. In the radial distribution functions, the peak around 2.9–3.0 Å is attributed to As scatterers. This peak appears in the FT for the adsorption of one equivalent

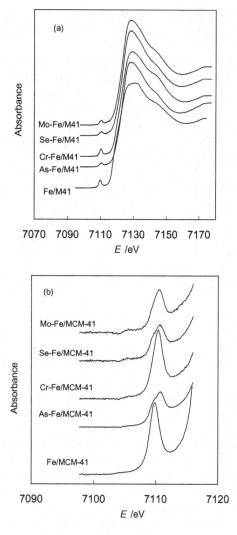

Figure 9. (a) Fe K-edge XANES spectra of Fe^{3+}/NN-MCM-41 and that after adsorbing arsenate, chromate, selenate and molybdate. (b) The preedge region expanded for comparison.

arsenate and it grows for the adsorption of two equivalent arsenates. However, the intensity for three equivalent adsorptions is almost the same as that for two. The coordination number of this bond is 0, 1.0 ± 0.4, 1.6 ± 0.5 and 1.6 ± 0.5 for before adsorption, adsorption of one, two and three equivalent arsenates, respectively. In contrast, as the adsorption progresses the

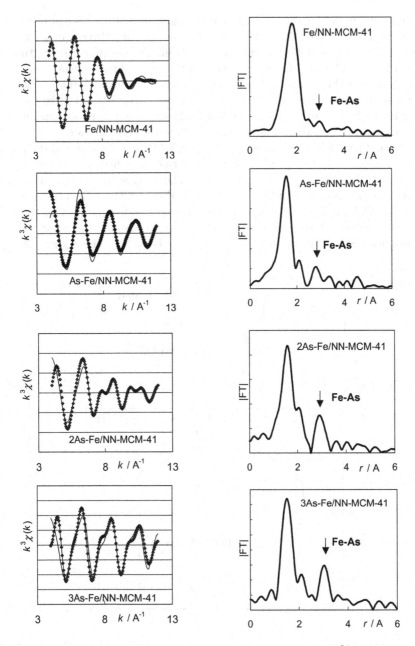

Figure 10. Fe K-edge EXAFS spectra and Fourier transforms of Fe^{3+}/NN-MCM-41 and that after adsorbing arsenate. The amount of arsenate adsorption was changed to control the coverage.

coordination number for the Cl shell decreases; found to be 2.1 ± 0.4 prior to adsorption, $1.2\pm0.4, 0.4\pm0.2$ and 0.2 ± 0.2, for one, two and three equivalent arsenates. The distance is unchanged at $0.227\,\mathrm{nm}$. The Fe–As bond length is constant at $0.327\,\mathrm{nm}$. This result suggests that two of the arsenate ions can be bound in the inner shell environment of Fe^{3+} even with full coverage of arsenate, where the stoichiometry is almost As:Fe = 3:1. Unlike the change in the Fe^{3+} centre, the EXAFS of As K-edge showed no significant change in the As-O bond after adsorption, both in the coordination number and the bond length, as shown in the $k^3\chi(k)$ EXAFS oscillations and their Fourier transforms in Figure 11. The result of the curve-fitting calculation is $N_{\mathrm{As-O}} = 3.9 \pm 0.4$ and $r_{\mathrm{As-O}} = 0.169\,\mathrm{nm}$. The tetrahedral coordination environment of arsenic is unlikely to be disturbed by the adsorption on Fe^{3+}/NN-MCM-41. The As-Fe length and the coordination number in the solid with As:Fe = 1:1 are $0.327\,\mathrm{nm}$ and 0.95, respectively, and these parameters are not changed in the adsorption with As:Fe = 2:1. The agreement

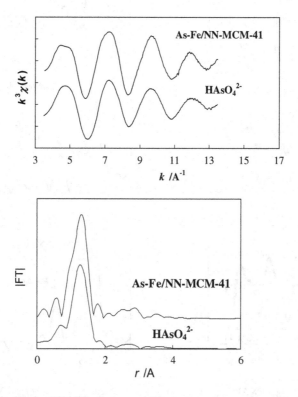

Figure 11. As K-edge EXAFS spectra and Fourier transforms of arsenate in water and on Fe^{3+}/NN-MCM-41. The molar equivalent arsenate is adsorbed.

of Fe-As and As-Fe distances from the Fe and As K-edge EXAFS spectra, respectively, provides the validity of the analysis.

As shown in the above discussion, one of the greatest advantages of mononuclear coordinated cations in a uniform organic layer is the possibility to monitor structural changes caused by adsorption. This is particularly important in the analysis of oxyanion adsorptions, because the oxygen shell of the toxic elements is not often distorted by adsorption, just like that of arsenate, and hence it is difficult to elucidate the structural factors that can explain the different adsorption behaviour. In contrast, the dissociation, bond formation, change in the symmetry, and so on, readily occur in the adsorption centres after oxyanion adsorption and their changes are closely related with the strength of adsorption. Spectroscopic analysis will be applied, more and more, to the adsorption structures on cation-anchored organic-inorganic hybrid materials, such as functionalized mesoporous silica.

Important Characteristics for Environmental Applications

Inhibition of Adsorption by Other Anions

The adsorption of oxyanions is hindered by the presence of other anions in solution. This is an important criterion *a priori* for the environmental applications. The inhibition by co-adsorption of Cl^- and SO_4^{2-} is particularly important, because these ions are widely found at $10^0 - 10^2$ ppm in freshwater. Adsorption with the same selectivity to the target oxyanion and other anions will lead to rapid saturation and the adsorbent will not be able to remove the toxic oxyanions sufficiently well to meet environmental regulations. This is not an exaggeration but a realistic problem, since it has been demonstrated that the retention of chromate is weaker than sulfate on γ-Al_2O_3.[42] The influence of the presence of phosphate and carbonate on the adsorption of arsenate on goethite has been analysed in detail in order to understand the adsorption mechanisms of arsenic species in the environment at a more realistic interface model than in a pure arsenate solution.[43-45]

The inhibition of arsenate adsorption on the EDA-propyl-functionalized MCM-41 (NN-MCM-41) that incorporates a transition metal cation has been explored by following the change in the uptake of arsenate in the presence of Cl^- and SO_4^{2-}.[31a] The ratio of the adsorption capacity to that without inhibitive anions is plotted against the initial concentration of the inhibitor, C_0, in Figure 12. Co^{2+}/NN-MCM-41 shows a nearly constant adsorption capacity with respect to the concentration of Cl^-.

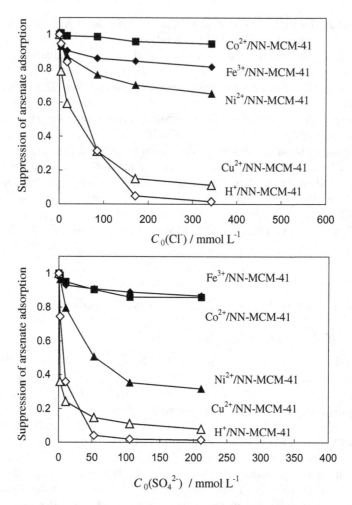

Figure 12. Inhibition of arsenate adsorption by the presence of chloride (upper) and sulfate (lower) evaluated by the change of adsorption capacity. The ordinates are the adsorption capacity in the presence of the inhibitor at an initial concentration of C_0 divided by that without the inhibitor. The adsorptions were carried out under the following conditions: initial concentration of arsenate: $12.2\,\text{mmol L}^{-1}$, adsorbent: $50\,\text{mg}$, volume of solution: $10\,\text{mL}$ and reaction temperature: $298\,\text{K}$.

The persistence of Fe^{3+}/NN-MCM-41 is ranked between those of Co^{2+} and Ni^{2+}/NN-MCM-41: a loss of 19% of adsorption when $C_0(Cl^-) = 342\,\text{mmol/L}$. The suppressions for Cu^{2+}/ and H^+/NN-MCM-41 are considerably larger than the other three cation-incorporated NN-MCM-41s; with $C_0(Cl^-) = 342\,\text{mmol/L}$, 89 and 98% of the adsorption is lost in Cu^{2+}/

and H^+/NN-MCM-41, respectively. A similar result is obtained as for the inhibition of arsenate by the presence of sulfate; the adsorption on Fe^{3+}/ and Co^{2+}/NN-MCM-41 is almost unaffected, whereas the Cu^{2+} and H^+ based adsorbents suffer from considerable inhibition.

Although the degree of inhibition by Cl^- and SO_4^{2-} varies among the cations, the average concentration of Cl^- and SO_4^{2-} in rivers throughout the world is 0.16–0.23 and 0.086–0.18 mmol/L, respectively.[46,47] When these concentrations are applied to the plots in Figure 12, the inhibition due to Cl^- and SO_4^{2-} is negligible. Thus, it is reasonable to assume that these anions have little effect in the environment. The strength of adsorption is another important factor as discussed above. The inhibition is more significant when the adsorption approaches saturation than when the coverage is low.[30,32] Since the arsenate concentration expected in the contaminated water will be 1×10^2 to 1×10^3 ppb, the inhibition by Cl^- and SO_4^{2-} will be much lower than in Figure 12, where the initial concentration of arsenate is 1700 ppm. From these considerations, the adsorption characteristics of arsenate on Fe^{3+} or Co^{2+}/NN-MCM-41 in the environment will remain unchanged, even in the presence of SO_4^{2-} and Cl^- below $1 \, mmol/L^{-1}$.

The lack of uniformity of the surface structure often causes problems in the extrapolation of the experimental results, especially when the adsorption is measured near saturation. The decrease in the distribution coefficient with the progress of adsorption in Figure 6 suggests that the surface contains "weak" and "strong" adsorption sites. In addition, we can see that the release of arsenate occurs more easily on H^+ sites than on Fe^{3+} and Co^{2+} sites. If a meaningful amount of H^+ sites exist in Fe^{3+}/ and Co^{2+}/NN-MCM-41, the exchange of adsorbed arsenate with Cl^- or SO_4^{2-} anions would occur in the same manner as in H^+/NN-MCM-41 in Figure 12. On the contrary, the absence of a rapid decrease in (arsenate adsorption)/(arsenate adsorption at $C_0 = 0$) near $C_0 = 0$ in Fe^{3+} and Co^{2+}/NN-MCM-41 demonstrates that the contribution of H^+ sites to the adsorption capacity in M^{n+}/NN-MCM-41, $(M^{n+} = Fe^{3+}$ and $Co^{2+})$ is negligible, even if such sites exist.

The separation of SeO_4^{2-} from SO_4^{2-} and of AsO_4^{3-} from PO_4^{3-} have been considered difficult and can be benchmarks for selective adsorption.[31b,32] Figure 13 shows the suppression of the adsorption capacity in the coadsorption systems of these anion combinations. The effect of the addition of sulphate on the adsorption of selenate is considerably greater than the suppressions in Figure 12. Even in the best adsorbent, Fe^{3+}/NN-MCM-41, more than 50% of adsorption capacity is lost with coexistence of 50 mmol L^{-1} of sulfate. Nevertheless, inhibition by the sulfate can be considered to be still limited for the following reason. As shown in Table 4, with the addition of 202 ppm of sulfate, suppression cannot be detected and the distribution coefficient remains beyond the detection limit (i.e. $K_d > 200000$) in the

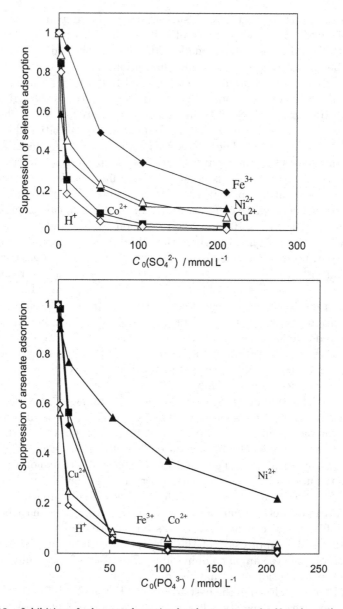

Figure 13. Inhibition of selenate adsorption by the presence of sulfate (upper) and inhibition of arsenate adsorption by the presence of phosphate (lower). The ordinates are
the adsorption capacity in the presence of the inhibitor at the initial concentration of
C_0 divided by that without the inhibitor. The central cations are indicated in the plots
and the substrate is NN-MCM-41. The adsorptions were carried out under the following conditions: initial concentration of selenate or arsenate: 12.2 mmol L^{-1}, adsorbent:
50 mg, volume of solution: 10 mL and reaction temperature: 298 K.

Table 4. Suppression of selenate adsorption ([adsorption of selenate in the presence of SO_4^{2-}]/[adsorption of selenate in the absence of SO_4^{2-}]) on cation incorporated NN-MCM-41.

Cation	SO_4^{2-} Initial ppm*	SeO_4^{2-} Initial ppm*	SeO_4^{2-} Final ppm*	SeO_4^{2-} Adsorption (mmol/g)	Suppression**	K_d***
Fe^{3+}	0	71.2	0	0.0997	—	> 200000
	202		0	0.0997	1.00	> 200000
	1014		5.5	0.0920	0.922	2381
	5071		36.1	0.0492	0.494	195
H^+	0	71.2	0	0.0997	—	> 200000
	202		14.2	0.0797	0.800	800
	1014		58.4	0.0180	0.181	44
	5071		68.2	0.00427	0.0428	9

N.B. *w/w ppm of oxyanions. ** = [adsorption of selenate in the presence of SO_4^{2-}]/[adsorption of selenate in the absence of SO_4^{2-}]. ***Distribution coefficient of selenate.

adsorption of selenate on Fe^{3+}/NN-MCM-41. With 1014 ppm of sulfate, about 8% suppression is observed though K_d diminishes to 2381, probably becoming less than 1% of the K_d observed in the adsorption without an inhibitor. However, since water containing 1000 ppm of SO_4^{2-} is not generally suitable for drinking as shown in Table 1, Fe^{3+}/NN-MCM-41 is qualified for the adsorbent that removes selenate from sulphate. On the other hand, 20% of the adsorption of selenate is lost and the distribution coefficient diminishes to 800 in the presence of 202 ppm sulfate, when the adsorption is carried out with H^+/NN-MCM-41. This result implies that protonated NN-MCM-41 is possibly disqualified as an adsorbent of selenate used in sulfate-containing water.

The order of resistance against inhibition by sulfate is Fe^{3+} > Cu^{2+} ∼ Ni^{2+} > Co^{2+} ∼ H^+ in selenate adsorption, which differs from that in arsenate adsorption, Fe^{3+} ∼ Co^{2+} > Ni^{2+} > Cu^{2+} ∼ H^+ (Figure 12). The difference in suppression implies that the transition metal cation at the adsorption site has a key importance for the prevention of interference by coexisting anions.

The suppression of arsenate adsorption by the inhibition of phosphate (Figure 13) is clearly larger than those by chloride and sulfate (Figure 12). Ni^{2+} has the most resistant capacity against this adsorption inhibitor. The largest decreases are observed among the four graphs in Figures 12 and 13. However, natural waters usually contains phosphorus whose concentration is lower than sulphate, as low as 20 ppb, which is a limiting factor for plant growth,[47] and, therefore, the effect of phosphate will be limited.

Recyclability of Used Adsorbents

Mesoporous silica frameworks, surface functional groups and adsorption sites are synthesized with more costly reagents than simple or modified natural adsorbents, such as grained goethite, pumice, γ-Al_2O_3, etc. The reactivation of used adsorbent is critically important for the demand in the potential market for reducing cost as well as emissions. When the adsorption is strong, reuse of the adsorbent is usually difficult because the regeneration of active sites inevitably includes desorption of the adsorbates. The removal may be accompanied by the destruction of the surface structure, resulting in the loss of adsorption capacity. It is demonstrated, in fact, that the desorption of arsenate from active carbon by treatment with a strong acid or base results in a significant loss of adsorption capacity.[49] The evaluation of the effectiveness of the recycling process can be done by analysing the elemental composition and adsorption characteristics of a regenerated adsorbent.

The arsenate adsorbed on Fe^{3+} incorporated NN-MCM-41 can be easily removed by washing with 1 M HCl at room temperature for 10 hours.[50] The results of the elemental analysis are summarized in Table 5. More than 99% of arsenate together with 94% of Fe are removed by this acid treatment, while the loss of amino groups is held at 23%. These percentages demonstrate that most of the arsenate is washed out with Fe^{3+} cation. The loss of iron signifies the destruction of the adsorption site, but does not mean the loss of the recovery method. Although the damage to the organic layer is not negligible, the re-incorporation of Fe^{3+} into the mesoporous silica after the removal of As recovers up to 82% of Fe^{3+} of the initial adsorbent level. The N/Fe ratio was nearly equal to 4 in this regenerated Fe^{3+}/NN-MCM-41, implying the recovery of the same coordination structure.

This recycling procedure works as well as or even better for the other oxyanions than for arsenate. The recovery of Fe^{3+} re-incorporation is 83, 90 and 96% using the same recycling treatment applied to chromate, selenate

Table 5.   Elemental analysis of arsenate-adsorbed Fe^{3+}/NN-MCM-41, that followed by treatment with hydrochloric acid and that by successive Fe^{3+} incorporation in addition.

	As* mmol	Fe^{3+}** mmol	N* mmol	N/Fe
After the adsorption**	1.50	0.51	1.98	3.9
After 1M HCl treatment	0.01	0.03	1.53	—
Fe^{3+} re-coordination	0.01	0.42	1.51	3.6

N.B. *The amount of the element per g-solid. **nearly full coverage.

Table 6. Arsenate adsorptions on "fresh" Fe^{3+}-incorporated NN-MCM-41 and "regenerated" Fe^{3+}-incorporated NN-MCM-41.

	Arsenate		Specific Adsorption (mmol/g)	K_d
	Initial (mg/L)	Final (mg/L)		
Fresh adsorbent	10.1	n.d.	0.01	$> 2 \times 10^5$
Regenerated adsorbent	10.1	n.d.	0.01	$> 2 \times 10^5$
Fresh adsorbent	1520	750	1.10	205
Regenerated adsorbent	1560	858	1.00	164
		As/Fe at the Saturation		
Fresh adsorbent		2.8		
Regenerated adsorbent		3.1		

N.B. The adsorptions were carried out under the following conditions: initial concentration of arsenate: 12.2 mmol L^{-1}, adsorbent: 50 mg, volume of solution: 10mL and reaction temperature: 298 K.

and molybdate adsorbed Fe^{3+}/NN-MCM-41, respectively. The difference is probably due to the amount of amino groups that remain after acid treatment: 83, 79 and 93% for chromate, selenate and molybdate adsorbed Fe^{3+}/NN-MCM-41, respectively.[50]

The arsenate adsorptions on the "fresh" and "regenerated" Fe^{3+}/NN-MCM-41 are compared in Table 6. It is clearly demonstrated in the table that the specific adsorption and distribution coefficient are sufficiently (80–100%) recovered. The adsorption capacity of regenerated adsorbent reaches 1.31 mmol/g,[50] which is equal to about 81% of the capacity of fresh adsorbent (226 mg/g = 1.61 mmol/g in Table 3); and, therefore, the As/Fe ratio does not change significantly as shown in Table 6, suggesting that the same kinds of adsorption site and adsorption mechanism are obtained in this regenerated adsorbent.

Together with the above results, the recycling of Fe^{3+}-incorporated NN-MCM-41 is illustrated in Figure 14. Nearly 81% of the initial capacity was regained through one cycle and the high efficiency of regeneration may serve to meet the cost and environmental demands in the future market. In the filtration solution of the "regeneration" process (the wash with HCl to remove arsenate and Fe^{3+}), the arsenate is dissolved due to the low pH. This species will probably be precipitated in the form of an insoluble ferric arsenate oxide, such as scorodite, by neutralizing the solution pH. In the other oxyanion adsorptions, such as chromate, selenate and molybdate adsorptions, 82, 86 and 85%, respectively, of adsorption capacity was recovered using regenerated Fe^{3+}/NN-MCM-41. The oxyanion/Fe

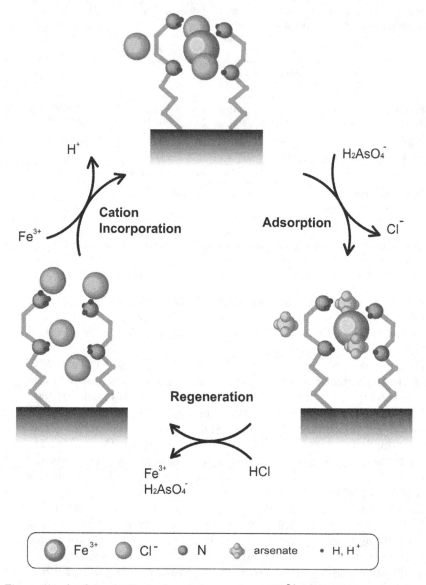

Figure 14. A schematic illustration of the recycling of Fe^{3+}/NN-MCM-41. This cycle includes the incorporation of Fe^{3+} by amino ligands, the adsorption of oxyanion, and the simultaneous removal of oxyanions and Fe^{3+} by acid treatment.

ratio remained almost the same as in the fresh material (1.9, 1.5 and 2.2, respectively). Consequently, the recycling mechanism in Figure 14 can be basically applied to the other oxyanion adsorptions.

Concluding Remarks

This chapter describes how amino-functionalized mesoporous silicas work as an adsorbent of toxic oxyanions, such as arsenate, chromate, selenate and molybdate. Their properties depend on the molecular structure of organic groups as well as the mesoporous structure. It is likely that the latter factor is closely related to the uniformity or the "real" density of the organic layer. The adsorption properties are influenced considerably by the cation on the amino group. An intense and large adsorption of arsenate is obtained on Fe^{3+}-incorporated EDA-propyl-functionalized MCM-41. The adsorption is selective even in the presence of chloride and sulfate. This adsorbent can be recycled by washing with hydrochloric acid and re-incorporation of Fe^{3+}.

The mesoporous adsorbents without organic functionalities, Fe, Al, Zn and La supported on SBA-15,[51] have been used for oxyanion adsorptions. When the element interacting with the oxyanions is highly dispersed and highly dense in a particular solid, a strong and capacious adsorption will be obtained. The platform used for the adsorption sites can be clay.[52]

The adsorption of cations on an adsorbent derived from mesoporous silica in a periodic structure has been investigated much more than that of anions. The typical example is the adsorption of mercury(II) on functionalized mesoporous silicas where 3-mercaptopropylsilane is used as a surface modifier.[53] It has been demonstrated that the Hg/S ratio depends on the kind of mesoporous silica,[54] being like arsenate adsorption on EDA-propyl functionalized mesoporous silicas as shown in Figures 2 and 3. The stoichiometry of the adsorption site — adsorbate ion is clearly important in the design of adsorbent, because the capacity is, roughly speaking, the product of the surface area, the surface density of the adsorption sites and this stoichiometry. The large capacity of Fe^{3+}/NN-MCM-41 for arsenate adsorption stems from the stoichiometry, Fe:As = 3:1, when the adsorption is saturated.

The reduction in the number of synthetic processes may be important. An interesting approach for reducing the processes is the combination of functionalization with amines, such as N, N-dimethyldecylamine (DMDA), with a pore expansion technique,[55] which is one of the most important techniques in the preparation of mesoporous silica. Since amine works as an anchoring site for transition metal cations, this "pore"-expanded mesoporous silica adsorbs toxic cations before extracting the surfactants. The adsorbent exhibits a high capacity: $106\,mg/(g\text{-solid})^{-1}$ of Cu^{2+}. Further development and application of this method are expected. The functionalization in association with templating, described above, is also a method combining templating and functionalization. These combined syntheses will become more and more important.

Acknowledgements

The author thanks his collaborators, Dr. Toshiyuki Yokoi (The University of Tokyo), and Prof. Takashi Tatsumi (Tokyo Institute of Technology). He is also grateful to the Ministry of Education, Culture, Sports, Science and Technology, Japan Society for the Promotion of Science, the Japan Securities Scholarship Foundation and the Japan Science and Technology Agency for financial support.

References

1. T. Yanagisawa, T. Shimizu, K. Kuroda and C. Kato, *Bull. Chem. Soc. Jpn.* **63**, 988 (1990).
2. C. T. Kresge, M. E. Leonowicz, W. J. Roth, J. C. Vartuli and J. S. Beck, *Nature* **359**, 710 (1992).
3. U. Ciesla and F. Schüth, *Microporous Mesoporous Mater.* **27**, 131 (1999).
4. J. Y. Ying, C. P. Mehnert and M. S. Wong, *Angew. Chem. Int. Ed.* **38**, 56 (1999).
5. (a) Q. Huo, D. I. Margolese, U. Ciesla, P. Feng, T. E. Gier, P. Sieger, R. Leon, P. M. Petroff, F. Schüth and G. D. Stucky, *Nature* **368**, 317 (1994); (b) Q. Huo, D. I. Margolese, U. Ciesla, D. G. Demuth, P. Feng, T. E. Gier, P. Sieger, A. Firouzi, B. F. Chmelka, F. Schüth and G. D. Stucky, *Chem. Mater.* **6**, 1176 (1994); (c) Q. Huo, R. Leon, P. M. Petroff and G. D. Stucky, *Science* **268**, 1324 (1995); (d) Q. Huo, D. I. Margolese and G. D. Stucky, *Chem. Mater.* **8**, 1147 (1996); (e) S. A. Bagshaw, E. Prouzet and T. J. Pinnavaia, *Science* **269**, 1242 (1995).
6. A. Tuel, *Microporous Mesoporous Mater.* **27**, 151 (1999).
7. (a) A. Corma, *Topics in Catal.* **4**, 249 (1997); (b) A. Corma, *Chem. Rev.* **97**, 2373 (1997).
8. G. E. Fryxell and J. Liu, *Adsorption at Silica Surface*, ed. E. Papirer (Marcel Dekker, New York, 2000), p. 665.
9. A. Sayari and S. Hamoudi, *Chem. Mater.* **13**, 3151 (2001).
10. H. van Bekkum, E. M. Flanigen, P. A. Jacobs and J. C. Jansen (eds.), For general information on zeolite, see *Introduction to Zeolite Science and Practice* (Elsevier, Amsterdam, 2001).

11. K. Kinoshita, For general information on active carbon, *Carbon* (Wiley-Interscience Publication, New York, 1988).
12. S. Shiraishi, *Carbon Alloys*, eds. E. Yasuda, M. Inagaki, K. Kaneko, M. Endo, A. Oya and Y. Tanabe (Elsevier, Amsterdam, 2003), Chap. 27.
13. (a) C. Y. Chen, H. X. Li and M. E. Davis, *Microporous Mater.* **2**, 17 (1993); (b) J. J. M. Kim, J. H. Kwak, S. Jun and R. Ryoo, *J. Phys. Chem.* **99**, 16742 (1995).
14. R. Nickson, J. McArhtur, W. Burgess, K. M. Ahmed, P. Ravenscroft and M. Rahman, *Nature* **395**, 338 (1998).
15. M. Karim, *Water Res.* **34**, 304 (2000).
16. H. Kondo, *Mizukankyo Gakkaishi* **20**, 6 (1997).
17. Y. S. Sen and C. S. Shen, *J. Water Pollut. Control Fed.* **36**, 281 (1964).
18. S. L. Chen, S. R. Dzeng, M. H. Yang, K. H. Chiu, G. M. Shieh and C. M. Wai, *Environ. Sci. Technol.* **28**, 877 (1994).
19. (a) R. D. Foust, Jr., P. Mohapatra, A.-M. Compton-O'Brien and J. Reifel, *Appl. Geochem.* **19**, 251 (2004); (b) A. E. Grosz, J. N. Grossman, R. Garrett, P. Friske, D. B. Smith, A. G. Darnley and E. Vowinkel, *Appl. Geochem.* **19**, 257 (2004); (c) J. R. Lockwood, M. J. Schervish, P. Gurian and M. J. Small, *J. Am. Stat. Assoc.* **96**, 1184 (2001); (d) J. M. Harrington, J. P. Middaugh, D. L. Morse and J. Housworth, *Am. J. Epidemiol.* **108**, 377 (1978); (e) J. Bundschuh, B. Farias, R. Martin, A. Storniolo, P. Bhattacharya, J. Cortes, G. Bonorino and R. Albouy, *Appl. Geochem.* **19**, 231 (2004).
20. M. M. Lawrence, *Environ. Sci. Technol.* **15**, 1482 (1981).
21. E. L. Tavani and C. Volzone, *J. Soc. Leather Technol. Chem.* **81**, 143 (1997).
22. H. Yoshitake, to be published.
23. A. P. Legrand (ed.), *The Surface Properties of Silicas* (John Wiley & Sons, Chichester, 1998), Chap. 3.
24. H. Yoshitake, *New J. Chem.* **29**, 1097 (2005).
25. H. Yoshitake, T. Yokoi and T. Tatsumi, *Chem. Mater.* **14**, 4603 (2002).
26. T. Yokoi, H. Yoshitake and T. Tatsumi, *J. Mater. Chem.* **14**, 951 (2004).
27. A. S. M. Chong, X. S. Zhao, A. T. Kustedjo and S. Z. Qiao, *Microporous Mesoporous Mater.* **72**, 33 (2004).
28. T. Yokoi, H. Yoshitake and T. Tatsumi, *Chem. Mater.* **15**, 4536 (2003).
29. Q. Zhang, K. Ariga, A. Okabe and T. Aida, *J. Am. Chem. Soc.* **126**, 988 (2004).
30. G. E. Fryxell, J. Liu, T. A. Hauser, Z. Nie, K. F. Ferris, S. Mattigod, M. Gong and R. T. Hallen, *Chem. Mater.* **11**, 2148 (1999).
31. (a) H. Yoshitake, T. Yokoi and T. Tatsumi, *Chem. Mater.* **15**, 1713 (2003); (b) H. Yoshitake, T. Yokoi and T. Tatsumi, *Bull. Chem. Soc. Jpn.* **76**, 2225 (2003).
32. T. Yokoi, Ms Engineering thesis, Yokohama National University, Yokohama, Japan, 2001.
33. S. D. Kelly, K. M. Kemner, G. E. Fryxell, J. Liu, S. V. Mattigod and K. F. Ferris, *J. Phys. Chem. B* **105**, 6337 (2001).
34. Y. Gao and A. Mucci, *Geochim. Cosmochim. Acta* **65**, 2361 (2001).
35. W. Zhang, P. Singh, E. Paling and S. Delides, *Minerals Eng.* **17**, 517 (2004).

36. T. F. Lin and J. K. Wu, *Water Res.* **35**, 2049 (2001).
37. S. Bang, M. Patel, L. Lippincott and X. Meng, *Chemosphere* **60**, 389 (2005).
38. A. N. Garg and P. N. Shukla, *Indian J. Chem.* **12**, 996 (1974).
39. R. L. Pecsok and J. Bjerrum, *Acta Chem. Scand.* **11**, 1419 (1957).
40. G. Gordon and R. K. Birdwhistell, *J. Am. Chem. Soc.* **81**, 3567 (1959).
41. T. E. Westre, P. Kennepohl, J. G. Dewitt, B. Hedman, K. O. Hodgson and E. I. Solomon, *J. Am. Chem. Soc.* **119**, 6297 (1997).
42. C. H. Wu, S. L. Lo and C. F. Lin, *Colloid Surf. A* **116**, 251 (2000).
43. Z. Hongshao and R. Stanforth, *Environ. Sci. Technol.* **35**, 4753 (2001).
44. J. Antelo, M. Avena, S. Fiol, R. Lopez and F. Arce, *J. Colloid Interf. Sci.* **285**, 476 (2005).
45. Y. Arai, D. L. Sparks and J. A. Davis, *Environ. Sci. Technol.* **38**, 817 (2004).
46. M. Meybeck, *Rev. Geol. Dyn. Geogr. Phys.* **21**, 215 (1979).
47. E. J. R. Conway, *Irish Acad., Proc. B* **48**, 119 (1942).
48. (a) C. D. Rail, *Groundwater Contamination: Sources, Control, and Preventative Measures.* Technomic Publishing Co., Inc.: Lancaster, PA, 1989, pp. 17; (b) E. A. Laws, *Aquatic Pollution*, 2nd edn. (John Wiley & Sons, Inc., New York, NY., 1993), pp. 148–149.
49. C. P. Huang and P. L. K. Fu, *Water Pollt. Control. Fed.* **56**, 233 (1984).
50. T. Yokoi, T. Tatsumi and H. Yoshitake, *J. Colloid Interf. Sci.* **274**, 451 (2004).
51. (a) M. Jang, J. K. Park and E. W. Shin, *Microporous Mesoporous Mater.* **75**, 159 (2004); (b) M. Jang, E. W. Shin, J. K. Park and S. I. Choi, *Environ. Sci. Technol.* **37**, 5062 (2003).
52. Y. Izumi, K. Masih, K. Aika and Y. Seida, *J. Phys. Chem. B* **109**, 3227 (2005).
53. X. Feng, G. E. Fryxell, L.-Q. Wang, A. Y. Kim, J. Liu and K. M. Kemner, *Science* **276**, 923 (1997).
54. L. Mercier and T. J. Pinnavaia, *Environ. Sci. Technol.* **32**, 2749 (1998).
55. (a) A. Sayari, Y. Yang, M. Kruk and M. Jaroniec, *J. Phys. Chem. B* **103**, 3651 (1999); (b) A. Sayari, *Angew. Chem. Int. Ed. Engl.* **39**, 2920 (2000).

Chapter 11

A Thiol-Functionalized Nanoporous Silica Sorbent for Removal of Mercury from Actual Industrial Waste

S. V. Mattigod, G. E. Fryxell and K. E. Parker

Pacific Northwest National Laboratory
Richland, Washington 99352, USA

Introduction

There are a number of existing technologies for mercury removal from water and wastewater. These include sulfide precipitation, coagulation/co-precipitation, adsorption, ion-exchange, and membrane separation. Recent reviews have included detailed discussions of the performance characteristics, advantages and disadvantages of these treatment methods.[1,2] Some of the proposed treatment technologies for RCRA metal and mercury removal include sulfur-impregnated carbon,[3] microemulsion liquid membranes,[4] ion-exchange,[5] and colloid precipitate flotation.[6] The sulfur-impregnated carbon weakly bonds a heavy metal, therefore, the adsorbed heavy metal needs secondary stabilization. In addition, a large portion of the pores in the active carbon is large enough for the entry of microbes that solubilize the mercury-sulfur compounds. Additionally, the RCRA metal loading on the carbon substrate is extremely limited.

The microemulsion liquid membrane technique consists of an oleic acid microemulsion liquid membrane containing sulfuric acid as the internal phase for mercury adsorption. However, this removal technology uses large volumes of organic solvent and involves multiple steps of extraction, stripping, de-emulsification, and finally mercury recovery by electrolysis. Also, the swelling of the liquid membrane reduces the efficiency of mercury extraction, and the slow kinetics of the metal adsorption requires long contacting times.

Typically, the ion exchange organic resins used for mercury removal are prone to oxidation leading to substantial reduction of resin life and their inability to reduce the mercury level to below the permitted level. Additionally, the mercury loading on organic resins is limited and the adsorbed mercury can be released into the environment if it is disposed of as a waste form because the organic substrates do not have the ability to resist microbial attack.

The removal of RCRA metal from water by colloid precipitate flotation has been reported to reduce mercury concentration from 0.16 ppm to about 0.0016 ppm. However, to remove colloids, this process involves adding hydrochloric acid (to adjust the wastewater pH), sodium sulfide and oleic acid solutions to the wastewater. Consequently, the treated water from this process needs additional processing to remove the residual additives. Also, the separated mercury needs further treatment to be stabilized as a permanent waste form.

During the last few years, we have designed and developed a new class of high-performance nanoporous sorbent materials for heavy metal removal that overcomes the deficiencies of existing technologies. These novel materials are created from a combination of synthetic mesoporous ceramic substrates that have specifically tailored pore sizes (2 to 10 nm) and very high surface areas ($\sim 1000\,\mathrm{m^2/g}$) with self-assembled monolayers of well-ordered functional groups that have high affinity and specificity for specific types of free or complex cations or anions. Typically, the nanoporous supporting materials are synthesized through a templated assembly process using oxide precursors and surfactant molecules. The synthesis is accomplished by mixing surfactants and oxide precursors in a solvent and exposing the solution to mild hydrothermal conditions. The surfactant molecules form ordered liquid crystalline structures, such as hexagonally ordered rod-like micelles, and the oxide materials precipitate on the micellar surfaces to replicate the organic templates formed by the rod-like micelles. Subsequent calcination to 500°C removes the surfactant templates and leaves a high surface area nanoporous ceramic substrate. Using surfactants of different chain lengths produces nanoporous materials with different pore sizes.

These nanoporous materials can be used as substrates for self-assembled monolayers of adsorptive functional groups that are selected to specifically adsorb heavy metals. Molecular self-assembly is a unique phenomenon in which functional molecules aggregate on an active surface, resulting in an organized assembly having both order and orientation. In this approach, bifunctional molecules containing a hydrophilic head group and a hydrophobic tail group adsorb onto a substrate or an interface as closely packed monolayers. The driving forces for the self-assembly are the intermolecular interactions (van der Waal's forces) between the functional molecules and the substrate. The tail group and the head group can be chemically-modified to contain certain functional groups to promote covalent bonding between the functional organic molecules and the substrate on one end, and the molecular bonding between the organic molecules and the metals on the other. Populating the head group with alkylthiols (which are well-known to have a high affinity for various heavy metals, including mercury) results in a functional monolayer which specifically adsorbs heavy metals. Using this technology, we synthesized a novel sorbent (thiol-SAMMS — thiol-self-assembled monolayers on mesoporous silica) for efficiently scavenging heavy metals from waste streams. Detailed descriptions of the synthesis, fabrication, and adsorptive properties of these novel materials have been published previously.[7-11] This chapter will summarize the effectiveness of thiol-SAMMS for mercury removal from two waste streams generated from a pilot-scale radioactive waste vitrification process. The treatment goal was to achieve residual mercury concentrations of ≤ 0.2 ppm as specified by Universal Treatment Standards 40CFR §268.38 (USEPA).

Experimental

The samples of waste streams tested were obtained from a pilot-scale study conducted at the Pacific Northwest National Laboratory. These waste streams result from a process designed to vitrify radioactive sludges that result from chemical separation of targeted actinide species. One of the waste streams (melter condensate) results from condensed liquid that is collected during the formation of waste glass in a melter. During this process, mercury contained in the radioactive sludge vaporizes and is collected in a condensate waste stream. A second waste stream results from periodic rinsing of a High Efficiency Mist Eliminator (HEME) filter with concentrated nitric acid. This highly acidic waste stream (HEME waste) typically contains over two orders of magnitude higher concentrations of mercury than the melter condensate waste stream.

Following filtration (0.45 μm), aliquots of each waste stream were analyzed using an ICP-OES (Inductively Coupled Plasma Optical Emission Spectrometer), and an IC (Ion Chromatograph). Mercury concentrations in these waste streams were measured by an ICP-MS (Inductively-Coupled Plasma Mass Spectrometer). The pH of the melter condensate sample was measured with a glass electrode and the acid content of HEME waste was determined by titrating an aliquot with 2.5 M sodium hydroxide solution.

Thiol-SAMMS used in these tests consisted of 3.5 nm MCM-41 material that was functionalized as described previously.[7-9] Because of the very high acidity, the HEME waste was preprocessed by neutralization with sodium hydroxide followed by filtration. This neutralization step also resulted in five-fold dilution of the sample. About 250 mL aliquot of this pretreated

HEME waste was initially equilibrated with 0.5 g of thiol-SAMMS material for about 4 hours. Following equilibration and the solid — liquid separation by filtration, the residual mercury concentration was measured by ICP-MS. Additional treatments were carried out by adding 0.5 g portions of thiol-SAMMS and sampling the solution until residual concentration of mercury was ≤ 0.2 ppm. For melter condensate tests, we added 0.02 g of thiol-SAMMS to an aliquot of 50 mL of the waste solution and equilibrated the mixture for 4 hours. After equilibrating the solid and liquid were separated by filtration, and the residual mercury concentration was measured by ICP-MS. Additional treatment with 0.02 g portions of thiol SAMMS was carried out until a residual mercury concentration of ≤ 0.2 ppm was achieved.

Results and Discussion

The data showed the principal component in HEME waste solution to be ∼ 9.15 M nitric acid. The mercury concentration in this waste stream was 732.7 mg/L (Table 1). Other dissolved components in the waste solution were Al, B, Ca, Fe K, Na Mg, Na, Si, and Zn in concentrations ranging from 6 to 500 mg/L. Minor dissolved components such as, Ba, Cd, Co, Cr, Cu, Mo, Ni, P, and Pb, were present in trace concentrations (< 3 mg/L).

Following neutralization (pH 6.9 SU), the principle component in HEME waste was ∼ 1.83 M $NaNO_3$ with a minor amount (∼ 0.002 M) of sodium borate. The waste after pretreatment was relatively concentrated with an ionic strength of 1.83 M. After applying dilution correction, the results (Table 2) indicate that the neutralization step removed a significant fraction (∼ 58%) of the mercury originally present in the HEME waste. The residual concentration of mercury after pretreatment was 61 mg/L. Analysis of the neutralized and filtered solutions indicate that this pretreatment step, also removed substantial fractions of Al (∼ 90%), Cr (∼ 87%), Fe (∼ 99%), Mn (∼ 81%), and Zn (∼ 82%) from the waste solution.

The melter condensate waste stream was alkaline in nature (pH 8.5 SU) with main dissolved components being ∼ 0.03 M sodium borate, ∼ 0.009 M sodium fluoride, and ∼ 0.003 M sodium chloride with minor amounts of sodium sulfate, sodium nitrate, sodium nitrite, and sodium iodide (Table 1). The ionic strength of this solution was ∼ 0.15 M which is an order of magnitude more dilute than the HEME Waste. The mercury concentration in this waste stream was measured to be 4.64 mg/L. Other dissolved components such as, Al, Ba, Ca, Cd, Co, C_2O_4 (oxalate), Cr, Cu, Fe, Mg, Mn, Mo, Ni, P, Pb, and Zn were present in trace concentrations (< 2 mg/L).

Results of the treatment of the HEME waste with thiol-SAMMS showed (Table 2) that the initial treatment removed ∼ 87% of mercury in solution

Table 1. Composition of Mercury-Containing Waste Streams.

Constituent	HEME Waste (mg/L)	Neutralized HEME (mg/L)	Melter Condensate (mg/L)	Constituent	HEME Waste (mg/L)	Neutralized HEME (mg/L)	Melter Condensate (mg/L)
Al	15.63	0.33	0.86	K	7.53	134.80	3.12
B	94.42	17.33	329.70	Mg	10.83	1.82	0.85
Ba	1.02	0.22	0.02	Mn	3.01	0.11	0.02
Ca	34.06	7.72	0.57	Mo	1.18	0.19	0.62
Cd	0.07	0.01	0.04	Na	510	42,146	522.00
Cl	—	—	105.00	Ni	1.03	0.16	0.08
Co	0.03	0.02	0.03	NO_3	567,300	113,460	39.00
C_2O_4	—	—	1.30	NO_2	—	—	40.00
Cr	2.26	0.06	1.67	P	0.39	0.13	0.65
Cu	0.28	0.07	0.08	Pb	0.78	0.13	0.23
Fe	26.55	0.07	0.68	Si	6.39	0.80	8.01
F	—	—	169.00	SO_4	—	—	75.00
Hg	732.7	61.0	4.64	Zn	5.99	0.21	0.25
I	—	—	25.00	pH	—	6.9 SU	8.5 SU

Table 2.   Mercury Removal from HEME Waste Solution.

Treatment	Residual Conc. (mg/L)	Cumulative Removal (%)
pH Adjusted & Filtered	61.00	—
0.5 g thiol- SAMMS	8.10	86.7
0.5 g thiol-SAMMS	4.20	93.1
0.5 g thiol-SAMMS	0.20	99.7

resulting in a residual concentration of 8.1 mg/L. Subsequent thiol-SAMMS addition reduced the mercury concentration to 4.2 mg/L. Adding an additional 0.5 g of thiol-SAMMS, for a total of 1.5 g per 250 mL of waste solution resulted in a cumulative reduction of mercury to 99.7% with the residual concentration of 0.2 mg/L. The cumulative mercury loading on the solid phase was calculated to be ~ 10.1 mg/g of thiol-SAMMS.

In the absence of strongly complexing ligands, mercury in neutralized HEME waste would exist mainly as hydrolytic species $[\mathrm{Hg(OH)}_2^0]$. Mercury therefore can adsorb on to the thiol functionality of SAMMS through two different mechanisms namely;

$$\mathrm{R\text{-}SH + Hg(OH)_2^0 = R\text{-}SHgOH + H_2O} \tag{1}$$

$$\mathrm{R\text{-}(SH)_2 + Hg(OH)_2^0 + R\text{-}S_2Hg + 2H_2O} \tag{2}$$

The adsorption reaction (1) indicates a monodentate bonding of mercury whereas, reaction (2) suggests a bidentate bonding of mercury to the thiol sites. Extended X-ray adsorption fine structure (EXAFS) data obtained on mercury loaded thiol-SAMMS has shown that bonding is typically bidentate in nature as indicated by the second reaction.[12]

We also calculated the selectivity (affinity) of thiol-SAMMS for adsorbing mercury from a high ionic strength waste solution. The selectivity (affinity) of a sorbent for a contaminant is typically expressed as a distribution coefficient (K_d) which defines the partitioning of the contaminant between sorbent and solution phase at equilibrium. Distribution coefficient is the measure of an exchange substrate's selectivity or specificity for adsorbing a specific contaminant or a group of contaminants from matrix solutions, such as waste streams. The distribution coefficient (sometimes referred to as the partition coefficient at equilibrium) is defined as a ratio of the adsorption density to the final contaminant concentration in solution at equilibrium. This measure of selectivity is defined as

$$K_d = \frac{(x/m)_{eq}}{c_{eq}} \tag{3}$$

where, K_d is the distribution coefficient (mL/g), $(x/m)_{eq}$ is the equilibrium adsorption density (mg of mercury per gram of thiol-SAMMS), and c_{eq} is the mercury concentration (mg/mL) in waste solution at equilibrium.

We computed a distribution coefficient (K_d) value of $\sim 5 \times 10^4 \, \text{mL/g}$ for mercury adsorption by thiol-SAMMS from HEME solution. These data showed that, with the combination of pretreatment (neutralization of the acid waste) and thiol-SAMMS treatment removed almost all ($\sim 99.9\%$) mercury originally present in very high concentration ($\sim 730 \, \text{mg/L}$) in this waste. This bench-scale test demonstrated the feasibility of effectively removing very high concentrations of mercury from strong acidic wastes by first neutralizing the waste, and then using the highly selective adsorptive properties of thiol-SAMMS to scavenge remaining mercury from the waste to meet the UTS limit of $\leq 0.2 \, \text{mg/L}$.

Data from the treatment of the melter condensate waste showed (Table 3) that the initial addition of 0.02 g of thiol-SAMMS removed nearly 85% of mercury in solution with a residual concentration of 0.72 mg/L. Adding an additional 0.02 g of thiol-SAMMS, for a total of 0.04 g per 50 mL of waste solution resulted in a cumulative reduction of mercury to 98.9% with the residual concentration of 0.05 mg/L. The cumulative mercury loading on the solid phase was calculated to be $\sim 5.6 \, \text{mg/g}$ of thiol-SAMMS. As compared to the HEME waste, the melter condensate waste contained halide ligands such as Cl^- and I^- which form strong complexes with dissolved mercury.[10,13] Based on the complexation constants listed,[10] we calculated that dissolved mercury in this waste existed mainly as a $HgOHI^0$ complex. Therefore, the bidentate adsorption mechanism for this complex can be represented by:

$$R\text{-}(SH)_2 + HgOHI^0 + R\text{-}S_2Hg + H_2O + H^+ + I^- \qquad (4)$$

We computed a distribution coefficient (K_d) value of $\sim 1 \times 10^5 \, \text{mL/g}$ for mercury adsorption by thiol-SAMMS from melter condensate solution. This test demonstrated that treatment of this alkaline waste solution with thiol-SAMMS can effectively remove mercury an order of magnitude below the UTS limit of $\leq 0.2 \, \text{mg/L}$ waste.

The data from these experiments showed that the presence of competing cations in the waste solutions did not significantly affect the high affinity of thiol-SAMMS for mercury in solution (K_d: 5×10^4 and $1 \times 10^5 \, \text{mL/g}$). Such selectivity and affinity in binding mercury by thiol-SAMMS can be explained on the basis of the hard and soft acid-base theory (HSAB)[14,16],

Table 3. Mercury Removal from Melter Condensate Waste Solution.

Treatment	Residual Conc. (mg/L)	Cumulative Removal (%)
Untreated	4.64	–
0.02 g thiol-SAMMS	0.72	84.5
0.02 g thiol-SAMMS	0.05	98.9

which predicts that the degree of cation softness directly correlates with the observed strength of interaction with soft base functionalities such as thiols (-SH groups). According to the HSAB principle, soft cations and anions possess relatively large ionic size, low electronegativity, and high polarizability (highly deformable bonding electron orbitals) therefore, mutually form strong covalent bonds. In these waste streams, all cations except lead and cadmium, are either hard or borderline cations therefore, were less likely to interact significantly with soft base moiety (-SH) of thiol-SAMMS. Because the relative degree of softness is in the order Hg $\gg$ Pb $\gg$ Cd,[17] mercury would preferentially bind to the thiol functionality. These tests showed that thiol-SAMMS can very selectively bind dissolved mercury from waste solutions containing varying concentrations of major and minor cations and anions. Data also showed that thiol-SAMMS sorbent can very effectively scavenge strongly complexed mercury from dilute to relatively concentrated waste matrices to meet the UTS limits for effluents.

Conclusions

Tests were conducted using a novel sorbent, thiol-functionalized nanoporous silica material, demonstrating its effectiveness in removing mercury from two waste streams. These waste streams originated from pilot-scale tests being conducted to refine the process of vitrifying radioactive sludges that result from chemical separation of targeted actinide species. Two waste streams resulting from this process (High Efficiency Mist Eliminator, HEME and melter condensate) contain mercury concentrations that ranged from ∼700 ppm to ∼5 ppm respectively. The data showed that thiol functionalized Self-Assembled Monolayers on Mesoporous Silica (thiol-SAMMS) was effective in reducing mercury concentrations in these two waste streams to meet a treatment limit of ≤ 0.2 ppm. These tests demonstrated that the thiol-SAMMS can very selectively (K_d: 5×10^4 and 1×10^5 mL/g) and effectively scavenge strongly complexed mercury from dilute to relatively concentrated waste matrices to meet the UTS limits for effluents.

Acknowledgements

Pacific Northwest National Laboratory is a multiprogram national laboratory operated for the U.S. Department of Energy by Battelle Memorial Institute under Contract DE-AC06-76RLO 1830.

References

1. M. A. Ebadian, M. Allen, Y. Cai and J. F. McGahan, Mercury contaminated material decontamination methods: Investigation and assessment, Final Report, Hemispheric Center for Environmental Toxicology, Florida International University, Miami, FL 33174 (2001).
2. United states Environmental Protection Agency, Aqueous Mercury Treatment — Capsule Report, EPA/625/R-97/004. Office of Research and Development, Washington, DC, 20460 (1997).
3. Y. Otani, H. Eml, C. Kanaoka and H. Nishino, *Environ. Sci. Technol.* **22**, 708 (1988).
4. K. A. Larson and J. M. Wiencek, *Environ. Progress* **13**, 253–262 (1994).
5. S. E. Ghazy, *Sep. Sci. Technol.* **30**, 933–947 (1995).
6. J. A. Ritter and J. P. Bibler, *Water Sci. Technol.* **25**, 165–172 (1992).
7. G. E. Fryxell, J. Liu, A. A. Hauser, Z. Nie, K. F. Ferris, S. V. Mattigod, M. Gong and R. T. Hallen, *Chem. Mat.* **11**, 2148–2154 (1999).
8. G. E. Fryxell, J. Liu and S. V. Mattigod, *Mat. Tech. Adv. Perf. Mat.* **14**, 188–191 (1999).
9. G. E. Fryxell, J. Liu, S. V. Mattigod, L. Q. Wang, M. Gong, T. A. Hauser, Y. Lin and K. F. Ferris and X. Feng, Environmental Applications of Interfacially-Modified Mesoporous Ceramics, *Proceedings of the 101st National Meetings of the American Ceramic Society* (1999).
10. S. V. Mattigod, X. Feng, G. Fryxell, J. Liu and M. Gong, *Sep. Sci. Tech.* **34**, 2329–2345 (1999).
11. J. Liu, G. E. Fryxell, S. V. Mattigod, T. S. Zemanian, Y. Shin and L. Q. Wang, *Studies Surf. Sci. Catalysis* **129**, 729–738 (2000).
12. K. M. Kemner, X. Feng, G. E. Fryxell, L.-Q. Wang, A. Y. Kim and J. Liu, *J. Synchrotron Rad.* **6**, 633–635 (1999).
13. D. Sarkar, M. E. Essington and K. C. Mishra, *Soil Sci. Soc. Am. J.* **64**, 1968–1975 (2000).
14. R. G. Pearson, *J. Chem. Educ.* **45**, 581–587 (1968).
15. R. G. Pearson, *J. Chem. Educ.* **45**, 643–648 (1968).
16. R. D. Hancock and A. E. Martell, *J. Chem. Educ.* **74**, 644 (1996).
17. M. Misono, E. Ochiai, Y. Saito and Y. Yoneda, *J. Inorg. Nucl. Chem.* **29**, 2685–2691 (1967).

Chapter 12

Amine Functionalized Nanoporous Materials for Carbon Dioxide (CO$_2$) Capture

Feng Zheng, R. Shane Addleman, Christopher L. Aardahl,
Glen E. Fryxell, Daryl R. Brown and Thomas S. Zemanian

Pacific Northwest National Laboratory
Richland, WA, USA

Introduction

Increasing levels of CO$_2$ concentration in the earth atmosphere and rising average global temperatures have raised serious concerns about the effects of anthropogenic CO$_2$ on global climate change. Meanwhile, most analyses project that fossil fuels will continue to be the dominant energy source worldwide until at least the middle of the twenty first century.[1] Significant reduction of the current level of CO$_2$ emission from the consumption of fossil-fuels is necessary to stabilize atmospheric concentration of CO$_2$. To this end, it is vital to capture and store CO$_2$ from large point sources of emission such as fossil-fuel power plants, cement production, oil refineries, iron and steel plants, and petrochemical plants. The focus of this chapter will be on the application of CO$_2$ capture technologies in relation to energy production from fossil fuels, as over one third of the world's CO$_2$ emissions from fossil-fuel use are attributed to fossil-fuel electric power-generation plants.[1] However, the same set of technologies will also be applicable to the other industrial processes mentioned above. In addition, the CO$_2$ capture technologies are also relevant to applications in space exploration and submarines, where the separation of CO$_2$ from air in contained breathing atmospheres is necessary.

In fossil-fuel based power plants, the chemical energy stored in coal, natural gas, oil, etc, is converted to thermal and mechanical energy by combustion and subsequently converted to electricity, with CO$_2$ as a primary byproduct of fossil fuel combustion. There are three competitive CO$_2$ capture technologies for fossil-fuel fired power plants: post-combustion capture, pre-combustion capture, and oxyfuel combustion. In post-combustion capture, CO_2 is captured from the flue gases after the fuel is burned in air. This system is applicable to retrofit existing power plants as well as new systems. However, the CO_2 concentration in flue gases is relatively low and the total flow rate is large because the nitrogen gas from combustion air significantly dilutes the flue gas stream. This makes it challenging to capture CO_2 from flue gas streams efficiently and economically. In pre-combustion capture, there is an initial fuel conversion step (usually gasification) and CO_2 is captured from the fuel gas. The advantage of this system is that the pressure and the CO_2 concentration of the fuel gas is higher, which makes separation easier. The flow rate of fuel gas streams is also smaller than that of flue gas streams as no nitrogen is introduced prior to combustion in air. The drawbacks of pre-combustion capture are two-fold: the flowsheet of the gasification and capture plant are inherently complex, and the technology is not nearly as proven as post-combustion capture. In oxyfuel combustion, fossil fuels are combusted in high purity oxygen to produce energy and a flue gas of essentially CO_2 and water, from which the CO_2 gas can be easily separated. The disadvantage here is the high energy cost associated with the air separation process that supplies the oxygen stream for consumption.

The selection of the separation processes for CO_2 capture from energy production depends on the gas stream conditions, which in turn depend on the selection of the overall CO_2 capture strategy. The CO_2 separation technologies currently available include gas/liquid scrubbing using physical or chemical absorbents as solvents, gas/solid scrubbing using sorbents by adsorption or absorption, cryogenic separation, and membrane separation. The topics covered in this chapter are those related to solid absorbents. Cryogenic separation and gas separation membranes are not discussed here as both of them would certainly deserve separate monologues on their own.

Physical solvents are nonreactive and the absorption and release of CO_2 are governed by Henry's law; in that, the concentration of CO_2 in the solvent is proportional to CO_2 partial pressure in the gas phase. Physical solvents thus work best when CO_2 partial pressure is high and more applicable to precombustion CO_2 capture. Commercial solvent processes such as Selexol^{TM} (polyethylene glycol dimethyl ethers) have an existing installation base in ammonia production, natural gas processing, and refinery operations. In the Rectisol^{TM} process, refrigerated methanol is used as solvents.

Unlike physical solvents where the absorption is driven by relatively weak intermolecular forces between the solvent and CO_2, chemical bonds are formed in chemical solvents when CO_2 is absorbed. Liquid amines such as monoethanolamine (MEA), diethanolamine (DEA), and methyldiethanolamine (MDEA) are chemical solvents that have been used in commercial plants to capture CO_2 from flue gases. Chemical solvents are more attractive for CO_2 capture at lower partial pressure. High energy consumption required for regeneration is one of the main disadvantages of liquid amine solvents.

Although liquid amine absorption is a mature technology for industrial CO_2 separation, significant improvements are needed in order to apply this concept to the extremely large-scale CO_2 capture in energy production. The major drawbacks of the current liquid amine technologies are the large amount of energy required for regeneration, corrosiveness of the amine solvents, and solvent degradation problems in the presence of oxygen. For example, the amine concentration in commercial amine process is limited to lower than 30% due to increasing corrosiveness of concentrated amine towards carbon steel. Therefore significant energy is consumed in the solvent regeneration step to heat and partially vaporize the water that makes up a large portion of the aqueous solvent, resulting in low energy efficiency of the overall process.

Solid sorbents include physical and chemical adsorbents. Examples of physical adsorbents are zeolitic materials and activated-carbons. Solid chemical absorbents include alkali carbonates, alkaline earth metal oxides, zirconate and silicate salts. These chemical absorbents can be used in bulk powder or supported forms. A different kind of solid absorbent incorporates organic amines in a porous support. This can be achieved by impregnating the pores with a bulk liquid or grafting ligands to the pore surfaces through chemical bonds. In this manuscript, these two approaches are referred to as supported amine and tethered amine sorbents, respectively. The CO_2 capture mechanism for both types of amine sorbents is similar to that of liquid amine solvents. Chemically-active solid sorbents have not been previously utilized commercially because the physical and economic performance was inferior. However, new materials under development may produce solid sorbents with performance superior to liquid systems with desirable economics.

Solid amine sorbents have promise to overcome some of the shortcomings of liquid amines. The dispersion, immobilization, and confinement of the amine functional groups into a porous solid support can result in a more stable, more mass transfer efficient, less toxic, and less corrosive material than the corresponding liquid amines. Solid amine sorbents allow a dry scrubbing process where the energy penalty associated with the evaporation of a large amount of water is avoided. Further, the amine functional groups can be tailored for lower regeneration energy requirement. The supports can be tailored independently for high stability and low mass transfer resistance.

In this chapter we will focus on solid amine sorbents (both supported and tethered amines) as the functional materials for CO_2 separation, and the design and performance of these structures is discussed. A detailed discussion of this topic is warranted and timely as increasingly more researchers are attracted to this area, as evidenced by the rapid growth of papers published in the last few years.[2-24]

Rational Design of Solid Amine Sorbents for CO_2 Capture

Desired Attributes of an Ideal CO_2 Sorbent

Adsorption capacity

In order to apply the solid amine sorbent approach to CO_2 capture in energy production processes, high CO_2 capture capacity is essential because the amount of CO_2 to be captured is extremely large and the capacity of the sorbents is one of the key drivers influencing the overall economic feasibility of the system.

Adsorption capacity is typically reported as the mass or moles of CO_2 adsorbed per mass of the sorbent, on a dry sorbent basis, at equilibrium with certain CO_2 concentration in the gas phase at specific temperature and pressure. When comparing literature data, care should be taken to compare capacities at the same gas phase conditions.

The equilibrium adsorption capacity is determined by the amine functionality, the type of supports, and the loading level of amine on the support. Depending on the method by which the amine functionality is imparted, there is generally an upper limit on the amine loading. For example, impregnation with liquid amine is limited by the pore volume while grafting amine groups on the surface is limited by the surface area of the support and the density of the anchor groups on the surface. In general we have found that the CO_2 capacity generally increases with higher amine loading; therefore, maximizing amine loading is desirable.

Selectivity

Bulk gas streams in a power plant contain N_2 and O_2 in the case of flue gases and H_2, CO, and CH_4 in the case of synthesis gas. Good solid sorbents should have high selectivity towards CO_2 over these bulk gas components. Additionally, there is a large presence of moisture or steam in both flue gas and synthesis gas. It is critical that the solid sorbents maintain high capacity for CO_2 in the presence of a significant amount of water vapor. Some physical sorbents such as certain zeolites have high CO_2 capacity, but the selectivity over water is poor and the sorbents need to be regenerated more frequently and at higher temperature than desired, unless the feed gas is dehumidified prior to CO_2 separation.[13,25] This increases energy cost and severely limits the use of such physical sorbents for CO_2 capture.

Gas streams from fossil fuel power plants also contain other acid gas contaminants such as SO_2, NO_2, and H_2S. An ideal CO_2 sorbent should exhibit minimal degradation of CO_2 capacity in the presence of these acid gases. Although the effect of the presence of water vapor on CO_2 capacity is commonly reported in the literature, data on the selectivity of solid amine sorbents over acid gas contaminants are few.[18,21,26]

Kinetics

The overall kinetics of CO_2 adsorption on a solid amine sorbent is influenced by the intrinsic reaction kinetics of the amine functional groups with CO_2 as well as by the diffusion of the gas phase through the sorbent structures. The amine functional groups can be selected to have reasonably fast kinetics. For example, it is known that better liquid amine absorbents can be formulated by having a fast reacting amine such as piperazine blended with another more strongly CO_2-binding amine.[27-30] The addition of the "promoter" amine improves the overall kinetics. A similar approach may be developed for solid amine sorbents but has not been reported. The porous supports of the solid amine sorbents can also be tailored to reduce mass transfer resistance. Large pore diameter, hierarchical pore structures, and cross connections of pore channels are all favorable for the enhancement of mass transport.

Adsorption is an exothermic process and is temperature-dependent. Large sorbent beds typically operate adiabatically and therefore the overall kinetics may also be influenced by the development of a thermal front traveling with the mass transfer zone. Ideally, high thermal conductivity of the sorbent materials is desirable. However, the choice of sorbent supports is usually limited by other considerations such as porosity, stability, ability to functionalize/impregnate, and of course cost.

Durability

Another key driver for the economics of CO_2 capture using solid sorbents is the durability of the sorbent. As the amount of sorbent required will be very large for fossil-fuel power plants, sorbent durability is critical to minimize sorbent makeup rate and to keep the process economically feasible.

While synthesis gas in precombustion capture is highly reducing, flue gas in post-combustion capture contains oxygen. The solid amine sorbents must be chemically stable in the flue gas. The operation temperature of the solid amine sorbents is limited by the regeneration temperature, typically from 100–140°C. At such temperatures, amine degradation is usually not a concern. However, amine loss to evaporation in supported amine sorbents is likely to occur with moderate heating. Therefore, the thermal stability of the sorbents should be well characterized.

Porous silica is among the commonly used solid supports for amine sorbents. Silica gel and mesoporous silica materials can undergo destruction of their pore structures under aggressive hydrothermal conditions. The long term effects of the presence of moisture or steam in the feed gas on the silica support for solid amine sorbents must be understood and minimized. Porous substrates with high hydrothermal stability are essential from a cost perspective.

Gas streams from fossil fuel conversion are quite dirty and typically contain acid gas contaminants such as SO_2 and NO_x in flue gases or H_2S in synthesis gas. If solid amine sorbents irreversibly bond to these contaminants, the CO_2 capture capacity is lost and the economical viability of using such solid amine sorbents is significantly reduced. However, if the acid gas contaminants are reversibly co-captured with CO_2, significant savings in cost can be realized. A good solid amine sorbent should be at least inert to the presence of relevant sulfur species.

Energy for regeneration

During sorbent regeneration, energy equivalent to the heat of adsorption needs to be supplied as desorption is an endothermic process. Thus, sorbents that bond with CO_2 weakly are desirable from an energy consumption perspective. In a temperature swing adsorption (TSA) process, the sorbent bed needs to be cycled between adsorption and desorption temperatures. Therefore a sorbent material that requires a small temperature differential is also advantageous. Chemisorption-based adsorbents typically require higher energy input to regenerate due to their higher heat of adsorption than that of the physical adsorbents such as zeolites or activated-carbons. However, physical adsorbents typically have poor selectivity for CO_2 in the presence of water. Thus, the overall economics will be a trade-off between the above considerations.

Tailoring Amine Functionalities

It has long been recognized that CO_2 reacts with ammonia and certain organic amines.[31–34] However, not all amines enter into this reaction with equal ease, and in fact, many do not react with CO_2 at all. These reactivity trends are a function of the nucleophilicity of the nitrogen lone pair, the steric environment surrounding the N atom, and the presence (or absence) of a proton on the amine. For example, typical primary alkyl amines (e.g. butyl amine) enter into this reaction with ease, while aniline does not. In the case of aniline, a combination of the inductive effect of the phenyl ring and conjugation of the π-system with the N atom's lone pair cause the amine to be insufficiently nucleophilic. Similar effects are in operation with various aryl amines (e.g. pyrroles, indoles, etc.), silazanes, and cyanamides. The steric environment surrounding the nucleophilic N atom is also important. Even though CO_2 is a small, linear molecule, steric congestion can impede the nucleophilic addition of the amine to the "carbonyl". This kinetic barrier also hinders the reversibility of the reaction. Thus, sterically congested amines (e.g. di-isopropylamine) are less desirable than are amines of similar molecular structures that are not so congested (e.g. n-hexylamine). Thirdly, it is important that the amine nucleophile have a proton to lose as the initially formed CO_2 adduct is not stable, and loss of a proton from the ammonium center creates a stable ureate salt. In the absence of this proton loss (e.g. with tertiary amines like triethylamine or pyridine), the CO_2 adducts are unstable and the equilibrium lies strongly on the side of the neutral amine and free CO_2. Thus, when tailoring the ideal nucleophilic center for this chemistry, alkyl amines are superior to aromatic amines, and primary alkyl amines are better than secondary alkyl amines.

However, it is also important to consider this chemistry from the perspective of the proton transfer reaction as well. Because of the spherical symmetry of the 1s orbital, and the lack of congesting substituents in the forward hemisphere of the proton's ligand field, proton transfer reactions tend to be quite facile and less susceptible to steric hindrance than other reactions. This means that the rate at which this proton is transferred (i.e. conversion of the unstable initial adduct to the stable ureate salt), and the position of this equilibrium will depend heavily on the basicity of the amine N atom that is removing the proton (assuming a primary amine of fixed identity for the initial nucleophilic addition). Alkyl amines are more basic than are aromatic amines, and tertiary amines are more basic than are secondary, which in turn are more basic than primary amines. Thus, basicity is a parameter that can be manipulated in order to "tune" the reversibility of CO_2 sorption.

Traditionally, amines have been used to absorb CO_2 in the liquid or amorphous polymer phases. Most commonly, a single amine is used to serve as both the nucleophile and the base in this chemistry. By employing an ordered self-assembled monolayer in a nanoporous ceramic architecture we have the luxury of being able to spatially organize these components within our system, and the ability to use different species for coordinating the CO_2 nucleophile with proton and the base. We also have the ability to incorporate both of these moieties into the same molecule, which allows us to "pre-organize" the transition state for the proton transfer reaction, facilitating both the forward and reverse reactions. Thus, suitably tethered diamines are particularly well-suited to this chemistry. For example, tethered ethylenediamine functional groups have two amine sites per group with the primary amine functioning as the nucleophile and the secondary amine as the base. This allows the formation of an intramolecular carbamate when a CO_2 molecule is captured by the ethylenediamine moiety, as shown in Scheme 1. The self-assembled monolayer also provides the ability to attach the greatest number of active sites per unit surface area, and still have all of them at the interface and easily accessible to the gas phase CO_2.

Tailoring Nanoporous Substrates

The ability to tailor the porosity and chemical functionality of a material is of critical importance to a range of technologies including microelectronics, catalysis, sensing and separations. Since chemical activity of a substance is frequently proportional to surface area, materials with high relative surface areas can be very valuable. However, the surface area is only useful to chemical processes if it is accessible and provides the desired functionality. The porosity of a material is therefore of importance since it provides access to the surface for chemical processes to occur. Many porous architectures prevent or retard mass transfer in a material, nullifying the advantages of high relative surface areas. Porous silica is a frequently utilized high surface area material since both the porosity and surface chemistry can be dictated to create the desired material functionality. In particular, functionalized nanoporous (or mesoporous) silica has recently been a material of significant interest.[35–45] Nanoporous silica's advantages over microporous

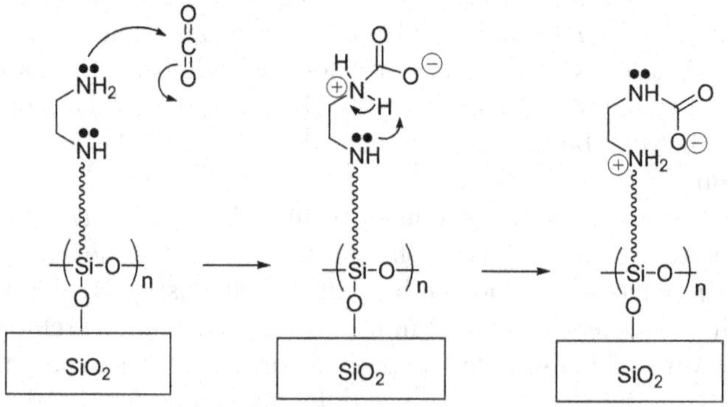

Scheme 1. Formation of intramolecular carbamate salt through the reaction of CO_2 with the surface-tethered EDA.

silica structures stem from its higher surface area and its open, frequently ordered, porosity.

Nanoporous silica materials can be created with large uniform pore structures, high specific surface area, and specific pore volumes. The nanostructure is imposed through the use of surfactants in siliceous sol-gel solutions that are used to create these materials. The surfactants can form ordered micellar structures in the sol-gel solution and as the sol-gel solution undergoes condensation, the solution's micellar structure is retained in the final solid form. Solution conditions can be controlled to produce cubic, hexagonal, lamellar, amorphous and other material phases. Selection of the surfactant and sol-gel condensation conditions enable tuning of the final silica material properties such as percentage of porosity, pore size, surface area, pore connectivity, and structural phase (i.e. hexagonal, cubic, amorphous). Selection of these parameters will dictate the material performance and necessitate trade-offs. Detailed discussions of sol-gel chemistry and mesoporous silica synthesis are available.[35-39]

For CO_2 capture, a material with the highest surface area possible is desired to provide the best capacity. However, this must be balanced with material stability and cost. As previously discussed, durability is a critical requirement for a CO_2 sorbent. Wall thickness must be maintained to a level that provides sufficient thermal and chemical stability. Pores should be large enough to allow facile mass transport (after functionalization) but not so large as to unacceptably reduce the surface area. Interpore connectivity is advantageous since it improves mass transport throughout the material. Ordered porosity can provide faster mass transfer (sorbent kinetics) but such materials are typically more expensive due to increased synthesis costs. Mesoporous materials that have been explored as nanoporous silica supports for CO_2 sorbents include MCM-41, MCM-48 and SBA-15. Templated silica pore diameters have been demonstrated from 2- to over 200 nm.[39,46] In some materials the pore diameter can be tuned by changing the surfactants and using pore expanders without the loss of ordered structure. For example, MCM-41 has a stable hexagonal structure with no pore interconnectivity, and the pore diameter can be tuned from 2 to 30 nm using established methods.[46] As one of the first mesoporous materials, MCM-41 has been shown to be very effective as a catalyst and sorbent support.[12,22,23,39,47,48] Templating with neutral amines instead of long chain cationic surfactants produces a more disordered hexagonal mesoporous silica (HMS) with somewhat better stability and pore connectivity than MCM-41.[5,47] MCM-48 has a three dimensional cubic phase structure with excellent pore interconnectivity that provides effective mass transport.[49,50] SBA-15 has a hexagonal structure with significant pore interconnectivity. The thicker walls provide SBA-15 greater stability than many nanoporous silica structures and the interconnected channels provide excellent mass transfer kinetics as well as resistance to pore plugging.[51,52] These properties make SBA-15 an excellent candidate structure for development of materials for chemical separations or transform processes.

A weakness with all silica structures is their chemical instability at a higher and lower pH conditions. At lower pH, the silica structure is typically stable but the alkyl silane bridge that allows functional groups to be attached is subject to acidic attack and degradation. At higher pH, the silica structure itself can undergo hydrolysis and even complete dissolution. The stability of mesoporous silica materials are very dependant upon their structure and any subsequent material modifications.[36,53,54] Typically, mesoporous silica exhibits stability between pH 3 and 7.5. The stability range can be extended by utilizing mesoporous structures with thicker walls, additives such as alumina to the silica, or coating with a high density cross-linked organosilane surface layer.[53,55–60] The typical thermal stability of calcined mesoporous structures is typically over 500°C. However, the stability of any organosilane functionalization is much lower, 100–300°C depending upon the ambient constituents and the chemistry employed on the surface.

Amine Functionalization

One of the principle advantages of nanoporous silica is its amenability to surface functionalization with different chemical moieties via a number of direct and indirect methods.[35,37,39,41] Silica functionalization methods include co-condensation, simple impregnations or surface deposition, and post-condensation covalent grafting.

Silica functionalization with co-condensation is accomplished by incorporation of the silanes with the desired chemical moieties into the original sol-gel solution before polymerization is initiated. During the sol-gel condensation reaction, the functional silane is covalently bound into the silica structure in a convenient one-pot synthesis approach. Because the functional organosilane is present during the sol-gel reaction, organic groups which could interfere with the condensation process or disrupt the micellar structure (such as organic acids) must be avoided. If such a group is desired, incorporation must be performed post-condensation via chemical conversion with established synthetic methods. The templating surfactant can be removed chemically, but this is not ideal for many applications, because the mesoporous structure has not been calcined (which would destroy the intercollated functional organosilane) and the silica continues to slowly undergo condensation resulting in an unstable structure. These structures are particularly vulnerable to hydrothermal degradation due to hydrolysis of the partially cross-linked silica structure.

More popular silica functionalization methods are the impregnation or surface deposition of the desired material directly into a fully calcined mesoporous material. Since the calcined mesoporous support material is fully condensed and cross-linked, many of the structural stability issues that arise with co-condensation functionalization are eliminated. This approach has been effectively applied for the functionalization of mesoporous materials with heterogeneous catalyst and absorbing materials, including amines for CO_2.[48]

Effective CO_2 sorbents have been created by deposition of polyethyleneimine (PEI) in the pores of mesoporous structures.[22,23] The synthetic methods and materials selected attempt to maximize amine loading (thereby increasing capacity) without plugging the pores (thereby reducing capacity and kinetics) within the materials. For polyamine surface deposition, materials with larger pores and good interpore connectivity have demonstrated superior performance.[7,49] CO_2 sorbents have also been created by impregnation of liquid amines into the pores of the mesoporous structures.[10,13] The high porosity of the mesoporous supports (typically over 50%) enable significant liquid loading of the sorbent. The open pore structure and short diffusion distances in the liquid phase provide excellent kinetics. However, the existence of a liquid phase places fundamental mass transfer limitation upon the system. Further, silica functionalization with liquid amine impregnation or surface deposition of polyamines produces a material where the active component is not covalently bound to the substrate. Consequetly, the functional component of the material can be removed from the sorbent via evaporative or elution processes, resulting in a reduction of sorbent performance. The operating conditions for such sorbents will be bound by the thermal, pressure and chemical conditions required to retain the active components within the solid support.

Covalent grafting of the chemically active moieties to the surface of calcined nanoporous silica is an attractive functionalization approach. It avoids the structural instabilities and chemical incompatibility issues inherent in the co-condensation approach since the silica structure is completely calcined before the desired organosilane species are introduced. Since the active moieties are covalently bound, grafting avoids the issues with evaporative or elution processes that may degrade sorbents materials created via impregnation or simple surface deposition. Consequently, covalent grafting of organosilanes post calcination is a very popular method for functionalization of nanoporous silica and has been demonstrated with a vast array of active groups for a range of applications.[35,37,41,43]

A number of groups have demonstrated that aminoalkoxysilanes grafted onto mesoporous silica supports results in effective CO_2 sorbents.[5–7,9,15,17–19,24,49] The selection of the terminal amine structure determines the complexation mechanisms and binding chemistries. The selection of the mesoporous silica support can strongly impact factors such as kinetics and capacity. Aminoalkoxysilanes grafted on MCM-48 and SBA-15 are particularly effective CO_2 sorbent materials. MCM-48 with monomeric aminopropyl grafting has excellent pore interconnectivity and has been demonstrated to be an effective support for CO_2 sorbents. While polymeric amines in MCM-48 showed high concentrations of sorbent sites, much of the pore structure was plugged resulting in reduced adsorption rates and capacities.[49] The thicker walls of SBA-15 sorbents provide greater stability than MCM-48 based materials while retaining good interconnected porosity. Several research groups have studied monomeric aminopropyl functionalized SBA-15 and found it to be an effective CO_2 sorbent.[6,7,15,17] Monomeric aminopropyl functionalized materials react at a 2:1 ratio with CO_2 (in the absence of water) and at a 1:1 ratio (captured as a bicarbonate salt) if sufficient water is present. EDA can react at a 1:1 mole ratio with CO_2 providing improved capacity (on a per ligand basis in the absence of water). Further, the principle complexation reaction for EDA-SBA15 is believed to be the formation of a carbamate salt which does not require the presence of water and has inherently fast reaction kinetics.[24,61] The capacity of EDA-SBA-15 reported to date compares favorably with previously reported aminopropyl-modified mesoporous silica. Furthermore, the adsorption capacity is not dependent upon the presence of water and the complexation kinetics is better.

Additional increases in sorbent capacity and kinetics are desirable and work is underway to provide materials with improved performance and economics. One way to improve performance is to increase the surface area of the support material. Undoubtedly there is room for incremental improvements but the specific surface area(s) of mesoporous silica cannot be significantly improved beyond the structures already used. However, many researchers have reported the loss of significant portions of the material surface area after installation of the amine moieties onto the surface, presumably resulting from pore plugging due to polymerization across the pores and not over the surface. Refinement of covalent grafting methods of relevant aminoalkoxysilanes to prevent vertical polymerization across the nanopore(s) should result in substantial improvements in both the capacity and kinetics. A direct approach to improving capacity is to increase the number of complexation sites per square nanometer of material. Such improvements also result in increased stability because the underlying substrate is protected by the functional monolayer. Ligand surface densities for EDA-SBA-15 have been estimated at or below 2.3 ligands per square nanometer.[15,17,24] Thiol-terminated monolayers on mesoporous silica have been reported as high as 6 silanes per square nanometer.[60] This suggests there can be significant improvements in CO_2 sorbent performance with increased amine site density. Refinement of covalent grafting methods to increase the density of ligands on the surface of mesoporous silica should result in improvements in both the capacity and stability. Improved surface ligand densities may be achieved with better silica surface preparation, multistep silanization processes, longer treatment times to enhance diffusion of precursors into pores, or more aggressive deposition conditions, i.e. intensive refluxing or the use of supercritical fluid solvents. Another approach to improving the performance of CO_2 sorbents may be to optimize the molecular environment around each ligand site to prevent competitive or interfering processes from reducing the chemical activity of complexation sites.

It should noted that the performance of aminoalkoxysilane modified nanoporous silica has met or exceeded the performance characteristic of commercially available solid phase CO_2 sorbents.[5–7,9,15,17–19,24,49] Furthermore, it should be clear that these new materials are not optimized and additional research will result in further improvements.

Sorbent Characterization and Performance Evaluation

Any sorbent materials must be well characterized because their physicochemical properties directly impact their performance in CO_2 adsorption and desorption. In addition, physical and chemical characterizations of the functionalized and un-functionalized materials are essential for development of improved synthesis techniques. The performance of the CO_2 sorbents is evaluated based on direct measurements of CO_2 uptake and release properties using various analytical methods.

One of the most critical properties of solid amine sorbents is the amine loading. This is best defined by the number of moles of amine functional groups per unit mass of the sorbent. There are several ways to determine the amine loading. The mass increase due to amine functionalization can be determined by gravimetric measurements.[12,13,15,18] Provided that the sorbent materials are properly outgassed to remove water and CO_2 that may have been adsorbed during synthesis or storage, accurate mass fraction of the amine moieties can be determined. The amine loading can be calculated from the amine mass fraction, the composition and molecular weights of the amine compounds. Direct elemental analysis of the functionalized sorbents can also provide the nitrogen mass fraction from which the amine loading is calculated.[4,5,19,49,62] X-ray photoelectron spectroscopy (XPS) can also be used to determine the atomic composition on the surface of amine-modified silica sorbents.[6-8] Because only the surface layer is probed by XPS, the nitrogen composition from XPS data cannot be converted to amine loading or surface amine concentration easily. The amine loading of porous silica sorbents can also be estimated from ^{29}Si solid state nuclear magnetic resonance (NMR) data.[24,63] The chemical environment of the silicon atoms is probed by ^{29}Si NMR and the abundance of Si-C bonds can be determined.

For tethered amine sorbents, the concentration of the amine groups on the sorbent surface, measured by the number of molecules per square nanometer, is useful when comparing the effectiveness of various synthesis methods. The surface amine concentration can be estimated from amine loading and the specific surface area of the sorbent material. The surface area of the sorbents is typically determined by nitrogen adsorption measurements using the Brunauer-Emmett-Teller method. The surface amine concentrations reported in the literature are typically between 1.0 to 3.0 molecules per square nanometer on porous silica supports.[5,15,19,24,49,63,64] Amine surface concentrations based on the surface area of both the unmodified and the functionalized substrates can be found in the literature. However, the surface area does change when the sorbent material is functionalized due to pore partial filling as well as plugging. Therefore one needs to be careful when comparing amine surface concentration values in the literature. It was proposed that the surface concentration based on the surface area of the unmodified substrate is a better indicator of the degree of functionalization.[5] In any case, it is important to characterize the surface area of the sorbent material both before and after the functionalization so that the amine surface concentration can be compared with other published data consistently.

The CO_2 adsorption and desorption properties such as capacity can be measured gravimetrically, typically using a thermogravimetric analysis (TGA) instrument.[4,5,12,13,18,22,23,26,49,65] When using TGA instruments that have differential scanning calorimetry capability, the heat of adsorption and the specific heat capacity of the sorbent can also be measured. Both properties are important for the analysis of the economics of the CO_2 capture process as they affect the energy consumption during the sorbent regeneration step. The thermal stability of the solid amine sorbents is also usually determined by thermogravimetric methods. TGA coupled with mass spectrometry of the effluent gas is particularly useful for measurements of desorption temperature of water and CO_2, as well as decomposition temperature of the incorporated amine moieties. Because the TGA method can be coupled with calorimetry and effluent gas analysis, and the typical sample size required for TGA analysis is only about a few tens of milligrams, it is a rather useful characterization tool for CO_2 sorbent studies. Static volumetric measurements similar to the nitrogen adsorption method for surface area measurements were also reported to gather CO_2 adsorption isotherm data.[18,19]

Another popular technique for studying CO_2 sorption is to use a fixed-bed reactor to make flow-through type measurements, where the sorption rate is determined from a mass balance based on gas phase flow rate and concentrations.[6–11,15,17–21] In a typical fixed sorbent bed experiment, a step change in the inlet CO_2 concentration of the sorbent bed results in a mass transfer front traveling down the bed during the adsorption step. The amount of CO_2 adsorbed by the sorbent bed at any moment equals the difference in the accumulated amount of CO_2 between what has entered and what has exit the bed up to that moment. Both quantities can be easily calculated by integration of the product of the total flow and the CO_2 concentration versus time. The variation in the total flow rate of the effluent stream due to adsorption or desorption can be compensated by simple mass balance considerations. Compared to the TGA method, the flow-through method allows better control of gas flow since TGA instruments typically have quite significant dead volume. The data by flow-through methods can be fitted to an adsorption column model for kinetics studies. The ability to obtain sorption kinetics data at conditions relevant to actual adsorption process is one of the main advantages of the flow through type measurements. The flow-through method also allows testing of different adsorption processes such as temperature swing adsorption (TSA) and pressure swing adsorption (PSA) and so on. One draw back of flow-through measurements is that the mass change of the sorbent is determined indirectly from concentration and flow rate measurements. Therefore, proper calibration of the concentration measurement instrument is critical, whether it is a gas chromatogram, a mass spectrometer, or other CO_2 detector. Baseline measurements with an empty sorbent bed are also necessary in order to eliminate the error in CO_2 adsorption calculation caused by the dead volume of the flow system.

Quantitative comparison of the CO_2 adsorption capacities reported in the literature is difficult as researchers have used different measurement techniques and different test conditions for CO_2 inlet concentration and adsorption temperature. Conversion of these capacity data to equivalent values at identical concentration and temperature is not practical as the adsorption isotherms are generally unknown other than for those discrete data points reported. Additionally, some capacity data were reported in units based on actual sorbent mass while others based on the mass of unfunctionalized starting materials. However, a qualitative comparison of the adsorption capacity is still valuable. In Table 1, CO_2 adsorption capacities reported in the literature were compiled. The raw data were expressed in various units such as weight percentage, milligrams, standard cubic centimeters, or millimoles of CO_2 per gram of the dry adsorbent. For the purpose of comparison the literature CO_2 capacity data were all converted to moles per kilogram.

The CO_2 capacity of the tethered amine sorbents at near ambient temperature is generally from 0.5 to 2.0 mol/kg. Comparable or higher capacities have been reported for the impregnated amine sorbents but their adsorption kinetics is slower than tethered amine sorbents, if the same solid support is used.[49] Although high amine loading, and therefore high CO_2 capacity, can be achieved by filling almost all available pore volume with a liquid amine, the adsorption rate is reduced as diffusion through liquid amine phase is much slower than through gas phase in open pores. A thorough analysis of the kinetics advantages of the different types of solid amine sorbents is not possible at this time because there are too few adsorption rate data available in the literature.

It is worth comparing the various solid amine sorbents in terms of amine efficiency, which is defined as the number of CO_2 molecules adsorbed for every nitrogen atom in the amine function groups. Therefore, the theoretical maximum of amine efficiency for carbamate formation is 0.5, i.e.

$$2R - NH_2 + CO_2 \longrightarrow R - NH_3^+ + R - NHCOO^- \tag{1}$$

When CO_2 is captured as bicarbonates, the theoretical maximum amine efficiency is 1, i.e.

$$R - NH_2 + CO_2 + H_2O \longrightarrow R - NH_3^+ + HCO_3^- \tag{2}$$

As can be seen in Table 1, the amine efficiency of the solid amine sorbents in the literature is typically lower than the theoretical values. Most synthetic amine sorbents contain defects that render a certain fraction of functional groups unavailable for reaction with CO_2. For example, polymerization of the amine silane precursors may result in plugged pores. The tethered amine groups can also form hydrogen bonds with surface silanol groups and become deactivated. Therefore, there are opportunities for improvements in sorbent performance through better synthesis techniques.

Generally, higher amine loading results in higher CO_2 adsorption capacity. For impregnated amine sorbents, the capacity increases with the amine loading until the loading is slightly over the level corresponding to complete pore filling.[13,22,23] Further increases in amine loading result in amine deposition outside the pores, which leads to reduced adsorption rates. At such high amine loading, the effective adsorption capacity after fixed adsorption time also decreases due to reduced adsorption rate even though the equilibrium capacity is expected to be higher with the added amine content. It was

Table 1. CO_2 capacity of solid amine sorbents reported in literature.

Amine Groups	Solid Supports	Capacity (mol/kg)	Amine Efficiency	Measurement Conditions[§]	Ref.
NH_2[†]	Silica Gel	0.62	0.67[∈]	100% CO_2, 23°C	19
NH_2[†]	SBA-15	1.75	—	10% CO_2/He 20°C	7
NH_2[†]	SBA-15	0.40	—	4% CO_2/He 25°C	6
NH_2[†]	HMS	1.59	0.69	90% CO_2/Ar 20°C	5
NH_2[†]	MCM-48	2.05	0.89	100% CO_2 25°C	18
	Silica Xerogel	1.14	0.67		
NH_2[†]	MCM-48	0.80	0.33	100% CO_2 25°C	49
Pyrrolidine[†]		0.30	0.20		
PEI[†]		0.40	0.08		
NH_2[†]	SBA-15	0.52	0.19	15% CO_2/N₂ 60°C	15
EDA[†]		0.87	0.21		
DETA[†]		1.10	0.22		
EDA[†]	SBA-15	0.79	—	50% CO_2/He	9
EDA[†]	SBA-15	0.45	0.08	15% CO_2/N₂ 25°C	24
PEI[‡]	MCM-41	4.89	—	100% CO_2 75°C	22
PEI[‡]	PMMA	0.84	—	2% CO_2/N₂	11
DEA[‡]	MCM-41	2.81	0.40	5% CO_2/N₂ 25°C	13
MEA[‡]	PMMA	0.83	0.14	1% CO_2/N₂ 20°C	10

[§]At 1 atm or ambient pressure and room temperature unless noted otherwise.
[∈]Average amine loading used based on a reported range.
[†]Anchored covalently to surface as propyl silanes derivatized with the following amine functional groups: PEI, polyethyleneimine; EDA, ethylenediamine; DETA, diethylenetriamine.
[‡]Immobilized by impregnation of the following liquid amines: PEI, polyethylene imine; DEA, diethanolamine; MEA, monoethanolamine.

proposed that the liquid amine incorporated within the pores reacts with CO_2 faster than in bulk liquid phase,[22,23] although the mechanism for this phenomenon has not been analyzed in detail. For tethered amine sorbents, it has been shown that the CO_2 concentration on the surface increased with surface amine concentration.[5] When the surface concentration of the adsorption sites was lower than approximately one nitrogen atom per square nanometer, the data suggested that physical adsorption was dominant.[5]

The effect of physical adsorption on overall adsorption capacity of the solid amine sorbents has not been well studied. The physical adsorption on unmodified supports is typically small compared to the capacity of the solid amine sorbents.[15,17,18,22,23] The heats of adsorption reported in the literature are consistent with strong adsorbate-surface interactions, i.e. greater than two to three times the heat of evaporation (see Table 2). For CO_2 the heat of evaporation at 0°C is 10.3 kJ/mol.[66] However, it is oversimplified to

Table 2. The heat of CO_2 adsorption of solid amine sorbents reported in literature.

Amine Groups	Solid Supports	Heat of Adsorption (kJ/mol)	Method	Ref.
NH_2[†]	Silica Gel	58.8 ± 4.2	Calculated from isosteres	19
NH_2[†]	HMS	60 ± 2	Measured by differential thermal analysis	5
EDA[†]	SBA-15	47.8	Calculated from desorption peak temperature and half width	9
PEI[‡]	PMMA[§]	94.6 ± 8.2	Measured by isothermal flow microcalorimetry	11

[†]Anchored covalently to surface as propyl silanes; EDA = ethylenediamine.
[‡]Impregnated polyethylene imine.
[§]Polymethyl methacrylate.

treat the interaction between CO_2 molecules and amine-modified surfaces as purely chemisorption. All of the reported CO_2 adsorption isotherms by solid amine sorbents show that the CO_2 capacity increases significantly with the CO_2 partial pressure.[9,13,18,19,22,24,49] This behavior is similar to that of physical adsorption. In strict chemisorption the sorbent surface is saturated rapidly even at low partial pressure and the equilibrium capacity is essentially independent of adsorbate partial pressure. Therefore contributions from intermolecular interactions between the CO_2 and amine-modified surfaces are also important to the overall CO_2 uptake/release behavior of solid amine sorbents. Hydrogen bonds of CO_2 molecules with tethered amine and residual surface silane groups are one example of the weaker interactions. A molecular dynamics study showed that there are distinctive structural "pockets" on amine-modified hexagonal mesoporous silica surface where the motion of CO_2 molecules is restricted.[3]

There are conflicting results in the literature regarding the effect of water on the CO_2 adsorption by solid amine sorbents. Some reported that when water was present the adsorption capacity increased significantly and the reason was thought to be the shift from carbamate formation to bicarbonate formation, which only requires 1 mole of amine to react with 1 mole of CO_2.[5,11,19,20] Others reported that the CO_2 adsorption capacity was not enhanced by the presence of water.[13,15,17,18,24] One group also reported that although the CO_2 adsorption capacity increased in presence of water, the rate of desorption decreased.[5] Obviously, more studies are needed to elucidate the role of water and to understand how sorbent synthesis can be improved to better control adsorption behavior. All existing studies show that the presence of water does not diminish the CO_2 adsorption capacity, so the presence of moisture in gas streams from energy production is not a problem for solid amine sorbents from a capacity point of view.

Gas streams from fossil fuel-fired power plants contain other acid gas such as SO_2, NO_x, or H_2S, depending on the fuel conversion process. Strong tolerance of solid amine sorbents to these acid gas contaminants is highly desirable. If solid amine sorbents are also reactive towards these acid gases, whether they can be regenerated easily after exposure is also critical. There are very few reports on the performance of solid amine sorbents with acid gas contaminants. One group of researchers showed that H_2S was adsorbed reversibly on amine-grafted MCM-48 mesoporous silica.[18] The regeneration temperature for H_2S was 60°C, which was lower than that of CO_2 at 75°C. The formation of $NH_3^+ HS^-$ group was evidenced by characteristic infrared absorption bands and the adsorption of H_2S was not affected by the presence of water. Another group reported that the adsorption of NO_x on polyethyleneimine-impregnated MCM-41 sorbents was irreversible under flue gas conditions.[21] In another study,[26] it was found that copolymers of styrene with pendent ethylenediamine, N,N,N'-trimethylethylenediame, and 1-methylpiperazine groups absorbed NO and NO_2 but not N_2O. The amine/NO adducts oxidized to ammonium nitrites and nitrosamines in the presence of oxygen, leading to loss of active amine sites. The NO_2/amine adducts underwent similar oxidative degradation and deactivation. On the other hand, the same amino-functional polymers exhibited thermally reversible adsorption of SO_2. The affinity for SO_2 was much higher than for CO_2 as SO_2 is a stronger Lewis acid.

Based on the above limited literature results, it appears that the utility of solid amine sorbents for CO_2 capture from synthesis gas will not be affected by the presence of trace H_2S contaminants. It is possible that H_2S can even be captured simultaneously with CO_2. For CO_2 capture from flue gas, the presence of NO_x will likely cause sorbent degradation and thus removal of NO_x prior to CO_2 capture is recommended. Simultaneous capture of SO_2 with CO_2 from flue gas seems possible using solid amine sorbents.

Economics

Solid based CO_2 sorbent systems for the most part have not been examined for cost-effectiveness in large scale applications. Such analysis is difficult because the unit operations required and their configuration are considerably different from typical liquid based scrubbing systems. Here, a preliminary analysis of the economics of using solid sorbents is presented to gain understanding of where the current level of development stands in comparison to commercial technologies such as liquid amine systems.

Methods and assumptions

The US EPA has published costing guidelines for carbon adsorbers used in the application of volatile organics capture.[67] This model has been used in combination with the TEAM model developed by Brown *et al.*[68] to make a rough estimate of levelized costs of carbon capture using amine-functionalized solids. Several assumptions had to be made to adapt these models to the problem at hand:

- It is envisioned that the desorbing period may be less than that required for adsorption; therefore, it is possible to be capturing CO_2 with several columns, while regenerating a lesser number of columns at any given time on a rotating basis.
- Flow of CO_2 through the adsorbing bed is terminated once a critical breakthrough concentration is reached. In this case, the adsorption cycle is ended when the integrated average capture fraction drops to 90%.
- Steam is used for desorption of CO_2 because it can quickly and more uniformly heat the bed to a given desorption temperature. It is also in good supply in a coal fired power plant. Steam is also used to carry the CO_2 out of the adsorber zone.
- Once desorption is complete, ambient air is blown through the bed to cool for another adsorption cycle. This air also serves to remove excess moisture from any steam condensation.
- Working capacity for CO_2 adsorption is assumed to be 5%. This is consistent with laboratory results for a feed of 15% CO_2 in a simulated coal flue gas.
- Adsorption is performed at 95°C and desorption is performed at 135°C. This temperature swing results in an estimated sensible heat of 325 Btu/lb CO_2.
- 1177 Btu/lb CO_2 is the desorption energy for a CO_2-liquid EDA complex; whereas, the desorption energy is 98 Btu/lb CO_2 for physical adsorption on zeolitic materials. Laboratory results show a combination of chemi- and physisorption, so it was assumed that the desorption energy was 357 Btu/lb CO_2 based on the ratio of chemi- to physisorption observed in lab tests.
- Steam consumption in the carrier role could dominate the steam demand. A rule of thumb for carbon beds is 3 to 5 lb of steam per pound of adsorbate or 3000 to 5000 Btu/lb CO_2.[67] It is possible to recover up to approximately 70% of this energy through recovery processes depending on the plant configuration; therefore, steam regeneration energy assumptions of 1000, 2500, and 4000 Btu/lb CO_2 were used.
- Steam cost is estimated at \$1.57/MMBtu based on the foregone electricity production.[69]

- The adsorbent material cost at large scale production is unknown; therefore, costs of \$5, \$20, and \$40/lb were used to understand sensitivity to this parameter.
- Economic life of the equipment assumed to be 30 years.
- Depreciable life of the equipment of 20 years.
- Economic life of the sorbent assumed to be five or 30 years.
- Depreciable life of the sorbent of three or 20 years.
- 9.3% discount rate.
- 3.1% inflation rate.
- 39.1% income tax rate.
- 2% property tax rate.

Economics results

The results of the analysis are presented in the form of two tables where Table 3 corresponds to a five-year sorbent life and Table 4 corresponds to a 30-year sorbent life.

From Tables 3 and 4, it is evident that for short sorbent lifetimes the cost of the sorbent is the dominant factor, and for long sorbent lifetimes the regeneration energy cost is a more significant factor. The relative order of magnitudes of cost per ton of CO_2 captured are in the neighborhood of estimates for liquid amine systems (\$20 to 70/ton) or somewhat higher.

Table 3. Levelized capture cost for five-year sorbent lifetime (\$/ton).

Sorbent Cost/Lb	Regeneration Energy Cost (Btu/lb CO_2)		
	1000	2500	4000
\$5	\$20.00	\$26.60	\$33.10
\$20	\$52.40	\$58.90	\$65.40
\$40	\$95.40	\$101.90	\$108.40

Table 4. Levelized capture cost for 30-year sorbent lifetime (\$/ton).

Sorbent Cost/Lb	Regeneration Energy Cost (Btu/lb CO_2)		
	1000	2500	4000
\$5	\$15.90	\$22.40	\$28.90
\$20	\$35.30	\$41.80	\$48.30
\$40	\$61.30	\$67.80	\$74.30

Economics conclusions

With the sorbent cost being a significant factor in the overall CO_2 capture cost, it is important to explore inexpensive substrates, e.g. silica gel, as the starting material for amine modifications. Replicating the amine-modified structures found in the mesoporous silica with tethered amine groups is possible when silica gel is used instead. In our own laboratory, using novel supercritical fluid processing techniques, we have successfully modified silica gels with amine functional groups. The above preliminary economics analysis also illustrates the importance of the lifetime of the solid sorbents. In currently published work, the durability of solid amine sorbents have not been tested adequately. Additional investigation of the stability of solid amine sorbents in CO_2 capture should be undertaken to obtain a better understanding of the deactivation process and its prevention. Provided that the uncertainty in sorbent cost and sorbent lifetime can be narrowed, a more detailed comparison of the solid amine sorbents with existing technologies is certainly highly desirable. While that is beyond the scope of this book chapter, it is our hope that this chapter can serve as useful resources for practitioners in this rapidly developing research field.

References

1. Working Group III of the Intergovernmental Panel on Climate Change *IPCC Special Report on Carbon Dioxide Capture and Storage*, Cambridge University Press, New York, 2005.
2. P. J. Branton, P. G. Hall, M. Treguer and K. S. W. Sing, Adsorption of carbon-dioxide, sulfur-dioxide and water-vapor by mcm-41, a model mesoporous adsorbent, *Journal of the Chemical Society-Faraday Transactions* **91**(13), 2041–2043 (1995).
3. A. L. Chaffee, Molecular modeling of HMS hybrid materials for CO_2 adsorption, *Fuel Processing Technology* **86**(14–15), 1473–1486 (2005).
4. G. P. Knowles, S. W. Delaney and A. L. Chaffee, Amine-functionalised mesoporous silicas as CO_2 adsorbents, *Nanoporous Materials IV Studies in Surface Science and Catalysis* **156**, 887–896 (2005).
5. G. P. Knowles, J. V. Graham, S. W. Delaney and A. L. Chaffee, Aminopropyl-functionalized mesoporous silicas as CO_2 adsorbents, *Fuel Processing Technology* **86**(14–15), 1435–1448 (2005).
6. A. C. C. Chang, S. S. C. Chuang, M. Gray and Y. Soong, In-situ infrared study of CO_2 adsorption on SBA-15 grafted with gamma-(aminopropyl)triethoxysilane, *Energy & Fuels* **17**(2), 468–473 (2003).
7. M. L. Gray, Y. Soong, K. J. Champagne, H. Pennline, J. P. Baltrus, R. W. Stevens, R. Khatri, S. S. C. Chuang and T. Filburn, Improved immobilized carbon dioxide capture sorbents, *Fuel Processing Technology* **86**(14–15),

1449–1455 (2005).

8. M. L. Gray, Y. Soong, K. J. Champagne, H. W. Pennline, J. Baltrus, Jr. R. W. Stevens, R. Khatri and S. S. C. Chuang, Capture of carbon dioxide by solid amine sorbents, *Int. J. Environmental Technology and Management* **4**(1/2), 82–88 (2004).

9. R. A. Khatri, S. S. C. Chuang, Y. Soong and M. Gray, Carbon dioxide capture by diamine-grafted SBA-15: A combined Fourier transform infrared and mass spectrometry study, *Ind. Eng. Chem. Res.* **44**(10), 3702–3708 (2005).

10. T. Filburn, J. J. Helble and R. A. Weiss, Development of supported ethanolamines and modified ethanolamines for CO₂ capture, *Ind. Eng. Chem. Res.* **44**(5), 1542–1546 (2005).

11. S. Satyapal, T. Filburn, J. Trela and J. Strange, Performance and properties of a solid amine sorbent for carbon dioxide removal in space life support applications, *Energy & Fuels* **15**(2), 250–255 (2001).

12. R. Franchi, P. J. E. Harlick and A. Sayari, A high capacity, water tolerant adsorbent for CO₂: Diethanolamine supported on pore-expanded MCM-41, *Nanoporous Materials IV Studies in Surface Science and Catalysis* **156**, 879–886 (2005).

13. R. S. Franchi, P. J. E. Harlick and A. Sayari, Applications of pore-expanded mesoporous silica. 2. Development of a high-capacity, water-tolerant adsorbent for CO₂, *Ind. Eng. Chem. Res.* **44**(21), 8007–8013 (2005).

14. N. Hiyoshi, K. Yogo and T. Yashima, Adsorption of carbon dioxide on modified mesoporous materials in the presence of water vapor, *Recent Advances in the Science and Technology of Zeolites and Related Materials, Pts A–C* **154**, 2995–3002 (2004).

15. N. Hiyoshi, K. Yogo and T. Yashima, Adsorption of carbon dioxide on aminosilane-modified mesoporous silica, *Journal of the Japan Petroleum Institute* **48**(1), 29–36 (2005).

16. N. Hiyoshi, K. Yogo and T. Yashima, Reversible adsorption of carbon dioxide on amine-modified SBA-15 from flue gas containing water vapor, *Carbon Dioxide Utilization for Global Sustainability* **153**, 417–422 (2004).

17. N. Hiyoshi, K. Yogo and T. Yashima, Adsorption of carbon dioxide on amine modified SBA-15 in the presence of water vapor, *Chemistry Letters* **33**(5), 510–511 (2004).

18. H. Y. Huang, R. T. Yang, D. Chinn and C. L. Munson, Amine-grafted MCM-48 and silica xerogel as superior sorbents for acidic gas removal from natural gas, *Ind. Eng. Chem. Res.* **42**(12), 2427–2433 (2003).

19. O. Leal, C. Bolivar, C. Ovalles, J. J. Garcia and Y. Espidel, Reversible adsorption of carbon dioxide on amine surface-bonded silica gel, *Inorganica Chimica Acta* **240**(1–2), 183–189 (1995).

20. X. Xu, C. Song, B. G. Miller and A. W. Scaroni, Influence of moisture on CO₂ separation from gas mixture by a nanoporous adsorbent based on polyethylenimine-modified molecular sieve MCM-41, *Ind. Eng. Chem. Res.* **44**(21), 8113–8119 (2005).

21. X. C. Xu, C. S. Song, B. G. Miller and A. W. Scaroni, Adsorption separation of carbon dioxide from flue gas of natural gas-fired boiler by a

novel nanoporous "molecular basket" adsorbent, *Fuel Processing Technology* **86**(14–15), 1457–1472 (2005).

22. X. C. Xu, C. S. Song, J. M. Andresen, B. G. Miller and A. W. Scaroni, Novel polyethylenimine-modified mesoporous molecular sieve of MCM-41 type as high-capacity adsorbent for CO_2 capture, *Energy & Fuels* **16**(6), 1463–1469 (2002).

23. X. C. Xu, C. S. Song, J. M. Andresen, B. G. Miller and A. W. Scaroni, Preparation and characterization of novel CO_2 "molecular basket" adsorbents based on polymer-modified mesoporous molecular sieve MCM-41, *Microporous and Mesoporous Materials* **62**(1–2), 29-45 (2003).

24. F. Zheng, D. N. Tran, B. J. Busche, G. E. Fryxell, R. S. Addleman, T. S. Zemanian and C. L. Aardahl, Ethylenediamine-modified SBA-15 as regenerable CO_2 sorbent, *Ind. Eng. Chem. Res.* **44**(9), 3099–3105 (2005).

25. E. S. Kikkinides, V. I. Sikavitsas and R. T. Yang, Natural-gas desulfurization by adsorption — feasibility and multiplicity of cyclic steady-states, *Ind. Eng. Chem. Res.* **34**(1), 255–262 (1995).

26. A. Diaf, J. L. Garcia and E. J. Beckman, Thermally reversible polymeric sorbents for acid gases — CO_2, SO_2 and Nox, *Journal of Applied Polymer Science* **53**(7), 857–875 (1994).

27. G. W. Xu, C. F. Zhang, S. J. Qin, W. H. Gao and H. B. Liu, Gas-liquid equilibrium in a CO_2-MDEA-H_2O system and the effect of piperazine on it, *Ind. Eng. Chem. Res.* **37**(4), 1473–1477 (1998).

28. G. W. Xu, C. F. Zhang, S. J. Qin and B. C. Zhu, Desorption of CO_2 from mdea and activated mdea solutions, *Ind. Eng. Chem. Res.* **34**(3), 874–880 (1995).

29. G. W. Xu, C. F. Zhang, S. J. Qin and Y. W. Wang, Kinetics study on absorption of carbon-dioxide into solutions of activated methyldiethanolamine, *Ind. Eng. Chem. Res.* **31**(3), 921–927 (1992).

30. H. Y. Dang and G. T. Rochelle, CO_2 absorption rate and solubility in monoethanolamine/piperazine/water, *Separation Science and Technology* **38**(2), 337–357 (2003).

31. E. Schering, *Chemisches Zentralblatt* **72**(II), 519 (1911).

32. E. Schering, German Patent 123,138, 1900.

33. F. Fichter and B. Becker, *Chemische Berichte* **44**, 3481 (1911).

34. V. Meyer and P. Jacobson, in *Lehrbuch der Organischen Chemie* (Leipzig, 1913), p. 1370.

35. A. Vinu, K. Z. Hossain and K. Ariga, Recent advances in functionalization of mesoporous silica, *Journal of Nanoscience and Nanotechnology* **5**(3), 347–371 (2005).

36. C. J. Brinker and G. W. Scherer, *Sol-Gel Science: The Physics and Chemistry of Sol-Gel Processing* (Academic Press: Boston, 1990).

37. G. E. Fryxell, Y. H. Lin, H. Wu and K. M. Kemner, Environmental applications of self-assembled monolayers on mesoporous supports (SAMMS), in *Nanoporous Materials III*, eds. A. Sayari and M. Jaroniec (Elsevier Science, Amsterdam, 2002), pp. 583–590.

38. M. S. Morey, A. Davidson and G. D. Stucky, Silica-based, cubic mesostructures: Synthesis, characterization and relevance for catalysis, *Journal of Porous Materials* **5**(3–4), 195–204 (1998).

39. J. Y. Ying, C. P. Mehnert and M. S. Wong, Synthesis and applications of supramolecular-templated mesoporous materials, *Angewandte Chemie-International Edition* **38**(1–2), 56–77 (1999).

40. A. Sayari, Catalysis by crystalline mesoporous molecular sieves, *Chemistry of Materials* **8**(8), 1840–1852 (1996).

41. K. Moller and T. Bein, Inclusion chemistry in periodic mesoporous hosts, *Chemistry of Materials* **10**(10), 2950–2963 (1998).

42. U. Ciesla and F. Schuth, Ordered mesoporous materials, *Microporous and Mesoporous Materials* **27**(2–3), 131–149 (1999).

43. A. Stein, B. J. Melde and R. C. Schroden, Hybrid inorganic-organic mesoporous silicates — nanoscopic reactors coming of age, *Advanced Materials* **12**(19), 1403–1419 (2000).

44. A. Sayari and S. Hamoudi, Periodic mesoporous silica-based organic — Inorganic nanocomposite materials, *Chemistry of Materials* **13**(10), 3151–3168 (2001).

45. A. Taguchi and F. Schuth, Ordered mesoporous materials in catalysis, *Microporous and Mesoporous Materials* **77**(1), 1–45 (2005).

46. A. Sayari, Y. Yang, M. Kruk and M. Jaroniec, Expanding the pore size of MCM-41 silicas: Use of amines as expanders in direct synthesis and postsynthesis procedures, *Journal of Physical Chemistry B* **103**(18), 3651–3658 (1999).

47. M. Dubois, T. Gulikkrzywicki and B. Cabane, Growth of silica polymers in a lamellar mesophase, *Langmuir* **9**(3), 673–680 (1993).

48. B. Zhou, S. Hermans and G. Somorjai, *Nanotechnology in Catalysis*, Vol. 2 (Kluwer Academic/Plenum Publishers, New York, 2004).

49. S. Kim, J. Ida, V. V. Guliants and J. Y. S. Lin, Tailoring pore properties of MCM-48 silica for selective adsorption of CO$_2$, *Journal of Physical Chemistry B* **109**(13), 6287–6293 (2005).

50. J. C. Vartuli, K. D. Schmitt, C. T. Kresge, W. J. Roth, M. E. Leonowicz, S. B. Mccullen, S. D. Hellring, J. S. Beck, J. L. Schlenker, D. H. Olson and E. W. Sheppard, Effect of surfactant silica molar ratios on the formation of mesoporous molecular-sieves — inorganic mimicry of surfactant liquid-crystal phases and mechanistic implications, *Chemistry of Materials* **6**(12), 2317–2326 (1994).

51. D. Y. Zhao, Q. S. Huo, J. L. Feng, B. F. Chmelka and G. D. Stucky, Nonionic triblock and star diblock copolymer and oligomeric surfactant syntheses of highly ordered, hydrothermally stable, mesoporous silica structures, *Journal of the American Chemical Society* **120**(24), 6024–6036 (1998).

52. D. Y. Zhao, J. L. Feng, Q. S. Huo, N. Melosh, G. H. Fredrickson, B. F. Chmelka and G. D. Stucky, Triblock copolymer syntheses of mesoporous silica with periodic 50 to 300 angstrom pores, *Science* **279**(5350), 548–552 (1998).

53. M. V. Landau, S. P. Varkey, M. Herskowitz, O. Regev, S. Pevzner, T. Sen and Z. Luz, Wetting stability of Si-MCM-41 mesoporous material in neutral, acidic and basic aqueous solutions, *Microporous and Mesoporous Materials* **33**(1–3), 149–163(1999).

54. A. Doyle and B. K. Hodnett, Stability of MCM-48 in aqueous solution as a function of pH, *Microporous and Mesoporous Materials* **63**(1–3), 53–57 (2003).

55. Y. D. Xia and R. Mokaya, On the hydrothermal stability of mesoporous aluminosilicate MCM-48 materials, *Journal of Physical Chemistry B* **107**(29), 6954–6960 (2003).

56. J. L. Zheng, Y. Zhang, Z. H. Li, W. Wei, D. Wu and Y. H. Sun, Hydrothermally stable mesoporous aluminosilicates with tubular morphology, *Chemical Physics Letters* **376**(1–2), 136–140 (2003).

57. M. Luechinger, L. Frunz, G. D. Pirngruber and R. A. Prins, Mechanistic explanation of the formation of high quality MCM-41 with high hydrothermal stability, *Microporous and Mesoporous Materials* **64**(1–3), 203–211 (2003).

58. D. R. Dunphy, S. Singer, A. W. Cook, B. Smarsly, D. A. Doshi and C. J. Brinker, Aqueous stability of mesoporous silica films doped or grafted with aluminum oxide, *Langmuir* **19**(24), 10403–10408 (2003).

59. Y. S. Ooi, R. Zakaria, A. R. Mohamed and S. Bhatia, Hydrothermal stability and catalytic activity of mesoporous aluminum-containing SBA-15, *Catalysis Communications* **5**(8), 441–445 (2004).

60. T. S. Zemanian, G. E. Fryxell, J. Liu, S. Mattigod, J. A. Franz and Z. M. Nie, Deposition of self-assembled monolayers in mesoporous silica from supercritical fluids, *Langmuir* **17**(26), 8172–8177 (2001).

61. A. Dibenedetto, M. Aresta, C. Fragale and M. Narracci, Reaction of silylalkylmono- and silylalkyldi-amines with carbon dioxide: Evidence of formation of inter- and intra-molecular ammonium carbamates and their conversion into organic carbamates of industrial interest under carbon dioxide catalysis, *Green Chemistry* **4**(5), 439–443 (2002).

62. A. R. Cestari and C. Airoldi, Diamine immobilization on silica gel through the sol-gel process and increase in the organic chain by using glutaraldehyde followed by ethylenediamine, *Langmuir* **13**(10), 2681–2686 (1997).

63. S. Huh, J. W. Wiench, J. C. Yoo, M. Pruski and V. S. Y. Lin, Organic functionalization and morphology control of mesoporous silicas via a co-condensation synthesis method, *Chemistry of Materials* **15**(22), 4247–4256 (2003).

64. G. S. Caravajal, D. E. Leyden, G. R. Quinting and G. E. Maciel, Structural characterization of (3-Aminopropyl)triethoxysilane-modified silicas by Si-29 and C-13 nuclear magnetic-resonance, *Analytical Chemistry* **60**(17), 1776–1786 (1988).

65. A. Diaf and E. J. Beckman, Thermally reversible polymeric sorbents for acid gases .3. CO_2-sorption enhancement in polymer-anchored amines, *Reactive & Functional Polymers* **27**(1), 45–51 (1995).

66. Compressed Gas Association, *Handbook of Compressed Gases*, 4th edn. (Kluwer Academic Publishers, Boston, 1999).

67. US Environmental Protection Agency, *EPA Air Pollution Control Cost Manual*, EPA/452/B-02-001, US EPA Office of Air Quality Planning and Standards: Research Triangle Park, NC, Jan, 02

68. D. R. Brown, D. A. Dirks, M. K. Spanner and T. A. Williams, *An Assessment of Methodology for Thermal Energy Storage Evaluation*, PNNL-6372, Pacific Northwest National Laboratory, Richland, WA, 87

69. *Integrated Environmental Control Model*, version 5.01, URL: http://www.iecm-online.com (Carnegie Mellon University: Pittsburgh, PA, 2004).

Nanomaterials that Enhance Sensing/Detection of Environmental Contaminants

Chapter 13

Nanostructured ZnO Gas Sensors

Huamei Shang and Guozhong Cao*

*Department of Materials Science and Engineering
University of Washington, Seattle, WA, 98195-2120, USA
gzcao@u.washington.edu

Introduction

A sensor is a form of transducer which converts a physical or chemical quantity into an electrical, optical or other measurable quantity. For example, a chemical sensor is intended to determine the composition and concentration of the relevant material via an electrical signal.[1] The specific needs for gas sensors which can detect and monitor, such as oxygen, flammable gases and toxic gases, have emerged to protect the environment and to monitor production processes. There are three types of solid state gas sensors currently in large scale use. They are based on solid electrolytes, on catalytic combustion, and on resistance modulation of semiconducting oxides.[2] For all sensing applications, the selectivity, sensitivity and rapid response are critical parameters. In addition, the compactness, low cost and easiness of integration of these sensors are all important for accommodating multiple sensors. Among these available gas sensors, semiconducting gas sensors are promising candidates for gas sensing development given their sensitivity to many gases of interest and the ability to fabricate them readily in many forms, such as single crystals, thin and thick films, and nanostructures, including nanoparticles, nanowires (or nanorods), and nanofilms, which are synthesized by various physical and chemical methods. These nanostructured materials used for gas sensors, in particular, have the advantage of fast response times due to short transport distance and high sensitivity due

---

*Corresponding author.

to the large surface area, as well as the potential for miniaturization via integration with IC-based technology leading to low power consumption.[3]

Various semiconducting materials have been investigated for applications in gas sensors, such as ZnO,[4] SnO_2,[5,6] WO_3,[7] Al_2O_3,[8] In_2O_3,[9,10] SiO_2,[11] V_2O_5,[12] Ga_2O_3,[13] TiO_2,[14] CdS,[15] ThO_2,[16] γ-Fe_2O_3,[17] CO_3O_4,[18] Ag_2O,[19] and MoO_3.[20] Among these materials, ZnO has been widely studied and is easily fabricated by both chemical and physical methods. The gas sensitive behavior of ZnO's electrical conductivity was first reported in 1954 by Heiland.[21] Since then, many fundamental investigations on the gas sensitive nature of single crystals and polycrystalline as well as ZnO films have been performed.[22-26] The most recent research has been devoted towards nanostructured oxides since reactions at grain boundaries and complete depletion of carriers in the grains can strongly modify the material's transport properties.[27-31]

Structures of ZnO

ZnO is a II-VI compound semiconductor with a wide direct band gap of 3.37 eV at room temperature. It has a stable hexagonal wurtzite structure in the space group $P6_3mc$ with lattice spacing a = 0.325 nm and c = 0.521 nm. The wurtzite structure of ZnO is composed of close packed lattice of oxygen anions with zinc cations filled in interstitial positions. All oxygen anions are tetrahedrally coordinated with zinc occupying one half of the tetrahedral sites in the oxygen HCP lattice achieving maximum cation separation.[32] Zinc d-orbital electrons hybridize with the oxygen p-orbital electrons to create the tetrahedral coordination.[33] Comparison of the X-ray Diffraction (XRD) patterns of the nanocrystals with the XRD pattern for bulk wurtzite ZnO suggests that the nanocrystals are formed in the wurtzite phase with similar lattice parameters. However, the XRD patterns of the nanocrystals are considerably broadened due to the very small size of these nanocrystals.[34]

The conduction characteristics of ZnO are primarily dominated by electrons generated by the oxygen vacancies and zinc interstitial atoms.[35] The chemical bonding between zinc and oxygen in ZnO is predominately ionic. Zinc gives away its $4s^2$ valence electrons which are in turn accepted by oxygen forming a complete $2p^6$ valence shell. If the lattice contains a zinc vacancy, then one oxygen atom is missing two electrons, or two oxygen atoms are each missing one electron. The latter may be considered as two individual $h^{\cdot}$ sites, this condition permits electronic conduction as acceptor states. Therefore, a zinc ionic vacancy acts as an acceptor such that unsatisfied bonds may accept electrons moving under external potential gradient. Conversely, oxygen vacancies will result in donor behavior as a result of unbound zinc electrons. The excess valence electrons of a zinc atom

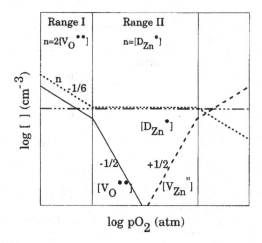

Figure 1. Brouwer diagram of ZnO showing a plateau in electron concentration at intermediate oxygen pressure and an increase in n at reduced pO_2. (Reprinted from G. W. Tomlins, J. L. Routbort and T. O. Mason, *J. Appl. Phys.* **87**, 117 (2000) with permission from AIP.)

may then be thermally excited into the conduction band and considered as donors. This form of intrinsic defect, which leads to n-type conduction, is located approximately 0.01 to 0.05 eV below the conduction band edge.[36] Brouwer diagram that best represents the defect structure of ZnO is shown in Figure 1.[37] The diagram shows a plateau in electron concentration at intermediate oxygen pressure and an increase in n at reduced p_{O_2}. The plateau is attributed to electroneutrality between electrons and a donor ion, while the upturn in n at low pO_2 is a result of an increase in doubly charged oxygen vacancies, which yields the $-1/6$ slope relationship.

Properties and Applications of ZnO

ZnO is of great interest for various photonic and electrical applications due to its unique physical and chemical properties, such as a wide band gap (3.37 eV), large exciton binding energy (60 meV) at room temperature, piezoelectricity, and surface chemistry sensitive to environment.[38] Applications of ZnO include light-emitting diodes,[39] diode lasers,[40] photodiodes,[41] photodetectors,[42] optical modulator wave guides,[43] photovoltaic cells,[44] phosphor,[45] varistor,[46] data storage,[47] and biochemical sensors.[48] Nanostructured ZnO, nanorods or nanowires in particular, has attracted intensive research, primarily for their large surface area for applications relying on heterogeneous reactions such as sensors and detectors,[47] their light confinement for nano-lasers,[40] and their enhanced freedom in lateral dimensions for more sensitive piezoelectric devices.[49]

Table 1.   Physical properties of wurtzite ZnO.

Properties	Value
Lattice constants (T = 300 K)	
a_0	0.32469 nm
c_0	0.52069 nm
Density	5.606 g/cm^3
Melting point	2248 K
Relative dielectric constant	8.66
Gap Energy	3.4 eV, direct
Intrinsic carrier concentration	$< 10^6$ cm^{-3}
Exciton binding energy	60 meV
Electron effective mass	0.24
Electron mobility (T = 300 K)	200 cm^2/Vs
Hole effective mass	0.59
Hole mobility (T = 300 K)	5–50 cm^2/Vs

Table 1 lists the basic physical properties of bulk ZnO.[50] It is worth noting that as the dimension of the semiconductor materials continuously shrinks down to nanometer or even smaller scale, some of their physical properties undergo changes known as the "quantum size effects". For example, quantum confinement increases the band gap energy of quasi-one-dimensional (Q1D) ZnO, which has been confirmed by photoluminescence.[51] Band gap of ZnO nanoparticles also demonstrates such size dependence.[52] X-ray Absorption Spectroscopy and Scanning Photoelectron Microscopy reveal the enhancement of surface states with the downsizing of ZnO nanorods.[53] In addition, the carrier concentration in Q1D systems can be significantly affected by the surface states, as suggested from nanowire chemical sensing studies.[54–57] Although the increase of the band gap in nanostructures is not favorable in most sensor applications, the change of surface states can have significant impacts on the sensing properties. Understanding the fundamental physical properties is crucial to the rational design of functional devices. Investigation of the properties of individual ZnO nanostructures is essential for developing their potential as the building blocks for future nanoscale devices. This chapter will review the up-to-date research progress on the development of ZnO nanostructures and their applications on gas sensing.

Synthesis of Nanostructured ZnO

ZnO nanostructures have attracted considerable attention for solid-state gas sensors with great potential for overcoming fundamental limitations due to their ultrahigh surface-to-volume ratio. There are a lot of nanostructures, such as nanoparticles, nanorods, nanowires, nanobelts, nanotubes, and nanofilms, that can be used for gas sensors. Various fabrication techniques have been established for the growth of ZnO nanostructures. Aqueous solution growth,[58] electrochemical deposition,[59–61] vapor-liquid-solid growth,[40] evaporation-condensation,[62] chemical vapor deposition,[63] carbothermal evaporation,[64] flux growth,[65] and template-based synthesis,[66] have all been reported to successfully grow ZnO nanostructures.

The early nanostructure of ZnO used for gas sensors is nanoparticles. The higher activity and fast response of the nanoparticle based sensors due to the enhancement of surface area at nanodimensions coupled with the possible low fabrication cost, ease of miniaturization, and the compatibility with microelectronic circuitry have provided a renewed interest in the gas sensing properties of metal oxide semiconductors.[67] It has been shown that particle sizes below 20 nm lead to higher values of the sensitivity and a more rapid response.[68,69] This has been explained not only by the increase of the specific reactive surface but also a complete depletion of the semiconductors as nanoparticle size approaches the thickness of the space-charge layer.[70] Another possibility is linked to the fact that the density of surface states induced by the chemisorbed oxygen species decreases with decreasing particle size, therefore leading to a lesser degree of Fermi-level pinning. Less Fermi-level pinning means that the surface barrier, and therefore the overall resistance, can undergo larger variations.[71] For fabricating nanoparticle-based gas sensors with superior characteristics and for understanding the size dependence of a gas sensing mechanism, the use of ZnO layers having a well-defined nanoparticle size is a primary requirement.[72] The size and uniformity of ZnO nanoparticles are closely related to sensor characteristics, so in practical sensor production the calcination process should be well-controlled to obtain uniform characteristics. The prevention of excess crystal growth is extremely important for achieving higher sensitivity and quick response of gas sensors. ZnO nanoparticles have been synthesized by vavious methods, including vapor decomposition, precipitation, and thermal decomposition.[73–78]

ZnO thin and thick films have also been investigated to be used for gas sensors, and they have been grown by many methods both in single and polycrystalline. The methods can be chemical vapor deposition,[79] electrochemical deposition,[80] pulse laser deposition (PLD),[81–83] atomic layer epitaxy (ALE) or molecular beam epitaxy (MBE),[84–86] and metal organic chemical vapor epitaxy (MOVPE)[87–89] and so on. The microstructural and physical properties of ZnO films can be modified by introducing changes into the procedure of its synthesis process. Especially for the application of ZnO as gas sensors, porous microstructure of the materials with controlled pore size is preferred. The sensitivity and response time of ZnO based sensors strongly depend on the porosity of the films. The grain size in the films also has noticeable effect on its gas sensing properties. The study of the influence of thickness on ppm level has also been done. It is concluded that thinner films of about 100 nm with low carrier concentration have better gas sensitivity than the thicker film with higher carrier concentration. The reason is that gas sensitivity is largest when the depletion region generated by chemisorbed oxygen extends through the sensor.[90] It should be noticed that many metal ions, such as Sn, Fe, Cu, In, and Al,[91,92] are doped in ZnO films to modify the microstructure and surface morphology as well as to increase the conductivity and result in an enhanced sensitivity. The microstructures are changed from non-oriented growth, for undoped films, to strongly (002) oriented, at intermediate (~ 1 at.%) doping level; and finally again to non-oriented and poor crystallinity, at high (> 3 at.%) doping level. The sensitivity of the films was studied in two steps: first as a function of their temperature (435 to 675 K) for a fixed concentration and secondly as a function of concentration (4 to 100 ppm) for a fixed temperature (675 K). A better sensitivity can be observed for Sn- and Al-doped films.

Furthermore, some mesoporous structures have also been investigated to be used for gas sensing either in the form of mesoporous films or by forming meso-pores in ZnO nanorods. Mesoporous ZnO with pore sizes from 2 nm to 50 nm are studied for the potential applications in gas sensing.[93,94] Normally, mesoporous structures are composed of amorphous materials and there are few reports of mesoporous structures based on crystalline materials. However, more and more research work on ZnO mesoporous materials are studied owing to their huge surface-to-volume ratio for the potential applications for solar cell, photocatalysis, and gas sensing.[95,96] Moreover, the synthesis of the crystalline ZnO 1D nanomaterials with mesoporous structure and well-defined size is developed, and it is expected that the mesoporous structure can increase the sensitivity of ZnO gas sensors greatly by increasing the surface-to-volume ratio.[97,98]

Most widely studied nanostructures are 1-dimensional ZnO, i.e. nanowires, nanorods, nanotube, and nanobelts. From the aspect of sensing performance, 1-dimensional ZnO is expected to be superior to its thin film counterpart.[56] Since their diameter is small and comparable to the Debye length, chemisorption induced surface states effectively affect the electronic structure of the entire channel, thus confer 1-dimensional ZnO higher sensitivity than thin film. Well-aligned arrays of ZnO nanorods were grown by vapor-phase process at high temperature on the single crystal substrates such as Si,[99] GaN,[100] and sapphire,[38] which have crystallographic similarity to ZnO. This method has limitations to the process of scaling up because of expensive single crystal substrates and high processing temperature. Aligned arrays of [001] ZnO nanorods on glass and silicon substrates have also been readily grown from aqueous solution with nanocrystal seeding;[101] the alignment of nanorods was achieved by evolution selection growth, i.e. the crystal orientation with the higher growth rate and perpendicular to the substrate surface will survive and continue to growth.[102] Well-aligned [001] ZnO nanorod arrays are also synthesized on ITO substrate from aqueous solution by electrochemical deposition or using a two-step growth process with electric field assisted nucleation. By using two-step process, a thin layer of ZnO on ITO substrate is first grown by electrochemical deposition, and then subsequently grow ZnO nanorod arrays with electrochemical deposited ZnO thin layer as substrate by spontaneous growth method. Low-temperature hydrothermal method has also been demonstrated to grow high quality ZnO nanowire arrays.[103] In the following section we will focus our discussion on the synthesis of nanorods and nanowires of ZnO.

Synthesis of ZnO Nanorods or Nanowires

Solution Based Spontaneous Growth

Two solution based approaches have been investigated in literature, one with, and the other without the use of a template. Spontaneous growth is a process driven by the reduction of Gibbs free energy or chemical potential. The reduction of Gibbs free energy is commonly realized by phase transformation or chemical reaction or the release of stress. For the formation of nanowires or nanorods, anisotropic growth is required, i.e. the crystal grows along a certain orientation faster than in other directions. Uniformly-sized nanowires, i.e. the same diameter along the longitudinal direction of a given nanowire, can be obtained when crystal growth proceeds along one direction, whereas no growth along other directions. In spontaneous growth, for given material and growth conditions, defects and impurities on the growth surfaces can play a significant role in determining the morphology of the final products.

According to the classic theory of nucleation and growth,[104] the free energy of forming stable nuclei on a substrate is determined by four factors: the degree of supersaturation S, the interfacial energy between the particle (c) and the liquid (l) σ_{cl}, the interfacial energy between the particle and the substrate (s) σ_{cs}, and the interfacial energy between the substrate and the liquid σ_{sl}:

$$\Delta G = -RT \ln S + \sigma_{cl} + (\sigma_{cs} - \sigma_{sl})A_{cs} \qquad (1)$$

where A_{cs} is the surface area of the particle.

Figure 2 is a schematic plot of the number of nuclei (N) as a function of degree of supersaturation (S), which is related to the concentration of the precursor and the solubility in the solution. Figure 2 suggests several regions for crystal growth. At a high concentration or a high temperature, homogeneous nucleation is dominant and precipitation in the bulk solution is the main mechanism. This region should be avoided if a controlled crystal growth is desired. The region slightly above the solubility line is the heterogeneous nucleation region. In this region heterogeneous nucleation on a substrate dominates and therefore it is possible to grow uniform nanostructured films and oriented 1-dimensional nanomaterials. Unfortunately, most solution synthesis is carried out at too high concentration so that undesirable precipitation dominates. As a result, oriented nanostructures were difficult to form. From Equation (1) and Figure 2, we can use

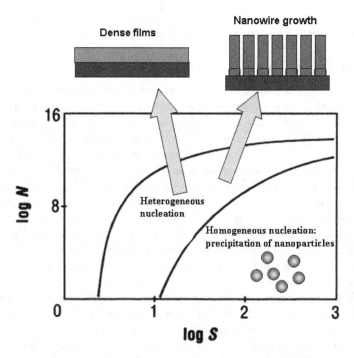

Figure 2. Number of nuclei generated in the solution as a function of the degree of supersaturation. The region for growing oriented nanostructured films and 1DNMs is indicated. (Reprinted from B. C. Bunker, P. C. Rieke, B. J. Tarasevich, A. A. Campbell, G. E. Fryxell, G. L. Graff, L. Song, J. Liu and J. W. Virden, *Science* **264**, 48 (1994) with permission from American Association for the Advancement of Science.)

the following rules to design a generalized solution synthesis method for oriented nanostructures:

(1) *Controlling the solubility of the precursors and the degree of supersaturation so that massive precipitation is not the dominating reaction.* Similar to VLS growth, the degree of supersaturation has a large effect on the growth behavior. A low degree of supersaturation promotes heterogeneous growth of 1DNMs, while a high degree of supersaturation favors bulk growth in the solution. Experimentally this is accomplished by reducing the reaction temperature, and by reducing the precursor concentrations as much as possible, while at the same time ensuring nucleation and growth can still take place. The formation of the new materials is characterized by the increase in cloudiness of the solution, which can be monitored by light scattering or turbidity measurement. A rapid increase in cloudiness is an indication of rapid precipitation and should be avoided.

(2) *Reducing the interfacial energy between the substrate and the particle.* In most cases, the activation energy for crystal growth on a substrate is lower than that required to create new nuclei from the bulk solution. Still, chemical methods can be used to further lower the interfacial energy. For example, self-assembled monolayers containing surface active groups were used to promote heterogeneous nucleation. The self-assembled monolayers have at least two functions: reducing the surface tension, and stimulate the nucleation of specific crystalline phases. A "biomimetic" approach was developed to prepare oriented nanostructured ceramic films.[104]

(3) *Controlling crystal growth.* Most crystalline materials, such as ZnO, have anisotropic crystalline structures and specific growth habits. If the growth along certain directions is much faster than other directions, nanowires or nanorodes can be produced. During crystal growth, different facets in a given crystal have different atomic density, and atoms on different facets have a different number of unsatisfied bonds (also referred to as broken or dangling bonds), leading to different surface energy. According to Periodic Bond Chain (PBC) theory developed by Hartman and Perdok,[105] crystal facets can be categorized into three groups based on the number of broken periodic bond chains on a given facet: flat surface, stepped surface, and kinked surface. The number of broken periodic bond chains can be understood as the number of broken bonds per atom on a given facets.

Vayssieres *et al.*, first introduced the concept "purpose-built materials" (Figure 3a) for growing oriented 1DNMs from solutions.[106–108] The same solution approach was extended to ZnO by decomposing Zn^{2+} amino complex.[107] By heating an equimolar zinc nitrate (0.1 M) and methenamine solution at 95°C, large arrays of oriented ZnO rods, about one to two μm in

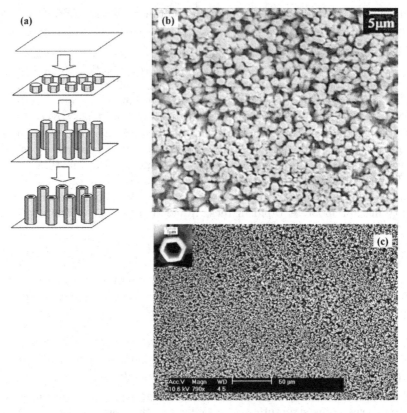

Figure 3. ZnO nanostructures from solution synthesis. (a) Schematics of growing oriented rods and tubes on a substrate. (b) Oriented ZnO microrods. (c) Oriented ZnO microtubes after aging. (Reprinted from L. Vayssieres, K. Keis, S.-E. Lindquist and A. Hagfeldt, *J. Phys. Chem.* **B105**, 3350 (2001) and L. Vayssieres, K. Keis, A. Hagfelt, S-E. Linquist, *Chem. Mater.* **13**, 4395 (2001) with permission from ACS.)

diameter, on various substrates, were produced (Figure 3b). Longer reaction time led to preferred dissolution of the ZnO rods on the metastable (001) polar surfaces, and produced hollow hexagonal microtubes (Figure 3c).[108]

Although the unseeded growth is successful in preparing a range of oriented 1DNMs, there is a need to better control the size, to reduce the dimension, to broaden the applicability of the solution-based approaches, and to control the density and the spatial distribution of the 1DNMs. A newly developed seeded growth has the promise to meet these challenges. In the seeded approach, nanoparticles are first placed on the substrates. Such nanoparticles are widely available through commercial sources and even can be readily prepared using techniques reported in literature. The crystal growth is carried out under mild conditions (low temperature and dilute concentration of the slat). Under these conditions, homogenous nucleation of new nuclei from the bulk solution is not favored and heterogeneous nucleation is prompted. Because the nanoparticles are the same as the materials to be grown, the low activation energy favored the epitaxial growth of the new 1DNMs from the existing seeds. This approach avoids the difficulty in separating the nucleation and growth steps. The size of the seeds and the density will to a large extend determine the size and the population density of the 1DNMs. The seeds can be deposited on the substrate using many mature techniques, such as dip coating, electrophoretic deposition, stamping, therefore making it possible to micropattern the 1DNMs for device applications. Furthermore, the surface characteristics of the substrate have a large effect on the interfacial energy and nucleation process. These effects are still not well understood despite of extensive investigation. The use of the seeds bypasses such complications and ensures that the materials produced are reliable and reproducible on different substrates.

Seeded growth was applied to prepare large arrays of ZnO nanorwires.[109–111] Uniform films up to wafer size were reported.[111] Three steps for growing oriented ZnO nanorods (Figure 4a) were revealed[112]:

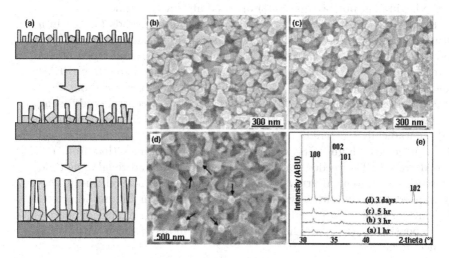

Figure 4. Mechanisms of growing oriented 1DNMs with seeded growth. (a) Illustration of growth controlled alignment of ZnO nanowires. (b) ZnO seed morphology after 30 minutes growth. (c) ZnO roughening after 1 hour of growth. (d) Initial nanorods growth after 3 hours. (e) XRD patterns at different growth stages, showing alignment at a later stage. (Reprinted from Z. R. Tian, J. A. Voigt, J. Liu, B. Mckenzie, M. J. Mcdermott, R. T. Cygan and L. J. Criscenti, *Nature Materials* **2**, 821 (2003) with permission from Nature Publishing Group.)

(1) deposition of crystal seeds on the substrate surface, (2) growth of randomly oriented crystals from the seeds, and (3) growth of aligned nanorods after extended reactions. In the early stages of growth, ZnO crystals grew along the fastest growth orientation, the $\langle 001 \rangle$ direction, and these crystals were not aligned. Figure 4b shows a SEM image of the ZnO nanoparticle seeds after 30 minutes of the crystal growth, with little sign of crystal growth. These particles mostly consist of rounded rectangular- and rod-shaped crystals with a wide size distribution. After one hour (Figure 4c) of the growth, surface roughness became visible on the ZnO seeds, indicating the initiation of the crystal growth. After three hours, short and faceted hexagonal rods were observed (Figure 4d), although most of these rods were not well aligned yet. However, as the rod-like crystals grew further, randomly oriented crystals began to overlap and their growth became physically limited as the misaligned nanorods began to impinge on neighboring crystals, giving rise to the preferred orientation of the film. Figure 4e confirmed that the ZnO crystals are hexagonal wurtzite structure ($P6_3mc$, a = 3.2495 Å, c = 5.2069 Å). The results from one to five hours are characteristic of randomly oriented ZnO powders, showing the (100) and (101) reflections as the main peaks. The (002) reflection was only significantly enhanced after extended growth, indicating that the ZnO were mostly randomly oriented at the beginning, but became $\langle 001 \rangle$ oriented after a long time growth.

Figure 5(a) shows large arrays of oriented ZnO nanorods formed in a 30 mL aqueous solution of 0.01 to 0.06 M hexamethylenetetramine (HMT) and 0.01 to 0.06 M $Zn(NO_3)_2$ at 60°C. The nanowires in uniformed arrays have a diameter of ~ 250 nm. The Scanning Electron Microscopy (SEM) images from the tilted samples suggest that these nanowires are about $3\,\mu m$ long and all stand up on the substrate (Figure 5b). A control study using no seeds shows that same synthetic conditions produced randomly oriented rods sporadically scattered on the substrates, about $1\,\mu m$ in diameter and $10\,\mu m$ in length (Figure 5c), or about three to four times that from the seeded growth. The seeded route is a straightforward and reliable method for synthesis of oriented arrays of the nanowires. This method can be easily applied to make patterned nanowires, as shown in Figure 5d. Here, the ZnO seeds were stamped onto a limited area using a ZnO nanocrystal suspension as the ink. The nanowires were mostly formed within this defined area.

Electrochemical Deposition (ECD)

Electrochemical deposition (ECD), also known as electrodeposition, has also been widely investigated to synthesize ZnO films and nanorod arrays.[113-115] ECD can be understood as a special electrolysis resulting in the deposition of solid material on an electrode. This process involves: (1)

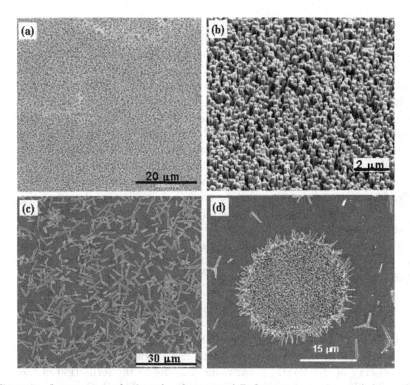

Figure 5. Large arrays of oriented and patterned ZnO nanowires using seeded growth. (a) A low magnification SEM image of large arrays of ZnO. (b) A SEM image from a tilted sample showing a good alignment. (c) ZnO microrods without seeds. (d) Preferred ZnO nanorod growth with stamped circular pattern of ZnO nanoseeds. (Reprinted from Z. R. Tian, J. A. Voigt, J. Liu, B. Mckenzie, M. J. Mcdermott, R. T. Cygan and L. J. Criscenti, *Nature Materials* **2**, 821 (2003) with permission from Nature Publishing Group).

oriented diffusion of charged growth species (typically positively charged cations) through a solution when an external electric field is applied, and (2) reduction of the charged growth species at the growth or deposition surface which also serves as an electrode. In general, electrochemical deposition is only applicable to electrical conductive materials such as metals, alloys, semiconductors, and electrical conductive polymers, since after the initial deposition, electrode is separated from the depositing solution by the deposit and the electrical current must go through the deposit to allow the deposition process to continue. Once little fluctuation yields the formation of small rods, the growth of rods or wires will continue, since the electric field and the density of current lines between the tips of nanowires and the opposing electrode are greater, due to a shorter distance, than that between two electrodes. The growth species will more likely be deposited onto the tip of nanowires, resulting in continued growth.

Epitaxial zinc oxide films were prepared on gallium nitride (0002) substrates by cathodic electrodeposition in an aqueous solution containing zinc salt and dissolved oxygen at 85°C.[114] The films have the hexagonal structure with the *c* axis parallel to that of GaN and the [100] direction in ZnO parallel to the [100] direction in GaN in the (0002) basal plane. The crystallographic quality appears to be very good, especially when considering the low temperature deposition and the relatively high growth rate conditions. Such results are related to the growth in the solution environment which, in comparison to the vapor phase, offers additional possibilities for atomic organization during deposition due to easy ionic exchanges between the surface and the solution through near equilibrium dissolution/precipitation processes.

Cao *et al.*, reported a soft and template-free electrochemical deposition method for preparing wafer-scale ZnO nanoneedle arrays on an oriented gold film coated silicon substrate.[115] It has been shown that the ZnO nano-needles possess single-crystal wurtzite structure and grow along the c-axis perpendicularly on the substrate. The {111}-oriented Au film/Si substrate results in the formation of {0001}-oriented ZnO nuclei on the film due to the small lattice mismatch between them. The oriented ZnO nuclei serve as seeds and grow preferentially along the c-axis due to the high surface free energy of the {0001} polar plane, leading to the formation of the ZnO nanoneedle array perpendicular to the substrate. This ZnO nanoneedle array is of good crystal quality.

ECD and solution-based spontaneous growth were also combined to grow well-oriented ZnO nanorod arrays.[116] In this method, [001] ZnO nanorods arrays on ITO substrates were fabricated by a two-step process: seeding and subsequent growth. Firstly, ITO substrates were placed in a 0.1 M of zinc nitrate aqueous solution for an initial growth or deposition. An external electric potential of 1.2 V was applied to ITO substrate, as a cathode, with a platinum plate as an anode for 20 minutes. The distance between two electrodes was kept at 1.5 mm. The ITO substrates with ZnO deposit were subsequently heat-treated at 500°C for 30 minutes in air. The ITO substrates with initial ZnO deposit after heat-treatment were placed in a mixture solution of 0.015 M zinc nitrate and 0.022 M methenamine at 60°C for 40 hours. $C_6H_{12}N_4$ is a growth directing agent widely used for the synthesis of ZnO nanorods.[58,61] Figure 6 compared XRD spectra and SEM images of ZnO nanorods grown with and without ECD assisted nucleation. One critical issue of this method is that the nucleation density of ECD deposition will affect the subsequent ZnO rod growth and alignment.[117]

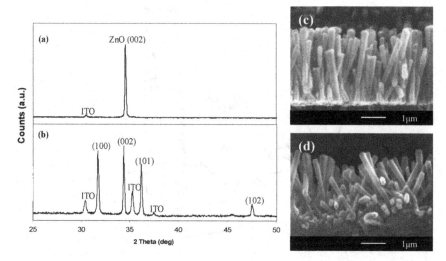

Figure 6. XRD and SEM of ZnO nanorod arrays grown from ECD assisted nucleation growth (a) XRD of ZnO nanorods grown with ECD assisted nucleation. (b) XRD of random nanorods grown without ECD assisted nucleation. (c) SEM cross-section of ZnO nanorods grown with ECD assisted nucleation. (d) SEM cross-section of ZnO nanorods grown without ECD assisted nucleation. (Reprinted from Y. J. Kim, H. M. Shang, and G. Z. Cao, *Proceedings of Materials Research Society Symposium* **879E**, Z4.1.1–Z4.1.6 (2005) with permission from MRS.)

Vapor-Liquid-Solid (VLS) Growth

In the VLS growth, a second phase material, commonly referred to as either impurity or catalyst, is purposely introduced to direct and confine the crystal growth on a specific orientation and within a confined area. Catalyst forms a liquid droplet by itself or by alloying with growth material during the growth, which acts as a trap of growth species. Enriched growth species in the catalyst droplets subsequently precipitates at the growth surface resulting in the one-directional growth. Wagner et al.,[118–120] first proposed the VLS theory over 40 years ago to explain the experimental results and observations in the growth of silicon nanowires, and Givargizov[121] further elaborated the experimental observations and models and theories developed regarding the VLS process.

In a VLS growth, the process can be simply described as sketched in Figure 7. The growth species is evaporated first, and then diffuses and dissolves into a liquid droplet. The surface of the liquid has a large accommodation coefficient, and is therefore a preferred site for deposition. Saturated growth species in the liquid droplet will diffuse to and precipitate at the interface between the substrate and the liquid. The precipitation

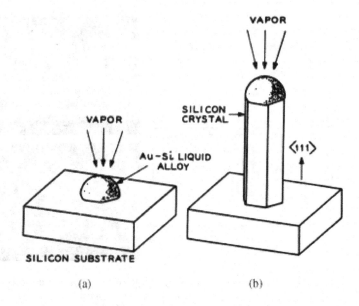

Figure 7.   Schematic showing the principal steps of the vapor-liquid-solid growth technique: (a) initial nucleation and (b) continued growth.

will follow first nucleation and then crystal growth. Continued precipitation or growth will separate the substrate and the liquid droplet, resulting in the growth of nanowires. For a perfect or an imperfect crystal surface, an impinging growth species diffuse along the surface. During the diffusion, the growth species may be irreversibly incorporated into the growth site (ledge, ledge-kink, or kink). If the growth species did not find a preferential site in a given period of time (the residence time) the growth species will escape back to the vapor phase. A liquid surface is distinctly different from a perfect or imperfect crystal surface, and can be considered as a "rough" surface. Rough surface is composed of only ledge, ledge-kink, or kink sites. As such, every site over the entire surface is to trap the impinging growth species. The accommodation coefficient is unity. Consequently, the growth rate of the nanowires or nanorods by VLS method is much higher than that without liquid catalyst. The enhanced growth rate can also be partly due to the fact that the condensation surface area for the growth species in the VLS growth is larger than the surface area of the crystal growth. While the growth surface is the interface between the liquid droplet and the solid surface, the condensation surface is the interface between the liquid droplet and vapor phase. Depending on the contact angle, the liquid surface area can be several times of the growth surface.

The size of nanowires grown by VLS method is solely determined by the size of the liquid catalyst droplets. To grow thinner nanowires, one can simply reduce the size of the liquid droplets. A typical method used to form small liquid catalyst droplets is to coat a thin layer of catalyst on the growth substrate and to anneal at elevated temperatures.[122] During annealing, catalyst reacts with the substrate to form a eutectic liquid and further ball up to reduce the overall surface energy. Au as a catalyst and silicon as a substrate is a typical example. The size of the liquid catalyst droplets can be controlled by the thickness of the catalyst film on the substrate. In general, a thinner film forms smaller droplets, giving smaller diameters of nanowires subsequently grown.

ZnO nanowires have been grown on Au-coated (thickness ranging from 2 to 50 nm) silicon substrates by heating a 1:1 mixture of ZnO and graphite powder to 900–925°C under a constant flow of argon for 5–30 minutes.[123] The grown ZnO nanowires vary with the thickness of the initial Au coatings as shown in Figure 8. For a 50 Å Au coating, the diameters of the nanowires normally range from 80 to 120 nm and their lengths are 10 to 20 μm. Thinner nanowires of 40 to 70 nm with lengths of 5 to 10 μm were grown on 30 Å Au-coated substrates. The grown ZnO nanowires are single crystal with a preferential growth direction of $\langle 001 \rangle$. The growth process of ZnO is believed to be different from that of elementary nanowires. The process involves the reduction of ZnO by graphite to form Zn and CO vapor at high temperatures (above 900°C). The Zn vapor is transported to and reacted with the Au catalyst, which would have already reacted with silicon to form a eutectic Au-Si liquid on silicon substrates, located downstream at a lower temperature to form Zu-Au-Si alloy droplets. As the droplets become supersaturated with Zn, crystalline ZnO nanowires are formed, possibly through the reaction between Zn and CO at a lower temperature. The above process could be easily understood by the fact that the reaction:

$$ZnO + C \leftrightarrow Zn + CO \tag{2}$$

is reversible at temperatures around 900°C.[124] Although the presence of a small amount of CO is not expected to change the phase diagram significantly, no ZnO nanowires were grown on substrates in the absence of graphite.

ZnO nanowire growth through the VLS mechanism also allows people to pattern these wires into a network.[123] Initial patterning experiments used a 300 mesh copper grid with a hexagonal pattern as a mask during the Au deposition. Most of the wires actually bridge the neighboring metal hexagons and form an intricate network. Figure 9 show SEM images of ZnO nanowire patterns. The nanowires at the edge have a diameter of about 50 to 200 nm, and can be over 50 μm long. This controlled growth of nanowires

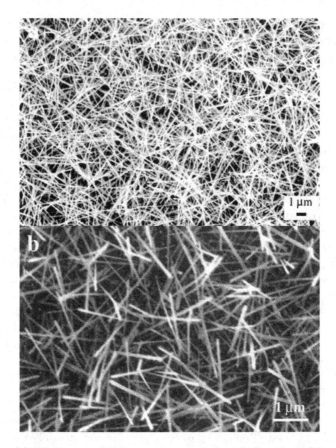

Figure 8. (a) SEM image of ZnO nanowires grown from ZnO and graphite powder in an argon flow on the surface of a silicon substrate coated with ∼50 Å thick Au film. (b) SEM image of ZnO nanowires grown under the same conditions as above but using a substrate coated with 30 Å thick Au film. (Reprinted from M. H. Huang, Y. Wu, H. Feick, N. Tran, E. Weber and P. Yang, *Adv. Mater.* **13**, 113 (2001) with permission from John Wiley and Sons.)

to form certain patterns and the eventual control of wire growth at certain locations may have implications for potential applications in nanoscale devices.

Evaporation–Condensation Growth

Evaporation–condensation process is also referred to as a vapor–solid (VS) process. Chemical reactions among various precursors may be involved to

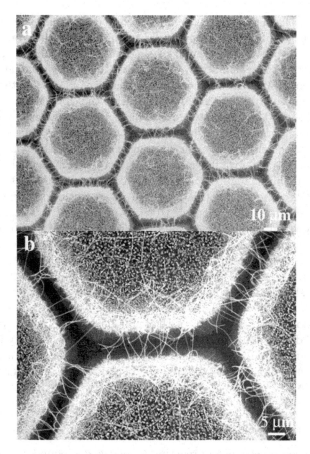

Figure 9. (a) SEM image of ZnO nanowire networks grown from the patterned Au islands. (b) SEM image of the same sample at a higher magnification. (Reprinted from M. H. Huang, Y. Wu, H. Feick, N. Tran, E. Weber and P. Yang, *Adv. Mater.* **13**, 113 (2001) with permission from John Wiley and Sons.)

produce desired materials. Nanowires and nanorods grown by evaporation-condensation methods are commonly single crystals with fewer imperfections. The formation of nanowires, nanorods or nanotubules through evaporation (or dissolution) — condensation is due to the anisotropic growth. Chemical reactions and the formation of intermediate compounds play important roles in the synthesis of various nanowires by evaporation-condensation methods. Reduction reactions are often used to generate volatile deposition precursors and hydrogen, water and carbon are commonly used as reduction agent.

Sears[125,126] was the first who demonstrated and explained the growth of nanowires by evaporation–condensation method. Recently, Wang and his co-worker[127] reported the growth of single crystal nanobelts of various semiconducting oxides simply by evaporating the desired commercially available metal oxides at high temperatures under a vacuum of 300 torr and condensing on an alumina substrate, placed inside the same alumina tube furnace, at relatively lower temperatures. The oxides include zinc oxide of wurtzite hexagonal crystal structure, tin oxide of rutile structure, indium oxide with C-rare-earth crystal structure, and cadmium oxide with NaCl cubic structure. We will just focus on the growth of ZnO nanobelts to illustrate their findings. Figure 10 shows the SEM and TEM pictures of ZnO nanobelts.[127] The typical thickness and width-to-thickness ratios of the ZnO nanobelts are in the range of 10 to 30 nm and five to ten, respectively. Two growth directions were observed: [0001] and [0110]. No screw dislocation was found throughout the entire length of the nanobelt, except a single stacking fault parallel to the growth axis in the nanobelts grown along [0110] direction. The surfaces of the nanobelts are clean, atomically sharp and free of any sheathed amorphous phase. Their further TEM analysis also revealed the absence of amorphous globules on the tip of nanobelts. The above observations imply that the growth of nanobelts cannot be attributed to either screw dislocation induced anisotropic growth, nor impurity inhibited growth, and is not due to the VLS mechanism. It seems worthwhile to note that the shape of nanowires and nanobelts may also depend on growth temperature.

Chemical Vapor Deposition (CVD)

CVD is the process of chemically reacting a volatile compound of a material to be deposited, with other gases, to produce a nonvolatile solid that deposits atomistically on a suitably placed substrate.[128] CVD process has been very extensively studied and very well documented[129–131] largely due to the close association with solid-state microelectronics. A variety of CVD methods have been developed, depending on the types of precursors used, the deposition conditions applied and the forms of energy introduced to the system to activate the chemical reactions. Metalorganic CVD (MOCVD), also known as organometallic vapor phase epitaxy (OMVPE) which differs from other CVD processes by the chemical nature of the precursor gases, has been recently developed to grow ZnO nanorod arrays.

Lu et al., have grown high quality epitaxial ZnO thin films on R-plane sapphire by MOCVD at low temperature (350°C–500°C).[132] They also have achieved the MOCVD growth of ZnO nanotips on various substrates[133]

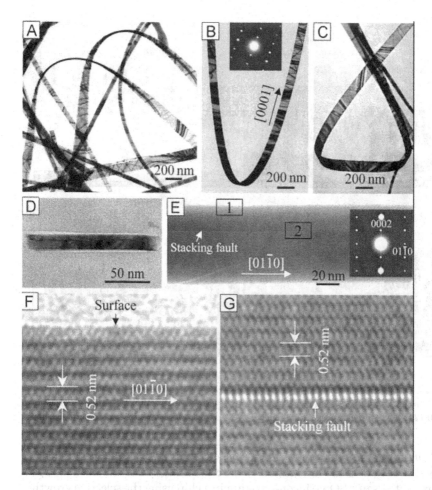

Figure 10.   SEM and TEM pictures of ZnO nanobelts. (A to C) TEM images of several straight and twisted ZnO nanobelts, displaying the shape characteristics of the belts. (D) Cross-sectional TEM image of a ZnO nanobelt, showing a rectanglelike cross section. (E) TEM image of a nanobelt growing along [010], showing only one stacking fault present in the nanobelt. (F) HRTEM image from box 1 in (E) showing a clean and structurally perfect surface. (G) HRTEM image from box 2 in (E), showing the stacking fault. (Reprinted from Z. W. Pan, Z. R. Dai and Z. L. Wang, *Science* **291**, 1947 (2001) with permission from American Association for the Advancement of Science.)

as shown in Figure 11. By controlling the ZnO initial growth (nucleation versus other growth mechanisms), columnar growth of ZnO with a high aspect ratio can be grown on Si, SiO_2/Si, GaN c-sappphire, and fused silica substrates. Nanotip arrays made with conductive ZnO are single crystalline,

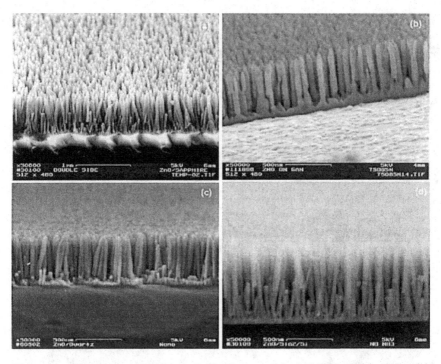

Figure 11.   SEM images of ZnO nanorod arrays grown by MOCVD on (a) Al₂O₃, (b) epi GaN, (c) fused silica, and (d) SiO₂/Si substrates. (Reprinted from S. Muthukumar, H. Sheng, J. Zhong, Z. Zhang, N. W. Emanaetoglu and Y. Lu, *IEEE Trans. Nanotech.* **2**, 50 (2003) with permission from IEEE.)

with uniform size and orientation and show good optical quality. Initially, they demonstrated the integration of epitaxial ZnO layer and ZnO nanotips arrays at the same SOS (silicon-on-sapphire) chip using the selective growth technology.

Yi and coworkers used catalyst-free MOCVD for growing ZnO nanorod and nanoneedle arrays.[63,134] In this method, no catalyst is employed for ZnO nanorod formation, which leads to preparation of high purity ZnO nanorods and easy fabrications of nanorod structures. After this research, related research of ZnO nanorod growth via MOCVD has been reported by other groups.[135–142]

Other Methods

ZnO nanostructures have been also synthesized by many other methods, and all of methods have provided the possibility of synthesis of well-aligned ZnO nanorod arrays with large surface area to be used for gas sensing and other applications. Heo *et al.*, reported on catalyst driven molecular beam epitaxy (MBE) of ZnO nanorods.[143] Okada *et al.*, succeeded in synthesizing ZnO nanorods by pulsed-laser deposition (PLD) without using any catalyst.[144] A novel and simple one-step solid state reaction in the presence of a suitable surfactant, triethanolamine (TEA), has been developed to synthesize uniform zinc oxide nanorods with an average diameter of 20 nm and length of 200 nm.[145] High-density, well-aligned wurtzite ZnO nanorod arrays with significantly different tip shapes were controllably fabricated by evaporating zinc powders at different temperatures.[146] Well-aligned single-crystalline wurzite ZnO nanowire array was successfully fabricated on an Al_2O_3 substrate by a simple physical vapor-deposition method at a low temperature of 450°C.[147] Wang *et al.*, used electrophoretic deposition to form nanorods of ZnO from colloidal sols.[148] Template-base method was also combined with other methods, such as electrodeposion, sol-gel and polymer assisted growth, to fabricated well-aligned ZnO nanorod arrays.[149,150] There are still many other methods that can be used to fabricate various ZnO nanostructures but are not included here. More details on ZnO fabrication can be found in some good books and review papers published recently.[27,55,96,151-153]

Nanostructured ZnO for Gas Sensing

The rapid development of modern industries raises a variety of serious environmental problems, for example, the release of various chemical pollutants, such as NO_x, SO_x, CO_2, volatile organic compounds (VOCs) and fluorocarbon, from industry, automobiles, and homes, into the atmosphere, resulting in global environmental issues, such as acid rain, the greenhouse effect, sick house syndrome, and ozone depletion. This atmospheric pollution can cause major disasters within a short period of time, since this type of pollution can diffuse rapidly over large areas. Since the kinds and quantities of pollution sources have also increased dramatically, the development of a method for monitoring and controlling these sources has become very important. Therefore, rapid and reliable detection systems are in great need so as to minimize the damage caused by atmospheric pollution. Solid-state gas sensors which are compact, robust, with versatile applications and a low cost are important devices for monitoring and controlling those pollutants by chemical reactions.[154] As one of the major materials for solid state gas sensor, bulk and thin films of ZnO have been investigated widely for gas sensing, and the typical detection gases are CO,[155] NH_3,[156] alcohol,[157] H_2.[158] Furthermore, it can be used not only for detecting the leakage of flammable gases and toxic gases but also for controlling domestic gas boilers.[159]

Gas Sensing Mechanisms of ZnO Semiconductor

N-type semiconducting materials such as SnO_2, ZnO, In_2O_3, and Fe_2O_3 have been known for the detection of flammable or toxic gases. The detection mechanism of them is straightforward: oxygen vacancies on metal-oxide surfaces are electrically and chemically active. These vacancies function as n-type donors, and often significantly increase the conductivity of oxide. Upon adsorption of charge accepting molecules at the vacancy sites, such as NO_2 and O_2, electrons are effectively depleted from the conduction band, leading to a reduced conductivity of the n-type oxide. On the other hand, molecules, such as CO and H_2, would react with surface adsorbed oxygen and consequently remove it, leading to an increase of conductivity. By measuring these conductance or resistance changes gas sensors can detect different gases. The effectiveness of gas sensors prepared from semiconducting oxides depends on several factors including the nature of the reaction taking place at the oxide surface, the temperature, the catalytic properties of the surface, the electronic properties of the bulk oxide and the microstructure.[2]

Although the detection mechanism of semiconductor sensors generally is based on the resistance changes, their response mechanisms may be a little different depending on detection gases. For the detection of oxygen, gas sensing element generally respond to changes in oxygen partial pressure at high temperatures (700°C) by exploiting the equilibrium between the composition of the atmosphere and the bulk stoichiometry. The relationship between oxygen partial pressure and the electrical conductivity of oxide may be represented by[160]:

$$\sigma = a\exp(-E_A/KT)P_{O2}^{1/N} \tag{3}$$

where σ is the electrical conductivity, A is a constant, E_A is the activation energy for conduction, P_{O2} is oxygen partial pressure and N is a constant determined by the dominant type of bulk defect involved in the equilibrium between oxygen and the sensor.

While for the detection of minority gases in air, semiconducting oxides are in an atmosphere of fixed oxygen partial pressure. Bulk changes on oxygen stoichiometry are not relevant to this type of sensing and the materials are normally held at temperatures in the range 300 to 500°C, where useful surface reactions proceed at a sufficient rate.

The relationship between sensor resistance and the concentration of deoxidizing gas can be expressed by the following equation over a certain range of gas concentration:

$$Rs = A[C]^{-\alpha} \tag{4}$$

Where, Rs = electrical resistance of the sensor, A = constant, $[C]$ = gas concentration, and α = slope of Rs curve. Due to the logarithmic relationship between sensor resistance and gas concentration, semiconductor type sensors have an advantage of high sensitivity to gas even at low gas concentration. The excellent stability and performance of the semiconductor type sensor provides maintenance-free, long-lived, and low-cost gas detection. Various sensors which have different sensitivity characteristics can be manufactured by selecting the most suitable combinations of sensing material, temperature and activity of sensor materials.

Grain Size and Structure Effects on Gas Sensing Performance

The sensitivity of nanostructured ZnO gas elements is comparatively high because of the grain-size effect.[161] For further development of high sensitivity ZnO gas sensors, understanding the correlation between the sensitivity and the microstructure of nano-grain ZnO gas elements is necessary. It is well-known that the sensing mechanism of ZnO belongs to the surface-controlled type. Its gas sensitivity is relative to grain size, geometry, surface state, oxygen adsorption quantity, active energy of oxygen adsorption, and lattice defects. Normally, the smaller its grain size, specific surface area and oxygen adsorption quantity, the higher its gas sensitivity is. In addition, the decrease in grain size of ZnO decreases its working temperature due to the increase of surface activity of ZnO. In general, working temperature of ZnO is 400 to 500°C, but that of nanometer ZnO made by emulsion is only 300°C.[161]

For ZnO nano-grain based gas sensors, the sensing properties are influenced not only by the microstructural features, such as the grain size, the geometry, but also by the connectivity between the grains.[162] In the case of neck–grain boundary-controlled sensitivity, a sensor's sensitivity will decrease with the increasing number of grain boundaries when the width of the space charge layer of the neck in air is comparable with the neck radius. On ZnO nano-grain's surface, the adsorbed oxygen extracts the conduction electrons from the near-surface region of the grain leading to the grain-boundary barriers and the neck barriers. Reaction of a reducing gas with the adsorbed oxygen results in its removal, and thereby, a reduction in the barrier height. The neck barrier controls the electron conducting channel through the neck[163] and the electron density in the space charge layer at the neck. Therefore the neck barrier determines the neck resistance. At the same time, the grain-boundary barrier determines the grain-boundary resistance. At a small grain size, the grain-boundary resistance is small and decreases slowly over the entire gas concentration range, therefore its total variation is also small. But the neck resistance decreases steeply and its total variation is large. The grain-boundary resistance may be much smaller than the neck resistance. Corresponding to the resistances, the grain boundary-controlled sensitivity of nano-grain ZnO gas elements is much lower than the neck-controlled sensitivity. It is notable that neck–grain boundary-controlled sensitivity is lower than neck-controlled sensitivity alone when the width of the space charge layer of the neck in air approaches the radius of the neck. This indicates that decreasing the number of grain boundaries will increase the sensitivity of nano-grain ZnO gas elements and implies that the grain-boundary resistance of nano-grain ZnO gas elements cannot be ignored though it may be much smaller than the neck resistance. When the gas concentration becomes high enough, the sensitivities begin to saturate. Ma *et al.*, studied the effects both of grain size and grain boundary on gas sensitivity.[162] Figure 12 shows that the smaller the grain size of nano-grain ZnO gas elements, the higher the neck-controlled sensitivity and the neck–grain boundary-controlled sensitivity are.[162]

Nanostructured ZnO for Gas Sensors

ZnO nanostructures have attracted considerable attention for solid-state gas sensors with great potential for overcoming fundamental limitations due to their ultrahigh surface-to-volume ratio. Besides ZnO nanoparticles, ZnO nanostructures which have been used for gas sensors include nanorods,

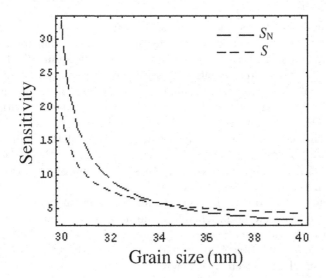

Figure 12.   The gas sensitivities decrease with increasing grain size for a nano-grain ZnO gas sensor. S_N is neck-controlled sensitivity and S is the neck-grain boundary-controlled sensitivity. (Reprinted from Y. Ma, W. L. Wang, K. J. Liao and C. Y. Kong, *J. Wide Band gap Mater.* **10**, 113 (2002) with permission from SAGE publications.)

nanowires, nanobelts, nanotubes. Furthmore, some mesoporous structures have also been investigated to be used for gas sensing either in the form of mesoporous films or by forming mesopores in ZnO nanorods. From the aspect of sensing performance, 1-dimensional ZnO, such as nanowires and nanorods, is expected to be superior to its thin film counterpart.[56] Since their diameter is small and comparable to the Debye length, chemisorption induced surface states effectively affect the electronic structure of the entire channel, thus confer 1-dimensional ZnO higher sensitivity than thin film.

ZnO nanowires and nanorods can be configured as either two terminal sensing devices or as FETs in which a transverse electric field can be utilized to tune the sensing property. Recently, Wan *et al.*,[164] fabricated ZnO nanowire chemical sensor using microelectromechanical system technology. Massive nanowires were placed between Pt interdigitating electrodes. The system exhibited a very high sensitivity to ethanol gas and a fast response time (within 10 seconds) at 300°C, showing a promising application for ZnO nanowire humidity sensors. Electrical transport studies show that ambient O_2 has considerable effect on the ZnO nanowires.[165,166] Fan *et al.*, discussed the relationship between oxygen pressure and ZnO nanowire FET performance.[165] It is shown that ZnO nanowires have fairly good sensitivity to O_2 (Figure 13a). In addition, it is observed that the sensitivity is a function of back gate potential, i.e. above gate threshold voltage of FET, sensitivity increases with decreasing gate voltage (Figure 13a inset). This implies that the gate voltage can be used to adjust the sensitivity range. Furthermore, a gate-refresh mechanism was proposed.[55] As demonstrated in Figure 13b, the conductance of nanowire can be recovered by using a negative gate potential larger than the threshold voltage. The gas selectivity for NO_2 and NH_3 using ZnO nanowire FET was also investigated under the gate refresh process, exhibiting a gas distinguishability function.

ZnO nanobelts have also attracted researchers' attention in gas sensing under atmospheric conditions. Nanobelts of ZnO with a rectangular cross section in a ribbon-like morphology, are very promising for sensors due to the fact that the surface-to-volume ratio is very high, the oxide is single crystalline, the faces exposed to the gaseous environment are always the same and the size is likely to produce a complete depletion of carriers inside the belt. Wen *et al.*, measured the NH_3 sensing characteristics.[167] Figure 14 shows the current response of the sensor to NH_3 at room temperature at a bias voltage of 5 V. The sensor is very sensitive to ammonia: when the sensor was exposed to 500 ppm NH_3, a current change as high as about 3.5-fold between the on and off states was observed, and this current response to NH_3 could be repeated many times without obvious change in signal intensity. The repeatability and stability of the sensor appear to be very good. For NH_3 sensors with ZnO, the change in current is mainly caused by the adsorption and desorption of NH_3 and H_2O molecules on the

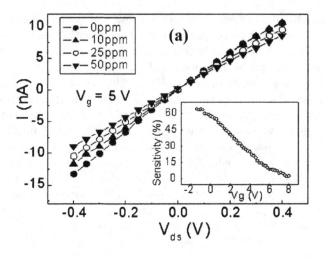

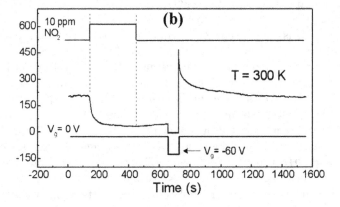

Figure 13.   (a) I-V curves of a ZnO nanowire under 50 ppm O_2. Inset: gate potential dependence of sensitivity under 10 ppm O_2. (Reprint from Ref. 76, Z. Fan *et al.*, *Appl. Phys. Lett.* **85**, 5923 (2004) with permission from American Institute of Physics). (b) The nanowire sensing response to 10 ppm NO_2 and the conductance recovery process assisted by a -60 V gate voltage pulse. (Reprint from Z. Fan *et al.*, *Appl. Phys. Lett.* **86**, 123510 (2005) with permission from American Institute of Physics.)

surface of the sensing materials. The high surface-to-volume ratio of our ZnO nanobelts sensor gives rise to high sensitivity to NH_3. In addition, the molecules are easy to adsorb on and desorb from the surface of nanobelts in contrast to the conventional thin-film sensor. Note that the exposed surfaces of ZnO nanobelts are (0001) for the most part, whereas a polycrystalline ZnO film exposes many different crystal planes on the surface, and this

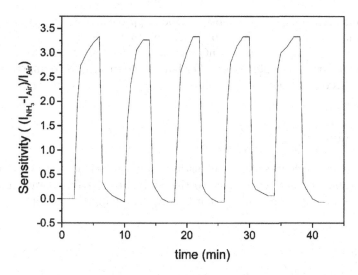

Figure 14. Current responses of the ZnO nanobelts sensors when the surrounding gas is switched between air and 500 ppm NH_3. (Reprinted from X. G. Wen, Y. P. Fang, Q. Pang, C. L. Yang, J. N. Wang, W. K. Ge, K. S. Wong and S. H. Yang, *J. Phys. Chem.* **B109**, 15303 (2005) with the permission from ACS.)

may contribute to different sensing properties. For conventional thin-film sensors, an elevated temperature is often used to facilitate molecular desorption. In nanobelts sensor, however, the available interbelt spaces allow the adsorbed molecules to desorb easily in the off state so that it shows a rapid response and good reversibility.

Conclusions

ZnO is a unique material and has attracted a lot of research for various applications. ZnO is a wide band gap semiconductor and its nanostructures are being explored for solar cells. ZnO is also a piezoelectric and its nanostructures can find applications in sensor and actuator or other micro- or nanao-electromechanical systems (MEMS or NEMS). The electrical conductivity of ZnO is sensitively dependent on its surface chemistry, and thus nanostructured ZnO is an ideal material for surface chemistry related sensors and detectors. To make nanostructured ZnO even more attractive from the engineering point of view is the ability of readily fabricating various well-controlled and high-quality nanostructures of single crystal ZnO at moderate processing conditions, and of being readily integrated to device fabrication. There is a myriad of publications available in open literature.

This chapter just focuses on the most common approaches for the synthesis of ZnO nanostructures and the applications of ZnO nanostructures in gas sensing. For sensing applications, ZnO nanowires or nanorods are the most preferred nanostructures, since they offer both large surface to volume ratio and easy collection of electrical signals.

Acknowledgements

H. M. Shang acknowledges the Joint Institute for Nanoscience (JIN) Graduate Fellowship funded by Pacific Northwest National Laboratories and University of Washington, and the Ford Motor Company Fellowship.

References

1. K. Ihokura and J. Watson, The Stannic Oxide Gas Sensor, CRC Press, Boca Raton (1994).
2. P. T. Moseley, *Meas. Sci. Technol.* **8**, 223 (1997).
3. Y. Min, H. L. Tuller, S. Palzer, J. Wöllenstein and H. Böttner, *Sens. Actuators B* **93**, 435 (2003).
4. T. Seiyama, A. Kato, K. Fujishi and M. Nagatami, *Anal. Chem.* **34**, 1502 (1962).
5. N. Taguchi, *Japanese Pat.* S45-38200 (1962).
6. S. C. Chang, Abst. Int. Seminar on Solid-state Gas Sensors (23rd WEH Seminar), 7 (1982).
7. P. J. Shaver, *Appl. Phys. Lett.* **11**, 255 (1967).
8. N. Taguchi, *Japanese Pat.* S50–30480 (1966).
9. J. C. Loh, *French Pat.* 1545292 (1967).
10. H. Steffes, C. Imawan, F. Solzbacher and E. Obermeier, *Sens. Actuators B* **68**, 249 (2000).
11. B. Bott, *British Pat.* 1374575 (1971).
12. C. Imawan, H. Steffes, F. Solzbacher and E. Obermeier, *Sens Actuators B* **77**, 346 (2001).
13. T. Schwebel, M. Fleischer, H. Meixner and C.-D. Kohl, *Sens. Actuators B* **49**, 46 (1998).
14. T. Tien, H. Stadler, E. Gibbons and P. Zacmanidis, *Am. Ceram. Soc. Bull.* **54**, 280 (1975).
15. M. C. Steele and B. A. Maclver, *Appl. Phys. Lett.* **28**, 687 (1976).
16. M. Nitta, S. Kanefusa, Y. Taketa and M. Haradome, *Appl. Phys. Lett.* **32**, 590 (1978).
17. S. Tao, X. Liu, X. Chu and Y. Shen, *Sens. Actuators B* **61**, 33 (1999).
18. J. R. Stetter, *J. Colloid Interface Sci.* **65**, 432 (1978).
19. N. Yamamoto, *Bull. Chem. Soc. Jpn.* **54**, 696 (1981).

20. C. Imawan, F. Solzbacher, H. Steffes and E. Obermeier, *Sens Actuators B* **64**, 193 (2000).
21. W. Heywang, Amorphe und polykristalline Halbleiter (Springer Verlag, Berlin, 1984), p. 204.
22. F. Lin, Y. Takao, Y. Shimizu and M. Egashira, *Sens. Actuators B* **25**, 843 (1995).
23. W. Göpel, *J. Vac. Sci. Technol.* **15**, 1298 (1978).
24. P. Esser and W. Göpel, *Surf. Sci.* **97**, 309 (1980).
25. W. Göpel, T. A. Jones, M. Kleitz, J. Lundstrom and T. Seiyama, Sensors Vol. 2 Chemical and Biochemical Sensors Part I (VCH, Weinheim, 1991).
26. W. Göpel, T. A. Jones, M. Kleitz, J. Lundstrom and T. Seiyama, Sensors Vol. 3 Chemical and Biochemical Sensors Part II (VCH, Weinheim, 1992).
27. Z. L. Wang, *Nanowires and Nanobelts: Materials, Properties and Devices* (Kluwer Academic Publishers, Boston, 2003).
28. G. Sberveglieri, G. Faglia, S. Groppelli, P. Nelli and A. Camanzi, *Semicond. Sci. Technol.* **5**, 1231 (1990).
29. M. Ferroni, V. Guidi, G. Martinelli, E. Comini, G. Sberveglierri, D. Boscarino and G. Della Mea, *J. Appl. Phys.* **88**, 1097 (2000).
30. V. E. Henrich and P. A. Cox, *The Surface Science of Metal Oxides* (Cambridge, 1994).
31. W. Göpel, *Sens Actuators A* **56**, 83 (1996).
32. W. D. Kingery, H. K. Bowen and D. R. UHlmann, Introduction to Ceramics, 2nd edn., John Wiley & Sons, Inc., New York (1976).
33. S. J. Pearson, D. P. Norton, K. Ip, Y. W. Heo and T. Steiner, *J. Vac. Sci. Technol. B* **22**, 932 (2004).
34. R. Viswanatha, S. Sapra, B. Satpati, P. V. Satyam, B. N. Dev and D. D. Sarma, *J. Mater. Chem.* **14**, 661 (2004).
35. F. A. Kröger, The Chemistry of Imperfect Crystals, Publishers Inc., Amsterdam (1964).
36. D. P. Norton, Y. W. Heo, M. P. Ivill, K. Ip, S. J. Pearton, M. F. Chisholm and T. Steiner, *Mater. Today* **7**, 34 (2004).
37. G. W. Tomlins, J. L. Routbort and T. O. Mason, *J. Appl. Phys.* **87**, 117 (2000).
38. M. H. Huang, S. Mao, H. Feick, H. Yan, Y. Wu, H. Kind, E. Weber, R. Russo and P. Yang, *Science* **292**, 1897 (2001).
39. N. Saito, H. Haneda, T. Sekiguchi, N. Ohashi, I. Sakaguchi and K. Koumoto, *Adv. Mater.* **14**, 418 (2002).
40. K. Keis, E. Magnusson, H. Lindstorm, S. E. Lindquist and A. Hagfelt, *Sol. Energ. Mater. Sol. Cells* **73**, 51 (2002).
41. J. Y. Lee, Y. S. Choi, J. H. Kim, M. O. Park and S. Im, *Thin Solid Films* **403**, 553 (2002).
42. S. Liang, H. Sheng, Y. Liu, Z. Hio, Y. Lu and H. Shen, *J. Cryst. Grow.* **225**, 110 (2001).
43. M. H. Koch, P. Y. Timbrell and R. N. Lamb, *Semicond. Sci. Tech.* **10**, 1523 (1995).

44. K. Keis, E. Magnusson, H. Lindstorm, S. E. Lindquist and A. Hagfelt, *Sol. Energ. Mater. Sol. Cells* **73**, 51 (2002).
45. S. J. Pearton, D. P. Norton, K. Ip, Y. W. Heo and T. Steiner, *Superlatt. Microstr.* **34**, 3 (2003).
46. Y. Lin, Z. Hang, Z. Tang, F. Yuan and J. Li, *Adv. Mater. Opt. Electron.* **9**, 205 (1999).
47. Y. C. Kong, D. P. Yu, B. Zhang, W. Fang and S. Q. Feng, *Appl. Phys. Lett.* **78**, 4 (2001).
48. Y. Cui, Q. Wei, H. Park and C. M. Lieber, *Science* **293**, 1289 (2001).
49. P. M. Martin, M. S. Good, J. W. Johnston, G. J. Posakony, L. J. Bond and S. L. Crawford, *Thin Solid Films* **379**, 253 (2000).
50. S. J. Pearton, D. P. Norton, K. Ip, Y. W. Heo and T. Steiner, *Prog. Mater. Sci.* **50**, 293 (2005).
51. X. Wang, Y. Ding, C. J. Summers and Z. L. Wang, *J. Phys. Chem. B* **108**, 8773 (2004).
52. L. M. Kukreja, S. Barik and P. Misra, *J. Cryst. Growth* **268**, 531 (2004).
53. J. W. Chiou, K. P. Krishna Kumar, J. C. Jan, H. M. Tsai, C. W. Bao, W. F. Pong, F. Z. Chien, M.-H. Tsai, I.-H. Hong, R. Klauser, J. F. Lee, J. J. Wu and S. C. Liu, *Appl. Phys. Lett.* **85**, 3220 (2004).
54. H. Chik, J. Liang, S. G. Cloutier, N. Kouklin and J. M. Xu, *Appl. Phys. Lett.* **84**, 3376 (2004).
55. Z. Fan and J. G. Lu, *Appl. Phys. Lett.* **86**, 123510 (2005).
56. A. Kolmakov and M. Moskovits, *Annu. Rev. Mater. Res.* **34**, 151 (2004).
57. Y. Zhang, A. Kolmakov, S. Chretien, H. Metiu and M. Moskovits, *Nano Lett.* **4**, 403 (2004).
58. L. Vayssieres, K. Keis, S. E. Lindquist and A. Hegfeld, *J. Phys. Chem. B* **105**, 3350 (2001).
59. M. Izaki and T. Omi, *Appl. Phys. Lett.* **68**, 2439 (1996).
60. Th. Pauporte and D. Lincot, *Appl. Phys. Lett.* **75**, 3817 (1999).
61. B. Cao, W. Cai, G. Duan, Y. Li, Q. Zhao and D. Yu, *Nanotechnology* **16**, 2567 (2005).
62. V. A. L. Roy, A. B. Djurisic, W. K. Chan, J. Gao, H. F. Lui and C. Surya, *Appl. Phys. Lett.* **83**, 141 (2003).
63. W. I. Park, D. H. Kim, S. W. Jung and G. C. Yi, *Appl. Phys. Lett.* **80**, 4232 (2002).
64. B. D. Yao, Y. F. Chan and N. Wang, *Appl. Phys. Lett.* **81**, 757 (2002).
65. X. Kong and Y. Li, *Chem. Lett.* **32**, 838 (2003).
66. Y. Li, G. W. Meng and L. D. Zhang, *Appl. Phys. Lett.* **76**, 2011 (2000).
67. W. Y. Chung, J. W. Lim, D. D. Lee, N. Miura and Y. Yamazoe, *Sens. Actuators B* **64**, 118 (2000).
68. C. Xu, J. Tamaki, N. Miura and N. Yamazoe, *Sens. Actuators B* **3**, 147 (1991).
69. M. K. Kennedy, F. E. Kruis, H. Fissan, B. R. Mehta, S. Stappert and G. Dumpich, *J. Appl. Phys.* **93**, 551 (2003).
70. H. Ogawa, M. Nishikawa and A. Abe, *J. Appl. Phys.* **53**, 4448 (1982).

71. C. Malagù, V. Guidi, M. Stefancich, M. C. Carotta and G. Martinelli, *J. Appl. Phys.* **91**, 808 (2002).
72. M. K. Kennedy, F. E. Kruis, H. Fissan and B. R. Mehta, *Rev. Sci. Instru.* **74**, 4908 (2003).
73. L. Spanhel and M. A. Anderson, *J. Am. Chem. Soc.* **113**, 2826 (1991).
74. E. A. Meulenkamp, *J. Phys. Chem. B* **102**, 5566 (1998).
75. A. Van Dijken, E. A. Meulenkamp, D. Vanmaekelbergh and A. Meijerink, *J. Phys. Chem. B* **104**, 1715 (2000).
76. M. Izaki and T. Omi, *Appl. Phys. Lett.* **68**, 2439 (1996).
77. E. M. Wong and P. C. Searson, *Appl. Phys. Lett.* **74**, 2939 (1999).
78. R. Liu, A. A. Vertegel, E. W. Bohannan, T. A. Sorenson and J. A. Switzer, *Chem. Mater.* **13**, 508 (2001).
79. S. Roy and S. Basu, *Bull. Mater. Sci.* **25**, 513 (2002).
80. F. Hossein-Babaei and F. Taghibakhsh, *Elec. Lett.* **36**, 1815 (2000).
81. S. Hayamizu, H. Tabata, H. Tanaka and T. Kawai, *J. Appl. Phys.* **80**, 787 (1996).
82. M. Okoshi, K. Higashikawa and M. Hanabusa, *Appl. Surf. Sci.* **154–155**, 424 (2000).
83. V. Craciun, J. Elders, J. G. E. Gardeniers and I. W. Boyd, *Appl. Phys. Lett.* **65**, 2963 (1995).
84. P. Yu, Z. K. Tang, G. K. L. Wong, M. Kawasaki, A. Ohotomo, H. Koinuma and Y. Segawa, *Solid State Commun.* **103**, 459 (1997).
85. P. Yu, Z. K.Tang, G. K. L. Wong, M. Kawasaki, A. Ohotomo, H. Koinuma and Y. Segawa, *J. Cryst. Growth* **184/185**, 601 (1998).
86. D. M. Bagnall, Y. F. Chen, Z. Zhu, T. Yao, M. Y. Shen and T. Goto, *Appl. Phys. Lett.* **73**, 1038 (1998).
87. S. Liang, C. R. Gorla and N. Emanetoglu, *J. Elec. Mater.* **27**, L72 (1998).
88. Y. Liu, C. R. Gorla, S. Liang, N. Emanetoglu, Y. Lu, H. Shen and M. Wraback, *J. Elec. Mater.* **29**, 60 (2000).
89. H. Yuan and Y. Zhang, *J. Cryst. Growth* **263**, 119 (2004).
90. N. G. Patel and B. H. Lashkari, *J. Mater. Sci.* **27**, 3026 (1992).
91. F. K. Sahn, Z. F. Liu, G. X. Liu, B. C. Shin and Y. S. Yu, *J. Korea. Phys. Soc.* **44**, 1215, (2004).
92. D. F. Paraguay, M. Miki-Yoshida, J. Morales, J. Solis and L. W. Estrada, *Thin Solid Films* **373**, 137 (2000).
93. B. O'Regan, V. Sklover and M. Grätzel, *J. Electrochem. Soc.* **148**, C498 (2001).
94. J. Jiu, K. Kurumada and M. Tanigaki, *Mater. Chem. Phys.* **81**, 93 (2003).
95. T. F. Jaramillo, S.-H. Baeck, A. Kleiman-Shwarsctein and E. W. McFarland, *Macromol. Rapid Commun.* **25**, 297 (2004).
96. Z. L. Wang, *Mater. Today* **7**, 26 (2004).
97. Y. H. Xiao, L. Li, Y. Li, M. Fang and L. D. Zhang, *Nanotechnology* **16**, 671 (2005).
98. X. D. Wang, C. J. Summers and Z. L. Wang, *Adv. Mater.* **16**, 1215 (2004).
99. Y. W. Zhu, H. Z. Zhang, X. C. Sun, S. Q. Feng, J. Xu, Q. Zhao, B. Xiang, R. M. Wang and D. P. Yu, *Appl. Phys. Lett.* **83**, 144 (2003).

100. M. Yan, H. T. Zhang, E. J. Widjaja and R. P. H. Chang, *J. Appl. Phys.* **94**, 5240 (2003).
101. L. Vayssieres, *Adv. Mater.* **15**, 464 (2003).
102. G. Z. Cao, J. J. Schermer, W. J. P. van Enckevort, W. A. L. M. Elst and L. J. Giling, *J. Appl. Phys.* **79**, 1357 (1996).
103. L. E. Greene, M. Law, J. Goldberger, F. Kim, J. C. Johnson, Y. Zhang, R. J. Saykally and P. Yang, *Angew. Chem. Int. Ed.* **42**, 3031 (2003).
104. B. C. Bunker, P. C. Rieke, B. J. Tarasevich, A. A. Campbell, G. E. Fryxell, G. L. Graff, L. Song, J. Liu and J. W. Virden, *Science* **264**, 48 (1994).
105. P. Hartman and W. G. Perdok, *Acta Cryst.* **8**, 49 (1955).
106. L. Vayssieres, N. Beermann, S.-E. Lindquist and A. Hagfeldt, *Chem. Mater.* **13**, 233 (2001).
107. L. Vayssieres, K. Keis, S.-E. Lindquist and A. Hagfeldt, *J. Phys. Chem. B* **105**, 3350 (2001).
108. L. Vayssieres, K. Keis, A. Hagfeldt and S.-E. Lindquist, *Chem. Mater.* **13**, 4395 (2001).
109. Z. Tian, J. A. Voigt, J. Liu, B. Mckenzie and M. J. Mcdermott, *J. Am. Chem. Soc.* **124**, 12954 (2002).
110. J. Liu, Y. Lin, L. Liang, J. A. Voigt, D. L. Huber, Z. R. Tian, E. Coker, B. Mckenzie and M. J. Mcdermott, *Chem.: A Europ. J.* **97**, 604 (2003).
111. L. E. Greene, M. Law, J. Goldberger, F. Kim, J. C. Johnson, Y. Zhang, R. J. Saykally and P. Yang, *Angew, Chem. Int. Ed.* **42**, 3031 (2003).
112. Z. R. Tian, J. A. Voigt, J. Liu, B. Mckenzie, M. J. Mcdermott, R. T. Cygan and L. J. Criscenti, *Nature Materials* **2**, 821 (2003).
113. M. Izaki and T. Omi, *Appl. Phys. Lett.* **68**, 2439 (1996).
114. Th. Pauporte and D. Lincot, *Appl. Phys. Lett.* **75**, 3817 (1999).
115. B. Cao, W. Cai, G. Duan, Y. Li, Q. Zhao and D. Yu, *Nanotechnology* **16**, 2567 (2005).
116. Y. J. Kim, H. M. Shang and G. Z. Cao, Accepted by *J. Sol-gel Sci. Technol.* (2005).
117. L. Vayssieres, *Int. J. Nanotechnology* **1**, 1 (2004).
118. R. S. Wagner and W. C. Ellis, *Appl. Phys. Lett.* **4**, 89 (1964).
119. R. S. Wagner, W. C. Ellis, K. A. Jackson and S. M. Arnold, *J. Appl. Phys.* **35**, 2993 (1964).
120. R. S. Wagner, in Whisker Technology, ed. A. P. Levitt (Wiley, New York, 1970).
121. E. I. Givargizov, Highly anisotropic crystals, D. Reidel, Dordrecht (1986).
122. E. I. Givargizov, *J. Vac. Sci. Technol. B* **11**, 449 (1993).
123. M. H. Huang, Y. Wu, H. Feick, N. Tran, E. Weber and P. Yang, *Adv. Mater.* **13**, 113 (2001).
124. D. R. Askkeland, *The Science and Engineering of Materials* (PWS, Boston, 1989).
125. G. W. Sears, *Acta Metal.* **3**, 361 (1955).
126. G. W. Sears, *Acta Metal.* **3**, 367 (1955).
127. Z. W. Pan, Z. R. Dai and Z. L. Wang, *Science* **291**, 1947 (2001).

128. M. Ohring, *The Materials Science of Thin Films* (Academic Press, San Diego, 1992).
129. K. F. Jensen and W. Kern, in *Thin Film Processes II*, eds. J. L. Vossen and W. Kern (Academic Press, San Diego, 1991).
130. K. L. Choy, *Prog. Mater. Sci.* **48**, 57 (2003).
131. P. Ser, P. Kalck and R. Feurer, *Chem. Rev.* **102**, 3085 (2002).
132. S. Muthukumar, N. W. Emanetoglu, G. Patounakis, C. R. Gorla, S. Liang and Y. Lu, *J. Vac. Sci. & Tech. A* **19**, 1850 (2001).
133. S. Muthukumar, H. Sheng, J. Zhong, Z. Zhang, N. W. Emanaetoglu and Y. Lu, *IEEE Trans. Nanotech* **2**, 50 (2003).
134. W. I. Park, G. C. Yi, M. Kim and S. J. Pennycook, *Adv. Mater.* **14**, 1841 (2002).
135. W. Lee, H. G. Sohn and J. M. Myoung, *Mater. Sci. Forum* **449**, 1245 (2004).
136. X. Liu, X. H. Wu, H. Cao and R. P. H. Chang, *J. Appl. Phys.* **95**, 3141 (2004).
137. K. Ogata, K. Maejima, S. Fujita and S. Fujita, *J. Cryst. Growth* **248**, 25 (2003).
138. K. Maejima, M. Ueda, S. Fujita and S. Fujita, *Japan. J. Appl. Phys.* **42**, 2600 (2003).
139. K. S. Kim and H. W. Kim, *Physica B* **328**, 368 (2003).
140. B. P. Zhang, N. T. Binh, Y. Segawa, K. Wakatsuki and N. Usami, *Appl. Phys. Lett.* **83**, 1635 (2003).
141. S. W. Kim, S. Fujita and S. Fujita, *Japan. J. Appl. Phys.* **41**, L543 (2002).
142. J. J. Wu and S. C. Liu, *Adv. Mater.* **14**, 215 (2002).
143. Y. W. Heo, V. Varadarajan, M. Kaufman, K. Kim, F. Ren, P. H. Fleming and D. P. Norton, *Appl. Phys. Lett.* **81**, 3046 (2002).
144. T. Okada, B. H. Agung and Y. Nakata, *Appl. Phys. A* **79**, 1417 (2004).
145. C. Jin, X. Yuan, W. Ge, J. Hong and X. Xin, *Nanotechnol.* **14**, 667 (2003).
146. N. Pan, X. Wang, K. Zhang, H. Hu, B. Xu, F. Li and J. Hou, *Nanotechnol.* **16**, 1069 (2005).
147. S. C. Lyu, Y. Zhang, C. J. Lee, H. Ruh and H. J. Lee, *Chem. Mater.* **15**, 3294 (2003).
148. Y. C. Wang, I. C. Leu and M. N. Hon, *J. Mater. Chem.* **12**, 2439 (2002).
149. Y. Li, G. S. Cheng and L. D. Zhang, *J. Mater. Res.* **15**, 2305 (2000).
150. B. B. Lakshmi, P. K. Dorhout and C. R. Martin, *Chem. Mater.* **9**, 857 (1997).
151. Z. R. Tian, J. A. Voigt, J. Liu, B. Mckenzie, M. J. Mcdermott, M. A. Rodriguez, H. Konishi and H. F Xu, *Nature Mater.* **2**, 821 (2003).
152. G.-C. Yi, C. Wang and W. I. Park, *Semicond. Sci. Technol.* **20**, S22 (2005).
153. G. Z. Cao, Nanostructures and nanomaterials: Synthesis, properties and applications (Imperial College Press, London, 2004).
154. D.-D. Lee and D.-S. Lee, *IEEE Sens. J.* **1**, 214 (2001).
155. H.-W. Ryu, B.-S. Park, S. A. Akbar, W.-S. Lee, K.-J. Hong, Y.-Jin Seo, D.-C. Shin, J.-S. Park and G.-P. Choi, *Sens. Actuators B* **96**, 717 (2003).
156. G. Sberveglieri, *Sens. Actuators B* **23**, 103 (1995).

157. G. S. Trivikrama Rao and D. Tarakarama Rao, *Sens. Actuators B* **55**, 166 (1999).

158. X. L. Cheng, H. Zhao, L. H. Huo, S. Gao and J. G. Zhao, *Sens. Actuators B* **102**, 248 (2004).

159. C. H. Kwon, H. K. Hong, D. H. Yun, K. Lee, S. T. Kim, Y. H. Roh and B. H. Lee, *Sens. Actuators B* **24/25**, 610 (1995).

160. P. Kofstad, Non-stoichiometry, diffusion and electrical conductivity in binary metal oxides (Wiley-Interscience, New York, 1972).

161. J. Q. Xu, Y. Q. Pan, Y. A. Shun and Z.-Z. Tian, *Sens. Actuators B* **66**, 277 (2000).

162. Y. Ma, W. L. Wang, K. J. Liao and C. Y. Kong, *J. Wide Band gap Mater.* **10**, 113 (2002).

163. C. Xu, J. Tamaki, N. Miura and N. Yamazoe, *Sens. Actuators B* **3**, 147 (1991).

164. Q. Wan, Q. H. Li, Y. J. Chen, T. H. Wang, X. L. He, J. P. Li and C. L. Lin, *Appl. Phys. Lett.* **84**, 3654 (2004).

165. Z. Fan, D. Wang, P. Chang, W. Tseng and J. G. Lu, *Appl. Phys. Lett.* **85**, 5923 (2004).

166. Q. H. Li, Q. Wan, Y. X. Liang and T. H. Wang, *Appl. Phys. Lett.* **84**, 4556 (2004).

167. X. G. Wen, Y. P. Fang, Q. Pang, C. L. Yang, J. N. Wang, W. K. Ge, K. S. Wong and S. H. Yang, *J. Phys. Chem. B* **109**, 15303 (2005).

Chapter 14

Synthesis and Properties of Mesoporous-Based Materials for Environmental Applications

Jianlin Shi*, Hangrong Chen, Zile Hua and Lingxia Zhang

Shanghai Institute of Ceramics, Chinese Academy of Sciences
1295 Ding-Xi Road, 200050, Shanghai, PR China
*jlshi@sunm.shcnc.ac.cn

The synthesis of nanoporous materials has attracted intensive attentions in the recent years for their wide-range of promising applications. Nanoporous materials are a large family of solid-state materials. According to the definition of the pore size by the International Union of Pure and Applied Chemistry (IUPAC),[1] porous materials can be divided into three classes: microporous materials (< 2 nm), mesoporous materials ($2 \sim 50$ nm), and macroporous materials (> 50 nm). Most or parts of these three classes of materials can be included in the so-called nanoporous materials ($1 \sim 100$ nm), which have found more and more important applications in the field of adsorption, separation, catalysis, supporting-materials, optic-electric devices, petroleum and chemical industries, and so on, owing to their well-defined pore structures, large surface areas and high thermal stability, and tunable pore surface characters, etc.[2-4]

Ordered mesoporous inorganic materials were first reported in 1992.[5,6] Such novel meso-structured materials templated by surfactant molecular self-assembly have found great potentials in many fields owing to their special pore characteristics. Several strategies for the fabrication of mesoporous silica materials have been extensively investigated over the past decade, and the approach to the synthesis of mesoporous silica has been extended to other metal oxides with various approaches. Herein we mainly focus on some typical synthesis approaches of mesoporous materials and mesoporous-based composites and their environmental catalysis processes.

Mesoporous Based Nanocomposites for Automotive Exhaust Treatments

Nowadays, automobiles are prevalent around the world as the most popular and necessary vehicles in our daily lives. About 50 million cars are produced every year, and over 700 million cars are used worldwide. The automotive exhaust has become an increasingly serious pollution source to our atmosphere. Emission regulations have been applied in European countries since 1988, and more rigorous regulations are being planned in both USA and Europe. Thus, the use of catalysts, especially the three-way catalysts (TWCs) for purifying exhaust gases, which contains mainly oxides of carbon (CO), oxides of nitrogen (NO_x), hydrocarbons (HC) and particulate matter, is necessary and indispensable in every vehicle. The automotive TWC system generally operates under a certain air-fuel ratio (A/F) range, which is controlled by an oxygen sensor device. Common three-way catalysts consist of noble metals, such as Pt, Rh, and Pd, as active sites; metal oxides as promoter materials; and a transition-alumina support. In order to obtain high conversion rates of CO, HC, and NO_x, the A/F ratio oscillations should be buffered. Therefore, CeO_2-based materials have been pervasively used as an oxygen storage material associated to the redox couple of Ce^{4+}/Ce^{3+}, i.e. ceria can provide oxygen for oxidizing CO and HC under rich A/F conditions and removes it from the exhaust gas phase for reducing NO_x under lean A/F. The addition of zirconia into the ceria lattice can significantly increase the oxygen storage capacity (OSC), as well as the thermal stability. This property is believed to be related to the displacement of oxygen sub-lattice around zirconium, leading to higher oxygen mobility and easier bulk reduction of the solid solution.[7] Therefore, CeO_2-ZrO_2 composite oxide has been widely used as automobile exhaust three-way catalysts in the last decade. A great deal of approaches has been reported to synthesis of high surface area CeO_2/ZrO_2 composite, such as co-precipitation,[8] sol-gel,[9] and microemulsion precipitation,[10] etc.

Synthesis of Mesoporous CeO₂ Catalysts and CO Catalytic Oxidation Property

Cerium oxide is an important rare earth oxide and has been widely investigated in the automotive exhaust purification, oxygen storage and release catalysis, and solid oxide fuel cell applications.[11,12]

Several novel strategies, including template methods, have been developed for the fabrication of mesoporous materials over the last years. The structures and properties of the templates play a very important role with respect to the properties of the replicated porous materials. The principle

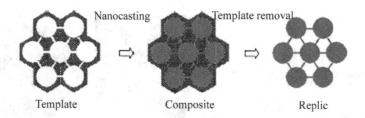

Figure 1.   Illustration of the nanocasting pathway to produce mesoporous oxides.[13]

of the nanocasting pathway with hard template mainly involves three main steps: (1) formation of the template, (2) the casting step and (3) removal of the template, as illustrated in Figure 1.[13] Generally, a three-dimensional pore network is necessary in the template to create a stable replica. Furthermore, the precursor should be easily converted into the desired composition with as little volume shrinkage as possible. And finally, the templates should be easily and completely removed to obtain the true replicas. Hereby, we illustrate a sample for the synthesis of ordered mesoporous CeO_2 using cubic Ia3d mesoporous silica KIT-6 as a hard template.

A typical mesoporous silica with cubic Ia3d symmetry (designated as KIT-6) was prepared according to the literature.[14] To obtain the surfactant-removed hard template (named as KIT-eo) with abundant silanol group on the interior mesopore surface, the surfactant was removed by solvent extraction using ethanol refluxing under stirring, followed by oxidization using 15 wt.% H_2O_2 solution under stirring at 310 K. The surfactant-removed product was filtered, washed with ethanol, and dried at 333 K in vacuum. A typical synthesis process of cerium oxide mesoporous material with uniform mesopore and fine particle morphology using the as-prepared KIT-eo of Ia3d symmetry as hard template was as follows[15]: First, 4.7 g of Ce(NO_3)$_3$· $6H_2O$ was dissolved in 20 mL of ethanol, then this solution was incorporated into 1 g of KIT-eo by the incipient wetness impregnation technique. After ethanol was evaporated at 333 K, the composite was calcined at 673 K in order to transfer the precursor into cerium oxide. The silica template in the CeO_2/SiO_2 composite was removed by washing with heated 1 to 2 M NaOH solution for three times. This template-free mesoporous CeO_2 array was collected by filtering, washed with water and ethanol, and dried at 373 K.

The small-angle X-ray Diffraction (XRD) pattern of the template-free mesoporous ceria was shown in Figure 2A, which shows the same characteristic Bragg diffraction peaks as the silica template belonging to the same symmetry (cubic Ia3d space group) with a slightly shrunk lattice space of a = 22.60 nm. The wide-angle XRD pattern of the mesoporous CeO_2 replica shows a little broader Bragg diffraction peaks, which can be indexed to a

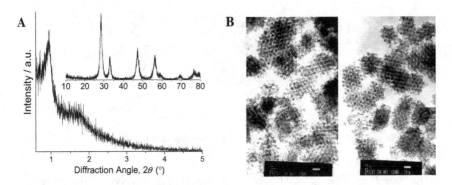

Figure 2. (A) Small- and wide-angle XRD patterns (inset) of ordered mesoporous ceria replica; (B) TEM images of the template-free ordered mesoporous ceria replica.[15]

cubic structure (space group Fm3m, JCPD No. 431002) with a lattice space of a = 0.5411 nm, indicating well-crystallized CeO_2 nanorod framework.

The Transmission Electron Microscope (TEM) images of the replica cerium oxide in Figure 2B show a well-ordered framework and a good replica of the silica template pore channels.

Compared with the reference sample CeO_2-D, which was obtained by directly calcining $Ce(NO_3)_3 \cdot 6H_2O$ at 823 K for four hours, the mesoporous ceria replica shows higher catalytic activity for the oxidation of CO to CO_2 at relatively low temperature. After further loaded with CuO, the composite can greatly enhance the CO oxidation reactivity. The conversions of CO over these samples were shown in Figure 3, which shows the strong effect of CuO loading on the CO conversions. The temperatures for 50% conversion of carbon monoxide (T_{50}) for the samples of CeO_2-D, and mesoporous ceria replica and CuO-loaded ceria replica are shown in the inset. The T_{50} of the cerium oxide replica (523 K) is significantly lower than that of the CeO_2-D (606 K), meaning that the activity of this replica CeO_2 is higher than that of the CeO_2-D. With the increase of the CuO loading amount, T_{50} is distinctively decreased, and the lowest T_{50} (389 K) of all catalyst samples is achieved at the 20% CuO loading, and then the T_{50} (400 K) increased slightly as the CuO loading amount increases to 30%.

Recently, Deshpande et al.[16] demonstrated a novel synthesis of thermally stable, mesoporous ceria using the sols of pure CeO_2 nanoparticles as building blocks in a block-copolymer-assisted assembly process. A hydrogenated polybutadiene–poly(ethylene oxide) block copolymer (PHB–PEO; $H[CH_2CH_2CH_2CH(CH_2 CH_3)]x(OCH_2CH_2)yOH$) was used as structure-directing agent. This synthesis pathway involves the addition of the CeO_2 nanoparticle sol in a mixture of ethanol and water to an alcoholic solution of the PHB–PEO block copolymer. Evaporation of the solvent induces the

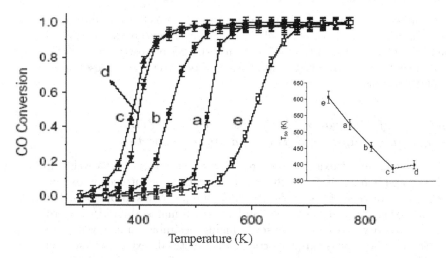

Figure 3. Percentage of CO conversion of $CO + O_2$ to CO_2 reaction and T_{50} (inset) value over ordered mesoporous ceria replica loaded with 0%(a), 10% (b), 20% (c), and 30% CuO. (e) is for a reference sample.[15]

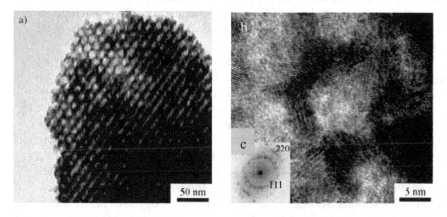

Figure 4. (a) TEM image of the sample calcined at 500°C for two hours; (b) HRTEM image of a $32 \times 32\,nm^2$ selected-area; and (c) the corresponding electron diffraction pattern.[16]

cooperative assembly of the nanoparticles and the PHB–PEO block copolymer micelles. By this approach, they have successfully assembled the nearly monodispersed and highly crystalline ceria nanoparticles (with a diameter of about 3 nm) into highly ordered 3D mesostructures. Furthermore, the mesopores are particularly large and show a high degree of ordering in terms of pore shape and 3D arrangement, as shown in Figure 4.[16] The

BET surface area obtained for the calcined sample was $87\,\mathrm{m^2/g}$ after calcination at $500°C$, and such relatively low specific surface area value is partly due to the high density of CeO_2 ($7.132\,\mathrm{cm^3/g}$). This mesostructure was stable upon calcination at $500°C$ and consists of relatively large mesopores of about $10 \sim 12\,\mathrm{nm}$.

Synthesis of Mesoporous Metal-Loaded CeO_2/ZrO_2 Composite Catalysts and Their Catalytic Properties

Ceria-zirconia mixed oxides are current targets of intensive researches for their successful applications in the latest generations of automotive exhaust three-way catalysts (TWCs) due to their unique redox properties, high oxygen storing/releasing capacity and high thermal stability.[17] Controlling the porosity of these catalysts is highly desirable, and it was noted that the chemical/physical properties and structural texture are strongly dependent on the preparation procedure. The use of high surface area mesoporous oxide supports may give rise to well-dispersed and stable noble-metal nanoparticles in the pore surface and thus can strongly influence their catalytic performances. Here we give an example of highly dispersive pure platinum nanoparticles, which was well-confined into the pore-channels of ordered mesoporous zirconia. This novel nanocomposite shows good catalytic activity.

The platinum-loaded mesoporous zirconia sample was synthesized by an ion-exchange method.[18] The as-synthesized mesophase zirconia containing surfactants was first post-treated by phosphoric acid to modify the inner pore surface and then, sulphuric acid and ethanol were used to extract the surfactants by an ion-exchange route. After the removal of the surfactants, the sucrose solution containing H_2PtCl_6 was filled into the pore channels by ion-exchange under vigorous agitation. After being aged and dried at $90°C$, the platinum can be confined and spontaneously *in-situ* reduced during thermal treatment at $773\,\mathrm{K}$. Figure 5A shows a representative High-resolutions Transmission Electron Microscopy (HRTEM) image recorded from platinum loaded ($0.5\,\mathrm{at\%}$) mesoporous zirconia sample. A large number of homogeneously dispersed metallic platinum nanoparticles in pore channels with uniform size can be directly observed. From the measurement of the Pt nanoparticles in these TEM images, an average particle size of $1.8 \pm 0.1\,\mathrm{nm}$ was determined. These confined platinum nanoparticles show high catalytic activity for the oxidation reaction of CO. The percent CO conversions over the prepared $0.5\,\mathrm{at\%}$ $Pt/M\text{-}ZrO_2$ sample and the reference Pt-loaded sample prepared by the traditional impregnation method are also shown in Figure 5B-b for comparison. The quickly accelerated CO conversion starts below $120°C$, and complete CO oxidation is achieved at

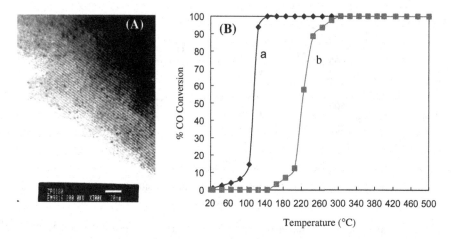

Figure 5. (A) Representative HRTEM image of sample Pt/M-ZrO$_2$ and (B) the percent CO conversion over: (a) 0.5 at% Pt/M-ZrO$_2$ for CO + O$_2$ reaction, and (b) the reference Pt-loaded mesoporous zirconia sample prepared by the traditional impregnation method.[18]

145°C over 0.5 at% Pt/M-ZrO$_2$ sample giving CO$_2$ as the product, which can ensure immediate activation of the catalyst on engine start-up, whereas, the same reaction under the identical conditions requires as high as 300°C for the reference sample.

After further incorporation with ceria, the temperature-programmed reaction profiles for the conversion of CO + NO into CO$_2$ + N$_2$ over such a composite sample CeO$_2$ (5wt%)-Pt(1at%)/M-ZrO$_2$ were shown in Figure 6. The mixed gases start their fast conversion into CO$_2$ + N$_2$ at 120°C, and reach their complete conversion at about 220°C, which indicates that this platinum-loaded and ceria-doped mesoporous zirconia material shows high and sustained catalytic activity for the conversion of NO + CO into CO$_2$ + N$_2$ at relatively low temperature.

In the recent years, hierarchically structured porous materials at multiple length scales have attracted much attention owing to their high surface areas and regular porosity at the nanometer scale, which can provide many novel properties and have important prospects in the practical industrial processes such as catalysis, adsorption, separation, and some biological applications.[19-21] In this kind of study, surfactants have been shown to be important in organizing the inorganic species into a variety of mesoporous forms via self-assembly or cooperative process. Thus, the combination of surfactant and the exo-templating methods, such as colloidal polymer latex, emulsions droplets, can allow the fabrication of hierarchical meso/macroporous architectures.[22] Yu and co-workers[23]

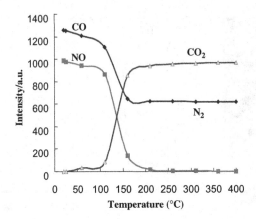

Figure 6.   Temperature-programmed reaction profiles for the conversion of CO + NO into $CO_2 + N_2$ reaction over sample $CeO_2(5\,wt\%)$-Pt$(1\,at\%)$/M-ZrO_2, (because the mass number of CO is the same as that of N2 ($m/z = 28$), their signal positions are at the same curve).[18]

demonstrated the synthesis of a thermally stable, hierarchically-ordered, meso/macrostructured, and palladium-loaded ceria-zirconia catalyst using a combination of surfactant and colloidal crystal templating methods. A typical synthesis process is as follows[1]: A 10 wt.% micellar solution of cetyltrimethyl ammonium bromide (CTAB) was prepared by dissolving the surfactant in water at room temperature under ultrasound irradiation. A meso/macro-structured sample with cerium content $Ce_xZr_{1-x}O_2$ (X = 0.6, denoted as MM60) was prepared from mixing an appropriate ratio of zirconium n-propoxide ($[Zr(OC_3H_7)_4]$) to ammonium cerium(IV) nitrate ($[(NH_4)_2Ce(NO_3)_6]$) in a minimum amount of n-propanol with ultrasound assisted dissolution. After homogenization, the resulting reddish-brown solution was slowly added to the surfactant solution with gentle stirring to disperse the droplets, and the pH of the solution was kept at 11 using ammonia. The obtained mass was transferred into a teflon-lined autoclave, and heated under static conditions at 80°C for 48 hours. The product was filtered and dried overnight and then calcined at 500°C to 800°C for the removal of the surfactant and crystallization. The reference sample prepared by co-precipitation method with a cerium content of 60% and was denoted as CP60. A 0.8 wt.% Pd in $Ce_xZr_{1-x}O_2$ was prepared by immersing the catalyst support in an aqueous solution of $PdCl_2$ under reduced pressure and ultrasound irradiation. The prepared sample shown in Figure 7[23] indicates that both macrochannels (A) and mesoporous structures (B,C) can be observed. The surface area and pore volume are 131.4 m²/g, and 0.14 cm³/g respectively, which are much higher than those of the reference

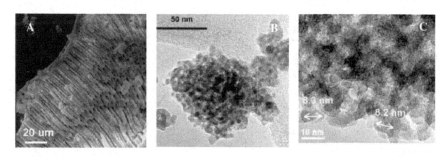

Figure 7.   SEM (A), and TEM images (B, C) with different magnifications for the Pd-MM60 after calcination.[23]

sample Pd-CP60. The catalytic actively profile of CO conversion over Pd-MM60 and the reference samples show that the T_{50} and T_{90} for Pd-MM60 are 65 and 75°C, respectively, which are much lower than those for Pd-CP60. This may be because that Pd-MM60 possesses mesoporous frameworks that provide a high surface area to volume ratio, and hence more active sites are available for CO oxidation. The poor activity of Pd-CP60 can be related to its structural and chemical properties. When calcined at 700°C, the BET surface area of Pd-CP60 decreases sharply to 20.6 m^2/g; in addition, Pd-CP60 annealed at 700°C consists of a mixture of cubic and tetragonal phases, such a phase inhomogeneity also strongly influences the oxidation/reduction behavior.

For the sake of energy saving, the ratio of air/fuel fed into internal combustion engines trends toward higher values than the stoichiometric 14.7, which gives rise to decreased NO reduction efficiency of three-way catalysts (TWCs) in such a fuel-lean condition. Therefore, development of a new generation of TWCs to improve the NO reduction activity under oxygen-rich condition is becoming urgent. Recently, Wang and co-workers reported[24] a surfactant-controlled synthetic method to obtain a nanophase of mesoporous ceria-zirconia solid solution by using cerium nitrate ($Ce(NO_3)_3 \cdot 6H_2O$) and zirconium oxychloride octahydrate ($ZrOCl_2 \cdot 8H_2O$) as precursors and a cationic surfactant, myristyltrimethylammonium bromide ($CH_3(CH_2)_{13}N(CH_3)_3Br$) as molecular template. The 3 wt.% $Pd/Ce_{0.6}Zr_{0.4}O_2$ catalyst was prepared by impregnating the mesoporous $Ce_{0.6}Zr_{0.4}O_2$ support, which was calcined at 800°C, with an aqueous solution of $Pd(NO_3)_2 \cdot 2H_2O$. The metal-supported catalyst was dried at 120°C for several hours and then calcined at 600°C for four hours. The surface area of the prepared sample by using such a method is 10 $\sim$ 15% higher than those prepared by a traditional co-precipitation method. Compared with pure ceria prepared by the same method with a crystallite size of 31 nm at 800°C, the prepared $Ce_{0.6}Zr_{0.4}O_2$ solid has much smaller crystallite size

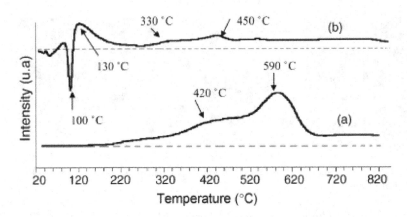

Figure 8.   TPR profiles of: (a) 3 wt.% $PdO/Ce_{0.6}Zr_{0.4}O_2$ and (b) $Ce_{0.6}Zr_{0.4}O_2$.[24]

of 11.4 nm after calcination at the same temperature, which indicates that zirconium addition to ceria can effectively inhibit the sintering and crystallite growth.

The temperature-program-reduction (TPR) method was applied to study the reduction properties of oxygen species in ceria-zirconia support and Pd/ceria-zirconia catalyst. The TPR profile of the ceria-zirconia calcined up to 800°C consisted of two hydrogen consumption peaks: one at 420°C and another at 590°C (Figure 8a).[24] The peak at the higher temperature (590°C) is ascribed to the reduction of bulk lattice oxygen and the one at 420°C is ascribed to the reduction of the surface oxygen species. On a pure CeO_2, two peaks are also observed in the TPR profile; however, the temperature for bulk oxygen reduction in ceria is as high as 900°C.[25] This result shows that addition of zirconium to ceria significantly lowers the oxygen reduction temperature, thus improving the reducibility of bulk oxygen. The TPR profile of the PdO/Ce-Zr-O shows a sharp, negative peak at about 100°C, which strongly indicates hydrogen generation during the TPR procedure (Figure 8b), which in turn was related to the formation/decomposition of a palladium hydride phase (b-PdHx). Compared to the TPR profile of the $Ce_{0.6}Zr_{0.4}O_2$ solid, the temperature of the peak maximum in the palladium-supported catalyst markedly shift to the low temperature range. In addition, a new reduction peak appeared at 130°C, which may associate with the hydrogen spillover from the metal to the support.

Catalytic tests of NO reduction by CO in an oxygen-excess condition showed that the $Pd/Ce_{0.6}Zr_{0.4}O_2$ catalyst had high activity for both CO oxidation and NO reduction, as can be seen in the Table 1.[24] The light-off temperature (T_{50}) that corresponds to 50% conversion of CO to CO_2 and

Table 1. Activity and selectivity of NO reduction by CO over 3 wt.% $Pd/Ce_{0.6}Zr_{0.4}O_2$ catalyst.[24]

Temperature (°C)	CO Conversion (%)	NO Conversion (%)	Selectivity (%)	
			N_2	NO_2
50	32	32	100	0
100	38	33	100	0
150	52	40	100	0
200	67	50	95	5
250	85	60	91	9
300	96	65	93	7
350	100	65	88	12
400	100	70	84	16

NO reduction was approximately 130 and 200°C, respectively. At cool start of reaction (below 150°C), NO reduction by CO with excess oxygen over the $Pd/Ce_{0.6}Zr_{0.4}O_2$ catalyst showed selectivity around 100% to N_2. However, a competition between NO reduction by CO and CO oxidation by O_2 was observed: at below 200°C, NO lowered CO oxidation activity, however, at reaction temperatures above 200°C, high activity of CO oxidation resulted in an inhibited effect on NO reduction. The NO presence may induce oxidation of the metallic Pd active sites, thus inhibiting the CO oxidation on Pd in the range of low reaction temperature. On the other hand, NO reduction was also strongly affected by CO oxidation in the oxygen-excess condition.

Electrocatalytic Performance of Mesoporous-Based Nanocomposites for Full Cell Reactions

Fuel cell-based automobiles have gained great attention in the recent years due to the growing public concern about air pollution and consequent environmental problems.[26] Polymer electrolyte membrane (PEM) fuel-cells and direct-methanol-fuel-cell (DMFC) as well as their accompanied environmental benefits have aroused intensive studies on fundamental and applied aspects for its potentials as clean and ideal power sources. To increase the catalytic activity of methanol electrooxidation, much effort has been devoted towards the development of new catalysts. In addition, the development of new carbon materials as supports to help achieve optimum catalytic performance has also induced great attentions, for the common carbon supports in practical use have broad pore size distributions and irregular structures.

Ordered Porous Carbons as Catalysts Supports in Direct-Methanol-Fuel-Cell

Direct-methanol-fuel-cells (DMFCs) are now being considered as candidate power sources for both portable power applications and electric vehicles as they provide high-energy output and high yield theoretically without harmful byproducts.[27,28] It is now widely accepted that highly dispersed platinum electrodes can provide higher activity and weaken the poisoning effect compared with bulk platinum. Therefore, high surface area and well-developed porosity are essential for a catalyst support to induce high catalytic activity. Except carbon black, which has been commercially used as carbon support, several different carbon materials such as mesostructure carbon, ordered porous carbon were reported as supports for the higher dispersion of the catalysts owing to the high surface area and well-defined porosity of these materials. A uniform porous carbon material has been recently reported by Yu and co-workers[29] via the colloidal crystal templates, as shown in Figure 9.

The silica spheres of 15 ~ 1000 nm in diameter were used as the colloidal crystal templates for mesoporous carbon spheres. With this approach,[29] the uniform ordered porous carbons of well-defined pore structures with tunable pore sizes and high surface area (from 450 to 1000 m^2/g) can be well-obtained by the carbonization of cross-linked phenol and formaldehyde, as can be observed in Figure 10A and 10B). Figure 10C shows the unit cell performance of direct-methanol-fuel-cell at 70°C by the porous carbon-supported catalysts as compared to those of the commercial E-TEK catalysts. The catalyst loadings were 3.0 mg/cm^2 for each of the supported Pt-Ru anode catalysts, and 5.0 mg/cm^2 for Pt black cathode catalyst, respectively. It can be seen that all the porous carbon-supported Pt-Ru

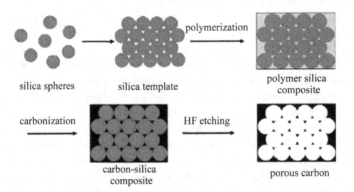

Figure 9.   Schematic synthesis procedure for uniform porous carbons via colloidal crystal template.[29]

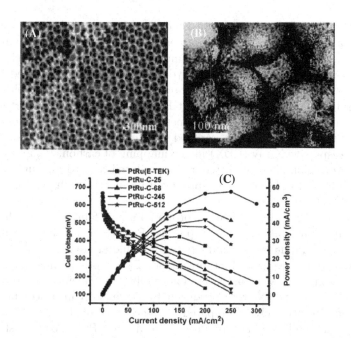

Figure 10. SEM (A) and TEM (B) images of the prepared ordered porous carbons, (C) Voltage and power density responses of porous carbon supported Pt50Ru50 alloy catalysts as compared to that of commercial catalyst (E-TEK) in direct methanol fuel cell. The DMFCs were operated at 70°C.[29]

catalysts exhibit much higher specific activity for methanol oxidation than the E-TEK catalyst by about 5 ∼ 43% in methanol oxidation activity, which was considered to be not only due to the higher surface areas and larger pore volumes, which allow higher degree of catalyst dispersion, but also to highly integrated and interconnecting pore systems with periodic order, which allow efficient gas diffusion. Thereinto, the porous carbon with a 25 nm pore diameter (PtRu-C-25) has the highest performance with a power density of $167 \, \text{mW/cm}^2$ at 70°C, which corresponds to a 43% increase as compared to that of the commercial catalyst.

As above referred, the recent discovery of ordered mesoporous silica provides suitable templates for the synthesis of carbons or other metal oxides with ordered mesoporous structures. Ordered mesoporous carbons such as CMK-1 and CMK-3 have been synthesized by using the three-dimensionally (3D) structured mesoporous silica as a hard template. These ordered mesostructured carbons have a regular array of uniform pores, large pore volume (1 ∼ 2 cm³/g), and high specific surface area (1300 ∼ 2000 m²/g). The first ordered mesoporous carbon (CMK-3)[30] with a faithful replica of the ordered silica has been successfully synthesized using SBA-15 silica as a template. The materials consisted of uniformly-sized carbon rods arranged in a hexagonal pattern. The mesostructured carbons are prepared by infiltrating the pore system with a suitable carbon precursor, such as sucrose, furfuryl, or a phenol-formaldehyde resin, and so on, followed by pyrolysis of the precursor. The silica framework can be dissolved either by NaOH solution or with HF. Pt, or Pt-Ru nanoparticles can be further loaded by impregnating the synthesized mesoporous carbon with a colloidal solution of Pt/Ru nanoparticles prepared by a ethylene glycol method.[31] The energy dispersive X-ray (EDX) image mapping of carbon, Pt, and Ru and the micro-region element distribution exhibited a good and uniform dispersion of Pt/Ru in the carbon support with a metal loading of 15.30% by mass.

Very recently, Yi and co-workers[32] proposed a new fabrication method for producing uniformly-sized mesoporous Pt-carbon catalysts using mesoporous Pt-alumina as a template with a metal source and using poly(divinylbenzene) as a carbon precursor. Two types of mesoporous Pt-alumina templates were prepared under different calcination conditions (PtAl-A and PtAl-N were produced by calcination in a stream of air or nitrogen, respectively). Compared to the conventional impregnation method, this method does not require an additional Pt deposition step on the support because the template itself serves as a metal source. The Pt-carbon catalysts retained a well-developed mesoporosity with a high surface area ($> 590\,\mathrm{m^2/g}$) and a large pore volume ($> 0.45\,\mathrm{cm^3/g}$). Furthermore, the mesoporous Pt-carbon (PtC) catalysts showed a higher metal dispersion and better electro-catalytic performance than the impregnated Pt-carbon (Pt/CMK-3) catalyst.[32]

Figure 11 shows the catalytic performance of the prepared Pt catalysts. Methanol electro-oxidation occurred at about 0.4 V in all Pt carbon catalysts, a typical feature for non-alloyed Pt-carbon catalysts.[33] The maximum current density observed at 0.6 V for the PtC catalysts (especially, PtC-N) was higher than that for the impregnated Pt-carbon (Pt/CMK-3) catalyst.

Mesoporous Precious-Metal Catalysts for Methanol Electrooxidation

The development and characterization of 'poison tolerant' catalysts for the oxidation of methanol has stimulated increased interest in the surface morphology of platinum and platinum-based catalysts. Mesoporous metallic systems appear to be a class of attractive electrocatalyts owing to their high specific surface area and periodic nanostructural texture. For example, platinum produced from the liquid crystal phase was characterized by a specific surface area of approximately $60\,\mathrm{m^2/g}$ (compared to $35\,\mathrm{m^2/g}$ for platinum

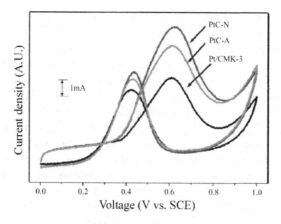

Figure 11. Cyclic voltammograms of PtCs and Pt/CMK-3 catalysts on a glass carbon disk electrode (working electrode) in H_2SO_4 (0.5 M) containing CH_3OH (1 M). The Pt loadings on working electrode were identical in all the cases (scan rate = 20 mV/s).[33]

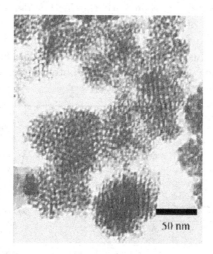

Figure 12. TEM image of mesoporous Pt nanoparticles.[34]

black) and large particle sizes, as shown in Figure 12.[34] The combination of high surface areas, uniform pore diameters, and large particles makes the metallic system considerably interesting for applications in catalysis, batteries, fuel cells, electrochromic windows, and sensors.[35] It was found that the intrinsic activity of platinum for oxygen reduction was more favorable compared to that of commercial platinum. Furthermore, the mesoporous platinum showed potential-dependent tolerance to CO poisoning and high electrocatalytic activity towards formic acid oxidation.[36]

In addition, the production of binary or ternary Pt-base alloys that combine high specific surface areas, controllable pore diameters and uniform distribution of metal species is expected to impact on a wide range of catalytic processes. Recently, a mesoporous Pt-Ru alloy has been successfully synthesized by chemical co-reduction of hexachloroplatinic acid and ruthenium trichloride within the aqueous domains of a normal topology hexagonal mesophase of oligoethylene oxide surfactants using Zn as the reductant.[36] The resultant material containing mesoporous structures has a high specific surface area and has been examined as an electrocatalyst towards the electrooxidation of methanol using cyclic voltammetry and chronoamperometry. Figure 13 shows the results of methanol catalytic oxidation. It can be seen that the oxidation current increases considerably at increased potential until a current peak at about 0.75 V. Further increases over the potential lead to a rapid decay in the oxidation current. Upon reversing the scan at 0.85 V, the current further decays until 0.80 V at which point a small peak is seen with the further decreases in potential. The current density of the forward-scan current peak is calculated to be $700\,mA/cm^2$, within a factor of two to three of the peak current density value assuming a diffusion-controlled irreversible six-electron oxidation reaction under the same conditions. When the potential scan is reversed at

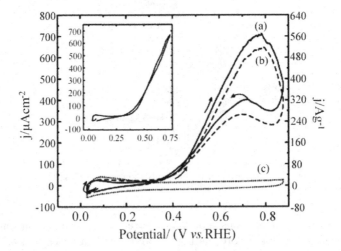

Figure 13.   Cyclic voltammograms for a mesoporous Pt-Ru electrode in 0.5 mol/dm H_2SO_4 solution containing 0.5 mol/dm methanol at 60°C (dE/dt = 50 mV/s). (a) first scan; (b) second scan; (c) cyclic voltammogram in the absence of any methanol. Inset: cyclic voltammogram under the same conditions over zero to 0.7 V.[36]

0.70 V, no peak or hysteresis loop in the current response is observed in the inset, which suggests that there is virtually no accumulation of poisons on the electrode surface when the potential is more negative than 0.70 V over the timescale of this experiment. It is believed[36] that methanol is oxidized mainly via a direct pathway to soluble products on the mesoporous Pt-Ru electrode and steady-state kinetics can be attained even at low potentials. The electrocatalytic activity of the mesoporous Pt-Ru towards methanol oxidation is more favorable compared to an ultrafine Pt-Ru electrocatalyst with similar bulk composition.

Synthesis of Mesoporous Titania and Its Photocatalytic Applications

Due to their low cost, strong redox power and high chemical/photocorrosion stability, titanium dioxide (TiO_2) based materials have been extensively used and further studied in the past decades to address the increasing air and water pollution problems through photocatalytic processing, especially for decomposing organic contaminants with low levels. Generally, under UV irradiation from either artificial or solar sources with energy larger than the material's band gap, electron-hole pairs are generated in the wide-band-gap semiconducting TiO_2 materials because of the electron transition from the valence band to the conduction band. As shown in Figure 14, in the pH 7 solution, the redox potentials for photogenerated holes (h^+) and photogenerated electrons (e^-) are +2.53 V and −0.52 V versus the standard hydrogen electrode (SHE), respectively.[38] Then interfacial electron-transfer

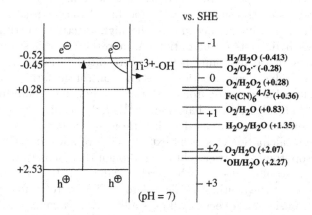

Figure 14. Schematic diagram showing the potentials for various redox processes occurring on the TiO_2 surface at pH 7.[38]

reactions with adsorbed organic compounds lead to their decomposing to CO_2 and H_2O through radical chain reactions or other mechanisms; and consequently purification of polluted air or wastewater. Since the catalytic reaction is a surface process, high surface area resulting from smaller particle size or porous structures means better catalytic efficiency. Therefore, with the discovery of ordered mesoporous materials,[5,6] recently the synthesis of titanium dioxide based mesoporous materials has also attracted great attention due to their higher surface areas and easy-recovery after reaction compared to that of the ordinary TiO_2 based materials; this probably leads to higher photocatalytic efficiency in the purification of polluted air or wastewater. In this part, preparation of mesoporous titanium dioxide with amorphous or crystallized frameworks and titanium dioxide containing mesoporous materials and their photocatalytic applications are presented.

Synthesis of Pure Titanium Dioxide Mesoporous Materials and Their Photocatalytic Applications

Mesoporous TiO_2 with amorphous framework

Towards mesoporous materials synthesis with surfactants as soft templates, the interactions among inorganic precursors, organic species, and solvent/co-solvent in a bulk solution cooperatively determine the ordering degree and symmetry of the final structures. Thus, since 1992, although ordered mesoporous silica materials with adjustable pore sizes and diversity of pore structures have been reported through proper choices of many process parameters,[5,6,39–41] (such as surfactant species, reactant molar ratios, pH values, reaction temperature and reaction time, and so on), the successful preparation of transition metal oxide based mesoporous materials, such as TiO_2, is difficult because of the high reactivity of titanium sources towards hydrolysis/condensation. As a result, poorly structured materials with a dense inorganic network were often formed. To lower the precursor hydrolysis/condensation rates and control the formation processes of non-silica based mesoporous materials, stabilizing agents such as acetylacetone,[42] glycolate,[43] triethanolamine,[44] alkyl phosphate[45] and primary amine[46] surfactants were used with little water or in the absence of it. Then through calcination or solvent extraction to remove the surfactant templates, mesoporous TiO_2 were obtained, though their structural ordering and thermal stability needs to be further improved. It is noted that at this stage most of the synthesized mesoporous TiO_2 were of amorphous frameworks or consisting of very small crystallites, which could hardly be recorded in X-ray Diffraction (XRD) or Electron Diffraction (ED) patterns. However, it is known that the basic requirements for photoactive TiO_2

Table 2. Characterization and reactivity results for titania materials.[45]

Sample	Surface Area (m^2/g)	Band Gap (eV)	Quantum Yield[a]
meso-TiO$_2$ (473 K)	712	3.19	0.0026
meso-TiO$_2$ (973 K)	90	3.25	0.0089
meso-TiO$_2$, extracted[b]	603	3.15	0.010
ns-TiO$_2$, as-synthesized		3.24	0.0094
ns-TiO$_2$ (873 K), 1 h	39		0.022
ns-TiO$_2$ (873 K), 3 h	36	3.22	0.038
ns-TiO$_2$ (873 K), 12 h	28	3.26	0.27
anatase (Aldrich)	9	3.28	0.41
Degussa P25	50	3.22	0.45

[a]Quantum yield is defined as the molecules of acetone formed per incident photon.
[b]Surfactant was extracted three times with 8% nitric acid in ethanol.

materials are their high crystallinity. The defects present in the amorphous or small crystallites are the recombination centers for the photogenerated electron-hole pairs, which would cause the decrease of quantum yields. Table 2 shows the structural properties and photocatalytic results of acetone for mesoporous TiO$_2$,[45] nano-sized TiO$_2$ prepared without surfactant template, and commercial TiO$_2$ powder Degussa P25, a standard titania P25. It is obvious that the quantum yields of the mesoporous samples (0.00xx) are several orders of magnitude lower than that of nano-sized TiO$_2$ (0.27) and of P25 (0.45), although the mesoporous one possesses the larger surface area. Thus, high-surface-area mesoporous titania with controllable crystalline framework is required to achieve its photocatalytic applications.

Mesoporous TiO$_2$ with crystallized framework

Recently, with the development of the so-called evaporation-induced self-assembly (EISA) process,[47] highly organized mesoporous TiO$_2$ materials with nano-sized anatase walls have been prepared reproducibly using poly(ethylene oxide) (PEO)-based templates.[48,49] The resulting materials not only possess high surface area and high thermal stability but also exhibit different and tailored mesostructural symmetry (wormlike, cubic *Im3m*, 2D-hexagonal *p6m*). Figure 15 is the schematic diagram for this whole process, and various characterization techniques accompanying with each step describe the time evolution of the mesostructure and consequently clarify the formation mechanism during EISA.[49] Starting from TiCl$_4$ ethanol solution, the Extended X-ray Absorption Fine Structure (EXAFS) and UV-Vis experiments confirmed the appearance of the intermediated product of metallic chloroethoxide precursors [TiCl$_{4-x}$(OEt)$_x$] with x $\approx$ 2. Further, to this highly acidic system, controlled quantities of water were slowly added to

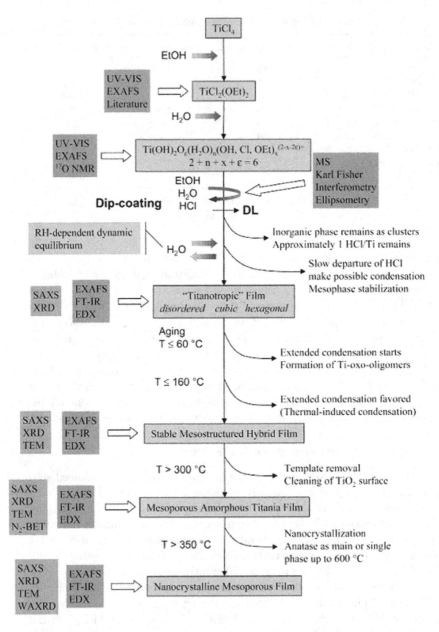

Figure 15. Sequential events in the formation of ordered mesoporous TiO₂ films.[49]

induce the hydrolysis of the inorganic moieties. In a certain range of molar ratios of m_{H_2O}/m_{Ti} (> 4), clear sols containing hydrolyzed hydrophilic moieties with proper polymerization degrees were formed. The dip-coating process was conducted in a circumstance with controlled relative humidity to prepared mesostructured TiO_2 films. *In situ* characterization with Mass Spectroscopy (MS), synchrotron radiation of the Small-Angle X-ray Scattering (SAXS) and interferometry reflected the mesostructure evolution (including the disordered film formation first and then the occurrence of the disorder-to-order transition) with fast evaporation of solvent ethanol and progressive concentration of water, HCl, and the other non-volatile components. Post-treatments under controlled atmosphere[50] or with gradual heating[49] would continuously enhance the condensation degree of TiO_2 framework with the slow elimination of water and HCl and eventually result in the formation of more thermally stable structures. After calcinated at higher than 300°C to remove surfactant template and thermally treated at elevated temperature (up to 600°C), mesoporous TiO_2 thin films with high surface area and crystallized frameworks were obtained (Figure 16). Using a similar process combined with more careful choice of surfactant species or thermal treating programs, crystalline mesoporous TiO_2 materials with larger pore size ($\sim 10\,nm$)[51] and higher thermal stability (up to 700°C)[52] have been achieved. Instead of common Ti source of chloride or alkoxide, "pre-synthesized" TiO_2 nanoparticles or other nano-building blocks were also considered to be used to alleviate the system sensitivity to the process conditions.[53-55] However, the structural ordering and the thermal stability of the resulted products needs to be further improved.

Considering the inorganic-inorganic interplay between two or more inorganic precursors, a new and general method for the preparation of non-siliceous mesoporous materials has been reported based on the "acid-base pair" principle.[56,57] To the synthesis of mesoporous TiO_2, according to the relative acidity and alkalinity of inorganic precursors on solvation, an "acid" precursor (titanium alkoxide) and a "base" counterpart (titanium chloride) were chosen. In the amphoteric solvents, such as ethanol or tetrahydrofuran, a self-adjusted sol-gel synthesis took place without any extra reagents. Moreover, since metal alkoxide served as extra oxygen donors, the Ti-O-Ti bridge partially originated from the condensation between Ti-Cl and Ti-OR and the cross-linkage of inorganic framework in the resulted materials were strengthened. After calcined at 350°C to remove the surfactant template, polycrystalline diffraction rings were observed in the ED pattern, which confirmed the semi-crystallized wall structures in the synthesized TiO_2. However, the material thermal stability at elevated temperature was not mentioned.

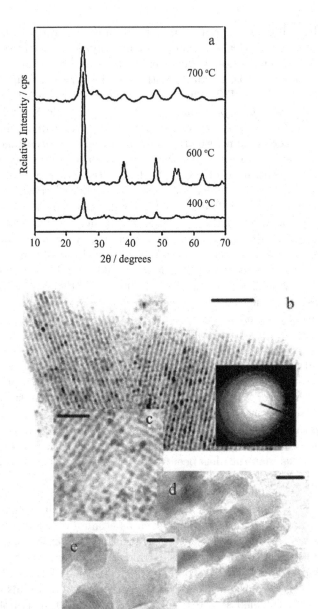

Figure 16. (a) XRD patterns of F127-tempalted TiO_2 mesoporous films calcined at the temperature indicated. (b–e) HRETM images of 600°C treated TiO_2 films of increased magnifications; the inset in part (b) is an ED pattern characteristic of anatase. Scale bars are (b) 100 nm, (c) 50 nm, (d) 10 nm, and (e) 5 nm.[49]

Photocatalytic behavior of mesoporous TiO_2
with crystallized framework

Under 365 nm UV irradiation, the photocatalytic activity of EISA-prepared mesoporous TiO_2 nanocrystalline films was evaluated by the photodecomposition of acetone,[58] a widely used industrial and domestic chemical. As shown in Table 3, the photocatalytic activity increases with the increase of thermal treatment temperature, which could be explained with the gradual crystallization of titania framework. For sample MT500 calcined at 500°C, two times higher specific photocatalytic activity than that of the conventional TiO_2 film could be obtained, which was resulted from the large surface area (85 m^2/g versus 9.1 m^2/g) and its three-dimensionally connected pore structures of the mesoporous catalysts.

Researchers studied the effect of solution conditions on the photocatalytic activity of the synthesized mesoporous titania.[50] As shown in Table 4, using the photodecomposition of rhodamine 6G as model reaction, the photocatalytic activity of mesoporous TiO_2 synthesized under different conditions had been compared. It is obvious that with the similar or even lower surface areas, TiO_2-CTABr series samples with larger pore size (> 3.0 nm) possess better photocatalytic activity than TiO_2-HDA-B-T sample with smaller pore size (2.0 nm), which reflects the accessibility of the anatase phase for rhodamine 6G and in turn determines the photocatalytic activity of the mesoporous titania. The same reasons could be used to explain higher photocatalytic activity of sample TiO_2-HDA-A-T than that of TiO_2-HDA-B-T, because under acidic conditions, the slower condensation of the titania precursor lead to larger pore-sized mesostructures and consequently better accessibility for photocatalysis.

Table 3.   Photocatalytic activity of mesoporous TiO_2 thin films sintered at different temperatures and their comparison with a conventional TiO_2 thin film[58,a].

Sample	Mass	Degradation[b]	Rate Constant K (min^{-1})	Specific Photoactivity	
	(mg)	Rate (%)		(mol/g/h)[c]	(mol/m^2/h)[d]
MT300	13.2	1.6	$2.6*10^{-4}$	$4.0*10^{-4}$	$3.8*10^{-4}$
MT400	12.5	9.9	$16.4*10^{-4}$	$2.6*10^{-3}$	$2.3*10^{-3}$
MT500	12.2	11.6	$19.4*10^{-4}$	$3.1*10^{-3}$	$2.7*10^{-3}$
MT600	11.8	7.9	$10.8*10^{-4}$	$1.8*10^{-3}$	$1.5*10^{-3}$
TiO_2-500[e]	8.8	4.8	$7.6*10^{-4}$	$1.9*10^{-3}$	$1.2*10^{-3}$

[a]The area of the substrate covered by the TiO_2 monolayer was 140 cm^2. [b]Average degradation rate of acetone after one hour of photocatalytic reaction. [c]Acetone degradation amount per unit mass catalyst after one hour of photocatalytic reaction. [d]Acetone degradation amount per unit film surface after one hour of photocatalytic reaction. [e]Data from Reference[59].

Table 4. Physical characterizations of NH_3-treated mesoporous titania synthesized with HAD and CTABr in Acidic (A) and Basic (B) medium. The calculated rate constants derived from the first-order photocatalytic reaction of rhodamine 6G photodecomposition.[50]

Sample[a]	S_{BET} (m^2/g)	V_p (cm^3/g)	D_{BJH} (Å)	k (min^{-1})
TiO$_2$-HDA-A-T	258	0.20	26.8	0.016
TiO$_2$-HDA-B-T	195	0.11	20.0	0.007
TiO$_2$-CTABr-A-T	340	0.28	34.0	0.029
TiO$_2$-CTABr-B-T	180	0.25	32.0 (broad)	0.029

[a]T stands for NH_3-treated. HDA and CTABr are the abbreviate of hexadecylamine and cetyltrimethylammoniumbromide, respectively.

Solvent effect during EISA process to prepare mesoporous TiO_2 materials was emphasized recently through several experiments.[60,61] Because of the higher hydrophobicity of 1-butanol than usually-used ethanol, the use of 1-butanol could enhance the microphase separation between the template and the inorganic precursor and allows for the formation of a robust, structurally well-ordered mesoporous TiO_2 material with nanocrystallized frameworks. It was further found that by just simply varying the solvent and co-solvent (methanol, ethanol, 1-butanol, or 1-octanol), stable mesoporous titania with pure anatase, pure rutile, bicrystalline (anatase and rutile) with controlled phase composition, and tricrystalline (anatase, rutile, and brookite) frameworks could be synthesized.[62] As different phase structure or phase mixtures of TiO_2 showed varied photocatalytic activity for different kinds of organic pollutants, this report would open up new directions of catalyst design for future TiO_2 based photocatalytic materials with higher quantum yields. As an example, mesoporous rutile TiO_2 powder with an average pore diameter of 2.6 nm and surface area of 174.5 m^2/g showed better photocatalytic results than that of nano-anatase titania with surface area of 313 m^2/g or nano-rutile titania with surface area of 115.6 m^2/g, when the gas-phase photooxidation of a mixture of benzene and methanol was used as the probe reaction.[63]

Photocatalytic activity of specially structured mesoporous TiO_2

Hierarchically macro/mesoporous titania is another interesting material for its photocatalytic application. The existence of additional macroporous structure not only accelerates the mass transport of organic pollutants to the catalytic titania phase and the discharge of photo-decomposed products, but also increase the photo-absorption efficiency and consequently the photocatalytic efficiency, because the extinction of light in semiconductors follows the exponential law $[I = I_0 \exp(-\alpha l)]$, where l is the penetration distance of the light and α is the reciprocal absorption length. Combining sol-gel chemistry, multiscale templating approaches, aerosol processing, and specific treatments, spheres of controlled diameter, made of a periodically-organized mesoporous titania crystalline network that surrounds spherical macropores have been prepared.[64] Just using single surfactant template (e.g. Brij56 or P123) with or without ultrasonic treatment,[65,66] hierarchical macro/mesoporous titania was also obtained (Figure 17), in spite of its decreased mesostructural ordering. Figure 18

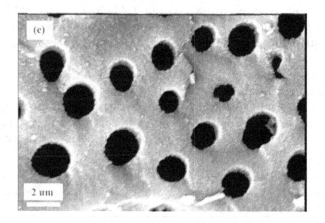

Figure 17.   SEM images of hierarchical macro/meso-porous TiO_2 calcined at 500°C.[66]

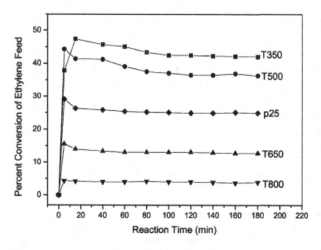

Figure 18.   Percent conversions of ethylene on hierarchical macro/meso-porous TiO_2 calcined at 350, 500, 650 and 800°C, and Degussa P25.[66]

shows the ethylene conversion on the series hierarchical macro/mesoporous TiO$_2$ and Degussa P25 with 254 nm UV irradiation.[66] The photocatalytic conversion rates with sample T350 and T500 are 42% and 37%, respectively, which are larger than $\sim$ 26% of commercial TiO$_2$ powder Degussa P25.

In addition to the phase structure, the photodecomposition catalytic activity of mesoporous TiO$_2$ on organic pollutants relies on its surface area as well. On the other hand, a novel photocatalytic activity of mesoporous TiO$_2$ on its structural advantage was reported recently, which was labeled as "stick-and-leave" transformation.[67] Photocatalytic transformation of various substrates could be found in Table 5. The letter "D" represents the distribution ratio of substrates in the solution and adsorbed on the catalysts and the "mTiO$_2$" and "nTiO$_2$" refer to mesoporous and non-porous TiO$_2$ materials, respectively. The transformation of benzene into phenol, one of the most difficult synthetic reactions, gives the typical results. Because benzene molecules are well-adsorbed on mTiO$_2$ ($D = 0.64$) and no phenol adsorption happens, it reduces the by-reaction rates and shows very high selectivity to phenol ($> 80\%$). In contrast, nTiO$_2$ catalysts gave very low product yield (1 to approximately 2%) and selectivity

Table 5. Photocatalytic transformation of various specially structured mesoporous TiO$_2$ substrates.[67]

Run	Catalyst	Reactant	D	Conv (%)	Product	D (%)	Yield (%)	Select (%)
1	nTiO$_2$(100)		0.17	90		0.01	39	35
2	nTiO$_2$(58)		0	82		0	28	34
3	mTiO$_2$(61)	11	0.11	89	1	0	61	72
4	nTiO$_2$(100)		0.10	91		0.01	16	18
5	nTiO$_2$(58)		0	87		0	11	13
6	mTiO$_2$(61)	14	0.12	90	6	0	65	72
7	nTiO$_2$(100)		0.05	60		0	26	23
8	nTiO$_2$(58)		0	71		0	12	17
9	mTiO$_2$(61)	9	0.09	84	4	0	75	89
10	nTiO$_2$(100)		0.24	26		0.01	2	8
11	nTiO$_2$(58)		0.28	16		0	1	6
12	mTiO$_2$(61)	16	0.64	23	1	0	19	83
13[b]	mTiO$_2$(61)		0.64	42		0	34	81
14[c]	mTiO$_2$(61)		0.64	10		0	8	80

[a]Reaction conditions were: reactants, 20 μmol; photoirradiation time, two hours; catalyst, 10 mg; buffered (pH 7) aqueous solution, 10 mL; temperature, 313 K. [b]Photoirradiation time, six hours. [c]Reactants, 0.5 mmol (which is not fully dissolved in the aqueous solution).

($< 10\%$), resulting from lower D value of benzene (approximately 0.2 times) and higher D value of phenol compared to that of $m\mathrm{TiO_2}$. Through enhancing the reaction selectivity, this process would address the environmental problem of chemical industries in the pollution sources.

Synthesis of Titanium Dioxide Containing Mesoporous Materials and their Photocatalytic Applications

Besides the above-mentioned pure $\mathrm{TiO_2}$ mesoporous materials, titanium dioxide containing mesoporous materials represent wider range for the choice of new photocatalysts, which includes nano-sized $\mathrm{TiO_2}$-particle-incorporated silica mesoporous materials,[68–72] mesostructured $\mathrm{TiO_2}$-other semiconductor (or noble metal) composites,[73,74] metal-ion-doped mesoporous $\mathrm{TiO_2}$ materials,[70,75] and surface-modified mesoporous materials,[76] and so on. In fact, in view of the complexity during the preparation of pure $\mathrm{TiO_2}$ mesoporous materials and the synergistic effect of structural and functional characters of different components in titanium dioxide containing mesoporous composites, the composite is more advantageous for practical applications.

TiO₂ supported on mesoporous silica

Using ordered mesoporous silica materials as supports, nano-sized $\mathrm{TiO_2}$ particles have been incorporated by many post-synthesis methods. Depending on the support structure, the size and amount of incorporated $\mathrm{TiO_2}$ particles could be varied. In comparison with ordinary anatase $\mathrm{TiO_2}$, they showed better photocatalytic activity towards the photodegradation of rhodamine 6G dye,[68] acetophenone,[72] and cyanide,[69] and so on. Figure 19 shows the comparison between kinetics of cyanide photooxidation in the presence of series $\mathrm{TiO_2}$-SBA-15 and $\mathrm{TiO_2}$-commercial $\mathrm{SiO_2}$ composite samples.[69] Although at lower $\mathrm{TiO_2}$ content both samples show similar activity, the rate constant increase of $\mathrm{TiO_2}$-SBA-15 sample is more significant with the content increase of $\mathrm{TiO_2}$. This is caused by the pore limitation of ordered SBA-15 supports, which controls the size and size distribution of incorporated $\mathrm{TiO_2}$ particles.

Mesoporous TiO₂ incorporated with metal particles or other metal Ions

Since the incorporated metals would enhance electron migration from the semiconductor surface and reduce the recombination probability of

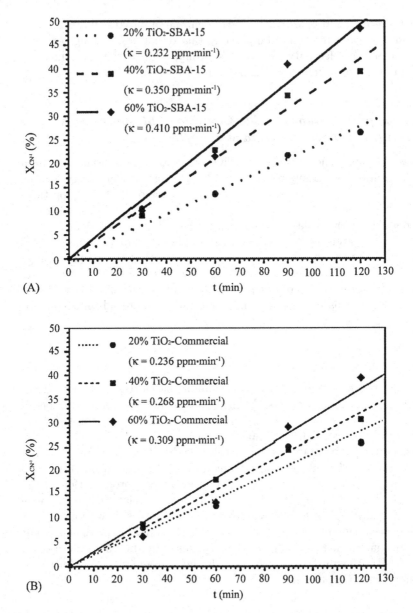

Figure 19.   Kinetics of cyanide photooxidation in the presence of (A) 20–60% TiO₂-SBA-15 samples; and (B) 20–60% TiO₂-commercial SiO₂ materials. Insets are the respective initial rate constants normalized to the titania content.[69]

photogenerated electrons and holes, silver nanoparticles incorporated meso-porous TiO_2 catalyst were prepared using wet impregnation followed by heat treatment, and its photocatalytic activity towards the decomposition of stearic acid were studied recently.[73] Figure 20 shows the photocatalytic activity of three anatase samples: 400°C calcined cubic mesoporous tita-nia with silver (red circles) or without silver (green triangles) and 600°C calcined mesoporous titania which has no longer mesostructured ordering (blue squares). It can be seen that Ag-containing catalysts has an appar-ent higher initial activity due to the photogenerated electron transfer from titania to silver. However, after 17 minutes, its photocatalytic activity is lower than that of the sample without silver. Possible reasons are suggested to be related to the stearic acid loading amount and deactivation of silver-containing sample.

La^{3+}-doped mesoporous titania with a highly crystallized framework and long-range order have been prepared by using nano-anatase particles as nano-building units.[75] In this process, La^{3+} plays two important roles. First, it works as the mesostructure stabilizer by restraining the growth of nano-anatase particles, especially in the direction perpendicular to wall. Second, it increases photocatalytic activity by enhancing the quantum yield. Through the degradation experiment of methyl orange dye under UV light irradiation, 0.25 at% of La^{3+} was determined as the optimal doping amount, at which the mesoporous titania had the highest photocatalytic activity (shown in Figure 21). With higher La^{3+} doping content, the decrease of photocatalytic activity may be caused by the accompanying increase of Ti^{3+}

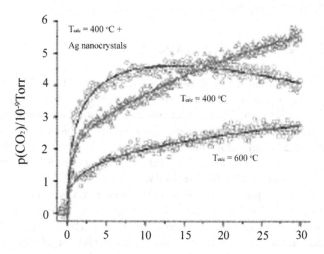

Figure 20.   Results from the photocatalytic experiments showing the generation of CO_2 as a function of time upon UV illumination of mesoporous titania with stearic acid.[73]

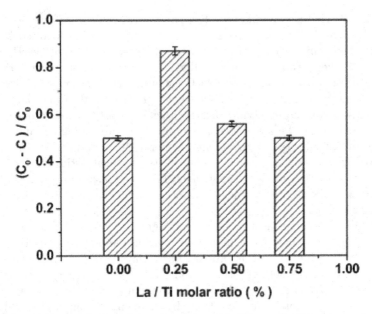

Figure 21. Photocatalytic activity of samples with different La/Ti molar ratio evaluated by the degradation of methyl orange solution illuminated by a UV source for 60 minutes.[75]

species, which act as photo-hole traps and enhance the recombination of photogenerated electrons and holes and consequently decrease the quantum yield of the catalytic process.

Mesoporous-Based Materials as Adsorbents

With the fast development of modern industry, varieties of toxic inorganic and organic chemicals as industrial wastes are discharged to the environment, causing serious water, air, and soil pollution. Adsorption is considered to be a safer and less expensive remedy to handle this problem than precipitation, ion exchange, and membrane filtration. An ideal adsorbent should have uniformly accessible pores, an interlinked pore system, a high surface area, and physical and/or chemical stability.

As stated above in this chapter, during the past few years, mesoporous materials are of great interest as adsorbents for heavy metals, dyes and harmful gases and so on, owing to their ultra high surface area, uniform and tunable pore diameter, and their tunable pore surface and/or pore wall framework characters with different functional groups for use as adsorbents. Furthermore, they allow molecular accessibility to large internal surface areas and volumes. Mesoporous adsorbents have shown good adsorption capacity and been proved to have great advantages (higher adsorption capacity or selectivity) over traditional adsorbents, e.g. resins, activated-charcoal, and so on.

Heavy and Other Toxic Metals

The removal of heavy metal ions from wastewater has been an extensive industrial research subject. Several kinds of materials, such as ion-exchange resins, activated-charcoal, and ion chelating agents immobilized on inorganic supports have been used as heavy metal adsorbents. However, these materials show many drawbacks because of their wide distribution of pore size, heterogeneous pore structure, low selectivity and low loading capacities for heavy metal ions. Research on environmental application of mesoporous silicas as novel heavy metal adsorbents has also been carried out recently. Mesoporous silicas such as MCM-41, HMS, SBA-15, and SBA-1 have been functionalized by several groups to make the materials able to interact strongly with metallic cations, and particularly mercury or metallic anions such as chromate and arsenate. There are two ways, including co-condensation and post-synthesis method, to introduce functional groups into the mesoporous matrix. In general, the former route leads to a uniform incorporation and dispersion of functional groups in their framework, while the latter one gives rise to functional layer on their pore wall surface. The latter leads to the loss of surface area and reduction of pore diameter and pore volume (as high as 60 to 84%).[77] The degree of pore volume reduction could be a result of the nature of the ligand, such as the size, and the ligand loading. Moreover, materials prepared by the immobilization method appear to have non-uniformly distributed functional groups, whereas the materials prepared by the former one-step sol-gel co-condensation method could obtain mesoporous functional adsorbents with high density and uniformly distributed functional groups and long life cycles.

In 1997, Liu and co-workers functionalized mesoporous silica with mercaptopropylsilane monolayers, which covalently bound to and closely packed on their pore surface with highest coverage up to 75%.[78,79] They used mercaptopropylsilyl functionalized large pore MCM-41 as adsorbents and obtained a remarkable adsorption capacity of 600 mgHg/g. They washed the mercury-loaded adsorbents with a concentrated HCl (12.1 M) solution to regenerate them and 100% mercury was removed and the adsorbents retained a capacity of 210 mgHg/g in the second cycle.[78] The selectivity of the adsorbents was tested in their research work. Mercier and co-workers used 3-mercaptopopyltrialkoxysilane (MPTS) modified HMS and MSU mesoporous silica samples by either post-grafting or co-condensation methods.[80–82] The authors compared the structure and adsorption properties of these two kinds of composites. Samples prepared by either one-step co-condensation or grafting method show high selectivity for Hg^{2+} and the Hg^{2+} adsorption capacity is proportional to the thiol content (Hg/S in mol = 1). Lee and co-workers synthesized a new kind of bifunctional porous silicas via the co-condensation of TEOS, MPTS and 3-aminopropyltrimethoxylsilane (APTS).[83] This adsorbent also has highly selective adsorption for Hg^{2+}, but the existence of amino groups decreases their metal adsorption capacities.

Jaroniec and co-workers reported a propylthiourea modified mesoporous silica for mercury removal.[84,85] This material was prepared via a two-step modification by attachment of an aminopropyl group and its subsequent conversion into a thiourea ligand. About 1.5 mmol/g of 1-benzoyl-3-propylthiourea groups was attached to the silica surface. The maximum loading of mercury ions from aqueous solution for this material was approximately. 1.0 g Hg^{2+}/g, double the capacity of the previously known thiol-containing samples. The authors claimed that the increase of adsorption capacity arose from the multifunctional nature and the deprotonation susceptibility of the 1-benzoyl-3-propylthiourea ligand. The adsorbent can be regenerated under mild conditions via washing with slightly acidified aqueous thiourea solution, and over 70% of the initial adsorption capacity retained.

Thioethers have the same affinity for heavy metal ions as thiols. Shi and co-workers prepared a thioether functionalized organic-inorganic mesoporous composites by a one step co-condensation of tetraethoxysilane (TEOS) and (1,4)-bis(triethoxysilyl)propane tetrasufide (BTESPTS) with tri-block copolymer poly(ethylene oxide)-poly(propylene oxide)-poly(ethylene oxide) (P123, $EO_{20}PO_{70}EO_{20}$) as template.[86] The new materials showed higher selectivity and adsorption capacity for Hg^{2+} (as listed in Table 6 and 7) than mesoporous adsorbents modified with

Table 6. Physichemical properties and Hg^{2+} adsorption capacity of the thioether functionalized organic-inorganic mesoporous composite samples.[86]

Sample	A_{BET}^{a} (m^2/g)	D_{BJH} (nm)	S Content (mmol/g)	Hg Adsorption Capacity (mg/g)	Hg/S
S2	634	5.02	1.50	627	2.08
S6	654	3.74	2.83	1280	2.25
S8	576	3.47	3.48	1860	2.66
S10	532	3.35	3.76	2330	3.09
S12	253	3.26	4.02	2540	3.15
S15	242	3.13	4.24	2710	3.19

^{a}A_{BET}, BET surface area; D_{BJH}, BJH pore diameter caculated from the desorption branch.

Table 7. The concentrations (ppm) of metal contaminants in wastewater solutions before and after treatment with thioether functionalized organic-inorganic mesoporous composite samples.[86]

Solution	Hg^{2+}	Pb^{2+}	Cd^{2+}	Zn^{2+}	Cu^{2+}
No treatment	10.76	13.07	12.43	6.79	8.78
After treatment with S2	3.80	12.85	12.35	6.76	8.75
After treatment with S6	0.34	12.64	12.32	6.76	8.74
After treatment with S12	0.23	12.47	12.23	6.51	7.16

3-mercaptopoyltrialkoxysilane (MPTS) or ordinary polythioether chelating resins with the equal amount of S content. Among them, sample S15 (15% BTESPTS in mol) shows an extraordinarily high adsorption capacity up to 2700 mg Hg^{2+}/g, which is about one order higher than that of ordinary polythioether chelating resins. The authors attributed these results to the special stereo-coordination chemistry of S with Hg^{2+} when the thioether groups had been incorporated into the mesoporous materials framework. This result also confirms the reactivity of bridging organic groups within the frameworks and its difference from that of terminal groups located in the channel space. A similar result was also reported by Mercier and co-workers with the substitution of ethylenediamine group for thioether group in the framework.[87] A preferential uptake for Cu^{2+} ions, compared with Ni^{2+} and Zn^{2+} ions, was observed.

Dai and co-workers improved the adsorption selectivity of specific cations by a novel surface imprinting method.[88] Two imprinting complex precursors $[Cu(apts)_xS_{6-x}]$ and $[Cu(aapts)_2S_2]^{2+}$, ($x = 3 - 5$; S = H_2O or methanol; apts = $H_2NCH_2CH_2CH_2Si(OMe)_3$; aapts = $H_2NCH_2CH_2NHCH_2CH_2CH_2Si(OMe)_3$) were post-grafted onto the mesoporous silica. The residue was washed with copious amount of 1 M HNO_3 to remove all the Cu^{2+} ions and any excess ligands. The Cu^{2+} adsorption capacity of these imprint-coated mesoporous composite reaches 0.1 mmol/g, which is ten times higher than that of samples prepared by the conventional grafting method as above mentioned. The authors attributed this to the stereochemical arrangement of ligands on the surface of mesopores, which optimizes the binding of targeted metal ions. However, this absorbant suffers from high cost and poor regenerability.

Most recently, Sayari and co-workers post-synthesized pore expanded MCM-41 silica using *N,N*-dimethyldecylamine and obtained material **A** with open pore structure and readily accessible amine groups.[89] However, the surface area of A was as low as 89 m^2/g due to the large molecules blocking the pore channels, which limited their adsorption capacities. The highest adsorption capacity was only 106 mg Cu^{2+}/g. After the selective removal of the occluded amine they obtained material **B**, a highly porous, hydrophobic material with large surface area and pore volume. Experiment found materials **B** to be a fast, sensitive, recyclable adsorbent for organic pollutants. This adsorption is still efficient in the presence of very dilute solutions of adsorbates.

Liu and co-workers reported the synthesis of a composite with Cu(II) ethyenediamine terminated silane immobilized on mesoporous silica and used it as an efficient and selective anion traps for both arsenate (AsO_4^{3-}) and chromate (CrO_4^{2-}).[90] These materials can remove arsenate and chromate almost completely from solutions containing more than 100 ppm toxic metals anions. The capacity of these materials is higher than 120 mg (anion)/g.[90]

Tavlarides and co-workers synthesized a high-capacity phosphonic acid functionalized adsorbent by co-condensation of tetraethoxysilane (TEOS) and trimethoxysilylpropyl diethylphosphonate (DEPPS).[91] They applied these composites to adsorb chromium(III) from aqueous solutions. Experiment shows an equilibrium chromium(III) adsorption capacity of 82 mg/g at pH 3.6 and rapid adsorption kinetics. Column adsorption at a high space velocity of 156 h^{-1} is achieved with a minimum effluent chromium concentration of 0.01 mg/L. The chromium-loaded adsorbent can be regenerated using 12 M HCl. Kinetics results show that near equilibrium is reached in as short as two minutes of contact time. Column tests show dynamic operation at high throughput with sharp breakthrough curves and effluent concentrations (< 0.02 mg/L) far below the USEPA MCL level (0.1 mg/L) of total chromium.

Yi and co-workers reported a post-hydrolysis method to prepare mesoporous alumina (MA) with a wide surface area ($307 m^2/g$), uniform pore size (3.5 nm), and a sponge-like interlinked pore system.[92] Batch experiment of these materials found that their maximum uptake for As (V) was 121 mg/g and was seven times higher than that of conventional active alumina (AA). The adsorption was also kinetically rapid with the complete adsorption in less than 5 hours as compared to about two days to reach half of the equilibrium value by AA. NaOH of 0.05 M was found to be the most suitable desorption agent. More than 85% of the arsenic adsorbed to the MA was desorbed in less than one hour. They also tested several other activated-alumina adsorbents with different pore properties and concluded that a uniform pore size and an interlinked pore system, rather than the surface area of the adsorbents, were the key factors.

Park and co-workers synthesized lanthanum impregnated SBA-15 materials and applied them to remove arsenate from water.[93] At the arsenate concentration of 0.667 mmol As/l, the adsorption capacity of 50% lanthanum-impregnated SBA-15 by weight (designated as La50SBA-15) was 1.651 mmol As/g (123.7 mg As/g), which was approximately ten and

14 times higher than La(III) impregnated alumina and La(III) impregnated silica gel, respectively.

Dyes and Other Organic Pollutions

There has been an increasing scientific concern about the hazardous effects of colored dyes in waste waters from dyeing process in textile industries. The release of the colored dyes into the ecosystem is a dramatic source of esthetic pollution, causting eutrophication and perturbations in aquatic life. Some azo dyes and their degradation products such as aromatic amines are highly carcinogenic. The traditional treatments (e.g. flocculation, sedimentation, and filtration) prove to be insufficient to purify a significant quantity of wastewaters. Adsorption technology has become one of the most effective technologies for the removal of dyes. Practically, adsorbents for dye removal require the following potential advantages: (1) a large accessible pore volume, (2) hydrophobicity, (3) high thermal and hydrothermal stability, (4) no catalytic activity, and (5) easy regeneration. In the past few years, mesoporous materials have become a promising adsorbent for dye removal.

Cetyltrimethylammonium bromide (CTAB)-modified mesoporous molecular sieve FSM-16 was tested as an adsorbent for acid dye (acid yellow, AY, and acid blue, AB) removal in comparison with as-prepared FSM-16 and activated-carbon (AC).[94] The hydrothermal modification of FSM-16 with CTAB remarkably narrowed the pore openings and therefore improved the acid dye sorption in the low concentration range (> 20 ppm) as compared with that of CTAB-free FSM-16 and AC.

Yeung and co-workers prepared ordered mesoporous silica (OMS) adsorbents by grafting amino- and carboxylic-containing functional groups onto MCM-41 for the removal of Acid blue 25 and Methylene blue dyes from wastewater.[95] The amino-containing OMS-NH_2 adsorbent has a large adsorption capacity (256 mg/g) and a strong affinity for the Acid blue 25. It can selectively remove Acid blue 25 from a mixture of dyes (i.e. Acid blue 25 and Methylene blue). The OMS-COOH is a good adsorbent for Methylene blue displaying high adsorption capacity (113 mg/g) and selectivity for the dye. However, the amount of dyes adsorbed is much less than that of available functional groups. Regeneration of these adsorbents was also operated by simple washing with alkaline or acid solution. Part of the functional groups and the adsorbed dyes could be recovered.

With regard to organic pollutants of aquatic ecosystems, phenols are harmful to living organisms even at ppb levels. A variety of recyclable inorganic adsorbents, e.g. activated-carbons, organically modified clays and layered double hydroxides (LDHs), have drawn much attention as promising adsorbents particularly for the efficient removal of organic pollutants from aqueous solutions. Mesoporous materials have attracted great interest for organic pollutants' removal because of their novel structure and chemical characters. As-synthesized mesoporous silicas were used for the adsorption of organic substrates.[96] However, these materials with filled channels would exhibit low adsorption capacity, and particularly low rates of adsorption because of mass transfer limitation within the crowded space.

Instead of using as-synthesized MCM-41, Inumaru and co-workers used calcined and surface-modified MCM-41via grafting alkyltriethoxysilane.[97] This needs the material hydrophobicity controllable with the alkyl chain length. The adsorption rate could also be optimized according to the amount of grafted material. Hanna and co-workers prepared a large pore MCM-41 using trimethylbenzene (TMB) as swelling agent, then removed the TMB by heating it in air at 115°C.[98] This endowed the hydrophobic material with an open structure, and thus improved its adsorption properties.

In 1998, Albanis and co-workers used mesoporous alumina aluminum phosphates (AAPs) to remove chlorinated phenols from aqueous solution.[99] The AAP material (with a ratio $P/Al = 0.6$) can adsorb 14.8% of 2,4-diehlorophenol, 27.1% of 2,4,6-trichlorophenol or 58.3% of pentachlorophenol, respectively. Most recently, Sayari and co-workers used a highly porous, hydrophobic material B with high surface area and pore volume to adsorb organic pollutants.[89] The initial capacities for adsorbing 4-chloroguaiacol and 2-6-dinitrophenol reached 95 and 110 mg/g, respectively, as shown in Table 8.

Humic acids are derived from the degradation of plants and microorganisms. They contain aliphatic, aromatic, and hydrophilic functional groups, such as carboxyl and phenolic groups. Since they cause color in all open water sources and potential health implications, their removal is an important task for improving the water quality. Liu and co-workers synthesized a new mesoporous organosilica material (ß-CD-Silica-4%) containing microporous ß-cyclodextrins (ß-CDs) by the co-condensation of a silylated

Table 8.   Adsorption of chloroguaiacol and dinitrophenol on material B.[89]

Pollutant	Concentration (g/L)	% Adsorbed Pollutant	Adsorbent Capacity (mg/g)
Chloroguaiacol	0.11	89.3	96.6
Chloroguaiacol	0.25	89.9	95.6
Chloroguaiacol	0.50	19.9	95.2
Dinitrophenol	0.08	98.7	115.5
Dinitrophenol	0.25	98.5	88.6

ß-CD monomer with tetraethoxysilane in the presence of cetyltrimethylammonium bromide.[100] Adsorption experiments showed that ß-CD-Silica-4% material removed up to 99% of humic acid from an aqueous solution containing 50 ppm of humic acid at a solution-to-solid ratio of 100 mL/g. However, the regeneration of these adsorbents is very difficult, after treatment with a concentrated NaCl solution, only 30% of the loaded humic acid could be removed.

Gases

Song and co-workers proposed a novel CO_2 "molecular basket" adsorbent using polyethylenimine (PEI) modified mesoporous materials.[101] At PEI loading of 50 wt.% in MCM-41-PEI, the highest CO_2 adsorption capacity of 246 mg/g-PEI was obtained, which is 30 times higher than that of the MCM-41 and about 2.3 times that of the pure PEI. Further, the adsorption/desorption processes of CO_2 for these composites were faster than that for pure PEI. These composites also have good selectivity for CO_2 and recycle stability. Most recently, they used these composites as a low-temperature H_2S adsorbent.[102] Polyethylenimine (PEI), which has a high density of amino groups to interact with H_2S, is therefore expected to show high H_2S adsorption capacity. More importantly, mesoporous MCM-41 acts also as a separation medium for the PEI nanoparticles, such nanoparticles are not likely to aggregate in the application process and therefore retain their original properties. The PEI in MCM-41 is highly dispersed and expected to show a higher adsorption efficiency for H_2S than bulk PEI particles. The H_2S adsorption by a commercial ZnO, which is used in practical applications, was also measured to compare with that of the new adsorbent. Higher H_2S adsorption capacity, lower breakthrough time and much lower operation temperature of the new adsorbent than commercial ZnO are achieved. Therefore, these composites can be a promising candidate for low-temperature adsorption of sulfur compounds from hydrocarbon fuels in solid oxide fuel cells (SOFCs). With regard to adsorbents and sensors using mesoporous materials for volatile organic compounds and other toxic gases, more references can be found in the following section.

Others

Ho and co-workers prepared mesoporous Ti oxohydroxide ($TiO_x(OH)_y$) using dodecylamine as template and studied their capacity for fluoride removal.[103] Zirconia and silica had been introduced into the mesoporous Ti oxohydroxide to enhance the ion-exchange capacity. Results showed that zirconia containing mesoporous Ti oxohydroxide exhibited the highest fluoride ion exchange capacity (about 1.2 mmol F^-/g), as it has the smallest particle size, with high uniformity among the mesoporous materials prepared.

Shin and co-workers studied the aluminum-impregnated mesoporous adsorbents for the phosphate removal from water.[104] They found that the Al-impregnated mesoporous materials showed faster adsorption kinetics (reached equilibrium in an hour) as well as higher adsorption capacities (as high as 862 μmol/g) as compared with activated alumina. The authors concluded from the study that: (a) the uniform mesopores of these adsorbents increase the diffusion rate of the adsorbate in the adsorption process, thus lead to the fast adsorption kinetics; and (b) the high phosphate adsorption capacities are attributed not only to the increase of surface hydroxyl density on Al oxide due to the well-dispersed Al component but also to the decrease in stoichiometry ratio of surface hydroxyl ions to phosphate by the formation of mono-dentate surface complexes.

Mesoporous-Based Materials as Sensors

Gas Sensors

Toxic volatile organic compounds emitted from building materials and furniture stay indoors, especially in new buildings, are suspected to be a cause of health hazard such as sick building or sick house syndrome. With the increasing number of biological and environmental issues, reliable, sensitive, selective, and user-friendly chemical sensors have become more and more attractive. A sensor includes a transducer (electrochemical, piezoelectrical, or optical) whose typical response is significantly altered by the presence of the chemical to be detected. Mesoporous materials, because of their novel structural properties, have become promising candidates not only for adsorbents but also for sensors. Their highly ordered, uniform and 3D cross-linked pore structure and large specific surface area can undoubtedly enhance the sensitivity of adsorption-type sensors. Sasahara and co-workers employed Pd-loaded mesoporous silica to fabricate a catalytic sensing material of an adsorption/combustion-type sensor for volatile organic compounds (VOCs).[105] Toluene was used as a typical example of VOCs with a pulse heating mode. A large response peak was observed by flash combustion of the toluene molecules adsorbed on the sensing material during the non-heating period and the integral response is nearly proportional to the specific surface area of the sensing materials. The response is stronger than that of Pd/γ-Al$_2$O$_3$ sensor. They also pointed out that thermal conductivity of the sensing material was another important factor to control the sensitivity of this type of sensors. Most recently, Floch

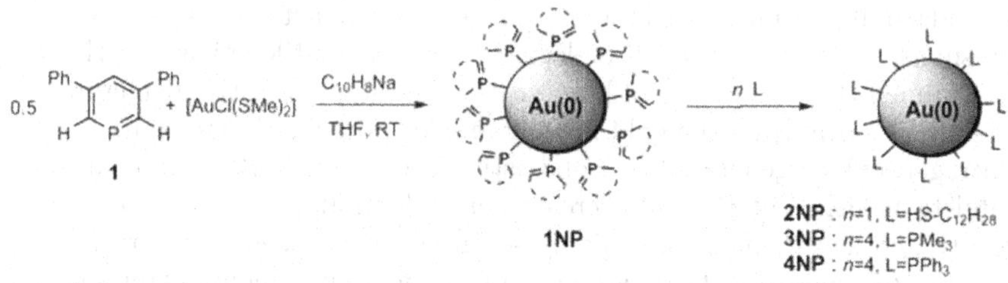

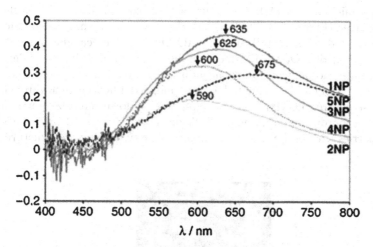

Figure 22. Synthesis diagram of nNP (n = 1 − 4).[106]

Figure 23. UV-Vis spectra recorded for nNP (n = 1–5). (The diffusing noise of the solution was simulated with an exponential function and subtracted from the recorded spectra in order to determine the peak maxima properly.[106]

and co-workers developed a selective chemical-sensor based on the plasmonic response of phosphinine-stabilized gold nanoparticles (NPs) hosted on periodically-organized mesoporous silica thin layers (Figure 22).[106] Gold NPs stabilized by phosphinine ligands were used. A subsequent substitute of phosphinine legands with thiol and phosphane results in a blueshift of the plasmon peak with a concomitant relative decrease of the peak intensity (Figure 23). Such signals show the possibility of the UV-Vis detection of a ligand-exchange reaction at the surface of gold nanoparticles in solution. Thus these nanoparticles can be used as a transducer. They immobilized these Au NPs in mesoporous silica thin films by a one-step evaporation-induced self assembly (EISA) approach, which can preserve both a good accessibility for the analyte to the detection center and good optical qualities. The properties of the phosphinine ligand enable a selective detection of

390 J. Shi et al.

thiols and small phosphanes, so the composite films can act as very sensitive and responsive sensors for small molecules, such as volatile thiols or PH3. It was claimed that the design of large scale or more-quantitative detectors is possible by optimising the gold loading and/or pore-size in the film.

Nitrogen oxides generated by combustion are not only dangerous and harmful to health, but they also create acid rain that destroys the environment. The development of a highly sensitive, responsive, and portable monitoring technique for these gases is an urgent requirement. Zhou and co-workers applied ordered mesoporous silica films to develop an NO_x gas sensor based on the surface photovoltage (SPV) system, using a metal-insulator-semiconductor (MIS) structure with mesoporous thin film as the insulator layer (Figure 24).[107,108] When the rear side of the semiconductor is irradiated by a LED beam, an AC photocurrent can be induced. The adsorption of NO_x on the mesoporous silica layers brings a change of the surface photocurrent on semiconductor layers. Thus the sensitivity to supplied NO_x gas is estimated by the induced photocurrent through the MIS structure. The large surface area and uniform nanosize pores of ordered mesoporous materials enable SPV devices to show good gas adsorption properties with better sensitivity and selectivity. Cubic pore structures

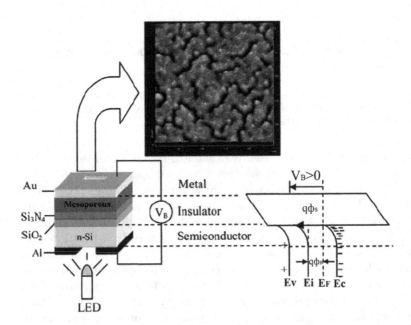

Figure 24.   Tin-modified mesoporous silica sensor with MIS structure and its energy band diagram. The AFM image of an electrode shows the porous layer of Au film.[108]

are a better choice than hexagonal ones because of their 3D cross-linked pore channels, which bring much higher gas sensitivity. Tin was added to the mesoporous silica thin films and detectable NO_2 concentration limits were lowered to 300 ppm at room temperature, as reported in one of their recent paper.[108] Hon and co-workers fabricated mesoporous WO_3 thin films as NO_2 gas sensors using Poly(alkylene oxide) block copolymer $EO_{100}PO_{64}EO_{100}$, F127, as a template. The film has nanocrystallite domains, porous structure with a relative surface area of $143 m^2/g$ after calcined at 250°C.[109] Upon exposure to NO_2, the electrical resistance of semiconducting mesoporous WO_3 thin films is found to increase dramatically. The mesoporous WO_3 thin film sensor shows a high sensitivity up to Rg/Ra = 226 (Rg and Ra are the electric resistance of the films in test gas and air, respectively) at 100°C for detecting 3 ppm NO_2. The one calcined at 250°C and operated at 35°C also shows a higher enough sensitivity of Rg/Ra = 23, the authors claimed it an unique NO_2 gas sensor fir its high sensitivity at such a low temperature.

Semiconductive mesoporous materials are of great interest as gas sensor materials. Hyodo and co-workers synthesized mesoporous SnO_2 and further improved their thermal stability by a phosphoric acid treatment process.[110,111] They found that thick film sensors prepared from the phosphoric acid (PA)-treated mesoporous SnO_2 powders of larger pore size showed better H_2 response, in comparison with the sensors prepared from untreated powder and from PA-treated mesoporous powder of smaller pore size. While semiconductor gas sensors exhibit high sensitivity (small change in gas composition causes dramatic change in resistance), their stability and selectivity still remain unsatisfactory for many applications. To solve this problem, Cabot and co-workers used Pd- and Pt-loaded mesoporous silica as a novel selective filter for a SnO_2-based gas sensors (Figure 25).[112]

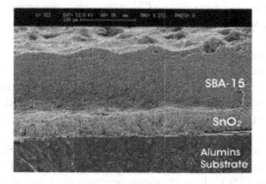

Figure 25.　A SEM cross-sectional view of a SnO_2 sensor with a mesoporous filter.[112]

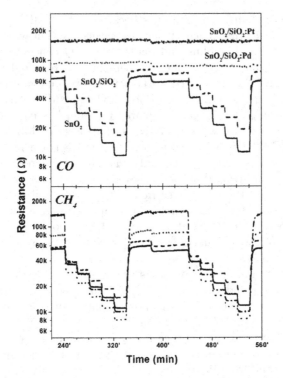

Figure 26. Transitory resistances of the SnO$_2$, SnO$_2$/SiO$_2$, SnO$_2$/SiO$_2$:Pd and SnO$_2$/SiO$_2$:Pt gas sensors when exposed to different concentrations of CO and CH: from 20 to 400 ppm and 200 to 4000 ppm, respectively. Two different relative humidity conditions are considered: 30 and 50%.[112]

Results indicate that SiO$_2$:Pd and SiO$_2$:Pt filters can eliminate response to CO without affecting the response to CH$_4$ when they are exposed to CH$_4$-CO mixtures (Figure 26). The large catalytically active surface area and the high resistivity of the mesoporous SiO$_2$ make it suitable as an efficient filter to improve the selectivity and stability of gas sensors.

Dyes are developed as oxygen sensors based upon the principle that oxygen is a powerful quencher of the electronically excited state of dyes. The key factors include the solubility of oxygen and the diffusion coefficient of oxygen in the host medium, and the excited state lifetime of the dye. Mesoporous materials were chosen as a kind of new candidate, which could adsorb sufficient dye to produce a strong photoluminescence signal and at the same time provide rapid access to oxygen molecules from the external atmosphere. Winnik and co-workers prepared mesoporous silica short rods.[113] Sensor films were prepared by adsorbing these particles at sub-monolayer coverage on a thin polymer film with a positively charged surface prepared by a layer-by-layer method. Various dyes were incorporated into the pores of the mesoporous silica particles when these films were dipped into solutions of dyes and the solvent was allowed to evaporate. Dyes incorporated into the mesoporous silica coated polymer films have a more intense fluorescence than those adsorbed to the layer-by-layer film itself. Therefore, composite films are promising oxygen sensors with a rapid response.

Other Sensors

Uranium and its salts, which can cause kidney damage and acute arterial lesions, are both radiologically and chemically toxic. Because concentrations of uranium compounds in contaminated groundwater are typically at very low levels, the accurate analyses become more difficult. Lin and co-workers reported a new uranium electrochemical sensor for uranium detection based on an adsorptive square-wave stripping votammetry technique.[114] This technique of pre-concentrating uranium chelates on an electrode surface prior to the detection step has been shown to be highly sensitive for the detection of trace uranium and its compounds. To prepare the sensor, an acetamide phosphonic acid ligand was immobilized via covalent bonding onto mesoporous silica MCM-41 and the ligand-bearing mesoporous silica was then embedded in a carbon graphite matrix. This sensor is mercury-free, solid-state, and has less ligand depletion than existing sensors. Using this functionalized mesoporous sensors, voltammetric responses for uranium detection are reported as a function of pH value, pre-concentration time, and aqueous phase uranium concentration. The uranium detection limit is 25 ppb after five minutes pre-concentration and improved to 1 ppb after 20 minutes pre-concentration. The relative standard deviations are normally less than 5%.

Martínez-Máñez and co-workers reported a new method for the determination of fluoride in water based on the specific reaction of fluorhydric acid with a MCM-41 solid functionalized with fluorescent or colourimetric signalling units.[115] The mesoporous MCM-41 solid is first functionalized with 3-aminopropyltriethoxysilane in refluxing toluene. Reaction of this amino-functionalized solid with 9-anthraldehyde (in absolute ethanol at 45°C; then with NaBH$_4$ in ethanol at room temperature 4-{2-[4-(dimethylamino)phenyl]diazenyl}benzoic acid (DCC–TsOH, 0°C) and lissamine rhodamine B sulfonyl chloride (in CH$_2$Cl$_2$, 0°C) results in the synthesis of sample S1, S2 and S3, respectively (Figure 27). Functionalized monoliths of S3 have been added to buffered solutions containing fluoride. The resulting solution colour was deepened with the increase of

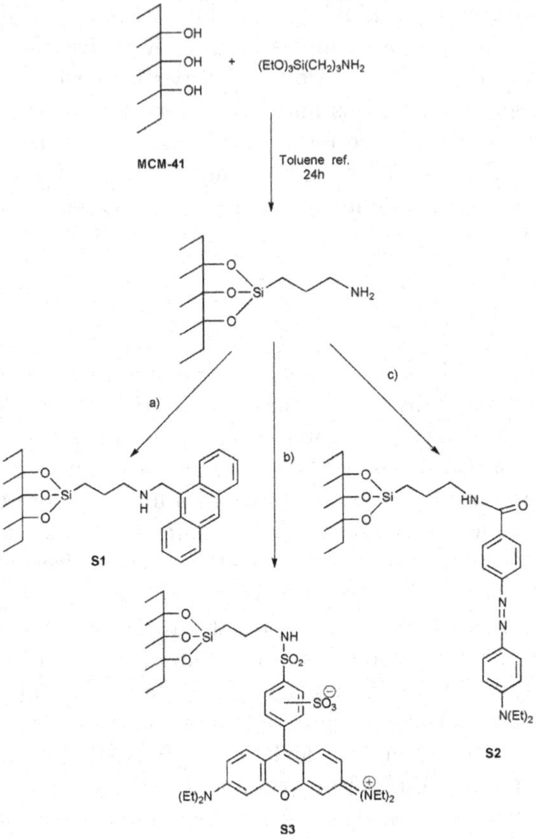

Figure 27. Synthesis of S1, S2 and S3 by the functionalization of an MCM-41 matrix firstly with 3-aminopropyltriethoxysilane and then with: (a) 9-anthraldehyde, (b) lissamine rhodamine B sulfonyl chloride, and (c) 4-{2-[4-(dimethylamino) phenyl]diazenyl}benzoic acid.[115]

the amount of fluoride. The good results obtained in the determination of fluoride content in toothpaste suggest its potential use as a practical method for the determination of fluoride anions in water samples.

Based on diffusion followed by an immobilizing reaction, Xue and co-workers first reported a new quantitative optical sensor for metal ions using amine-grafted mesoporous silica monoliths with $H_2N(CH_2)_3Si-(OMe)_3$.[116] The sensor unifies two processes including slow diffusion of the metal ions to the binding sites and fast metal-ligands(ML_n) complexation. When the ligands have been saturated and diffusion of the Cu^{2+} reaches a balance to the metal solution, the adsorbance of the complex ML_n could be observed spectroscopically. They studied its response properties for Cu^{2+} in detail and gave a mathematical model for such a quantitative sensor.

Summary

As a special kind of nanomaterials, nanoporous materials have attracted a lot of research attentions in recent years, among which ordered mesoporous materials are one of the most attractive focusing points due to its unique characteristics such as extraordinary high surface area, uniform pore size distributions, and so on. To meet the ever increasing needs for environmental protections and environment-friendly processes, mesoporous materials have been expected to play more and more important roles in the fields of environment related areas due to its high catalytic, adsorbing and sensing performances. In this chapter, we have reviewed the recent progresses in the synthesis and properties of various kinds of mesoporous materials and mesoporous-based nanocomposites by chemical assembly approaches on nanoscale. Emphasis has been focused on the environment-related catalytic, adsorbing and sensing performances of these materials.

Part I focuses mainly on the catalytic performances for automobile exhaust treatments of mesoporous zirconia based three way catalysts. The metal-loaded composite catalysts show significantly lower ignition temperature for catalysis and higher stability and lower noble metal consumption as compared to the common noble metal-loaded zirconia/ceria powders. The reason is the high dispersion of noble metal particles in the pore channels, which provide abundant active sites for the exhaust conversion and prevent the aggregation of metal particles. The mesoporous materials, especially conductive mesoporous carbons, are promising as electrode reaction catalysts for polymer electrolyte membrane cells. After the catalytically active site loading in mesoporous carbons, the composites show high potential for future applications in such clean energy power sources, as pointed out in part II. Another active research field is the photo-catalysis of titania-based materials for the decomposition of harmful organics. As introduced in part III, mesoporous titania has also attracted much attention due to its high surface area. Main attention has been paid to the preparation of phase structure controllable mesoporous titania, which leads to attractive photo-catalytic performances. The last two parts (part IV and V) deal with the adsorbing and sensing performances of mesoporous-based composites, such as mesoporous organic/inorganic hybrids. Again, owing to the very high surface area of the pore channels and its tunable wall and pore surface chemical properties, the materials illustrates very promising applications for the detection and removal of harmful substances such as heavy metal ions, toxic gases, organic pollutants, and so on.

Further research work is expected on the synthesis of nano/meso-porous materials with controlled structures and/or architectures, and their properties, especially the performances for environment applications. The quantity production of these advanced materials with high reproducibility and low cost need to be explored in the future for the practical applications as well.

References

1. K. S. W. Sing, D. H. Evereet, K. H. W. Haul, L. Moscou, R. A. Pierotti and J. Rouquerol, *Pure Appl. Chem.* **57**, 603–619 (1985).
2. J. Y. Ying; C. P. Mehnert and M. S. Wong, *Angew. Chem. Int. Ed.* **38**, 56–77 (1999).
3. F. Schuth and W. Schmidt, *Adv. Mater.* **14**, 629–638 (2002).
4. A. Taguchi and F. Schuth, *Micro. Meso. Mater.* **77**, 1–45 (2005).
5. C. T. Kresge, M. E. Leonowicz, W. J. Roth, J. C. Vartulli and J. S. Beck, *Nature* **359**, 710 (1992).
6. J. S. Beck, J. C. Vartuli, W. J. Roth, M. E. Leonowicz, C. T. Kresge, K. D. Schmitt, C. T.-W. Chu, D. H. Olson, E. W. Sheppard, S. B. McCullen, J. B. Higgins and J. L. Schlenker, *J. Am. Chem. Soc.* **114**, 10834 (1992).
7. P. Fornasiero, E. Fonda, R. Di. Monte, G. Vlaic, J. Kaspar and M. Graziani, *J. Catal.* **187**, 177 (1999).
8. C. Bozo, N. Guilhaume, E. Garbowski and M. Primet, *Catal. Today* **59**, 33 (2000).
9. P. Fornasiero, N. Hickey, J. Kaspar, C. Dossi, D. Gava and M. Graziani, *J. Catal.* **189**, 326 (2000).
10. A. P. Oliveira and M. L. Torem, *Powder Technol.* **119**, 181 (2001).
11. H. Shinjoh, *Journal of Alloys and Compounds* 2005 (In Press), *Available online 20 July 2005.*
12. I. M. Hung, H. P. Wang, W. H. Lai, K. Z. Fung and M. H. Hon, *Electrochimica Acta* **50**, 745–748 (2004).
13. A. H. Lu and F. Schuth, *C. R. Chimie* **8**, 609–620 (2005).
14. F. Kleitz, S. H. Choi and R. Ryoo, *Chem. Commun.* 2136 (2003).
15. W. H. Shen, X. P. Dong, Y. F. Zhu, H. R. Chen and J. L. Shi, *Micro. Meso. Mater.* **85**, 157–162 (2005).
16. A. S. Deshpande, N. Pinna, B. Smarsly, M. Antonietti and M. Niederberger, *Small* **1**, 313–316 (2005).
17. M. Ozawa, *Journal of Alloys and Compounds* 275–277, 886–890 (1998).
18. H. R. Chen, J. L. Shi, J. N. Yan, Z. L. Hua, H. G. Chen and D. S. Yan, *Advanced Materials* **15**(13), 1078–1081 (2003).
19. D. B. Kuang, T. Breaesinski and B. Smarsly, *J. Am. Chem. Soc.* **126**, 10534 (2004).
20. A. Corma, P. Atienzer, H. Garcia and J. Y. C. Ching, *Nature* **3**, 394 (2004).
21. H. R. Chen, J. L. Gu, J. L. Shi, Z. C. Liu, J. H. Gao, M. L. Ruan and D. S. Yan, *Adv. Mater.* **17**, 2010–2014 (2005).

22. T. Sen, G. J. T. Tiddy, J. L. Casci and M. W. Anderson, *Angew. Chem. Int. Ed.* **42**, 4649 (2003).
23. C. Ho, J. C. Yu, X. C. Wang, S. Lai and Y. F. Qiu, *J. Mater. Chem.* **15**, 2193–2201 (2005).
24. L. F. Chen, G. Gonzalez, J. A. Wang, L. E. Norena, A. Toledo, S. Castillo and M. M. Pineda, *Applied Surface Science* **243**, 319–328 (2005).
25. F. Fajardie, J. F. Tempere, J. M. Manoli, G. D. Mariadassou and G. Blanchard, *J. Chem. Soc. Faraday Trans.* **94**, 3727 (1998).
26. Z. Q. Mao, *Chin. J. Power Sources* **27** 179–182 (2003).
27. W. Chrzanowski and A. Wieckowski, *Langmuir* **14**, 1967 (1998).
28. H. D. Dinh, X. Ren, F. H. Garzon, P. Zelenay and S. Gottesfeld, *J. Electroanal. Chem.* **492**, 222 (2000).
29. G. S. Chai, S. B. Yoon, J. S. Yu, J. H. Choi and Y. E. Sung, *J. Phys. Chem. B* **108**, 7074–7079 (2004).
30. R. Ryoo, S. Joo, M. Kruk and M. Jaroniec, *Adv. Mater.* **13**, 677 (2001).
31. Y. Wang, J. Ren, K. Deng, L. Gui and Y. Tang, *Chem. Mater.* **12**, 1622 (2000).
32. H. Kim, P. Kim, J. B. Joo, W. Y. Kim, I. K. Song and J. Yi, *Journal of Power Sources*, 2005 (In Press), Available online 1 September 2005.
33. K. W. Park, J. H. Choi and Y. E. Sung, *J. Phys. Chem. B* **107**, 5951 (2003).
34. J. H. Jiang and A. Kucernak, *Journal of Electroanalytical Chemistry.* **533**, 153–165 (2002).
35. G. S. Attard, P. N. Bartlett, N. R. B. Coleman, J. M. Elliott, J. R. Owen and J. H. Wang, *Science* **278**, 838 (1997).
36. J. H. Jiang and A. Kucernak, *Journal of Electroanalytical Chemistry.* **543**, 187–199 (2003).
37. J. M. Orts, E. Louis, L. M. Sander and J. Clavilier, *Electrochim. Acta* **44**, 1221 (1998).
38. A. Fujishima, T. N. Rao and D. A. Tryk, *J. Photochem. Photobiol. C: Photochem. Rev.* **1**, 1 (2000).
39. Q. Huo, D. I. Margolese, U. Ciesla, P. Feng, T. E. Gler, P. Sieger, R. Leon, P. M. Petroff, F. Shüth and G. D. Stucky, *Nature* **368**, 317 (1994).
40. D. Zhao, J. Feng, Q. Huo, N. Melosh, G. H. Fredrickson, B. F. Chmelka and G. D. Stucky, *Science* **279**, 548 (1998).
41. P. T. Tanev and T. J. Pinnavaia, *Science* **267**, 865 (1995).
42. D. M. Antenolli and J. Y. Ying, *Angew. Chem. Int. Ed. Engl.* **34**, 2014 (1995).
43. D. Khushalani, G. A. Ozin and A. Kuperman, *J. Mate. Chem.* **9**, 1491 (1999).
44. S. Cabrera, J. El-Haskouri, A. Beltrán-Portier, D. Beltrán-Portier, M. D. Marcos and P. Amorós, *Solid State Sci.* **2**, 513 (2000).
45. V. F. Stone Jr. and R. J. Davis, *Chem. Mater.* **10**, 1468 (1998).
46. D. M. Antonelli, *Microporous Mesoporous Mater.* **30**, 315 (1999).
47. Y. Lu, R. Ganguli, C. A. Drewien, M. T. Anderson, C. J. Brinker, W. Gong, Y. Guo, H. Soyez, B. Dunn, M. H. Huang and J. I. Zink, *Nature* **389**, 364 (1997).

48. E. L. Crepaldi, G. J. de A. A. Soler-Illia, D. Grosso and C. Sanchez, *New J. Chem.* **27**, 9 (2003).

49. E. L. Crepaldi, G. J. de A. A. Soler-Illia, D. Grosso, F. Cagnol, F. Ribot and C. Sanchez, *J. Am. Chem. Soc.* **125**, 9770 (2003).

50. E. Beyers, P. Cool and E. F. Vansant, *J. Phys. Chem. B* **109**, 10081 (2005).

51. B. Smarsly, D. Grosso, T. Brezesinski, N. Pinna, C. Boissière, M. Antonietti and C. Sanchez, *Chem. Mater.* **16**, 2948 (2004).

52. D. Grosso, G. J. de A. A. Soler-Illia, E. L. Crepaldi, F. Cagnol, C. Sinturel, A. Bourgeois, A. Brunet-Bruneau, H. Amenitsch, P. A. Albouy and C. Sanchez, *Chem. Mater.* **15**, 4562 (2003).

53. Y. K. Hwang, K. Lee and Y. Kwon, *Chem. Commun.* 1738 (2001).

54. M. S. Wong, E. S. Jeng and J. Y. Ying, *Nano Letters* **1**, 637 (2001).

55. G. J. de A. A. Soler-Illia, E. Scolan, A. Louis, P. Albouy and C. Sanchez, *New J. Chem.* **25**, 156 (2001).

56. B. Tian, H. Yang, X. Liu, S. Xie, C. Yu, J. Fan, B. Tu and D. Zhao, *Chem. Commun.* 1824 (2002).

57. B. Tian, X. Liu, B. Tu, C. Yu, J. Fan, L. Wang, S. Xie, G. D. Stucky and D. Zhao, *Nature Mater.* **2**, 159 (2003).

58. J. C. Yu, X. Wang and X. Fu, *Chem. Mater.* **16**, 1523 (2004).

59. J. C. Yu, J. G. Yu and J. C. Zhao, *Appl. Catal. B-Environ.* **36**, 31 (2002).

60. S. Haseloh, S. Y. Choi, M. Mamak, N. Coombs, S. Petrov, N. Chopra and G. A. Ozin, *Chem. Commun.* 1460 (2004).

61. S. Y. Choi, M. Mamak, N. Coombs, N. Chopra and G. A. Ozin, *Adv. Funct. Mater.* **14**, 335 (2004).

62. H. Luo, C. Wang and Y. Yan, *Chem. Mater.* **15**, 3841 (2003).

63. Y. Li, N. Lee, E. Lee, J. S. Song and S. Kim, *Chem. Phys. Lett.* **389**, 124 (2004).

64. D. Grosso, G. J. de A. A. Soler-Illia, E. L. Crepaldi, B. Charleux and C. Sanchez, *Adv. Funct. Mater.* **13**, 37 (2003).

65. L. Zhang and J. C. Yu, *Chem. Commun.* 2078 (2003).

66. X. Wang, J. C. Yu, C. Ho, Y. Hou and X. Fu, *Langmuir* **21**, 2552 (2005).

67. Y. Shiraishi, N. Saito and T. Hirai, *J. Am. Chem. Soc.* **127**, 12820 (2005).

68. B. J. Aronson, C. F. Blanford and A. Stein, *Chem. Mater.* **9**, 2842 (1997).

69. R. van Grieken, J. Aguado, M. J. López-muñoz and J. Marugán, *J. Photochem. Photobiol. A: Chem.* **148**, 315 (2002).

70. E. P. Reddy, B. Sun and P. G. Smirniotis, *J. Phys. Chem. B* **108**, 17198 (2004).

71. E. P. Reddy, L. Davydov and P. G. Smirniotis, *Appl. Catal. B: Environ.* **42**, 1 (2003).

72. Y. Xu and C. H. Langford, *J. Phys. Chem. B* **101**, 3115 (1997).

73. M. Andersson, H. Birkedal, N. R. Franklin, T. Ostomel, S. Boettcher, A. E. C. Palmqvist and G. D. Stucky, *Chem. Mater.* **17**, 1409 (2005).

74. M. H. Bartl, S. P. Puls, J. Tang, H. C. Lichtenegger and G. D. Stucky, *Angew. Chem. Int. Ed.* **43**, 3037 (2004).

75. S. Yuan, Q. Sheng, J. Zhang, F. Chen, M. Anpo and Q. Zhang, *Microporoua Mesoporous Mater.* **79**, 93 (2005).

76. Y. Yang, Y. Guo, C. Hu, Y. Wang and E. Wang, *Appl. Catal. A: General* **273**, 201 (2004).
77. A. Walcarius, M. Etienne and B. Lebeau, *Chem. Mater.* **15**, 2161 (2003).
78. X. Feng, G. E. Fryxell, L. Q. Wang, A. Y. Kim, J. Liu and K. M. Kemmer, *Science* **276**, 923–926 (1997).
79. J. Liu, X. D. Feng, G. E. Fryxell, L. Q. Wang, A. Y. Kim and M. L. Gong, *Adv. Mater.* **10**, 161–165 (1998).
80. L. Mercier and T. J. Pinnavaia, *Adv. Mater.* **9**, 500–503 (1997).
81. J. Brown, L. Mercier and T. J. Pinnavaia, *Chem. Commun.* 69–70 (1999).
82. J. Brown, R. Richer and L. Mecier, *Microporous. Mesoporous. Mater.* **37**, 41–48 (2000).
83. B. Lee, Y. Kim, H. Lee and J. Yi, *Micro. Meso. Mater.* **50**, 77–90 (2001).
84. V. Antochshuk and M. Jaroniec, *Chem. Commun.* 258–259 (2002).
85. V. Antochshuk, O. Olkhovyk, M. Jaroniec, I. Park and R. Ryoo, *Langmuir* **19**, 3031–3034 (2003).
86. L. X. Zhang, W. H. Zhang, J. L. Shi, Z. L. Hua, Y. S. Li and J. N. Yan, *Chem. Commun.* 210–211 (2003).
87. K. Z. Hossain and L. Mercier, *Adv. Mater.* **14**, 1053–1056 (2002).
88. S. Dai, M. C. Burleigh, Y. Shin, C. C. Morrow, C. E. Barnes and Z. Xue, *Agew. Chem. Int. Ed.* **38**, 1235–1239 (1999).
89. A. Sayari, S. Hamoudi and Y. Yang, *Chem. Mater.* **17**, 212–216 (2005).
90. G. E. Fryxell, J. Liu, T. A. Hauder, Z. Nie, K. F. Ferris, S. Mattigod, M. Gong and R. T. Hallen, *Chem. Mater.* **11**, 2148–2154 (1999).
91. K. H. Nam and L. L. Tavlarides, *Chem. Mater.* **17**, 1597–1604 (2005).
92. Y. Kim, C. Kim, I. Choi, S. Rengaraj and J. Yi, *Environ. Sci. Technol.* **38**, 924–931 (2004).
93. M. Jang, J. K. Park and E. W. Shin, *Microporous Mesoporous Mater.* **75**, 159–168 (2004).
94. M. M. Mohamed, *J. of Colloid and Interface Sci.* **272**, 28–34 (2004).
95. K. Y. Ho, G. McKay and K. L. Yeung, *Langmuir* **19**, 3019–3024 (2003).
96. Y. Miyake, T. Yumoto, H. Kitamura and T. Sugimoto, *Phys. Chem. Chem. Phys.* **4**, 2680–2684 (2002).
97. K. Inumaru, Y. Inoue, S. Kakii, T. Nakano and S. Yamanaka, *Phys. Chem. Chem. Phys.* **6**, 3133–3139 (2004).
98. K. Hanna, I. Beurroise, R. Denoyel, D. Desplantier-Giscard, A. Galarneau and F. J. De Renzo, *Colloid Interface Sci.* **252**, 276–283 (2002).
99. T. G. Danis, T. A. Albanis, D. E. Petrakis and P. J. Pomonis, *War. Res.* **32**(2), 295–302 (1998).
100. C. Liu, N. Naismith and J. Economy, *J. of Chromatography A* **1036**, 113–118 (2004).
101. X. Xu, C. Song, J. M. Andresen, B. G. Miller and A. W. Scaroni, *Microporous Mesoporous Mater.* **62**, 29–45 (2003).
102. X. Xu, I. Novochinskii and C. Song, *Energy Fuels* **19**(2), 2214–2215 (2005).
103. L. N. Ho, T. Ishihara, S. Ueshima, H. Nishiguchi and Y. Takita, *J. Colloid Interface Sci.* **272**, 399–403 (2004).

104. E. W. Shin, J. S. Han, M. Jang, S. H. Min, J. K. Park and R. M. Rowell, *Environ. Sci. Technol.* **38**, 912–917 (2004).

105. T. Sasahara, A. Kido, H. Ishihara, T. Sunayamaa and M. Egashira, *Sensors Actuators B* **108**, 478–483 (2005).

106. F. Goettmann, A. Moores, C. Boissière, P. L. Floch and C. Sanchez, *Small* **6**, 636–639 (2005).

107. T. Yamada, H. S. Zhou, H. Uchida, M. Tomita, Y. Ueno, T. Ichino, I. Honma, K. Asai and T. Katsube, *Adv. Mater.* **14**, 812–815 (2002).

108. B. Yuliarto, H. S. Zhou, T. Yamada, I. Honma, Y. Katsumura and M. Ichihara, *Anal. Chem.* **76**, 6719–6726 (2004).

109. L. G. Teoh, Y. M. Hon, J. Shieh, W. H. Lai and M. H. Hon, *Sensors Actuators B* **96**, 219–225 (2003).

110. Y. Shimizu, T. Hyodo and M. Egashira, *J. Eur. Ceram. Soc.* **24**, 1389–1398 (2004).

111. Y. Himizua, A. Jonoa, T. Hyodob and M. Egashira, *Sensors Actuators B* **108**, 56–61 (2005).

112. A. Cabot, J. Arbiol, A. Cornet, J. R. Morante, F. Chen and M. Liu *Thin Solid Films* **436**, 64–69 (2003).

113. B. H. Han, I. Manners and M. A. Winnik, *Chem. Mater.* **17**, 3160–3171 (2005).

114. W. Yantasee, Y. Lin, G. E. Fryxell and Z. Wang, *Electroanalysis* **10**, 870–873 (2004).

115. A. B. Descalzo, D. Jiménez, J. E. Haskouri, D. Beltrán, P. Amorós, M. D. Marcos, R. Martínez-Máñez and J. Sotoa, *Chem. Commun.* 562–563 (2002).

116. D. L. Rodman, H. Pan, C. W. Clavier, X. Feng and Z. Xue, *Anal. Chem.* **77**, 3231–3237 (2005).

Chapter 15

Electrochemical Sensors Based on Nanomaterials for Environmental Monitoring

Wassana Yantasee, Yuehe Lin and Glen E. Fryxell

Pacific Northwest National Laboratory
Richland, Washington 99352, USA

Introduction

Current Stage of Electrochemical Sensors for the Analysis of Toxic Metals

Environmental monitoring especially of toxic metal ions in surface and sub-surface water sources as well as occupational monitoring of worker exposure to toxic metals presently rely on the collection of discrete liquid samples for subsequent laboratory analysis using techniques such as ICP-MS and AAS. Sensors that are field-deployable and able to measure part-per-billion (ppb) or nanomolar levels of toxic metal ions will reduce time and costs associated with environmental and occupational monitoring of hazardous metal species. Electrochemical sensors based on adsorptive stripping voltammetry (AdSV) appear to be a very promising technique that offers desired characteristics such as field-deployability, specificity for targeted metal ions, enhanced measurement frequency and precision, robustness, inexpensiveness, and infrequent regeneration of sensor materials.[1-3]

Adsorptive stripping voltammetry (AdSV) usually involves selective preconcentration of metal ions at an electrode surface, followed by the quantification of the accumulated species by a voltammetric method that generates favorable signal-to-background ratio. Preconcentration of metal ion species at an electrode surface prior to the detection step allows sensitive analyses of the metal species that may be present at extremely low levels in contaminated groundwater, wastewater, or body fluids. Conventionally, the preconcentration of metal species has been done at mercury-based electrodes by forming amalgam of the metal ions, which have issues related to the use and disposal of toxic mercury and the mechanical instability of mercury electrodes (i.e. mercury drop electrodes), making them unfit for routine field applications.[4,5] To develop mercury-free sensors, functional ligands have been immobilized on electrode surfaces for the preconcentration of trace metal ions by employing the specific binding properties of the ligands towards the target metal ions without applying a potential. Immobilization of ligands at the electrode surface can be done by coating of monolayer[6-9] or polymeric films[10-12] on the electrode surfaces, or embedding of suitable functional ligands in a conductive porous matrix.[13-20] The number of functional groups on the monolayer films is often limited and the stability and durability of the films may be poor. Polymeric films are often affected by shrinking and swelling because of the changes in solution pH or electrolyte concentration.[21,22] Thus conductive carbon paste mixed with functional ligands is more widely used. However, the ligands in these sensors are in loose association or physical contact with the carbon paste materials, thus degradation of the sensor occurs over time as a result of depletion of ligand-bearing materials unless certain solvents in which the ligands are insoluble are employed.[19]

Nanostructured Silicas and Carbon Nanotubes (CNT) in Electrochemical Sensing of Toxic Metals

Nanostructured materials, defined as materials that have at least one dimension smaller than 100 nm, are increasingly popular in the development of electrochemical sensors for toxic metal ions because they permit the fabrication of miniature sensing devices that are sensitive, compact, low-cost, low-energy-consuming, and easily integrated into field-deployable units. Two nanomaterials that are the fastest growing in electrochemical sensing field are ordered mesoporous silicas (OMSs) and carbon nanotubes (CNTs) which will be the subjects of this review. Figure 1 shows the structures of (a) ordered mesoporous silica with parallel hexagonal pore structures, and (b) open-end multi-wall CNT in comparison with ordered pyrolytic graphite.

Nanostructured silica materials

The uses of silica-based organic-inorganic materials, including silicas coated with inorganic layers supporting catalysts, and silica gels grafted with organic groups or immobilized with enzymes, as electrode modifiers were widely studied in 1990s and have been nicely reviewed by Walcarius.[23] Also

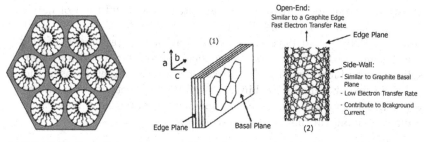

(a) MCM-41 with self-assembled monolayers   (b) ordered pyrolytic graphite (1) and open-end multi-wall CNT (2)

Figure 1.   Structures of (a) MCM-41 with self-assembled monolayers (SAMMS), (b-1) highly ordered pyrolytic graphite, and (b-2) open-end multi-walled CNT, from Ref. 135, reproduced by permission of The Royal Society of Chemistry.

in the past decade, crystalline nanoporous silica, MCM-41, having hexagonal arrays of regular pore structure have been synthesized by the surfactant templating process.[24] Attributed to the rigid, open-parallel pore structure, the MCM-41 significantly improves accessibility of analytes in and out of the pores when compared with amorphous silicas.[25] Amorphous materials are not desirable for electrochemical applications since most adsorption of analytes on the functional groups residing inside the pores as well as electrochemical transformations are diffusion-controlled.[26] Unmodified ordered-mesoporous silicas have been dispersed in carbon paste matrix in order to utilize the reactivity of their silanol groups for the preconcentration of Cu^{2+},[27,28] Hg^{2+},[29] and Hg^{2+}, Ag^+, Pb^{2+}, and Cu^{2+}.[30] However, only metal hydroxides of mercury(II) are soluble enough to reach and react with silanol groups in the pH range of 4 to 7.[29] Other metal species that are not hydroxylated can only bind with silanolate groups, which are formed in significant amounts only at above pH 7. Because both number of immobilized functional groups and the structure of the organically modified silicates are believed to affect the voltammetric responses of metal ions,[27,31] installation of well-designed organic functional groups in the pores of ordered mesoporous silicas opens a new frontier for a new class of electrochemical sensors.

Organically-modified ordered mesoporous silica (OMS) materials have been used for a decade as sorbent materials, yet their use in electrochemical sensors is relatively new. Few work has employed electrochemistry methods to study the properties of such materials, including cyclic voltammetry to study the voltammetric responses of electroactive species,[28,29,31–35] potentiometric monitoring of pH to study the protonation kinetics,[36] or steady-state current method to study the uptake[36] and diffusion of metal ions[37,38] of various mesoporous silicas. At the Pacific Northwest National Laboratory (PNNL), we have exploited the attractive properties of the organically modified OMSs in electrochemical sensors for metal ions that are solid-state and mercury-free.[39-46] Preconcentration of metal ions at the electrode surface by utilizing the binding affinity and selectivity of the functionalized ordered mesoporous silicas at open-circuit potential enables their voltammetric quantitations at ultra-trace levels.

Carbon nanotubes

There have been enormous interests in exploiting CNTs in electrochemical sensors since they were first introduced in 1991.[47] CNTs are distinguished according to their structural properties[48]: a single-wall CNT (SWCNT) consists of a single graphitic sheet rolled into a cylinder (with 1 to 2 nm o.d. and several microns in length), and multiwall CNT (MWCNT) consists of graphitic sheets rolled into closed concentric tubes (with 50 nm o.d. and microns in length), each separated by van der Waals forces to have a gap of 3.4 Å. The electron transfer rate of the electrode is dominated by the CNT surface structure. Multi-wall CNTs have the side-wall structure similar to the graphite basal plane of highly ordered pyrolytic graphite (HOPG) and the open-end similar to the edge-plane of HOPG (Figure 1b), thus the electron transfer rate along the open ends of the CNTs is normally faster than their side-walls.[49]

In the past decade, researchers have emphasized on the use of anti-interference layers or artificial electron mediators for improving the selectivity of amperometric biosensors. Because of the electrocatalytic properties of CNTs, the artificial mediators to shuttle electrons between the enzymes and the electrodes are not needed, thereby eliminating the dependence on the mediators and enhancing the reproducibility. CNTs also minimize the surface fouling of biosensors, thus imparting higher stability onto these devices. With these advantages, CNTs have been investigated more extensively for electrochemical sensors for biomolecule than for metal ions. With CNTs, the overvoltages for the oxidation of hydrogen peroxide[50,51] and nicotinamide adenine dinucleotide (NADH) is reduced,[50,52] allowing low-potential detections of glucoses,[50,51,53] organophosphorous compounds,[54] and alcohols.[50] Nonetheless, our work and others have shown that CNTs are also very appealing for the development of mercury-free sensors and CNT nanoelectrode array for trace metal ion analysis. Given that we have previously reviewed the use of CNTs in electrochemical biosensors,[55,56] this review article will focus only its use for the analysis and sensing of metal ions.

SAMMS-Based Electrochemical Sensors: Principles and Configurations

Metal Ion Preconcentration at SAMMS Materials

For electrochemical sensors, successful preconcentration of trace metal ions (in μg/L or ppb) present in complex matrices requires that the sorbent meet a number of important criteria, including (a) high selectivity for target metals, (b) high loading capacity, (c) fast sorption kinetics, (d) excellent stability, and (e) ability to be easily regenerated. Pacific Northwest National Laboratory (PNNL) has been a leader in developing a new class of nanostructured sorbents, the self-assembled monolayer on mesoporous supports (SAMMS) by installation of various organofunctional moieties (Figure 2a for example) on MCM-41. Initially aimed at facilitating environmental cleanup of the complex nuclear/chemical waste, the materials are designed based on stereochemistry of the ligand, size of the chelation cavity, and hardness/softness of the metal ion and ligand field to selectively sequester a specific target species, including lanthanides,[57,58] actinides,[59-61] heavy and transition metal ions,[21,62-65] radiocesium,[66] radioiodide,[67] and oxometallate anions.[68] SAMMS materials are highly efficient sorbents that meet all the above requirements; their large adsorption capacity and multi-ligand chelation ability enhances their binding affinity and stability, their open

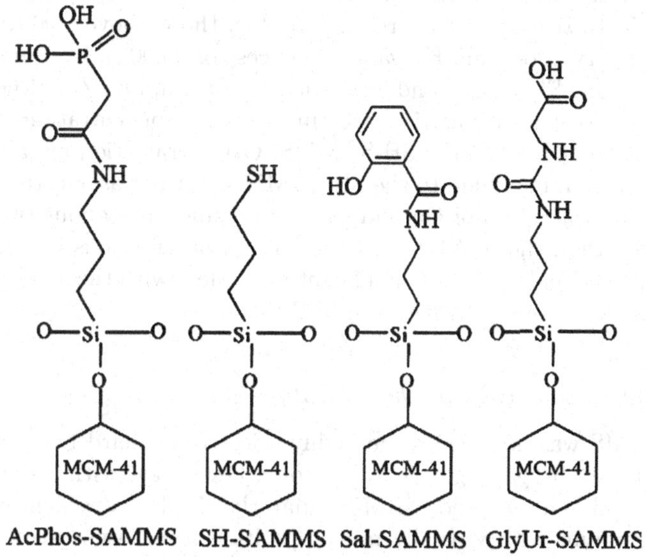

AcPhos-SAMMS SH-SAMMS Sal-SAMMS GlyUr-SAMMS

Figure 2. SAMMS materials with various organosilanes.

parallel-pore structure with pore size of 5.0 nm allows for easy diffusion of analytes into the nanoporous matrix, and their rigid ceramic backbone prevents pore closure due to solvent swelling resulting in fast sorption kinetics. Equilibrium adsorption on SAMMS is normally reached within minutes. The high surface area of silica substrate $(1,000\,m^2/g)$ and the monolayer self-assembly technique afford a high functional group density up to tenfold higher than simple functionalization methods.[41,65] With their superior properties over conventional sorbents, they are highly promising for the preconcentration of metal ions at electrochemical sensors. Figure 2 shows schematics of four SAMMS materials that have been investigated in our laboratory first as sorbent materials and later as electrode modifiers including thiol (SH)-SAMMS, acetamide phosphonic acid (Ac-Phos) SAMMS, glycinyl urea (Gly-Ur) SAMMS, and salicylamide (Sal) SAMMS.

Thiol (SH)-SAMMS

SH-SAMMS was designed to be an excellent sorbent for lead and mercury.[21,62] With thiol coverage of up to 82%, the SH-SAMMS has a loading capacity for Hg of 0.64 g Hg/g.[21] In the presences of Ag, Cr, Zn, Ba, and Na, the SH-SAMMS has demonstrated the mass-weighted distribution coefficients (K_d) of Hg and Pb in the order of 10^5 mL/g at neutral pH.[62] The higher the K_d values, the more effective the sorbent material is at sequestering the target species at trace concentrations. In general, K_d values above 500 mL/g are considered acceptable, those above 5,000 mL/g are considered very good, and K_d values in excess of 50,000 mL/g are considered outstanding.[60] Background ions, such as Na, Ba, and Zn, although present at high concentrations (i.e. 350 times higher concentrations than Hg and Pb), did not bind to the SH-SAMMS. Other transition metals like Cd, Cu, Ag, and Au, may bind to the SH-SAMMS, but not as effectively as Hg and Pb. The K_d values of Cu and Cd, for instance, are about twenty-five times lower than those of Hg. The presence of other anions (i.e. CN^-, CO_3^{2-}, SO_4^{2-}, PO_4^{3-}) also did not significantly interfere with the adsorption of lead and mercury ions onto the SH-SAMMS.[64]

Acetamide phosphonic acid (Ac-Phos) SAMMS

Ac-Phos SAMMS was designed to be highly selective for hard Lewis acids like actinide ions[69] by pairing a "hard" anionic Lewis base with a suitable synergistic ligand. The Ac-Phos SAMMS material displays excellent selectivity for plutonium with virtually no competition from large excess of a wide variety of metal cations (e.g. Ni, Pb, Cd, Hg, Cu, Cr, Ca, and Na) and anions (phosphate, sulfate, and citrate).[60] In absence of actinides, the Ac-Phos SAMMS containing phosphonic acids is a good sorbent material for heavy and transition metal ions such as Cu, Pb, and Cd; the K_d values are in the order of 10^4 mL/g in excess of 4000-fold by mole of acetate buffer and at solution per solid ratio of 200 mL/g.[65] It can uptake 99% of 2 mg/L Cd^{2+} within a minute at a solution to solids ratio of 200 mL/g.[65]

Glycinyl-Urea (Gly-Ur) SAMMS

Gly-Ur SAMMS contains carboxylic acid and amide carbonyl groups that are arranged in a suitable fashion with large chelating cavity and may bind to metal ions by many different schemes.[41] From pH 4.6 to 6.4 (Table 1), Gly-Ur SAMMS has very good affinity for Pb^{2+} and Hg^{2+} and moderate affinity for Cu^{2+}: the K_d values are in the order of 10^4 mL/g for Pb and Hg and 10^3 mL/g for Cu even in excess of 50-fold by mole of Ca^{2+} and 10,000-fold by mole of $NaNO_3$ per metal ions.[41] At pH 2, the K_d dropped significantly suggesting that the regeneration of the sorbent materials is very feasible using an acid wash. After ten cycles of washing in 0.5 M HCl and reuse, Gly-Ur-SAMMS showed no sign of material degradation and the binding affinity of europium remained unchanged.[60]

Salicylamide (Sal) SAMMS

Sal-SAMMS contains stabilized phenolic group that is in resonance with the adjacent amide carbonyl group, thereby conveniently providing an easy anchoring point for metal ions.[57] Because of the size its ligand cavity, Sal-SAMMS prefers to sequester smaller lanthanide cations such as Lu and Eu over the larger lanthanides like La as demonstrated by the K_d values in Table 1.

Table 1. Distribution coefficients (K_d) for Gly-Ur-SAMMS at initial metal ions of 0.5 mM in 0.1 M $NaNO_3$ (Ref. 41) and for Sal-SAMMS at initial lanthanides of 2 ppm in 0.1 M $NaNO_3$ (Ref. 57) both with solution/solids of 200.

Gly-Ur-SAMMS			
Metal	pH 2.0	pH 4.6	pH 6.4
Cu	6	3438	2111
Hg	2023	22274	17563
Pb	23	39971	38071
Sal-SAMMS				
Cation	pH = 1	pH = 2.5	pH = 4.5	pH = 6.5
La	0	0	17	7870
Nd	0	0	18	> 100,000
Eu	0	129	17100	48000
Lu	6	2	20	> 100,000

Sensors Based on SAMMS-Conductive Materials

Carbon is a versatile electrode material that can undergo various chemical and electrochemical modifications to produce suitable surfaces for high electrode responses. Carbon electrodes have a wide useful potential range, especially in the positive direction, because of the slow kinetics of carbon oxidation. Conductive carbon paste and graphite ink provide easy ways to construct electrodes for many decades, but their use in conjunction with OMSs as electrode materials are still limited. At PNNL, thiol,[45] acetamide phosphonic acid,[39,42,43,46] glycinyl urea,[41] and salicylamide[40] self-assembled mesoporous silica materials have been incorporated into carbon paste electrodes (CPEs) or screen-printed carbon electrodes (SPCEs) for the electroanalysis of many metal ions important for environmental and occupational monitoring. Figure 3 shows various configurations of electrodes and sensors that have been investigated in our laboratory, while Table 2 summarizes our work and others using different types of ordered mesoporous silica materials in electrochemical sensors for metal ions. The lower detection limits of electrochemical sensors employing electrolytic deposition are often estimated from their signal-to-noise ratio (e.g. 3S/N). For SAMMS-based electrodes that preconcentrate metal ions at open-circuit, detection limits are obtained experimentally as a function of the preconcentration time.

Detection principles

Unlike other classes of electrode modifiers that are intrinsically conductive such as zeolites,[70-72] metal oxide,[73] and carbon nanotubes, SAMMS materials are electronic insulators. To use SAMMS materials in electroanalysis, they must be (a) embedded into the conductive matrix like graphite paste or carbon ink, or (b) in close contact with the conductive surface. On the contrary, since SAMMS particles are not conductive, the high surface of the material does not contribute to the charging current for the sensors, leading to the low background current for the measurements.

The overall analysis involved a two-step procedure: a preconcentration step at open circuit, followed by medium exchange to a pure electrolyte solution for the voltammetric quantification. During the preconcentration step, metal ions (e.g. M(II)) are accumulated on SAMMS, which are embedded on the electrode surface, by complexation with the functional groups (e.g. -SH) on SAMMS:

$$SiO_2 ---- SH + M(II) \leftrightarrow SiO_2 ---- S\text{-}M(II) \qquad (1)$$

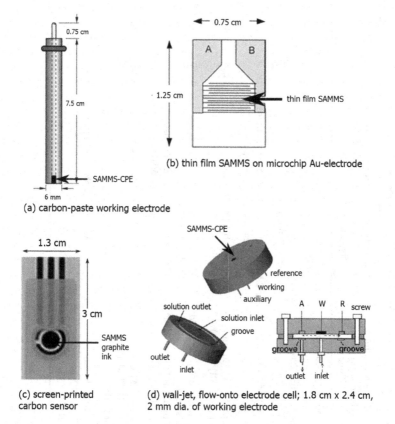

Figure 3. Four electrode configurations used at PNNL, from Ref. 46, reproduced by permission of Elsevier.

The detection step consists of electrolysis and stripping. In the electrolysis step, accumulated M(II) are desorbed in an acidic medium, then cathodically electrolyzed at a negative potential (i.e. $-1.0\,V$) for a certain period (i.e. 60 seconds) to convert metal ion (M(II)) to elemental metal M(0):

$$SiO_2 \text{----} S\text{-}M(II) + H^+ \leftrightarrow SiO_2 \text{----} SH + M(II), \tag{2}$$

$$M(II) + 2\,e^- \leftrightarrow M(0) \tag{3}$$

In the detection step, the elemental metal M(0) is subsequently detected by an anodic stripping voltammetry technique:

$$M(0) - 2e^- \leftrightarrow M(II) \tag{4}$$

Table 2.   Ordered mesoporous silica materials in electrochemical sensors for metal ions

Mesoporous Silicas	Electrode Type	Metal Ions	Detection Limits (Preconcentration Time)	Linear Range (Preconcentration Time)	Reproducibility (% RSD)	Figure	Ref.
SH-SAMMS	CPEs	Simultaneous Hg/−Pb	0.5 ppb Pb^{2+}/3 ppb Hg^{2+} (20 min), 10 ppb Pb^{2+}/20 ppb Hg^{2+} (2 min)	10–15 ppb Pb^{2+} (2 min), 2–1600 ppb Hg^{2+} (2 min)	< 5%	3a, 4	45
Gly-Ur-SAMMs	CPEs	Simultaneous Cu/Pb/Hg	1 ppb Pb^{2+} (5 min)	2.5–50 ppb pb^{2+} (2 min)	3.5% (6 samples), 50 ppb	3a, 5	41
Ac-Phos-SAMMS	CPEs	U(VI)	1 ppb U(VI) (20 min) 25 ppb U(VI) (5 min)	25–500 ppb U(VI) (5 min)	< 5%	3a, 6	43
Ac-Phos-SAMMS	CEPs	Simultaneous Cu/Pb/Cd	0.5 ppb Cu^{2+}/Pb^{2+}/Cd^{2+} (20 min) 10 ppb Cu^{2+}/Pb^{2+}/Cd^{2+} (2 min)	10–2000 ppb Cu^{2+}/Pb^{2+}/Cd^{2+} (2 min)	5% (7 samples), 12% (5 sensors), 50 ppb	3a, 7	42
Monsil	CPEs	Hg^{2+}, Ag^+, Pb^{2+}, Cu^{2+}	2×10^{-4} M Cu^{2+}, 5×10^{-4} M Pb^{2+}, 7×10^{-4} M Ag^+	NA	6% (5 samples)	NA	30
MCM-41	CPEs	Cu^{2+}, Hg^{2+}	3×10^{-8} M Cu^{2+}, 5×10^{-8} M Hg^{2+}, (3S/∧ N)	NA	5% (6 samples), 1×10^{-5} M	NA	27–29

Table 2. (*Continued*)

Mesoporous Silicas	Electrode Type	Metal Ions	Detection Limits (Preconcentration Time)	Linear Range (Preconcentration Time)	Reproducibility (% RSD)	Figure	Ref.
Propylammonium-MCM-41	CPEs	Cu^{2+}	1×10^{-5} M Cu(II) (1 min)	NA	NA	NA	80
SO_3H-MCM-41	CPEs	Cu^{2+}	1×10^{-5} M Cu(II)	NA	NA	NA	81
Thin-film SH-SAMMS	Microchip Au array	Pb^{2+}	25 ppb Pb^{2+} (30 min), 250 ppb Pb^{2+} (5 min)	250-5000 ppb Pb^{2+} (5 min)	< 5%	3b, 8	44
Thin-film mercaptopropyl-MPS	Gold, platinum, GC	Hg^{2+}	1×10^{-5} M Hg^{2+} (1 min)	NA	NA	NA	93
Ac-Phos-SAMMs	CPEs embedded in wall-jet, flow-onto cell	Pb^{2+}	1 ppb Pb^{2+} (3 min)	1–25 ppb Pb^{2+} (3 min)	2.5% (7 samples), 10 ppb	3d, 9	46
Ac-Phos-SAMMS	SPCEs	Pb^{2+}	2.5 ppb Pb^{2+} (5 min)	2.5–500 ppb Pb^{2+} (5 min)	5% (6 samples), 10% (5 sensors), 500 ppb	3c, 10	39
Sal-SAMMS	SPCEs	Eu^{3+}	10 ppb Eu^{2+} (10 min)	75–500 ppb Eu^{3+} (5 min)	10% (5 samples), 10% (5 sensors), 100 ppb	3c; 11	40

From a voltammogram, the peak location is used to identify metal ions (e.g. Hg at 0.3 V, Cd at -0.8 V, and Pb at -0.5 V, approximately), while the response (either peak area in volt-ampere or peak height in ampere) is a function of metal ion concentration in the sample. After the metal preconcentration (step (1)), the detection step is performed in a clean medium, thus the interference only affects the preconcentration step and not the detection step unlike other types of electrochemical sensors where the stripping steps are performed in the same solution.

Factors affecting the performance of the SAMMS modified electrodes were investigated. Table 3 summarizes typical operating conditions. It is worth noting that most of the operating conditions (e.g. electrode compositions, electrolysis/stripping media conditions, electrolysis time) for metal ion detection obtained using SH-SAMMS-CPEs can be used effectively with the CPEs modified with other SAMMS materials with slight modifications. The most sensitive and reliable electrode contains about 10 to 20 wt.% SAMMS in carbon paste electrodes or 10 wt.% SAMMS in screen-printed carbon electrodes. The protocol is a sequence of a two to five minutes preconcentration period, a 60 seconds electrolysis period of the preconcentrated species at a negative potential in 0.2 to 0.5 M acid solution, followed by square wave anodic stripping voltammetry to a positive-potential direction also in the same acid solution. The binding between SAMMS and metal ions is reversible, therefore the SAMMS-CPEs can be easily regenerated without damaging the ligand monolayer by desorption of the preconcentrated species in an acidic solution. By choosing an appropriate acid solution, the electrodes usually are ready to be used again without preconditions. We found that SAMMS-based electrodes can be reliable even after 80 consecutive runs with no need for surface renewal.

SH-SAMMS-CPEs

Figure 4(a) shows simultaneous detections of Pb^{2+} at -0.5 V and Hg^{2+} at 0.3 V, while Figure 4(b) shows their linear calibration curves measured simultaneously at a SH-SAMMS modified carbon paste electrode.[45] The large differential potential (ΔE) of 0.8 V enables Pb^{2+} and Hg^{2+} to be detected simultaneously without interference with each other. After two minutes of preconcentration, the area of each anodic peak of Pb^{2+} and Hg^{2+} was proportional to their concentrations in the range of 10 to 1500 ppb Pb^{2+} and 20 to 1600 ppb Hg^{2+}, respectively. Because of the extremely large surface area of SAMMS and the self-assembly chemistry, SAMMS contains a high loading density of the functional group, which minimizes the competition for the binding sites among the metal ions and enables simultaneous detection of the metal ions without losing signal intensity.

Table 3. Typical operating parameters for the voltammetric measurements of metal ions at SAMMS-based electrochemical sensors

Mesoporous Silicas	Electrode Type	% SAMMS	Metal Ions	Preconcentration*	Electrolysis	Stripping	Anodic Peak* Location
SH-SAMMS	CPE	20%	Hg/Pb	2 min	-1.0 V, 60s in 0.2 M HNO_3	-1.0 V to 0.6 V	-0.55 V (Pb), 0.25 V (Hg)
Ac-Phos-SAMMS	CPE	20%	Cu/Pb/Cd	2 min	-1.0 V, 60s in 0.2 M HNO_3	-1.0 V to 0.4 V	-0.47 V (Pb), -0.1 V (Cu), 0.23 V (Hg)
Gly-Ur-SAMMs	CPE	12.5%	Cu/Pb/Hg	2 min	-0.8 V, 60s in 0.3 M HNO_3	-0.8 V to 0.6 V	-0.47 V (Pb), -0.1 V (Cu), 0.23 V (Hg)
Ac-Phos-SAMMS	CPE	20%	U	U in acetate buffer (pH 5), 5 min	-0.8 V, 60s in 0.2 M HNO_3	-0.8 V to 0.4 V	-0.37 V
Thin-film-SH-SAMMS	Au-electrode array	100%	Pb	5 min	-1.0 V, 60s in 0.1 M HNO_3	-1.0 V to -0.4 V	-0.48 V
Ac-Phos-SAMMS	CPEs embedded in wall-jet, flow-onto cell	10%	Pb	360 µL flowed at 2µL/s	-1.0 V, 70s in 0.3 M HCl	-0.8 V to -0.2 V	-0.64 V
Ac-Phos-SAMMS	SPCEs	10%	Pb	3 min	-1.0 V, 120s in 0.3 M HCl	-0.7 V to -0.49 V	-0.6 V
Sal-SAMMS	SPCEs	10%	Eu	8 mL, Eu in acetate (pH 4.6), 5 min	-0.9 V, 60s in 20 µL of 0.1 M NH_4Cl (pH 3.5)	-0.95 V to -0.4 V	-0.72 V

*Unless specified otherwise, electrodes were immersed in 15 mL of stirred metal ions in DI water (pH 5.5) under open circuit.

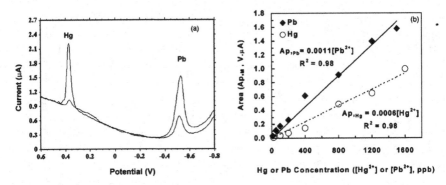

Figure 4. Simultaneous voltammetric detection of Hg^{2+} and Pb^{2+} at SH-SAMMS-CPEs: (a) anodic peaks, and (b) linear responses with a function of their solution concentrations, from Ref. 45, reproduced by permission of The Royal Society of Chemistry.

Gly-Ur-SAMMS-CPEs

Figure 5 shows the voltammetric response (current) of Pb as a function of (a) preconcentration time and (b) Pb solution concentration at a Gly-Ur-SAMMS modified carbon paste electrode.[41] A linear response current was obtained from one to at least 15 minutes of the preconcentration period (at open circuit) in 25 ppb Pb^{2+} solution and from 2.5 to 50 ppb Pb^{2+} in the solution after two minutes of preconcentration period. The percent relative standard deviation (%RSD) of 6 consecutive measurements of 50 ppb Pb^{2+} was found to be 3.5. The same electrode can also be used for simultaneously detection of lead, copper, and mercury as shown in Figure 5c where the peak are found at $-0.47\,V$, $-0.06\,V$, and $0.23\,V$, respectively.

Ac-Phos-SAMMS-CPEs for uranium

Uranium is one of the most important actinides at numerous nuclear sites. Most recent voltammetric detection of uranium has been made successfully at sub-nanomolar level by employing electrolytic accumulation of uranyl species in complexes with chelating agents like catechol,[74] oxine,[75] chloranilic acid,[76] cufferon,[77] and salicylidenimine[78] at mercury drop electrodes or cufferon[79] at Bi-film electrode. Because of its excellent affinity and selectivity for actinide ions, the Ac-Phos-SAMMS was used to modify a carbon paste electrode for uranium detection[43] which will eliminate the potential interferences from heavy and transition metal ions existing at higher concentrations in most wastes.

Uranium in aqueous solution will form complexes with ligands that vary as a function of solution pH. Speciation can greatly affect the voltammetric

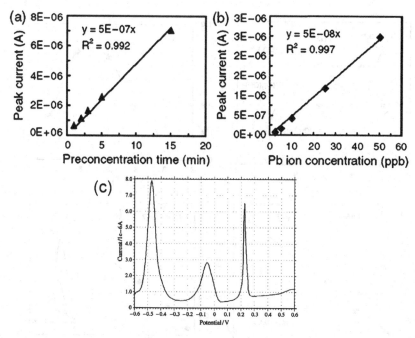

Figure 5. Voltammetric detection of metal ions at Gly-Ur-SAMMS-CPEs: (a) linear response of 25 ppb Pb^{2+} as a function of preconcentration time, (b) after two minutes preconcentration as a function of Pb^{2+} solution concentration, and (c) simultaneous detection of Pb^{2+}, Cd^{2+}, and Hg^{2+}, from Ref. 41, reproduced by permission of the American Scientific Publishers.

responses of uranium (matrix effect). The type and concentration of anions, the concentration of uranium, and the pH of the solution are factors that determine which uranium complex will be preferentially formed in the solution and adsorbed on the SAMMS. Sensors that rely on electrodeposition of metal-ligand complexes on electrode surfaces before detecting the metal within the same solution are often affected by the presence of competing ligands in both deposition and detection steps. Using SAMMS-based electrodes with medium exchange in the detection step after uranium preconcentration can overcome the matrix effect and ligand interference.

The pH of the solution was predicted to affect the distribution of uranyl complexes with acetate (used as the buffer), hydroxyl, and carbonate ligands as shown in Figure 6(a). In the pH range of 2 to 5, the dominant uranyl species have positive or neutral charges (e.g. UO_2^{2+}, $UO_2(Ac)^+$), which bind stronger to the phosphonic acid on Ac-Phos-SAMMS as pH increases, leading to increased voltammetric response (Figure 6b). Above pH 6, uranyl

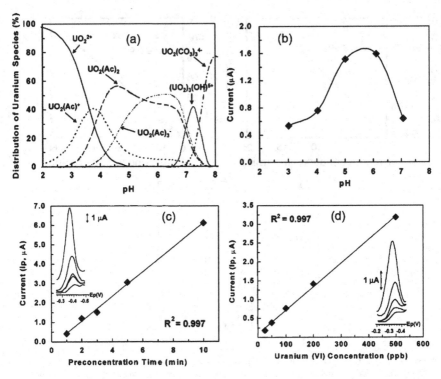

Figure 6. Voltammetric detection of U(VI) at Ac-Phos-SAMMS-CPEs: (a) predicted speciation of 500 ppb U(VI) in 0.05 M Na-acetate, (b) voltammetric responses of 500 ppb U(VI) after three minutes preconcentration as a function of pH, (c) linear response of 500 ppb U(VI) at pH 5 as a function of preconcentration time, and (d) linear response after five minutes preconcentration as a function of U(VI) concentration, from Ref. 43, reproduced by permission of Wiley-VCH.

species with negative charge appear to cause repulsive forces, resulting in low uranium adsorption and low voltammetric responses. After a five-minute-preconcentration period of U(VI) in acetate buffer (pH 5), the current at −0.37 V was proportional to the preconcentration time (Figure 6c) and U(VI) concentration in the range of 25 to 500 ppb (Figure 6d). The uranium detection limits improved significantly with longer preconcentration time (Table 2).

Ac-Phos-SAMMS-CPEs for heavy metal ions

Ac-Phos-SAMMS has been used to modify a carbon paste electrode for simultaneous detection of cadmium (Cd^{2+}), copper (Cu^{2+}), and lead

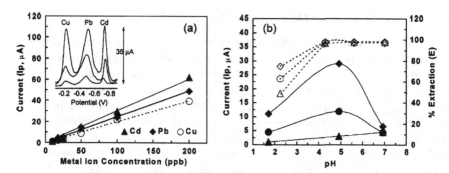

Figure 7. Simultaneous detection of metal ions at Ac-Phos-SAMMS-CPEs: (a) voltammetric responses of $Pb^{2+}/Cd^{2+}/Cu^{2+}$ after two minutes preconcentration as a function of their solution concentrations, and (b) responses of 50 ppb metal ions as a function of pH (two minutes preconcentration), from Ref. 42, reproduced by permission of Elsevier.

(Pb^{2+}) in the absence of actinides.[42] Figure 7(a) shows that the linear voltammetric responses of Cu^{2+}, Pb^{2+}, and Cd^{2+} after two minutes of preconcentration in the metal ion concentrations range of 10 to 200 ppb. The signals of Cu were not affected by the presence of Pb^{2+} and Cd^{2+} in the concentration range studied, owing to the much higher loading capacity of the SAMMS than the total moles of metal ions. Above 200 ppb, the detection of each Cu^{2+}, Pb^{2+}, and Cd^{2+} started to interfere with each other because of the close proximity of their signals (i.e. $\Delta E_{Cu-Pb} = 0.3\,V$, $\Delta E_{Pb-Cd} = 0.25\,V$). The sample pHs were found to exert a significant but predictable effect on the preconcentration process. In Figure 7(b), the voltammetric responses of Cu, Pb, and Hg followed the adsorption isotherms in the pH range of 2 to 6. Above pH 6, while the adsorption isotherms still remained at maximum, the peak currents were negligible due to the formation of metal hydroxide complexes that may be sparingly soluble and not adsorb on the electrode surface.[31] Such information about speciation can be obtained only at electrochemical sensors and not by ICP-MS which yield only the total metal concentrations.

In addition to our work summarized above, the Walcarius group has developed carbon paste electrodes modified with propylammonium-[80] and mercaptopropyl-[81] functionalized-mesoporous silica materials for voltammetric detection of Cu^{2+} (Table 2). To exploit the amine ligands on the propylammonium moieties, a five-minute electrochemical pre-treatment at $-1.3\,V$ was performed in order to deprotonate the N-bearing group.[80] In the other work, mercaptopropyl functionalized mesoporous silica was oxidized by H_2O_2 to first create SO_3H functional groups.[81]

Sensors Based on SAMMS Thin-Film

Mesoporous silica films with pore sizes of up to 100 Å have been synthesized from spin-casting of silica sol gels (normally consisting of silica precursor, acid solution, organic solvent, and water) by a surfactant-templating process in which the pores are formed upon removal of the surfactant by calcinations.[82–86] The precise design and control of pore structure, pore size/volume, and pore orientation of the silica films can be achieved by using structure-directing agents, such as the non-ionic surfactant Pluronic F-127 ($EO_{106}PO_{70}EO_{106}$). By controlling the film thickness (e.g. changing the composition of the precursor solution, spin rate, and calcination conditions), a continuous, defect-free, mesoporous silica thin film is produced.[44] Spin-casted silica thin films have been found to be stable in pure water and acidic electrolyte solutions, but unstable in basic aqueous solutions due to the dissolution of the mesoporous silica framework.[35] Because electrodes modified with these silica films cannot be used in basic solutions, the silanol groups, inherent to the nanoporous silica, are not useful in binding with positively charged metal ions.[27] Therefore, without functional ligands attached, the spin-cast films of mesoporous silicas are often used as molecular sieves in electrodes by either blocking or improving the voltammetric responses of the analyte species.[35] To remove this limitation, we have used a self-assembly technique to attach organic thiol functional groups onto the mesopore surfaces of the silica thin film that was previously "spin-casted" and calcined on a surface of microchip-based gold electrode array, as shown in Figure 3b. This two-step approach has also been used by the Sanchez group.[87,88] Several surface characterization methods were performed on the mesoporous silica thin film before and after thiol functionalization. TEM measurement of the SH-FMS film shows short-range ordered mesopores. The nitrogen adsorption/desorption isotherms of the untreated SiO_2 thin film are of type IV with a clear H_2 type hysteresis loop, which is consistent with the film's large ratio of pore size to pore spacing[89,90] and cubic arrangement of pores.[91] From BET analysis, the calcined film had a large primary pore diameter of 7.7 nm, a porosity of 60%, and a high BET surface area of 613 m^2/g. The Fourier-Transform Infrared (FTIR) spectra indicate that organic thiol monolayers were successfully immobilized inside the nanopores of the mesoporous silica. After the thiol immobilization, increased in the film's refractive index suggests the decrease in film porosity from 66% to 23%. The similarity of the X-ray Diffraction (XRD) patterns before and after thiol functionalization indicates that the structure of mesoporous film remained unchanged after the thiols attachment.

  With the SH-SAMMS thin film-modified electrode array, the preconcentration of Pb can be performed in slightly acidic electrolyte solutions or

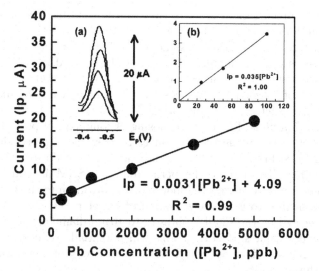

Figure 8. Voltammetric response of Pb^{2+} after five minutes preconcentration at SH-SAMMS thin-film on a microchip-Au-array, insets show (a) the Pb voltammograms, and (b) responses after 30 minutes preconcentration, from Ref. 44, reproduced by permission of the Royal Society of Chemistry.

neutral non-electrolyte solutions[44] which corresponded to previous batch sorption study of SH-SAMMS in a powder form.[62] In non-basic solutions, the mesoporous silica films are more likely to remain stable for a long period of time. The voltammetric responses increased linearly with Pb^{2+} concentrations ranging from 250 to 5000 ppb after five minutes of preconcentration and from 25 to 100 ppb after 30 minutes of preconcentration (Figure 8 and Table 2). The large dynamic range for lead signals is a result of the high loading capacity of the SAMMS thin-film in comparison to that of conventional chemically self-assembled thin-film electrodes that often has smaller linear range. Through the silica-surfactant self-assembly process, conductive additives or binders that are often used to mix with SAMMS to make an electrode can be eliminated. Thus the electrodes can be constructed in a manner that is highly reproducible which is very important for any two-dimensional surface reaction to be successful.[92] The SAMMS thin-film can be made to adhere to a wide variety of surfaces. The microchip-based SH-SAMMS electrodes also have low maintenance requirements (no activation and regeneration were required) and can be integrated into microfluidic devices for field applications.

In a different approach, the Walcarius group has recently used a one-step technique by spin-casting of a sol-gel mixture containing both alkoxysilane and organosilane precursors in order to coat thin-films of mercaptopropyl-mesoporous on gold, platinum, and glassy carbon electrodes.[93] The surfactant was removed by a solvent extraction instead of calcination. The electrodes were used to accumulate 10^{-5} M of Hg(II) in 0.1 M HNO$_3$ at open circuit prior to its stripping voltammetry in 3 M HCl (Table 2).

Improvement of sensor performances may be accomplished by manipulating the pore structure, size and orientation as well as the thickness of the SAMMS thin-films. Partial pore-closing or long axes of pores being oriented parallel to the plane of the film caused by the exposed side of the film being dried first.[94] may often cause slow metal diffusion into the SAMMS thin-films. To increase the permeability of SAMMS thin-films, PNNL researchers[95] have engineered hierarchical porosity into the films by using cellulose nitrate to generate a system of larger pores (15 to 30 nm diameter) connected by the smaller pores (2 to 10 nm diameter) generated using traditional organic surfactants. This will potentially reduce the preconcentration time and increase the detection sensitivity of metal ions at the SAMMS-thin-film sensors.

Field-Deployable SAMMS-Based Sensors

Once the sensors based on SAMMS were successfully evaluated and optimized in batch experiments using electrode configuration as in Figure 3a, two sensor platforms have been investigated for field applications: (1) a portable analyzer employing a stripping voltammetry technique at a wall-jet, flow-onto electrode cell (Figure 3d), and programmable sequential injection, and (2) reusable screen-printed carbon electrodes (SPCEs) (Figure 3c). Lead (Pb) has been used as the model metal for evaluating the two sensor platforms developed from Ac-Phos SAMMS modification of electrodes because: (a) Pb is one of our most concerned metals due to its high toxicity and common occurrence of Pb poisoning from environmental and occupational exposures especially in children,[96,97] and (b) Pb has excellent binding capacity, selectivity, and rate[65] on Ac-Phos SAMMS.

Automated Portable Analyzer Based on Flow-Injection/ Stripping Voltammetry

To develop the next-generation analyzers that are portable, fully-automated, and remotely-controllable, we have integrated a sequential injection analysis (SIA) method with the nanostructured electrochemical sensors. Since it was first introduced in 1990, SIA is gaining popularity because of its economical use of samples and reagents, robustness,

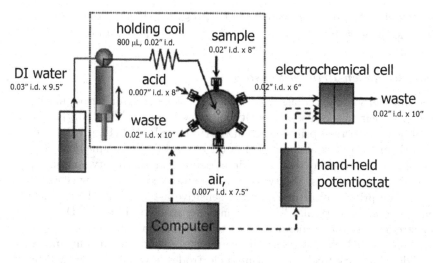

Figure 9. Portable metal ion analyzer consisting of programmable sequential injector and wall-jet, flow-onto cell having an Ac-Phos-CPE as the working electrode, from Ref. 46, reproduced by permission of Elsevier.

inexpensiveness, and simple design of instrumentation.[74,98,99] To detect Pb at low ppb level, carbon paste modified with Ac-Phos SAMMS was embedded in a very small wall-jet (flow-onto) electrochemical cell (Figure 3d) in connection with the SIA system as shown in Figure 9.[46] Microliters of samples and reagents are injected into the system by a programmable sequential flow technique. The portable system yielded linear calibration curve in the useful range of 1 to at least 25 ppb of Pb(II). The automation increases the reproducibility by eliminating human errors and batch-to-batch variation, thus increasing the measurement reproducibility; the % RSD of seven consecutive measurements of Pb^{2+} was reduced from 5% in batch experiments to 2.5 using the automated system.

The effect of interferences on the Pb detection was evaluated. An ionic species may be considered as the interference if: (a) it can out-compete the target metals for the binding sites on SAMMS during the preconcentration step and (b) once preconcentrated on SAMMS its peak response can overlap that of the target metals in the stripping step. Organic molecules and anions can also act as interference if it can out-compete SAMMS for the target metal ions. From batch competitive sorption experiments,[65] Ac-Phos-SAMMS has an affinity for metal ions in the decreasing order of $Pb^{2+} > Cu^{2+} > Mn^{2+} > Cd^{2+} > Zn^{2+} > Co^{2+} > Ni^{2+} \sim Ca^{2+} \gg Na^+$. This corresponds to the results obtained at an Ac-Phos-SAMMS sensor that large excess of 100-fold (by mole) of Ca, 70-fold of Zn, Ni, and Co, and 10-fold of Mn did not interfere with Pb signals.[46] The electrode was reliable for at least 90 measurements over five days of operation.

Reusable Screen-Printed Sensors

Disposable sensors for the assay of toxic metal ions are gaining popularity because of their ease-of-use, simplicity and, low costs.[100–106] Of all the disposable sensors, screen-printed carbon electrodes (SPCEs) (Figure 3c) coupled with an adsorptive stripping voltammetry (AdSV) technique have been increasingly investigated due to their measurement sensitivity, simplicity during field applications, ability to be mass-produced at very low costs.[107] Most screen-printed electrodes for the sensitive assay of metal ions have been based on mercury film,[108–111] or mercury oxide particles.[107] Disposal of electrodes containing mercury leads to occupational and environmental heath concerns which may result in future regulation of the mercury-based electrodes. Mercury-free screen-printed electrodes have been developed by employing gold,[112] silver,[113] or bare carbon electrodes by applying suitable reduction potential to accumulate the target metals (e.g. Cu and Pb).[104,114] However, the sensitivity, reliability, and cost competitiveness of such electrodes are yet to reach those of the mercury based electrodes. Although chemically modified screen-printed electrodes have been developed by using ligands such as 1-(2-pyridylazo)-2-naphthol[103] or calixarene,[102] the ligand modification has been done *ex situ* by drop coating technique, followed by the complexation of the ligands and Pb^{2+} in ammonia buffer. Thus they often have the disadvantages of not being user-friendly, short electrode life time, and poor inter-electrode reproducibility.

Ac-Phos-SAMMS-SPCEs

We have modified screen-printed carbon electrodes (SPCEs) with Ac-Phos SAMMS for Pb detection.[39] A screen-printed sensor consists of the built-in three electrode system: screen-printed SAMMS-graphite ink mixture (10 wt. % SAMMS) as a working electrode and screen-printed carbon as counter electrode, and Ag/AgCl as the reference electrode, situated on a 1.3 cm × 3 cm × 0.5 mm plastic substrate (Figure 3c). Similar to SAMMS-CPEs, preconcentration of Pb^{2+} at SAMMS-based sensors can be accomplished at open circuit potential without electrolyte and solution degassing. Linear calibration curve was found in the range of zero to at least 100 ppb Pb^{2+} after five minutes of preconcentration as shown in Figure 10(a). Cadmium, lead, and copper can also be detected simultaneously at the screen-printed sensors (Figure 10b). Even when being mixed with graphite ink, SAMMS on the screen-printed

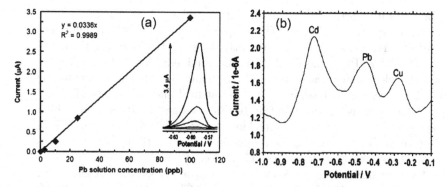

Figure 10. Metal ion detection after five minutes preconcentration at Ac-Phos-SAMMS-SPCEs: (a) linear response of Pb^{2+} as a function of solution concentration, and (b) simultaneous detection of 90 ppb Cd^{2+}/18 ppb Pb^{2+}/18 ppb Cu^{2+}, from Ref. 39, reproduced by permission of Elsevier.

sensors still adsorbed Pb well within a short period, which is attributed to the rigid, open-parallel mesopores and the suitable interfacial chemistry of SAMMS that allow easy access to hydrated metal ions.[65] The SAMMS-SPCEs can be reused for tens of measurements with minimal degradation, thereby enabling the establishment of the calibration curve and lowering the costs compared to single-use SPCEs. Reproducibilty (%RSD) was found to be 5% for a single sensor (six measurements) and 10% for five sensors. The inter-electrode reproducibility can be improved through the precision of manufacturing of the sensors, in which SAMMS modification is done *in situ*.

Sal-SAMMS-SPCEs

Rare earth cations like lanthanides are difficult to electrochemically reduce to elemental forms; the voltammetric detection limits of europium (Eu) in perchlorate solutions after being directly reduced at a boron-doped diamond electrode were only in millimolar level.[115] In order to detect Eu at nanomolar level, researchers have electrolytically accumulated Eu in the presence of thenoyltrifluoroacetone[116] or salicylic acid[117] in aqueous solutions at the mercury drop electrodes. The preconcentration of lanthanides at an electrode surface by non-electrolytic methods are more preferable.[118,119] Nafion-coated electrodes have been used to preconcentrate Eu^{3+} at open circuit potential, but the Eu detection limits by normal voltammetric methods are still in micromolar level since Eu^{3+} is only weakly incorporated by ion-exchange into Nafion coating[10,11] and is completed by NH_4^+ used as the electrolyte.[120]

Eu are often present in nuclear wastes and used as Am(III) mimic. It is a very similar in size to U(IV), Np(IV), and Pu(IV).[58] Having considered its significance, we have recently developed sensors for Eu based on colloidal-Au-SPCEs modified with salicylamide-SAMMS. In acidic solutions, the electroanalysis of trace Eu is highly challenging because the Eu^{3+}/Eu^{2+} reduction peak can be found at fairly negative potential (e.g. -1.2) where the hydrogen evolution reaction occurs resulting in very high background.[115] Ugo *et al.*[10] and Moretto *et al.*[120] found that NH_4Cl improved the reversibility of the reduction of rare earth metals at Nafion-modified glassy carbon electrodes. Therefore, after Eu^{3+} preconcentration with Sal-SAMMS-graphite ink on a screen-printed electrode (Figure 3c), Eu^{3+} was converted to Eu^{2+} by applying $-0.9\,V$ for 60 seconds in $20\,\mu L$ of NH_4Cl solution (pH 3.5), followed by the stripping voltammetry in the same solution yielding an Eu^{3+}/Eu^{2+} oxidation peak at $-0.72\,V$. The peak current was a linear function of Eu^{3+} in the samples as shown in Figure 11. After five to ten minutes of preconcentration at open circuit potential, ppb levels of Eu can be detected. Ten-fold higher concentrations of Lu and Nd did not interfere with Eu adsorption, while that of La reduced the Eu peak current by 15%. As the concentration of NH_4Cl increased from 0.05 M to 0.2 M, the Eu responses also increased, suggesting NH_4^+ did not compete

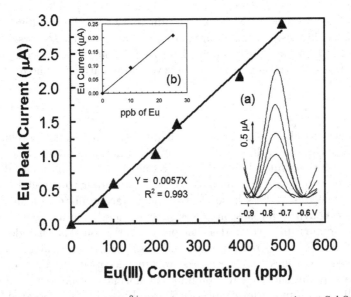

Figure 11.　Linear response of Eu^{3+} after five minutes preconcentration at Sal-SAMMS-SPCEs (with built-in colloidal Au), insets (a) the corresponding voltammograms, and (b) the linear response after ten minutes preconcentration, from Ref. 40, reproduced by permission of the Royal Society of Chemistry.

with Eu for the binding sites of Sal-SAMMS, unlike at the Nafion-film electrodes. Other lanthanides and actinides are less likely to interfere with Eu sorption on Sal-SAMMS at this Eu concentration range because the large capacity of SAMMS will minimize the competitive binding.[42] The Sal-SAMMS sensors had large working pH of the Eu solutions from 2 to 6.5 although the K_d measured from batch sorption was very small at pH below 2.5 (Table 1). This is perhaps due to the local pH effect; the ligand field is buried in a graphite ink matrix of the SPCEs which may restrict the transport of H^+ into such a matrix causing the effective pH at the Sal-ligand interface to be more moderate than it is out in the bulk solution. Because the close proximity of Pb peak (found at -0.6 V) to Eu peak, Pb at much higher concentration (e.g. four-fold by mole) was found to reduce Eu peak current by 40%, while the influence of Pb was negligible at one-to-one molar ratio of Pb and Eu. If needed, separation of Pb from lanthanides prior to the detection may be easily done using materials like Ac-Phos SAMMS or Gly-Ur SAMMS.[57] Nevertheless, Sal-SAMMS-SPCEs provide a simple, reproducible, and mercury-free method for Eu detection.

CNT-Based Electrochemical Sensors

Various Methods for Preparation of CNT-Based Sensors

Although CNTs are relatively new in analytical fields, their unique electronic (electron transfer rate similar to edge-plane graphite), chemical (biocompatibility and ability to be covalently functionalized), and mechanical properties (three times stronger than steel) make them extremely attractive for chemical and biochemical sensors.[121,122] Various methods have been used to prepared CNT-based sensors: (a) casting of CNT thin films, from the suspensions of CNTs in solvents,[51,52,123–129] such as (DMF),[128] acetone,[129] Nafion,[51,126] dihexadecyl hydrogen phosphate (DHP),[127] and sulfuric acid,[52] prior to being coated on electrode surfaces, (b) using CNTs as paste electrodes or electrode composites,[130–133] and (c) using aligned CNT as electrode substrates.[53,134–141] With specific to metal ion sensors (summarized in Table 4), dihexadecyl hydrogen phosphate (DHP)[127] and Nafion[126] have been used to disperse MWCNTs under ultrasonication prior to being drop-coated on glassy carbon electrodes. The CNT film enables the development of mercury-free electrodes that can detect from 10^{-9} to 10^{-6} M of Cd and Pb.

Most CNT-based sensors take advantage of the bulk properties of CNTs, including increased electrode surface area,[142] fast electron transfer rate,[143] and good electrocatalitivity in promoting electron-transfer reactions of many important species.[55,56] Using CNTs as nanoelectrodes have been

Table 4.   Carbon nanotubes in electrochemical sensors for metal ions

Electrode Configuration	Electrode Substrate	Metal Ions	Detection Limits (Preconcentration Time)	Linear Range (Accumulation Time)	Figure	Ref.
Hg-film aligned CNT array	Au	Pb	1 ppb Pb^{2+} (3 min)	2-100 ppb Pb^{2+} (3 min)	12	139, 145, 154, 155
Bi-film aligned CNT array	Au	Simultaneous Cd/Pb	0.04 ppb Pb^{2+} (from 3S/N)	0.5-8 ppb Pb^{2+} (2 min)	12, 13	135
MWNT/Nafion film	Glassy carbon	Cd	1×10^{-6} M (4 min)	1×10^{-6} M 4×10^{-6} M (4 min)		126, 127
MWN/DHP film	Glassy carbon	Simultaneous Cd/Pb	6×10^{-9} M Cd, 4×10^{-9} M Pb (5 min)	2.5×10^{-8} M-1 $\times 10^{-5}$ M Cd 2×10^{-8} M to 1×10^{-5} M Pb		127, 128

increasingly explored since conventional macroelectrodes (having diameter in millimeters), such as glassy carbon and carbon paste, are known to have slow mass transport. Using nanoelectrodes (having diameter in nanometers) can enhance mass transport[138,139,144]; as electrodes decrease in size, radial (3-dimensional) diffusion becomes dominant and results in fast mass transport and fast electron transfer. The high diffusion rate at nanoelectrodes enables the study of fast electrochemical and chemical reactions.[146] Nanoelectrodes also have higher responsiveness (or higher mass sensitivity) than macroelectrodes, attributed to their lower background (charging) currents.[145] Additionally, they are less influenced by solution resistance due to lower ohmic drop.[139] Despite its advantages, a single nanoelectrode offers extremely low capacitive current (in pico-amperes), thereby requiring expensive signal amplifiers. To solve this issue, nanoelectrode array consisting of millions of nanoelectrodes have been developed at PNNL in collaboration with the Boston College in order to provide magnified signals without the need for a signal amplifier.

High-density aligned CNT electrode arrays (i.e. dense CNT forest) have been reported to show fast electron transfer and electrocatalytic characteristics,[143,147] but they do not maintain the properties of individual nanoelectrodes due to the overlapping of their diffusion layers.[148-152] To make each carbon nanotube on the array work as an individual nanoelectrode, the spacing must be sufficiently larger than the diameter of the nanotubes. A nanoscale in size of the electrode and the million numbers of electrodes will result in an improved signal-to-noise ratio (and hence improved detection limits).

Fabrication of Aligned CNT-Nanoelectrode Array

The fabrication procedure for low-site density aligned CNT-nanoelectrode array has been refine.[138,139] Figure 12(a) shows the schematic of the fabrication of the CNT-nanoelectrode array. Briefly, Ni nanoparticles were first electrodeposited on a Cr-coated silicon (Si) substrate with an area of 1 cm^2. The aligned CNT arrays with low-site density were subsequently grown from those nickel (Ni) nanoparticles by plasma-enhanced chemical vapor deposition. To take advantage of a faster electron transfer rate at the open-ends of CNTs, and to eliminate the background generated from the side-wall of CNTs, an Epoxy passive layer was used to block the side-wall of the carbon nanotube.[139] Specifically, an Epon epoxy resin 828 passivation layer (7 to 9 μm) was coated on the aligned CNT array to preserve the CNTs on the silicon substrate surface using a standard spin-coating technology. The protruding parts of CNTs are broken and removed by ultrasonication in water. Figures 12(b) and 12(c) show the SEM images of aligned CNT array with site density of 2×10^6 per cm^2 and after coating with the Epoxy resin

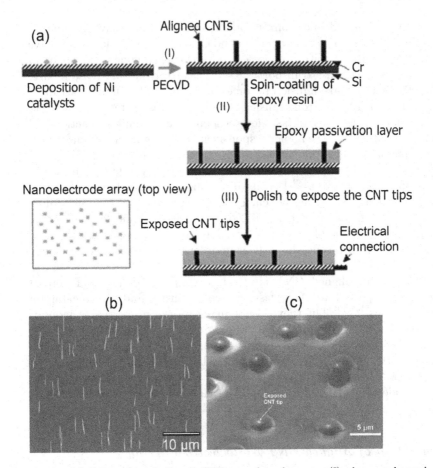

Figure 12. (a) Fabrication of aligned CNT-nanoeletrode array: (I) plasma-enhanced vapor deposition of CNTs on top of Ni nanoparticles created by electrolytic-deposition, (II) spin-coating of Epoxy resin, and (III) polishing to expose the tips of CNTs, and SEM images (b) after step (I): showing CNT density of $2 \times 10^6 \, cm^2$, and (c) after step III: coating with the Epoxy resin and polishing to expose the CNT tips. (a) is from Ref. 56, reproduced with permission of *Frontiers in Bioscience*, and (b) and (c) from Ref. 139, reproduced with permission of Wiley-VCH.

and polishing to expose the CNT tips (white dot), respectively. The CNTs are apart from their nearest neighbors for at least 5 μm, while the diameter of each nanotube are about 50 to 80 nm. From these low site density CNTs, the nanoelectrode arrays consisting of up to millions of CNTs, each serving as individual nanoelectrodes, was fabricated by adding the electrical connection on the CNT-Si substrate.

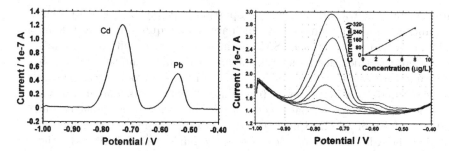

Figure 13.   Voltammetric responses of metal ions at aligned CNT-nanoelectrode array: (a) voltammograms of 5 ppb Cd^{2+} and 5 ppb Pb^{2+} after 2 minutes deposition at -1.2 V in the presence of 500 ppb Bi^{3+}, (b) linear response of Cd^{2+} with a function of solution concentration, same other conditions, from Ref. 135, reproduced by permission of The Royal Society of Chemistry.

Applications of CNT-Based Sensors for Metal Ion Monitoring

With Hg-coated CNT-nanoelectrode array, linear relationship between the Pb signals and the Pb concentration in the solutions ranging from 1 to 100 ppb (μg/L) has been obtained.[139] Because Hg is highly toxic and not suitable for field-deployable use, the relatively benign bismuth, Bi(III), has been evaluated as a Hg substitute.[79,135,153,154] Bi-based electrodes performed as well as Hg-based electrode for Cr(VI) quantitation after the accumulation of Cr(VI)-diethylenetriamine pentaacetic acid (DTPA) chelate.[153,155] Highly responsive voltammetric analysis of Cd and Pb at the Bi-based CNT-nanoelectrode array was obtained.[135] Figure 13(a) shows well-defined peaks of 5 μg/L of Cd and Pb obtained after only two minutes of accumulation, in which Bi was accumulated *in situ* with the target metals at -1.2 V. Figure 13(b) shows a linear calibration curve of Cd achieved at a very useful concentration range of 0.5 to 8 μg/L of Cd.

Conclusions and Future Work

This article is a review of the work relevant to the two fastest growing nanomaterials in electrochemical sensing of metal ions: organically modified ordered mesoporous silicas (OMSs) and carbon nanotubes (CNTs). Nanostructured self-assembled monolayers on mesoporous supports (SAMMS) materials are highly effective as electrode modifiers; they can be either mixed with conductive materials or spin-casted as a thin-film on electrode surface. The interfacial chemistry of SAMMS can be fine-tuned to selectively preconcentrate the specific metal ions of interest. The functional groups on SAMMS materials enable the preconcentration to be done of target metal ions without mercury, supporting electrolytes, applied potential, and solution degassing, all of which are often required in conventional stripping voltammetric sensors. SAMMS materials grafted with four organic functionalities (thiol, acetamide phosphonic acid, salicylamide, or glycinyl-urea) have been used successfully at PNNL in the electrochemical sensors for heavy and transition metals (cadmium, copper, lead, and mercury), actinide uranium, and lanthanide europium. The high loading capacity and high selectivity of SAMMS materials are desirable for metal ion detection based on the AdSV technique because they minimize the competition for the binding sites of the non-target species, thereby reducing the interferences and preserving the signal intensity of the target metal ions. The covalent bonding between the functional groups and the surface of silica prevents sensors from degradation over time due to depletion of ligand-bearing materials. The reversible binding between metal ions and ligands on SAMMS enables many successive uses of SAMMS-based electrodes with virtually no additional work required for surface regeneration.

Since it was first introduced in 1991, CNTs have been widely investigated for electrochemical sensors of many important biomolecules because of their electrocatalytic and antifouling properties, biocompatibility, high surface, and mechanical strength. For trace metal analysis, CNT thin-film created by drop-coating of CNT-solvent suspensions on electrode surfaces has been explored in order to develop mercury-free sensors by exploiting the bulk properties of the CNTs. Array of low-site-density aligned carbon nanotubes has been grown on metal substrates by a non-lithographic method. Each CNT serves as a nanoelectrode which normally has greater mass transfer rate and higher mass sensitivity than conventional macroelectrodes. The array of millions of CNT nanoelectrodes provides magnified voltammetric signals for trace metal ions without the need for a signal amplifier.

Our future work will focus on using other SAMMS materials in electrochemical sensors of other pollutants. Interfacial chemistry and electrochemistry of metal species on surfaces of SAMMS-based electrodes will be an on-going study: this fundamental knowledge is needed for predicting how the sensors will perform in real wastes that consist of many interferences and ligands at a broad spectrum of pH levels. Using the SAMMS-based sensors with biological samples for occupational monitoring of toxic metal ions is also desirable. To overcome the electrode fouling by proteins in biological samples, new metal sensors are being developed at PNNL from the composite of SAMMS as the metal ion preconcentrator and CNT as the conductive matrix. In another approach, to avoid the use of hydrophobic binders and improve the mass transfer of the metal ions in samples to the SAMMS surface, superparamagnetic SAMMS is being developed to first bind with metal ions in liquid phase, followed by immobilization of metal-bound magnetic SAMMS on the electrode surface *via* magnetic or electromagnetic force. The results will be reported in due course.

Acknowledgements

This work was supported by DOE-EMSP and NIH/NIESH/1R01 ES010976-01A2. Pacific Northwest National Laboratory (PNNL) is operated by Battelle Memorial Institute for the U.S. Department of Energy (DOE). The research was performed in part at the Environmental Molecular Sciences Laboratory (EMSL), a national scientific user facility sponsored by the DOE's Office of Biological and Environmental Research and located at PNNL. The authors specially thank Dr. Charles Timchalk, Prof. Zhifeng Ren, Dean A. Moore, Brad J. Busche, Dr. Guodong Liu, Yi Tu, and Xiaohong Li for their contributions.

References

1. Y. Lin, C. A. Timchalk, D. W. Matson, H. Wu and K. D. Thrall, *Biomed. Microdevices* **3**, 331 (2001).
2. Y. Lin, R. Zhao, K. D. Thrall, C. Timchalk, W. D. Bennett and M. D. Matson, *Proc. Soc. Opt. Eng. (SPIE)* **3877**, 248 (1999).
3. J. Wang, Y. Lin and L. Chen, *Analyst* **118**, 277 (1993).
4. J. Wang, J. Lu, D. D. Larson and K. Olsen, *Electroanal* **7**, 247–250 (1995).
5. J. Wang, J. Wang, B. Tian and M. Jiang, *Anal. Chem.* **69**, 1657–1661 (1997).
6. N. Muskal and D. Mandler, *Curr. Separations* **19**, 49–54 (2000).
7. I. Turyan, M. Atiya and D. Mandler, *Electroanalysis* **13**, 653 (2001).
8. I. Turyan and D. Mandler, *Anal. Chem.* **66**, 58–63 (1994).
9. I. Turyan and D. Mandler, *Anal. Chem.* **69**, 894–897 (1997).
10. P. Ugo, B. Ballarin, S. Daniele and A. Mazzocchin, *J. Electroanal. Chem.* **291**, 187–199 (1990).
11. P. Ugo, B. Ballarin, S. Daniele and A. Mazzocchin, *J. Electroanal. Chem.* **324**, 145 (1992).
12. N. Akmal, H. Zimmer and H. B. Mark, *Anal. Letters.* **24**, 1431–1443 (1991).
13. R. P. Baldwin, J. K. Christensen and L. Kryger, *Anal. Chem. Commun.* **58**(8), 1790–1798 (1986).
14. T. H. Degefa, B. S. Chandravanshi and H. Alemu, *Electroanalysis* **11**, 1305–1311 (1999).
15. V. S. Ijeri and A. K. Srivastava, *Anal. Sci.* **17**, 605–608 (2001).
16. M. F. Mousavi, A. Rahmani, S. M. Golabi, M. Shamsipur and H. Sharghi, *Talanta.* **55**, 305–312 (2001).
17. Z. Navratilova, *Electroanalysis* **3**(8), 799–802 (1991).
18. B. Ogorevc, X. H. Cai and I. Grabec, *Anal. Chim. Acta* **305**, 176–182 (1995).
19. S. V. Prabhu, R. P. Baldwin and L. Kryger, *Anal. Chem.* **59**(8), 1074–1078 (1987).

20. I. G. Svegl, M. Kolar, B. Ogorevc and B. Pihlar, *Fresenius J. Anal. Chem.* **361**, 358–362 (1998).

21. X. B. Chen, X. D. Feng, J. Liu, G. E. Fryxell and M. Gong, *Sep. Sci. Technol.* **34**, 1121–1132 (1999).

22. H. Ju, Y. Gong and H. Zhu, *Anal. Sci.* **17**, 59 (2001).

23. A. Walcarius, *Chem. Mater.* **13**, 3351–3372 (2001).

24. C. T. Kresge, M. E. Leonowicz, W. J. Roth, J. C. Vartuli and J. S. Beck, *Nature.* **359**, 710–712.

25. A. Walcarius, M. Etienne, S. Sayen and B. Lebeau, *Electroanalysis* **15**, 414 (2003).

26. A. Walcarius, *C. R. Chimie.* **8**, 693–712 (2005).

27. A. Walcarius, C. Despas and J. Bessière, *Anal. Chim. Acta* **385**, 79–89 (1999).

28. A. Walcarius, C. Despas, P. Trens, M. J. Hudson and J. Bessière, *J. Electroanal. Chem.* **453**, 249–252 (1998).

29. A. Walcarius and J. Bessière, *Chem. Mater.* **11**, 3009–3011 (1999).

30. A. M. Bond, W. Miao, T. D. Smith and J. Jamis, *Analytica Chim. Acta* **396**, 203–213 (1999).

31. A. Walcarius, N. Luthi, J. L. Blin, B. L. Su and L. Lamberts, *Electrochim. Acta* **44**, 4601–4610 (1999).

32. J. F. Diaz, K. J. Balkus, F. Bedioui, V. Kurshev and L. Kevan, *Chem. Mater.* **9**, 61 (1997).

33. J. F. Diaz, F. Bedioui, E. Briot, J. Devynck and K. J. Balkus, *Mater. Res. Soc. Symp. Proc.* **431**, 89 (1996).

34. Y.-X. Jiang, N. Ding and S.-G. Sun, *J. Electroanal. Chem.* **563**, 15 (2004).

35. C. Song and G. Villemure, *Micropor. Mesopor. Mater.* **44**(Sp. Iss.), 679–689 (2001).

36. A. Walcarius, M. Etienne and B. Lebeau, *Chem. Mater.* **15**, 2161 (2003).

37. M. Etienne, B. Lebeau and A. Walcarius, *New J. Chem.* **26**, 384 (2002).

38. A. Walcarius and C. Delacôte, *Chem. Mater.* **15**, 4181 (2003).

39. W. Yantasee, L. A. Deibler, G. E. Fryxell, C. Timchalk and Y. Lin, *Electrochem. Commun.* **7**, 1170–1176 (2005).

40. W. Yantasee, G. Fryxell and Y. Lin, *Analyst* **131**, 1342 (2006).

41. W. Yantasee, G. E. Fryxell, M. M. Conner and Y. Lin, *J. Nanosci. and Nanotech.* **5**, 1537–1540 (2005).

42. W. Yantasee, Y. Lin, G. E. Fryxell and B. J. Busche, *Analytica Chim. Acta* **502**, 207–212 (2004).

43. W. Yantasee, Y. Lin, G. E. Fryxell and Z. Wang, *Electroanalysis* **16**, 870–873 (2004).

44. W. Yantasee, Y. Lin, X. Li, G. E. Fryxell, T. S. Zemanian and V. V. Viswanathan, *Analyst* **128**, 899–904 (2003).

45. W. Yantasee, Y. Lin, T. S. Zemanian and G. E. Fryxell, *Analyst* **128**, 467–472 (2003).

46. W. Yantasee, C. Timchalk, G. E. Fryxell, B. P. Dockendorff and Y. Lin, *Talanta.* **68**, 256–261 (2005).

47. S. Iijima, *Nature.* **354**, 56–58 (1991).

48. R. Saito, G. Dresselhaus and M. S. Dresselhaus, *Physical Properties of Carbon Nanotubes* (Imperial College Press, London, 1998).
49. J. M. Nugent, K. S. V. Santhanam, A. Rubio and P. M. Ajayan, *Nano Lett.* 1, 87 (2001).
50. J. Wang and M. Musameh, *Anal. Chem.* 75, 2075–2079 (2003).
51. J. Wang, M. Musameh and Y. Lin, *J. Am. Chem. Soc.* 125, 2408–2409 (2003).
52. M. Musameh, J. Wang, A. Merkoci and Y. Lin, *Electrochem. Commun.* 4, 743–746 (2002).
53. Y. Lin, F. Lu, Y. Tu and Z. F. Ren, *Nano Lett.* 4, 191–195 (2004).
54. Y. Lin, F. Lu and J. Wang, *Electroanalysis* 16, 145–149 (2003).
55. Y. Lin, W. Yantasee, F. Lu, J. Wang, M. Musameh, Y. Tu and Z. F. Ren, Biosensors based on carbon nanotubes, *Dekker Encyclopedia of Nanoscience and Nanotechnology* (Marcel Dekker, New York, 2004), pp. 361–374.
56. Y. Lin, W. Yantasee and J. Wang, *Frontiers in Bioscience* 10, 492–505 (2005).
57. G. E. Fryxell, H. Wu, Y. Lin, W. J. Shaw, J. C. Birnbaum, J. C. Linehan, Z. Nie, K. Kemner and S. Kelly, *J. Chem. Mat.* 14, 3356–3363 (2004).
58. W. Yantasee, G. E. Fryxell, Y. Lin, H. Wu, K. N. Raymond and J. Xu, *J. Nanosci. Nanotech.* 5, 527–529 (2005).
59. J. C. Birnbaum, B. Busche, Y. Lin, W. Shaw and G. E. Fryxell, *Chem. Communications* 13, 1374 (2003).
60. G. E. Fryxell, Y. Lin, S. Fiskum, J. C. Birnbaum, H. Wu, K. Kemner and S. Kelly, *Environ. Sci. Technol.* 39, 1324–1331 (2005).
61. Y. Lin, S. K. Fiskum, W. Yantasee, H. Wu, S. V. Mattigod, E. Vorpagel and G. E. Fryxell, *Environmental Science and Technology* 39, 1332–1337 (2005).
62. X. D. Feng, G. E. Fryxell, L. Q. Wang, A. Y. Kim, J. Liu and K. Kemner, *Science.* 276, 923–926 (1997).
63. G. E. Fryxell, Y. Lin, H. Wu and K. M. Kemner, in *Studies in Surface Science and Catalysis*, eds., A. Sayari and M. Jaroniec, Environmental applications of self-assembled monolayers on mesoporous supports (SAMMS), Vol. 141 (Elsevier Science, 2002), pp. 583–590.
64. G. E. Fryxell, J. Liu, S. V. Mattigod, L. Q. Wang, M. Gong, T. A. Hauser, Y. Lin, K. F. Ferris and X. Feng, in *Ceramics Transactions*, eds. G. T. Chandler and X. Feng, Environmental Issues and Waste Management Technologies in the Ceramic and Nuclear Industries, Environmental Applications of Interfacially Modified Mesoporous Ceramics; The American Ceramic Society, Vol. 107 (2000), pp. 29–37.
65. W. Yantasee, Y. Lin, G. E. Fryxell, B. J. Busche and J. C. Birnbaum, *Sep. Sci. Technol.* 38, 3809–3825 (2003).
66. Y. Lin, G. E. Fryxell, H. Wu and M. Engelhard, *Environ. Sci. Technol.* 35, 3962–3966 (2001).
67. S. V. Mattigod, G. E. Fryxell, R. J. Serne, K. E. Parker and F. M. Mann, *Radiochimica Acta* 91, 539–545 (2003).

68. G. E. Fryxell, J. Liu, T. A. Hauser, Z. Nie, K. F. Ferris, S. Mattigod, M. Gong and R. T. Hallen, *Chem. Mater.* **11**, 2148–2154 (1999).

69. R. G. Pearson, *J. Am. Chem. Soc.* **85**, 3533–3539 (1963).

70. A. Walcarius, T. Barbaise and J. Bessiere, *Analytica Chimica Acta* **340**, 61–76 (1997).

71. S. Kilinc Alpat, U. Yuksel and H. Akcay, *Electrochem. Commun.* **7**, 130–134 (2005).

72. C. Bing, G. Ngoh-Khang and C. Lian-Sai, *Electrochimica Acta* **42**, 595–604 (1997).

73. X. Cui, G. Liu, L. Li, W. Yantasee and Y. Lin, *Sensor Letters* **3**, 16–21 (2005).

74. J. F. van Staden and R. E. Taljaard, *Talanta.* **64**, 1203 (2004).

75. J. Wang, R. Setiadji, L. Chen, J. Lu and S. Morton, *Electroanal* **4**, 161 (1992).

76. S. Sander, W. Wafner and G. Henze, *Anal. Chim. Acta* **349**, 93 (1997).

77. J. Wang and R. Setiadji, *Anal. Chim. Acta* **264**, 205 (1992).

78. M. B. R. Bastos, J. C. Moreira and P. A. M. Farias, *Anal. Chim. Acta* **408**, 83–88 (2000).

79. L. Lin, S. Thongngamdee, J. Wang, Y. Lin, O. A. Sadik and S.-Y. Ly, *Anal. Chim. Acta* **535**, 9–13 (2005).

80. S. Sayen and A. Walcarius, *J. Electroanal. Chem.* **581**, 70–78 (2005).

81. V. Ganesan and A. Walcarius, *Langmuir* **20**, 3632–3640 (2004).

82. I. A. Aksay, M. Trau, S. Manne, I. Honma, N. Yao, L. Zhou, P. Fenter, P. M. Eisenberger and S. M. Gruner, *Science* **273**, 892–898 (1996).

83. H. Yang, N. Coombs, I. Sokolov and G. A. Ozin, *Nature* **381**, 589–592 (1996).

84. H. Yang, A. Kuperman, N. Coombs, S. MamicheAfara and G. A. Ozin, *Nature* **379**, 703–705 (1996).

85. D. Y. Zhao, J. L. Feng, Q. S. Huo, N. Melosh, G. H. Fredrickson, B. F. Chmelka and G. D. Stucky, *Science* **279**, 548–552 (1998).

86. D. Y. Zhao, Q. S. Huo, J. L. Feng, B. F. Chmelka and G. D. Stucky, *J. Am. Chem. Soc.* **120**, 6024–6036 (1998).

87. F. Cagnol, D. Grosso and C. Sanchez, *Chem. Commun.* 1742 (2004).

88. L. Nicole, C. Boissiere, D. Grosso, A. Quach and C. Sanchez, *J. Mater. Chem. Commun.* **15**, 3598 (2005).

89. Q. Huo, R. Leon, P. M. Petroff and G. D. Stucky, *Science* **268**, 1324–1327 (1995).

90. V. Luzzati, H. Delacroix and A. Gulik, *J. Phys. II* **6**, 405–418 (1996).

91. D. Zhao, P. Yang, N. Melosh, J. Feng, B. F. Chmelka and G. D. Stucky, *Adv. Mater.* **10**, 1380–1385 (1998).

92. J. J. Gooding, V. G. Praig and E. A. H. Hall, *Anal. Chem.* **70**, 2396–2402 (1998).

93. M. Etienne and A. Walcarius, *Electrochem. Commun.* **7**, 1449–1456 (2005).

94. D. Grosso, F. Cagnol, G. J. d. A. A. Soler-Illia, E. L. Crepaldi, H. Amenitsch, A. Brunet-Bruneau, A. Bourgeios and C. Sanchez, *Adv. Fun. Mater.* **14**, 309 (2004).

95. R. E. Williford, G. E. Fryxell, X. S. Li and R.S. Addleman, *Micropor. Mesopor. Mat.* **84**, 201–210 (2005).
96. W. L. Roper, V. Houk, H. Falk and S. Binder, Preventing Lead Poisoning in Young Children. Centers for Disease Control and Prevention, Atlanta, GA (1991)
97. G. A. Wagner, R. M. Maxwell, D. Moor and L. Foster, *J. Am. Med. Assoc.* **270**, 69–71 (2000).
98. J. Růžička, G. D. Marshall and G. D. Christian, *Anal. Chem.* **62**, 1861 (1990).
99. J. Růžička and G. D. Marshall, *Anal. Chim. Acta* **273**, 329 (1990).
100. I. Palchetti, C. Upjohn, A. P. F. Turner and M. Mascini, *Anal. Lett.* **33**, 1231–1246 (2000).
101. K. C. Honeychurch, D. M. Hawkins, J. P. Hart and D. C. Cowell, *Talanta.* **57**, 565–574 (2002).
102. K. C. Honeychurch, J. P. Hart, D. C. Cowell and D. W. M. Arrigan, *Sens. Actuators, B* **77**, 642–652 (2001).
103. K. C. Honeychurch, J. P. Hart and D. C. Cowell, *Anal. Chim. Acta* **431**, 89–99 (2001).
104. K. C. Honeychurch, J. P. Hart and D. C. Cowell, *Electroanalysis* **12**, 171–177 (2000).
105. K. C. Honeychurch and J. P. Hart, *Trends Anal. Chem.* **22**, 456–469 (2003).
106. Q. G. Q. Health, Cadmium. Queensland Government, Canberra (2002), pp. 1–4.
107. J.-Y. Choi, K. Seo, S.-R. Cho, J.-R. Oh, S.-H. Kahng and J. Park, *Anal. Chim. Acta* **443**, 241–247 (2001).
108. D. Desmond, B. Lane, J. Alderman, M. Hill, D. W. M. Arrigan and J. D. Glennon, *Sens. Actuators, B* **48**, 409–414 (1998).
109. J. Wang and T. Baomin, *Anal. Chem.* **64**, 1706–1709 (1992).
110. J. Wang, J. Lu, B. Tain and C. Yarnitsky, *J. Electroanal. Chem.* **361**, 77–83 (1993).
111. J. Wang, B. Tian, V. B. Nascimento and L. Angnes, *Electrochim. Acta* **43**, 3459–3465 (1998).
112. P. Masawat, S. Liawruangrath and J. M. Slater, *Sens. Actuators, B* **91**, 52–59 (2003).
113. J.-M. Zen, C.-C. Yang and A. S. Kumar, *Anal. Chim. Acta* **464**, 229–235 (2002).
114. K. Z. Brianina, N. F. Zakharchuck, D. P. Synkova and I. G. Yudelevich, *J. Electroanal. Chem.* **35**, 165 (1972).
115. S. Ferro and A. D. Battisti, *J. Electroanal. Chem.* **533**, 177–180 (2002).
116. M. Mlakar and M. Branica, *Anal. Chim. Acta* **247**, 89 (1991).
117. M. Mlakar, *Analytica Chim. Acta* **260**, 51–56 (1992).
118. M. Zhang and X. Gao, *Anal. Chem.* **56**, 1917 (1984).
119. X. Gao and M. Zhang, *Anal. Chem.* **56**, 1912 (1984).
120. L. M. Moretto, J. Chevalet, G. A. Mazzocchin and P. Ugo, *J. Electroanal. Chem.* **498**, 117–126 (2001).
121. R. H. Baughman, A. Zakhidov and W. A. De Heer, *Science* **297**, 787 (2002).

122. Q. Zhao, Z. Gan and Q. Zhuang, *Electroanalysis* **14**, 1609 (2002).

123. J. Wang, M. Li, Z. Shi, N. Li and Z. Gu, *Anal. Chem.* **74**, 1993–1997 (2002).

124. J. Wang, M. Li, Z. Shi, N. Li and Z. Gu, *Electroanal* **14**, 225–230 (2002).

125. Z. Wang, J. Liu, Q. Liang, Y. Wang and G. Luo, *Analyst* **127**, 653–658 (2002).

126. Y.-C. Tsai, J.-M. Chen, S.-C. Li and F. Marken, *Electrochem. Commun.* **6**, 917–922 (2004).

127. K. Wu, S. Hu, J. Fei and W. Bai, *Analytica Chim. Acta* **489**, 215–221 (2003).

128. H. X. Luo, Z. J. Shi, N. Q. Li, Z. N. Gu and Q. K. Zhuang, *Anal. Chem.* **73**, 915–920 (2001).

129. F. H. Wu, G. C. Zhao and X. W. Wei, *Electrochem. Commun.* **4**, 690 (2002).

130. V. G. Gavalas, S. A. Law, J. C. Ball, R. Andrews and L. G. Bachas, *Anal. Biochem.* **329**, 247–252 (2004).

131. M. D. Rubianes and G. A. Rivas, *Electrochem. Commun.* **5**, 689–694 (2003).

132. F. Valentini, S. Orlanducci, M. L. Terranova, A. Amine and G. Palleschi, *Sens. Actuat. B* **100**, 1170–125 (2004).

133. J. Wang and M. Musameh, *Analyst* **128**, 1382–1385 (2003).

134. M. Gao, L. M. Dai and G. Wallace G, *Electroanal* **15**, 1089–1094 (2003).

135. G. Liu, Y. Lin, Y. Tu and Z. Ren, *Analyst* **130**, 1098–1101 (2005).

136. K. P. Loh, S. L. Zhao and W. D. Zhang, **13**, 1075–1079 (2004).

137. W. C. Poh, K. P. Loh, W. D. Zhang, T. Sudhiranjan, J. S. Ye and F. S. Sheu, *Langmuir* **20**, 5484–5492 (2004).

138. S. G. Wang, Q. Zhang, R. Wang and S. F. Yoon, *Biochem. Biophys. Res. Commun.* **311**, 572–576 (2003).

139. S. G. Wang, Q. Zhang, R. Wang, S. F. Yoon, J. Ahn, D. J. Yang, J. Z. Tianm, J. Q. Li and Q. Zhou, *Electrochem. Commun.* **5**, 800–803 (2003).

140. J. Wang, G. Liu and R. Jan, *J. Am. Chem. Soc.* **126**, 3010 (2004).

141. J. Gooding, R. Wibowo, J. Liu, W. Yang, D. Losic, S. Orbons, F. Mearns, J. Shapter and D. Hibbert, *J. Am. Chem. Soc.* **125**, 9006 (2003).

142. Y. Tu, Z. P. Huang, D. Z. Wang, J. G. Wen and Z. F. Ren, *Appl. Phys. Lett.* **80**, 4018–4020 (2002).

143. V. Menon and C. Martin, *Anal. Chem.* **67**, 1920 (1995).

144. D. Arrigan, *Analyst* **129**, 1157–1165 (2004).

145. X. Yu, D. Chattopadhyay, I. Galeska, F. Papadimitrakopoulos and J. Rusling, *Electrochem. Commun.* **5**, 408 (2003).

146. R. Feeney and S. P. Kounaves, *Electroanalysis* **12**, 677 (2000).

147. S. Fletcher and M. D. Horne, *Electrochem. Commun.* **1**, 502 (1999).

148. H. J. Lee, C. Beriet, R. Ferrigno and H. H. Girault, *J. Electroanal Chem. Commun.* **502**, 138 (2001).

149. J. Li, H. T. Ng, A. Cassell, W. Fan, H. Chen, Q. Ye, J. Koehne, J. Han and M. Meyyappan, *Nano Lett.* **3**, 597 (2003).

150. M. E. Sandison, N. Anicet, A. Glidle and J. M. Cooper, *Anal. Chem.* **74**, 5717 (2002).

151. Y. Tu, Y. Lin and Z. F. Ren, *Nano Lett.* **3**, 107–109 (2003).
152. Y. Tu, Y. Lin, W. Yantasee and Z. Ren, *Electroanalysis* **17**, 79–84 (2005).
153. L. Lin, N. S. Lawrence, S. Thongngamdee, J. Wang and Y. Lin, *Talanta.* **65**, 144–148 (2005).
154. J. Wang, J. Lu, S. B. Hocevar and B. Ogorevc, *Electroanalysis* **13**, 13 (2001).

Chapter 16

Nanomaterial Based Environmental Sensors

Dosi Dosev*, Mikaela Nichkova† and Ian M. Kennedy*

*Department of Mechanical and Aeronautical Engineering,
†Department of Entomology, University of California,
Davis, CA 95616, USA

Nanomaterials offer unique properties that can be exploited in novel environmental sensors. We consider both inorganic and organic materials that are synthesized as spherical particles, tubes, rods, belts and wires. The electrical, magnetic, chemical and optical properties of these materials are used to detect the presence of environmental toxins or other undesirable vapors or liquids at the particle's surface. We review the application of nanomaterials primarily to biosensors that use a biological recognition element such as an antibody that is bound to the surface; we also include a discussion of physical gas sensors that are based on nano-scale materials.

Introduction

Environmental science and technology frequently require the measurement and detection of the constituents of soil, water and air. Measurements of these components of the environment are often necessary first steps in the development and validation of models of environmental processes; measurements are also very important during the monitoring of processes in the environment, particularly with regard to monitoring the transport of contaminants, and detection of toxins like pesticides. Air quality modeling requires high quality data on the concentrations of air contaminants. Typical components of the air of interest are those compounds that are often emitted by combustion sources such as engines. The air pollution that is created from these emissions is well-known to have an adverse impact on health.[1,2] Gas phase species such as carbon monoxide, unburned hydrocarbons, oxides of nitrogen, ozone, and most significantly in recent times, particulate matter, are known to be associated with adverse effects on human health. Over the years, many bench-top analyzers have been developed for detecting these compounds in the ambient air. Those technologies are well established but the instruments are fairly large and expensive. Their size and cost do not permit the deployment of such systems in large-scale arrays of sensors. The thrust of recent sensor research is directed towards the development of smaller, cheaper, portable devices.

The detection of toxins in ground water and in soil is very important in assessing threats to human health from seepage of waste from contaminated soils, and in monitoring the transport of toxins through the ground water and eventually consumption by humans. In general, well-established techniques such as liquid and gas chromatography are used to analyze samples that are returned to the laboratory. This method of analysis is slow and expensive. It is certainly not amenable to the deployment of large-scale arrays for detecting the transport of toxins from contaminated sites. The contaminants of soil and water that are of common interest include industrial pollutants such as polyaromatic hydrocarbons (PAH), chlorinated solvents and metal-containing compounds. In some cases, the materials may be widely dispersed, as is the case with dioxins.

The detection of trace compounds in the environment faces problems that are unique to environmental science and technology. First, the detection must take place within a complex matrix of many compounds, all of which may interfere with the detection process. Oxygen, which is generally ubiquitous in environmental settings, can interfere with measurements by quenching fluorescence. Sensors that are based on immuno assay techniques, that typically use antibodies as bio-recognition elements, can also suffer from interference from other compounds in the sample due to non-specific binding. In this case, compounds other than the target compounds are bound to the antibody and are detected as the sample analyte. Over many years, methods have been devised to improve the sensitivity and specificity of antibodies for application to immunoassays. However, in environmental samples there may be many closely-related chemical compounds that could give rise to this problem. Diminished specificity can be a particular problem if molecularly-imprinted polymers (MIPs) are used as substitutes for antibodies; the use of MIPs has been advocated by several research groups.[3–7]

The monitoring of environmental sites generally requires the deployment of many sensors in a network.[8] For example, monitoring the transport of contaminated waste from soils or water requires a fairly large area to be instrumented with sensors. For this reason, environmental applications typically require sensors to be both rugged and relatively cheap. The latter requirement is exacerbated by budgets for environmental clean-up research and technology that are often smaller than the funding that is available for biomedical research. Sensors that are deployed in the field need to be simple enough to be operated by minimally-trained technicians, and to involve fairly straightforward technology. They need to be able to withstand a harsh environment, to operate off a self-contained power supply, and ideally to interface to a network with telemetry that can transmit data back to a base station.

Not all sensors that are employed in environmental research and technology need to be used in the field — many important sensing technologies can be used in the laboratory. Research into the exposure of human populations to environmental toxins is typically done using laboratory instrumentation. However, many samples are collected from a large population.[9] In this case, the technology needs to be able to handle a large number of samples in a reasonable time. Consequently, high throughput technologies become very important in handling large numbers of samples. The requirement for high throughput suggests that it might be desirable to work with small volumes, and to be able to tolerate minimal clean-up or preparation prior to analysis.

Working with very small volumes of sample indicates the need for detection methods with high sensitivity. Recent advances in nanotechnology may be able to provide new analytical approaches that meet the new requirements for ultra sensitive detection.[10–18] The development of quantum dots offers a good example of nanotechnology applied to biosensors.[19–25] Quantum dots are very small particles made of semiconductor materials such as cadmium sulfide, cadmium selenide, zinc sulfide, as well as other compounds.[26–29] When the semiconductor materials are synthesized into very small particles of the order of 10 nanometers in diameter, the quantum confinement effect serves to shift their bulk phase optical emission towards the blue end of the spectrum. Hence, it is possible to synthesize a wide range of particles of different emission wavelengths. This technology has found commercial application with several companies worldwide and is being applied widely in biotechnology. The use of quantum dots as labels in techniques such as immunoassays in environmental samples may provide some advantages: narrow emission characteristics may enable discrimination against broad background fluorescence; the lack of photo-bleaching allows persistent excitation; and strong luminescence offers an excellent signal to noise ratio.

Considerable effort has been channeled into the development of new nanostructured surfaces for application to optical biosensors. Nanotechnology can be used to create novel surfaces on which to carry out conventional detection. Surface Enhanced Raman Scattering (SERS) has been coupled with nanostructured silver[30] and gold surfaces[31,32] to gain additional signal strength. The characteristics of nanoparticles that are used for SERS may be important,[33] including the refractive index of the materials and the size of the particles. It has been shown that Surface Enhanced Raman Scattering (SERS) is much more effective on a surface that provides nanoscale features or roughness. In fact, some researchers have found that fractal aggregates of nanoparticles are actually more effective then single nanoparticles by themselves. Other surfaces may contain porous features[34–38] that enable selective capture of analytes within the sample, immobilization of the biological detection elements within the porous matrix, or immobilization of gold nanoparticles for SERS.[39] The matrix serves to protect the detection elements, and screens out many confounding compounds in the sample.

Over the past decade, a great deal of attention has been given to the development of novel materials such as carbon nanotubes, and zinc oxide nanowires and nanobelts. Carbon nanotubes are particularly attractive for sensor applications because their conductivity may change as materials or compounds are adsorbed onto the surface. Zinc oxide is a semiconductor material that also exhibits piezoelectric behavior. When it is synthesized in the form of nanowires and nanobelts, the change in its electrical or piezo-electric properties can be used to detect the presence of adsorbents on the surface. Both of these very interesting materials and morphologies are being applied to the development of sensors, biological and electrochemical.

Novel nano-scale materials are beginning to offer a range of new functionalities and possibilities for environmental sensing. There are many examples of nano-scale materials applied to biosensing in general, with a primary emphasis on applications in biomedical and biological sciences. There are fewer examples of applications to environmental problems. This disparity is being redressed to some extent by research over the past decade. We expect to see many more applications of nano-engineered materials in environmental sensors in the future. This review will cover some of the fundamentals of nano-scale materials applied to sensing. Examples of successful applications of these materials to novel environmental sensors will be discussed.

Sensor Technologies

We begin our discussion of sensor technologies with a review of the fundamentals of biosensors that make use of a biological molecule for capture and recognition of specific compounds in a mixture.

Detecting Element for Biosensors

A *biosensor* is defined as an analytical device that consists of a biological component (enzyme, antibody, receptor, DNA, cell, and so on) in

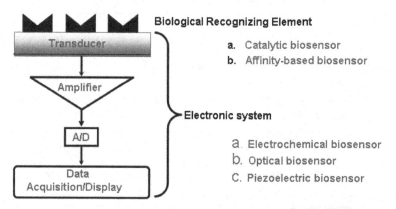

Figure 1. General configuration of biosensors. The biological recognizing element (enzyme, antibody, receptor, DNA, cell, etc.) is in intimate contact with a physical transducer that converts the biorecognition process into a measurable signal (electrical, optical, etc). Catalytic biosensors have enzyme as biorecognition element and affinity biosensors are based on antibodies, DNA, cells or other type of receptor.

intimate contact with a physical transducer that converts the biorecognition process into a measurable signal (electrical, optical, and so on) as seen in Figure 1. Biosensors are commonly classified as immunosensors, enzymatic (catalytic), non-enzymatic receptor, whole-cell (microbial sensors) or nucleic acid (DNA) biosensors, according to the biological recognition element.

Immunosensors are affinity-based sensors designed to detect the direct binding of an antibody (Ab) to an antigen (Ag) to form an immunocomplex at the transducer surface. Depending on the transducer technology employed, immunosensors can be divided into three principal classes: optical, electrochemical and piezoelectrical. They are based on the principles of solid-phase immunoassays, where the immuno-reagent (Ab or Ag) is immobilized on a solid support, so that the interaction takes place at the solid-liquid interface.

Antibodies are globular proteins produced by the immune system of mammals as a defense against foreign agents (antigen, Ag). The structure of the antibody molecules is usually typified by the immunoglobulin G (IgG) subclass, which is the most commonly used in immunochemical application. The IgGs (molecular weight of 150 kDa) are composed of four polypeptide chains: two identical heavy (H) and two identical light (L) chains (50 kDa and 23 to 25 kDa, respectively) interlinked by disulphide bridges (see Figure 2).[40] The region that carries the antigen binding sites is known as the Fab fragment, and the constant (crystallized) region that is involved in immune regulation is termed Fc. Both H and L chains are

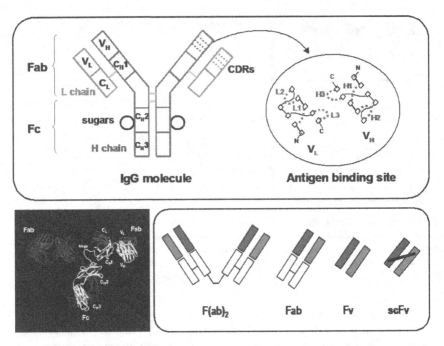

Figure 2. Antibody structure. Constant domains: C_L, C_H1, C_H2 and C_H3; variable domains: V_L and V_H. Antigen binding site: CDR (hypervariable regions): L1, L2, L3, H1, H2, H3; constant framework: black line (from Harlow and Lane[40] with permission)

divided into constant (C) and variable (V) domains based on their amino acid sequence variability. The most important regions of the antibody with regard to the Ab-Ag binding interaction are the variable regions, consisting of the association of the VH and VL domains. Within each of these domains there are three distinct areas of even higher sequence variability, known as complementary determining regions (CDRs), bounded by four framework regions (FR). The CDRs of the VH and VL domains form the antigen-binding site. Their amino acid sequence and spatial conformation determine the binding specificity and binding strength (affinity) of the Ab molecule to the Ag molecule. The interaction between Ab and Ag is reversible and it is stabilized by electrostatic forces, hydrogen bonds, hydrophobic and van der Waals interactions.[41] At equilibrium, the affinity constant Ka is defined as:

$$Ka = \frac{[AbAg]}{[Ab] \cdot [Ag]} \tag{1}$$

Antibody fragments can be generated by enzymatic or chemical degradation and by genetic engineering.[42] These include the F(ab)2; Fab; Fd fragments; Fv, the smallest fragment required for complete Ag binding; single-chain scFv fragments, in which Fv is stabilized by a flexible amino acid linker; and others (see Figure 2).

Immunosensors can be either direct (where the immunochemical reaction is directly determined by measuring the physical changes induced by the formation of the immune complex) or indirect (where a sensitive detectable label is combined with the Ab or Ag of interest). As environmental contaminants are often small-sized molecules, most of the devices that are described in the literature perform indirect measurements by using competitive immunoassay configurations and/or labels such as enzymes, fluorescent chemicals, or electrochemically active substances (see Figure 3). The sensitivity and specificity of an immunosensor are determined by the same characteristics as in other solid-phase immunoassays, namely, the affinity and specificity of the binding agent, and the background noise of the detection system (transducer). Challenges still encountered in biosensor development include fabrication, immobilization of onto a transducer, effective signal generation, miniaturization and integration.

The majority of biosensor research is currently directed towards clinical applications, but a variety of immunosensors for environmental applications have also been developed in the last years.[43–45] Both types of optical and electrochemical transducers have been demonstrated to provide detection limits in the low parts-per-billion to the high parts-per-trillion range. Research in the Ab field is still growing; future developments will make use of recombinant fragments with desired characteristics, better defined chemical structure, stability, and so on. As the orientation of the binding molecule (recognition element) after being immobilized onto a transducer is likely to

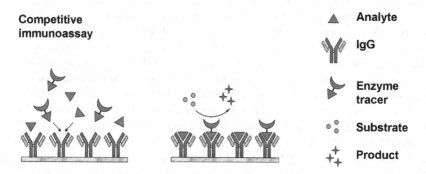

Figure 3. A Schematic illustration of competitive immunoassay. The analyte (target compound) competes with the enzyme tracer (competitor with similar molecular structure) for the antibodies (IgG) immobilized on the surface. The amount of tracer is detected by the enzymatic reaction which converts the substrate to a measurable product. The signal is reversibly proportional to the concentration of the analyte in the sample.

affect the sensitivity, it is desirable to equip the recombinant Ab fragments with suitable tags to facilitate their orientation at the sensor surface.[46]

Even though not based on biological elements, *molecularly imprinted polymer (MIP)*-based sensors mimic the biological activity of antibodies, receptor molecules, etc.[47] MIPs combine highly selective molecular recognition, comparable to biological systems, with the typical properties of polymers such as high thermal, chemical and stress tolerance, and an extremely long shelf life without any need for special storage conditions.[48] However, until now the affinities achieved by MIP biosensors are not comparable to those reached by antibodies. MIP optical biosensors have been developed for detection of pesticides[49] and chemical agents[50] in water.

Immunosensors are limited by problems with matrix effects that arise from the non-specific binding of extraneous compounds from the sample or matrix to the antibody. The regeneration of the antibodies following an assay is also problematical. Implementation and commercialization of immunosensor technology is slow. With further improvement in the fabrication of miniaturized biosensors that are simple, rapid, portable, cost-effective, and with the ability to regenerate, it is anticipated that they will provide a powerful tool for environmental and biological monitoring.

DNA biosensors

Nucleic acid hybridization is the underlying principle of DNA biosensors. A single stranded DNA molecule immobilized on a sensory surface is able to seek out, or hybridize to its complementary strand in a sample (Figure 4) by binding their complementary bases, i.e. C to G and A to T.[51]

The parallel analysis of large numbers of DNA fragments can be provided by DNA chips or arrays.[52] Currently, most DNA chips use optical detection of fluorescence-labeled oligonucleotides. Despite the dominance of the optical systems, advances in electrochemical detection are quite promising. Besides the voltametric detection of redox intercalators, new developments have focused on the mediated oxidation of guanidine within the DNA, the amplification of the hybridization event by an enzyme label and impedance analysis, and the electron transport through the DNA double helix.[52] The ssDNA (single strand DNA) probes (see Figure 4) have gained considerable interest because of their importance in the early diagnosis of diseases, such as cancer, hypercholesteremia, etc. They are used for the detection of microbial and viral pathogens (i.e. HIV, *Crytosporidium, E. coli, Giardia, Mycobaterium tuberculosis*, and so on) in combination with Polymerase Chain Reaction (PCR) units, detection of mutated genes (gene *p*53 for tumor suppression is mutated in malign tumors), and analysis of gene sequences. However, for the monitoring of carcinogenic low

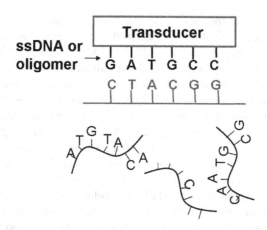

Figure 4. A Schematic illustration of ssDNA probe based on the specific DNA hybridization reaction. The ssDNA or oligomer is immobilized on the transducer surface and it hybridizes with the complementary DNA from the sample. (From Zhai, Cui and Yang[51] with permission.)

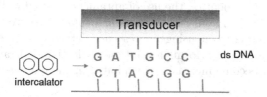

Figure 5. A Schematic illustration of DNA receptor (dsDNA probes). The double-stranded DNA is immobilized on the surface and small organic compounds intercalate selectively into it.

molecular weight compounds — DNA intercalators (endocrine disruptors, polyaromatic hydrocarbons,[53] aromatic amines, drugs,[54] organophosphate pesticides,[55] and so on) — a second type of DNA biosensors referred to as a DNA receptor (double strand or ds DNA probe) is applied. In this approach, biosensors monitor the interaction of small pollutants with affinities for DNA with the immobilized dsDNA layer (Figure 5). These biosensors may therefore be used as general indicators of pollution when integrated in a panel of tests, since they can give rapid and easy to evaluate information on the presence of such compounds.

Aptasensors

Aptamers are artificial nucleic acid ligands (RNA or DNA) that can be generated *in vitro* from vast populations of random sequences; they are designed to bind to amino acids, drugs, proteins and other molecules by so-called Systematic Evolution of Ligands by Exponential enrichment (SELEX).[56,57] Aptamers are ideal candidates for use as biocomponents in biosensors (aptasensors), possessing many advantages over state-of-the-art affinity sensors. Aptamers have a number of advantages compared to antibodies such as smaller size, rationally designed binding-affinity and -specificity, higher temperature stability, no need for immunization of animal hosts, and a higher degree of purification, thus eliminating batch to batch antibody variation. Furthermore, the functional groups of aptamers can be altered for the direct or indirect covalent immobilization on biochips or other support materials, resulting in highly ordered receptor layers. Regeneration of aptamers is easier than antibodies, due to the simple and quite stable structure, which is a key factor that permits repeated use, economy and precision of biosensors. Development of aptasensors is being boosted by using optical and acoustic methods to analyze biological phenomena in solution in real time or by immobilizing the aptamer onto a solid support. Aptasensors for the detection of thrombin, anthrax spores, and proteins have been reported.[57] Despite the wealth of literature detailing the use of aptamers in therapeutics, the use of aptamers in *in vitro* diagnostics is a field that is still in its infancy with the main focus on the detection of the aptamer-target binding event using fluorescence detection. However, with the advent of automated platforms for aptamer selection, it is anticipated that aptamers will become increasingly exploited in the biosensor field.

Microbial (whole-cell) biosensors

Microbial (whole-cell) biosensors can monitor the metabolism of cells by the measurement of pH, O_2 consumption, CO_2 production, redox potential, electric potential on nervous system cells or bioluminescence in bacteria. Viable microbial cells (bacteria and yeasts) have been genetically modified to produce a recombinant organism that exhibits a number of important traits, e.g. expression of cellular degradative enzyme, specific binding proteins, and a reporter enzyme such as bacterial luciferase which is induced in the presence of the target analytes (antibiotics, pesticides, benzene, toluene and xylene).[58] Nerve cells growing on array structures have also been implemented in the development of chips and sensors.[59]

Catalytic (enzymatic) biosensors

Catalytic biosensors rely on the enzyme-catalyzed conversion of a non-detectable substrate into an optically or electrochemically detectable product or vice versa (see Figure 6).[58,60] This process allows the detection of substrates, products, inhibitors and modulators of the catalytic reaction. Several enzyme-catalyzed reactions involve the production or consumption

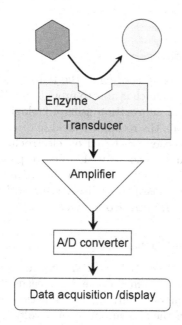

Figure 6. A Schematic illustration of catalytic biosensors. A non-detectable substrate is converted into a electrochemically or optically detectable product by the enzyme immobilized on the sensor surface.

of a detectable product of low molecular weight species, such as O_2, CO_2, and ions. The most frequently employed enzymes are oxidoreductases and hydrolases.[61] The activity of cholinosterases is inhibited by a variety of organophosphorous and carbamate insecticides. Tyrosinase is used to detect contamination caused by phenolic compounds. On the other hand, various pollutants such as hydrazines, atrazine, cyanide, and diethyldithiocarbamtes act as tyrosinase inhibitors. An improvement in sensitivity of the enzymatic biosensors can be achieved by the use of recombinant enzyme mutants.[62]

Physical Sensors

It is not necessary to incorporate a biological recognition element in a sensor device; physical adsorption onto a surface can be sufficient to change the conductivity, for example, of a thin film. The change is detected electrically and is related to the concentration of a particular analyte. Physical adsorption lacks the high degree of specificity that is provided by the biological or biomimetic recognition elements that were discussed above. Physical adsorption finds most application to gas sensors.

Signal Transduction

Electrochemical Transducers

An electrochemical biosensor, according to the IUPAC definition, is a self-contained integrated device, which is able to provide specific quantitative or semi-quantitative analytical information using a biological recognition element (biochemical receptor); the receptor is held in direct contact with the transduction element.[63] Biosensors based on electrochemical transducers have the advantage of being economic with a fast response; the possibility of automation allows application in a wide number of areas.[64] The electrochemical biosensors can be classified as either **conductimetric, impedimetric, potentiometric or amperometric.**

Conductimetric and impedimetric biosensors

These immunosensors are essentially based on the same physical principle and one or another parameter can be measured. Conductimetry describes the measurement of a current as a function of applied voltage while impedance refers to the measurement of a voltage as a function of an applied current. In both cases the conductimetric or impedimetric properties are influenced by a sensing layer placed between a set of two electrodes. For example, in the case of conductimetric immunosensors[60] the measurement is based on the use of an electroconductive polymer doped with analyte, to which specific antibodies have been immobilized. A direct competitive immunoassay produces a change in the conductimetric properties of the polymer.

Impedimetric biosensors have been used for the detection of bacteria and have been applied to sanitation microbiology.[65] These biosensors are based on the principle that microbial metabolism results in an increase in both conductance and capacitance, causing a decrease in the impedance. Impedance is usually measured by a bridge circuit. Often a reference module is included to measure and exclude non-specific changes in the test module. The reference module serves as a control for temperature changes, evaporation, changes in amounts of dissolved gases and degradation of culture medium during incubation.

Potentiometric transducers

Potentiometric sensors incorporate a membrane or surface that is sensitive to a desired species and that generates an electrical potential proportional to the logarithm of the concentration of the active species, measured in relation to a reference electrode. The potentiometric devices can measure changes in pH and ion concentration. It is possible to use transistors as electronic signal amplifiers coupled to ion sensitive electrode (ISE), and such transistors are called ISFET. These biosensors are based on the immobilization of a biologically active material, usually enzymes, antigens or antibodies, or on a membrane, or on the surface of a transducer such as an ISE that responds to the species formed in the enzymatic reaction or to the formation of an antigen-antibody immunocomplex.

In the case of potentiometric immunosensors, a change in potential occurs after the specific binding of the antibody or the antigen to the immobilized partner. Proteins in aqueous solution are poly-electrolytes and consequently the electrical charge of an antibody can be affected by binding the corresponding antigen. The difference in electrical potential is measured between the working electrode where the specific antibody has been immobilized and a reference electrode. A main disadvantage of this system is that variations in the potential due to the antibody–antigen interaction are small (1 to 5 mV) and therefore the reliability and sensitivity of the analysis are limited by background effects.[60]

The research in this field has been aimed at getting better limits of detection and selectivity of the ISE, with the aim of supplying the necessary requirements for its application in industry. New developments include sensor arrays, new ionophores, improvement of the detection limit and new electrodes for miniaturization.[66,67]

Ion-sensitive field-effect transistors (ISFETs) are an important subset of potentiometric sensors. An ISFET is composed of an ion-selective membrane applied directly to the insulated gate of the FET. When such ISFETs are coupled with a bio-catalytic or bio-complexing layer, they become biosensors, and are usually called either enzyme (ENFETs) or immunological (IMFETs) field effect transistors. Operating properties of ENFET- and IMFET-based devices are strongly related to those of the ISE-based biosensors. The importance of the ISFET can be attributed to its capacity for miniaturization and the possibility of exploiting the processes of microelectronics in its manufacture.[68]

Amperometric transducers

Amperometric sensors measure the electrical current produced by a chemical reaction of an electroactive species with an applied potential, which is related to the concentration of the species in solution. The amperometric sensor is faster, more sensitive, precise and accurate than the potentiometric ones. Therefore, it is not necessary to wait until thermodynamic equilibrium is obtained; the response is a linear function of the concentration of the analyte. However, the selectivity of the amperometric devices is only governed by the redox potential of the electroactive species present. Consequently, the electric current measured by the instrument can include the contributions of several chemical species. The first amperometric biosensor[69] for glucose analysis using the glucose oxidase enzyme with the Clark oxygen electrode was based on monitoring of the oxygen consumption. The formation of the reaction product or consumption of reagent can be measured to infer the analyte concentration. These biosensors are referred to as the first generation. Amperometric biosensors modified with mediators are referred to as second generation biosensors. Mediators are redox substances that facilitate the electron transfer between the enzyme and electrode. The direct enzyme-electrode coupling, or mediator-free biosensors based on direct electron transfer mechanisms, are referred to as third generation. In this case, the electron is directly transferred from the electrode to enzyme and to the substrate molecule (or vice versa). In this mechanism, the electron acts as a second substrate for the enzymatic reactions and leads to the generation of a catalytic current. The substrate transformation (electrode process) is essentially a catalytic process.[70,71]

Piezoelectric transducers

Piezoelectric transducers find application as immunosensors. In these devices, an antigen or antibody is immobilized on the surface of a crystal.[72] The interaction of these elements with the analyte is highly specific and can be monitored through the oscillation of the immersed crystal in a liquid which will produce a modification of mass in the crystal, measurable by means of the modification of the frequency of oscillation. Immunosensors based on the wave acoustics principle, among others types, can be used for detection of pathogenic microorganisms, gases, aromas, pesticides, hormones and other compounds.[73–75]

Optical Transduction Devices

We begin with a general description of the devices that have been developed and implemented as platforms for biosensors, without restricting the discussion to approaches that involve nanoscale materials.

The binding of a target molecule to a receptor can be detected by an interrogating light beam, or by an intrinsic light output from the products of the binding reaction. Optical interrogation can be implemented in several different modes. For example, is possible to detect subtle differences in the refractive index in a medium at the surface of a sensor on which antibodies have been immobilized. When an antigen-antibody interaction takes place, the refractive index of the medium becomes slightly higher; this change can be measured using a variety of different optical techniques that are sensitive to small changes in refractive indexes. Fluorescence techniques are very commonly used in biosensors. In this case, molecules are labeled with fluorophores that are excited by a light source; the resulting fluorescence is measured by a photodetector. Spectroscopy can be used to detect the presence of an analyte by evaluating the optical spectrum directly; the Raman scattering signal that is observed at a shifted wavelength that is unique to the target compound can be used in a detection method. Spontaneous Raman scattering is normally quite weak. However, nanoscale materials are able to provide a significant enhancement in the Raman signal. In all cases, the optical interrogation will lead ultimately to an electrical signal that provides a calibrated measure of the extent of binding to receptors, and hence of the amount of an analyte that is present in a sample.

Waveguide sensors

Several variants of biosensors are based on the application of the evanescent wave that appears at the interface between two media that have different refractive indexes. The evanescent wave at the interface can be used in conjunction with waveguides, with surface plasmon waves, and with resonant mirrors. In several configurations of biosensors, the evanescent wave is utilized within a micro channel that is filled with sample fluid. The capture or binding of analyte molecules by the biological detection element at the sensor surface is detected through techniques that rely on refractive index changes.

Another family of biosensors uses the fluorescent output from a reporter molecule on a target molecule as the detection mode. Evanescent waves may also be used in this case to excite the labeling fluorophores, as seen in Total Internal Reflection Fluorescence (TIRF) devices. Antibodies that are raised against a specific analyte are immobilized on the top of a waveguide. Fluid that contains previously labeled analyte molecules, along with unlabeled sample analyte, is added to the system. Competition for binding to the antibodies takes place. A larger concentration of sample analyte will lead to a lower fluorescence signal as fewer of the labeled molecules are bound to antibodies. The evanescent waveguide field will only excite fluorophores that are in the immediate vicinity of the surface, thereby avoiding the excitation of more distant unbound fluorophores which can contribute to unwanted background signal.

Several configurations have been used for evanescent field detection, including planar waveguides and optical fibers, with signal transduction implemented via fluorescence or interferometry. Sapsford *et al.*,[76] have provided a thorough review of fluorescence planar waveguide biosensors;

Taitt and Ligler[77] have reviewed developments in evanescent wave optical fiber biosensors; Campbell and McCloskey[78] have reviewed interferometric biosensors.

SPR

An interface between two optically transparent media can be useful not only in the format of Total Internal Reflection (TIRF) spectroscopy, but also in application through a phenomenon known as Surface Plasmon Resonance (SPR). These surface waves occur at a metal — dielectric interface. The propagation of the plasmon wave is sensitive to the refractive index of the dielectric medium, and this provides its utility in terms of sensing. The use of SPR for biosensors has been thoroughly reviewed by Homola and colleagues.[79,80]

Surface Plasmon Resonance devices find application in two basic areas: direct sensing of an analyte in samples, and the fundamental study of the kinetics or thermodynamics of biophysical and biochemical reactions. In the first case, binding of an analyte to sensing elements at the surface is detected through the change in the refractive index. The second type of application takes advantage of the fact that surface plasmon resonance biosensors do not require labeling of analyte or detecting element, as is required in fluorescence-based detection schemes. Therefore, it is possible to measure the kinetics, or time dependence, of reactions between an analyte and sensing element in the biosensor.

Interferometric sensors

Because the phase of light changes as the refractive index in a medium changes, variations in refractive index caused by binding of analyte molecules can be detected through interference effects. Interferometric biosensors have been fabricated on waveguides where one leg of the interferometer is the sensing part, and a second waveguide channel carries a reference beam. As the refractive index in the detection leg changes, the phase change of light, when recombined with the reference signal, produces interference patterns that can be recorded and interpreted in terms of concentrations of analytes. The use of interferometry for biosensors is reviewed in detail by Campbell et al.[78]

Interferometric sensors have been used in a number of environmental applications, including competitive immunoassays for the detection of the herbicide, atrazine;[81,82] in this case the limit of detection was about 100 ng/L. In other cases, nanoscale materials have been adopted as part of the sensor. Tinsley-Bown et al.,[83] used a porous silicon structure with pore sizes greater than 50 nm for an immunoassay. Ryu *et al.*,[84] also used a porous silicon interferometer to detect beta-galactosidase that was released from recombinant *E. coli*.

Resonant mirror devices

The evanescent field above a waveguide can be enhanced when the incident light enters the waveguide at a special angle and is reflected back by a mirror. A resonant condition occurs at this angle, and the device is referred to as a resonant mirror. One of the earliest descriptions of the resonant mirror concept in the literature is due to Cush *et al.*[85] The concept has been commercialized in the IAsys biosensor. The principles and applications of resonant mirror devices have been thoroughly reviewed by Cooper[86] and by Kinning and Edwards.[87]

The device is made of two basic parts: a low refractive index coupling layer, and a high refractive index waveguide layer. An evanescent field is established above the waveguide; it is used to detect changes in the refractive index of the medium that is immediately adjacent to the liquid — waveguide interface. Binding of an analyte to a sensing element such as an antibody will change the refractive index where the evanescent field is propagating. This will affect the angle at which an incident ray of light will resonate within the cavity created by the waveguide. The incident beam is scanned over a range of angles in order to determine when the maximum intensity is observed at the detector. Polarizers are used so that only light that is propagating through the resonant structure is detected, thereby blocking extraneous light from the detector.

Spectroscopy

Although labeling with a fluorophore is standard procedure in biosensors, in some cases it may be advantageous to carry out label-free detection. This can be achieved by making use of unique spectral signatures that can be obtained from biomolecules through processes such as Raman Scattering. A well-known method that uses this technique is so called Surface Enhanced Raman Scattering (SERS). The normal spontaneous Raman signal is quite weak but its strength can be enhanced if detection is carried out on gold or silver surfaces that support Raman resonances. The surface enhanced resonant Raman process gives rise to a considerable increase in the Raman cross-section, and therefore in the limit of detection. The nanoscale features of the surfaces are important in determining the extent of the signal enhancement. SERS has been used in basic studies of biochemistry[88] and recently for DNA detection.[89,90]

D. Dosev, M. Nichkova and I. M. Kennedy

Raman spectroscopy can also be carried out in small volumes through the use of optical tweezers to control samples.[91,92] With this technique, it has been shown that is possible to trap cells and other biomolecules within the confocal region of a micro Raman spectroscopy system. The Raman signature of the biomolecules is unique and allows identification of particular cells and compounds without additional labeling. The equipment to carry this out is rather more complex than simple lab-on-a-chip devices but it offers considerable promise for label-free detection.

Nanoscale Materials for Sensors

Quantum Dots

Quantum dots (QDs) represent an early implementation of optically-active nanoscale materials. They are made of semiconductor materials with electrons that have energy levels typical of the particular semiconductor material. The exciton Bohr radius represents the average physical separation between excited electron-hole pairs for each material. If the size of a semiconductor crystal becomes as small as the size of the material's exciton Bohr radius, then the electron energy levels become discrete. This phenomenon is called quantum confinement and under such conditions the material is termed a quantum dot. Due to the quantum confinement effect, the absorbance peak of quantum dots and their emission maxima shift to higher energies (shorter wavelengths) with decreasing particle size.[93]

Since "naked" QDs are susceptible to photo-oxidation, they need to be capped by a protective shell of an insulating material or wide-bandgap semiconductor, structurally matched with the core material. Core-shell geometries where the nanocrystal (typically from 2 to 8 nm in diameter) is encapsulated in a shell of a wider band gap semiconductor have increased fluorescence quantum efficiencies ($> 50\%$) and greatly improved photochemical stability. The emission quantum yield of these particles increases with increasing thickness of the shell and can reach values of up to 0.85.[31]

In contrast to the organic fluorophores that are conventionally used in biotechnology, quantum dots absorb light over a very broad spectral range. This makes it possible to optically excite quantum dots with different emission spectra (colors) by using a single excitation laser wavelength, which enables simultaneous detection of multiple markers in biosensing and assay applications.[29] Quantum dots are made in CdSe-CdS core-shell structures that emit in the visible region from approximately 550 nm (green) to 630 nm (red). Near-infrared optical emission has been achieved with other material combinations, such as InP and InAs. It is necessary to provide a hydrophobic, chemisorbed surface layer to facilitate bioconjugation via adsorption or covalent binding. Widely used approaches for this purpose include the treatment of the QDs with tri-n-octylphosphine oxide (TOPO) and hexadecylamine; several alternative approaches have also been proposed.[27] Currently, quantum dots with a variety of properties are available commercially.

Carbon Nanotubes (CNTs)

Carbon nanotubes (CNTs) are graphitic sheets that are curled up into seamless cylinders. During the past few years they have attracted a great deal of attention, and have been subject to intense research associated with their possible application as carbon-based electronic devices. Carbon nanotubes are classified into two main types: multi-walled (MWCNTs) and single walled (SWCNTs). The earliest observations were of micron-long multi-walled nanotubes in 1991.[94] Shortly after that, the synthesis of single-walled tubes was reported.[95] Their diameters could be controllably changed via variations in the growth parameters.

SWCNT possess a cylindrical nanostructure (with a high aspect ratio), formed by rolling up a single graphite sheet into a tube. The atoms are located using a pair of integers (n, m) and the lattice vector $\vec{C} = n\vec{a}_1 + m\vec{a}_2$ as shown in Figure 7.[96]

A tube can be classified using the pair of integers n and m by viewing the rolling up of the sheet as the "placement" of the atom at 0,0 on the atom at n, m. Hence, different diameter tubes and helical arrangements of hexagons can arise by changing n, m. An example is shown in Figure 8.[96]

MWCNT consist of several layers of graphene cylinders that are concentrically nested like rings of a tree trunk (with an interlayer spacing of 3.4 Å).[97,98] Figure 9 compares TEM images of SWCNT and MWCNT.[94] While SWCNTs typically have diameters ranging from 0.7 nm to 3 nm, they can be several microns in length. They exhibit unusual mechanical properties such as high strength and stiffness — the Young's modulus and tensile strength of nanotubes are higher than those of any other material.[99,100]

The unique electrical properties of carbon nanotubes make them extremely attractive for the task of chemical sensors, in general, and electrochemical detection, in particular.[101] The electronic properties of CNTs are sensitive to the structure of the tube. Carbon nanotubes can be metallic or semiconducting, depending on n, m. For example, if $n - m$ is three times an integer, the carbon nanotube has an extremely small band gap, and at room temperature, it has metallic behavior. For $n = m$, the tubes are metallic; and for other values of $n - m$, the tubes behave as semiconductors with a band gap.

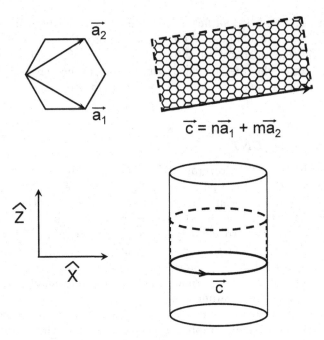

$$\vec{c} = n\vec{a}_1 + m\vec{a}_2$$

Figure 7.   Structure of an n,m carbon nanotube. The carbon atoms are at the vertices of the hexagons. (From Cohen[96] with permission.)

Inorganic Nanotubes, Wires and Belts

In addition to carbon nanotubes, inorganic nanotubes and nanowires are exciting increasing interest in many areas.[102,103] Like carbon nanotubes, inorganic nanotubes consist of long hollow cylinders that typically have diameters of about 10 nm and wall thicknesses of about 1 nm.[104–107] Nanowires and nanorods are flexible or stiff solid cylinders that are much longer than their diameters. They provide a much greater surface area for sensing applications than the area offered by spherical particles.

These unusual morphologies can be synthesized from a wide variety of materials. Nanotubes have been made from semi-conductor metal oxides such as ZnO and TiO_2[108,109] from sulfides, from phosphides,[110] and other materials.[104] Nanorods have also been formed from gold and silver.[111] In general, the synthesis routes follow one of two paths: a gas phase synthesis in which precursors decompose at high temperature to reform as wires and rods; or a solution phase process at low temperature. Methods for the synthesis of these materials are described in several recent reviews.[104–106,109,112]

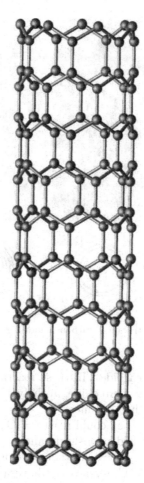

Figure 8. Ball and stick model for a nanotube. The balls represent carbon atoms. (From Cohen[96] with permission.)

Rao *et al.*,[104] describe several of these methods. The so-called carbo-thermal route makes use of the reaction of metal oxide powders in a reducing environment with carbon nanotubes that supply a source of reactive carbon. This method produces nanowires and nanorods of metal oxides such as SiO_2, InO, ZnO, and under the right conditions of excess nitrogen, can produce nitrides of these metals and oxynitrides. One of the simple methods of synthesis of inorganic nanowires employs the carbothermal route. The method essentially involves heating a metal oxide with an adequate quantity of carbon in an appropriate atmosphere. For example, ammonia provides the atmosphere for the formation of nitrides. An inert atmosphere and a

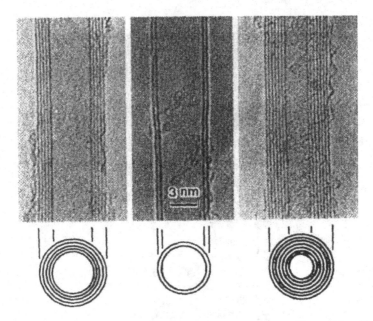

Figure 9. TEM images of multi-walled carbon nanotubes. (From Iijima[94] with permission.)

slight excess of carbon yield carbides. In these reactions, carbon helps to form an oxidic species, usually a sub oxide, in the vapor phase, which then transforms to the final crystalline product.

Rao *et al.*,[104] showed that they could react Ga_2O_3 powder with multi-walled carbon nanotubes to yield nanosheets and nanorods. Figure 10(a) shows a SEM image of Ga_2O_3 nanowires that were obtained by this reaction at 1100°C in Argon gas. The nanowires have a diameter of around 300 nm with lengths extending to tens of microns. The nanowires exhibit good photoluminescence characteristics.[104] Nanowires of other oxides such as SiO_2 (Figure 10b) have also been prepared by reacting the corresponding metal/metal oxide with carbon in an inert atmosphere.[104] The reaction of NH_3 with a mixture of SiO_2 gel and carbon nanotubes or activated carbon in the presence of a Fe catalyst yielded nanowires of silica.[104] Using silica gel heated with activated-carbon at 1360°C in NH_3 yielded SiC nanowires (Figure 10c). Silicon nanowires have been formed from Si powders or solid substrates with a similar approach (Figure 10d).[104]

Many synthesis techniques have been reported that use the solution phase or so-called soft chemistry synthesis of nanorods, tubes and wires. Solution phase chemistry can produce many of the same materials as the gas phase route, including metal oxides as such as ZnO,[113,114] indium tin oxide

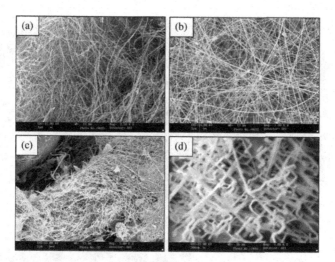

Figure 10. SEM images of (a) Ga_2O_3 nanowires obtained by the reaction of Ga_2O_3 with MWNTS at 1100°C in 60 sccm Ar; (b) Si34 nanowires synthesized by heating silica gel and MWNTs in an NH3 atmosphere; (c) SiC nanowires obtained by the reaction of silica gel and activated carbon in NH3 at 1360°C for seven hours and (d) Si nanowires synthesized using the carbon-assisted method on a Si substrate. (From Rao *et al.*,[104] with permission.)

(ITO)[115] and others. One of the novel methods for the synthesis of metal oxide nanotubes employs gels derived from low molecular weight organic compounds. The gel fibers are coated with oxidic materials followed by the dissolution of the gel in a suitable solvent and by calcination. Figure 11 shows Transmission Electron Microscopy (TEM) images of (a) gel fibers; and nanotubes of (b) titania, (c) ZnO and (d) $ZnSO_4$ obtained by coating the respective precursors on the hydrogel fibers, followed by the removal of the template. The use of hydrogels enabled Rao *et al.*,[104] to avoid the use of alkoxides, which are difficult to handle due to their moisture sensitivity. The gel fibers have diameters between 8 to 10 nm with lengths extending to several hundreds of nanometers. Nanotubes of titania were obtained by the reaction of the titanium alkoxide with the gel fibers, followed by the removal of the gel fiber template. The nanotubes generally have inner diameters comparable to the diameter of the gel fibers. ZnO nanotubes were obtained using zinc acetate as the precursor; $ZnSO_4$ nanotubes were obtained by a similar process.

Inorganic nanowires find their greatest application in sensing through the modification of their electrical properties as compounds adsorb to their surface. Slight changes to their conductivity can be detected

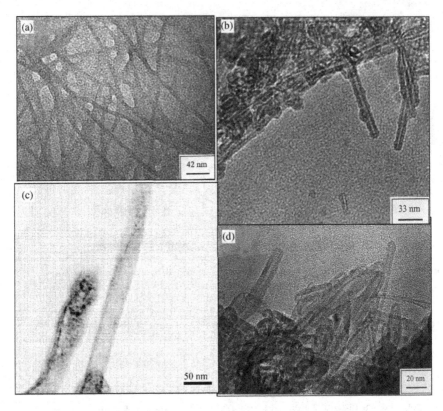

Figure 11.   TEM images of (a) gel fibers and nanotubes of (b) titania, (c) ZnO and (d) ZnSO₄ obtained by coating the respective precursors on the hydrogel fibers, followed by the removal of the template. (From Rao *et al.*,[104] with permission.)

following adsorption. The difficulty lies in attaching electrodes to individual nanowires and reading out the results.

Colloidal Gold and Colloidal Silver

Nanoparticles of noble materials such as gold and silver have been used for centuries for coloring of glass and silk. The high conductivity of the noble metals supports the presence of a plasmon absorption band. This absorption band results when the incident photon frequency is in resonance with the collective excitation of the conductive electrons of the particle. This effect was termed "Localized Surface Plasmon Resonance" (LSPR). The resonant frequency of the LSPR band is strongly size dependent. Spherical gold particles of 10 to 40 nm diameter are red (resonance frequency at about 600 nm). On the other hand, irregularly shaped, and bigger particles turn blue (resonance frequency about 450 nm). The full visible spectrum may be covered by gold particles with diameters from 60 to 200 nm. Silver nanoparticles in the same size ranges behave in a similar way. Today, colloidal gold finds application in non-linear optics, supramolecular chemistry, molecular recognition and the biosciences.[31,116]

Lanthanide-Based Nanoparticles

Lanthanides or rare earths metals comprise all elements bearing the atomic numbers 58 to 71 in the periodic table. Some of the lanthanide ions (e.g. Eu(III), Dy(III), Tb(III) and Sm(III)) have well determined excitation and emission spectra when they are incorporated into appropriate environments such as chelate molecules or crystal lattice of oxide. Chelates are organic complexes that serve as antennae which transmit the excitation energy to the lanthanide ion.[117] Oxides like Gd_2O_3 of Y_2O_3 provide ideal hosts for doping with luminescent lanthanide ions;[118] the crystal lattice of the oxide permits efficient excitation and emission of the electronic transitions of their f-electrons (Figure 12). The lanthanide emission is narrow, exhibits a large Stokes shift, and yields long luminescence lifetimes (from several μs to ms).

The use of nanoparticles containing luminescent lanthanide ions enables multiple luminescent centers to be incorporated into a single nanoparticle. Single polystyrene nanoparticle of 100 nm may contain several thousand chelate molecules, possibly with different lanthanide ions. The amount of doping into a Gd_2O_3 nanoparticle is even greater. Nevertheless, if the doping exceeds a certain concentration (e.g. Eu^{3+} into Gd_2O_3), the average distance between two Eu^{3+} ions will become too short and they will tend to quench each other instead of emitting light. The direct consequence of this so-called concentration quenching is a decrease in the luminescent lifetime.

The long lifetime of optimally doped nanoparticles offers the opportunity for so-called time-gated measurements: after the sample is excited with a short pulse of UV light the emission signal is accumulated for a suitable time interval with a time delay on the order of several 100 ns. Therefore, interference by short-lived background emission and scattering from the sample can be minimized as seen in Figure 13.

Magnetic Nanoparticles

Magnetic nanoparticles, especially magnetite (Fe_3O_4) nanoparticles, are promising candidates for biomolecule tagging, imaging, sensing, and separation.[15,119,120] Magnetite is a common magnetic iron oxide that has

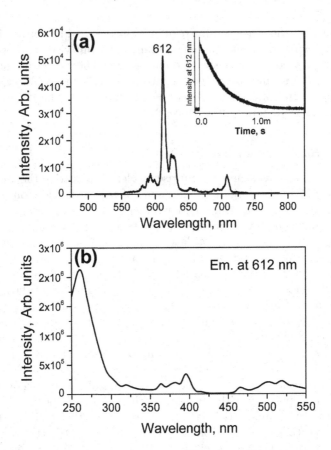

Figure 12. (a) Emission spectrum and 1 ms lifetime and (b) excitation spectrum of Eu^{3+} doped into Gd_2O_3 nanoparticles

a cubic inverse spinel structure with oxygen forming a face-centered cubic (fcc) structure and Fe cations occupying interstitial tetrahedral sites and octahedral sites. The electrons can hop between Fe^{2+} and Fe^{3+} ions in the octahedral sites at room temperature, thus giving the magnetite properties of half-metallic material. Using chemical precipitation methods, different nanometer-scale MFe_2O_4 materials (M = Co, Ni, Mn, Mg, etc) can be synthesized and used in studies of nanomagnetism.[121]

For magnetic particles below a certain critical diameter, a single nanoparticle cannot support more than one domain. The critical diameter generally falls in the 10 to 100 nm range.[122] In a single-domain cluster, the atomic magnetic moments are coupled via exchange interactions to form a large net cluster moment — the cluster is called a super-paramagnet.

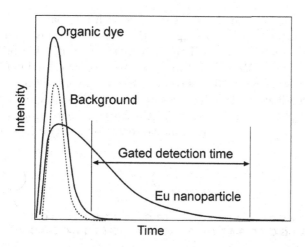

Figure 13. Time-gated detection with long-lifetime nanophosphors avoids the short-lifetime background fluorescence from a sample (background shown as dashed line). An organic dye has a short lifetime.

After surface coating and functionalization, the magnetic nanoparticles can be dispersed into water, forming water-based suspensions. From a suspension, the particles can interact with an external magnetic field and be positioned to a specific area, facilitating magnetic resonance imaging for medical diagnosis and cancer therapy based on heating with an external AC magnetic field. These applications require that the nanoparticles are superparamagnetic with sizes of a few nanometers and a narrow size distribution so that the particles have uniform physical and chemical properties.

The Application of Nanomaterials in Sensors

Optical Transduction Using Nano-scale Materials

Nano-scale materials offer unique opportunities for optical transduction in biosensors due to the particular interaction between light and matter at the nano-scale. Light can be either absorbed or emitted.

Absorbance based sensors

When gold nanoparticles form aggregates, the light absorption spectrum of the aggregate depends strongly on the average distance between the particles within the aggregate. As the interparticle distances in these aggregates decrease to less than approximately the average particle diameter, the color becomes blue. On the other hand, aggregates with interparticle distances substantially greater than the average particle diameter appear red.[123,124] Several research groups have used this phenomenon to develop biosensor detection schemes for DNA,[125–129] for studying biomolecular interactions in real time,[130] and for immunoassays. A refractive index sensor based on colloidal gold-modified optical fiber was reported by Cheng and Chau.[131]

As an example, Figure 14 describes the basic principle for detection of DNA using gold nanoparticles. Mercaptoalkyloligonucleotide-modified

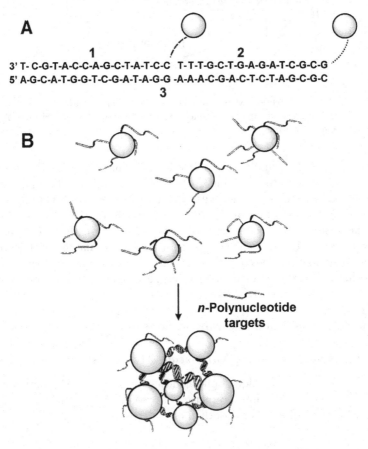

A

1   **2**
3' T- C-G-T-A-C-C-A-G-C-T-A-T-C-C T-T-T-G-C-T-G-A-G-A-T-C-G-C-G
5' A-G-C-A-T-G-G-T-C-G-A-T-A-G-G -A-A-A-C-G-A-C-T-C-T-A-G-C-G-C
3

B

n-Polynucleotide
targets

Figure 14. (A) Hybridization of two mercaptoalkyl-oligonucleotide-modified Au nanoparticles and polynucleotide target. (B) Schematic representation of the concept for generating aggregates signaling hybridization of nanoparticle-oligonucleotide conjugates with oligonucleotide target molecules. The nanoparticles and the oligonucleotide interconnects are not drawn to scale, and the number of oligomers per particle is believed to be much larger than depicted. (From Elghanian[125] with permission.)

Au nanoparticles are used as reporter groups. Hybridization results in the binding of an oligonucleotide probe to the target sequence, and in the formation of an extended polymeric network in which the reporter units are interlocked.[125]

Localized Surface Plasmon Resonance (LSPR) excitation in silver and gold nanoparticles produces strong extinction and scattering spectra that in recent years have been used for important sensing and spectroscopy applications. Several ultra-sensitive nanoscale optical biosensors based on LSPR have been reported. LSPR nanobiosensors for the biotin-streptavidin system have been developed using gold nanoparticles self-assembled on a functionalized glass surface[130,132] and using triangular silver nanoparticles directly fabricated on the glass by nanosphere lithography.[133] Colloidal gold-enhanced SPR was developed for ultra-sensitive detection of DNA hybridization.[134] Use of the Au nanoparticles tags leads to a greater than ten-fold increase in angle shift, corresponding to a more than 1000-fold improvement in sensitivity for the target oligonucleotide as compared to the unamplified event and makes it possible to conduct SPR imaging on DNA arrays with a sensitivity similar to that of traditional fluorescence based methods for DNA hybridization.

Luminescence-based sensors

Gold nanoparticles are also applied as quenchers for fluorescent labels in a new class of optical biosensors for recognizing and detecting specific DNA sequences.[25] Oligonucleotides labeled with a thiol group at one end and a fluorescent dye at the other end were attached to the surface of gold nanoparticles, spontaneously forming arch-like conformation (see Figure 15). Binding of target molecules resulted in a conformation change and this restored the fluorescence of the quenched fluorophore. The biosensor developed on this basis was able to detect single-base mutations in a homogeneous format.

Recently, the same principle was employed for the development of biosensors for the detection of neurotoxic organophosphates that have found widespread use in the environment for insect control.[55]

A large variety of environmental sensors are based on the photoluminescence quenching in porous silicon and silicon nanocrysrals due to adsorption of different molecules on their surface. The advances in this type of sensors were reviewed in detail by Shi *et al.*[135] After different treatments of Si nanocrystals surface, their photoluminescence was found to be sensitive to a variety of organic solvents, amines, aromatic compounds, metal ions and corrosive, toxic and explosive gases.

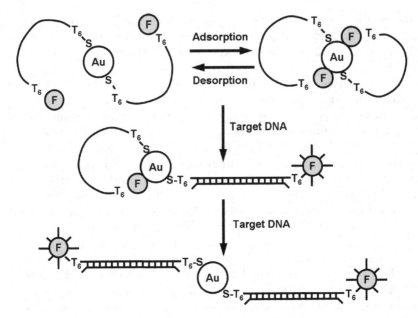

Figure 15. Nanoparticle-based probes and their operating principles. Two oligonucleotide molecules (oligos) are shown to self-assemble into a constrained conformation
on each gold particle (2.5 nm diameter). A T6 spacer (six thymines) is inserted at both
the 3′- and 5′-ends to reduce steric hindrance. Single-stranded DNA is represented by
a single line and double-stranded DNA by a cross-linked double line. In the assembled
(closed) state, the fluorophore is quenched by the nanoparticle. Upon target binding, the
constrained conformation opens, the fluorophore leaves the surface because of the structural rigidity of the hybridized DNA (double-stranded), and fluorescence is restored. In
the open state, the fluorophore is separated from the particle surface by about 10 nm.
See text for detailed explanation. Au, gold particle; F, fluorophore; S, sulfur atom. (From
Maxwell et al.,[25] with permission.)

Quantum dots are very attractive for biolabeling due to properties
such as excellent brightness, narrow and precise tunable emission, negligible photo-bleaching, fairly high quantum yields, and photostability.[93,136]
Although quantum dots have proven to be suitable labels for biological
imaging, their application in quantitative biosensors is still in an early stage.
A biosensing detection scheme was developed that utilizes donor–acceptor
energy transfer between QDs and receptors for conducting recognition-
based assays.[137] In this configuration, the luminescence of antibody-coated
quantum dot was quenched due to specific binding with quencher-labeled
antigen. Gerion et al.,[28] have employed quantum dots as fluorescent labels
for DNA microarray analysis while Geho et al.,[138] reported the use of quantum dots as detection elements for protein microarrays.

Silica nanoparticles doped with organic fluorescent dyes were synthesized by Tapec *et al.*[139] They exhibited very good photostability and minimal dye leakage. The silica matrix of the nanoparticles permits different methods of surface biomolecular modification for biosensor and bioanalysis applications. Application of dye-doped silica particles for the detection of glutamate was demonstrated.

Nanoparticles of Eu_2O_3 have shown promising application as luminescent tags in immunoassay for detection of atrazine.[140] Recently, a microarray format for an immunoassay using improved Eu-doped Gd_2O_3 nanoparticles was reported.[141] The excellent photostability of these particles allows the samples to be observed and analyzed under UV excitation for prolonged time periods. Their sharp emission spectrum facilitates easy separation of the signal from the background fluorescence. Detection of phenoxybenzoic acid (a generic biomarker of human exposure to pyrethroids) was demonstrated using these nanoparticles.[141] In this configuration, an internal fluorescent standard was introduced by making use of the emission from the nanoparticle as a measure of the number of antibodies that were attached to its surface; the detection of analyte was reported by a second fluorophore of a different wavelength. Miniaturization of a conventional immunoassay offers the possibility for building portable biosensor devices; the lanthanide oxide nanoparticles offer the opportunity to build new detection schemes using their unique luminescent properties such as long lifetime,[142,143] excellent photostability[141] and sharp emission spectra.

Electrochemical Biosensors Using Nanomaterials

Nanomaterials can be used in a variety of electrochemical biosensing schemes. New nanoparticle-based signal amplification and coding strategies for bioaffinity assays, along with carbon nanotube molecular wires, have been proposed during the last few years. Gold, silver and metal oxide nanoparticles have been widely used in biosensors because of their large relative surface area and good biocompatibility.

Enzyme-based electrochemical biosensors

An extremely important challenge in amperometric enzyme electrodes is the establishment of efficient electron transfer between the redox active site of the enzyme (protein) and the electrode surface. The possibility of direct electron transfer between enzymes and electrode surfaces could pave the way for superior reagentless biosensors, as it obviates the need for co-substrates or mediators and allows efficient transduction of the biorecognition event. The conductive properties of some nanomaterials, such as carbon nanotubes and metal nanoparticles, suggested that they could mediate the electrical communication between the redox enzyme and the electrode. The ability of CNT to promote electron transfer reactions is attributed to the presence of edge plane defects at their end caps. CNT associated with carbon paste electrodes or glassy carbon electrodes show excellent eletcrocatalytic activity toward the oxidation of ascorbic acid, dopamine, phenols, tryptophan, carbohydrates, and the reduction of hydrogen peroxide and nitrite ions.[144,145]

The structural alignment of enzymes on the ends of CNT organized as an array on a conductive surface was accomplished by Willner's group.[146] An array of perpendicularly oriented carbon nanotubes on a gold electrode was fabricated by covalently attaching carboxylic acid functionalized carbon nanotubes to a cystamine-monolayer-functionalized gold electrode. The amino-derivative of the FAD cofactor (flavin adenine dinucleotide) was covalently coupled to the carboxylic groups at the free ends of the aligned CNT (see Figure 16).

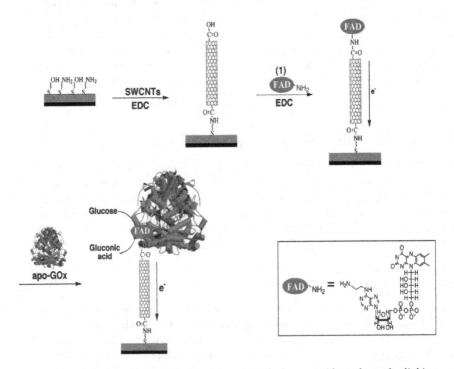

Figure 16. Assembly of CNT electrically contacted glucose oxidase electrode: linking the reconstituted enzyme, on the edge of the FAD-functionalized CNT, to the electrode surface. (From Patolsky[146] with permission.)

Such enzyme reconstitution on the end of the CNT represents an extremely efficient approach for plugging an electrode into glucose-oxidase (GOx). Electrons were thus transported across distances greater than 150 nm with the length of the CNT controlling the rate of electron transport. An interfacial electron transfer rate constant of 42 s^{-1} was estimated for a 50 nm long CNT.

The anodic detection of dehydrogenase and oxidase enzymes at ordinary electrodes is often hampered by the large over-voltage encountered during their oxidation. The greatly enhanced redox activity of hydrogen peroxide[147] and NADH[148] at CNT-modified electrodes addresses these over-voltage limitations. Further improvement of the detection of the enzymatically-liberated hydrogen peroxide was accomplished by deposition of platinum[145] and palladium[149] nanoparticles onto CNT. Co-axial nanowires, consisting of a concentric layer of polypyrrole uniformely coated onto aligned CNT, have provided a template for making glucose sensors with a large amount of electrochemically entrapped GOx in the ultrathin polypyrrole film.[150] Similarly, amperometric detection of organophosphorus pesticides and nerve agents was performed using a screen-printed biosensor based on co-immobilized acetocholine esterase, choline esterase and CNTs.[151]

The excellent electrocatalytic properties of metal nanoparticles (compared to bulk metal electrodes) can also enhance amperometric enzyme electrodes. Gold nanoparticles were used as "electrical nanoplugs" for the alignment of GOx on conducting supports and wiring its redox center.[152] An improved amperometric biosensing of glutamate was achieved by a transducer built on dispersed iridium nanoparticles ($d = 2$ nm) in graphite-like carbon.[153] Ferrocene-doped silica nanoparticles conjugated with a biopolymer chitosan served as an electron transfer mediator for the determination of glucose in rat brain.[154] ZrO_2 nanoparticles/chitosan composite film was used as an immobilization matrix in an acetylcholinestarse biosensor for the determination of pesticide in real vegetable samples.[36] Glassy carbon electrodes with a film of mercury-doped silver nanoparticles was used for cysteine detection in urine samples.[155] HRP-modified nanoelectrodes formed by arrays of gold nanotubes offered enhanced sensitivity compared to a classical gold macroelectrode.[156]

Bioaffinity (immuno and DNA) electrochemical biosensors

Nanomaterials can be used as quantitation tags or as amplification carriers in a variety of bioaffinity electrochemical formats. By analogy to fluorescence-based methods, several electrochemical detection methods have been pursued in which DNA sequences and antibodies have been labeled with electroactive nanoparticles. The characteristic electrochemical response of the nanoparticle reporter signals the hybridization event or the immuno-reaction, respectively.

Since CNTs exhibit a high conductivity and reveal very high area-to-weight ratio, they have been used as a support for the immobilization of biomolecules and further used in immuno- and DNA sensors. For example, immunosensing systems with electrochemiluminescent detection have been designed using CNTs as supports of the immuno-recognition systems.[157] DNA-functionalized carbon nanotubes arrays were constructed, hybridized with the complementary DNA and the redox label daunomycin was intercalated into the double-stranded DNA-carbon nanotube assembly and detected by differential pulse voltammetry.[158] In another configuration, amino-functionalized DNA covalently bound to aligned CNTs on a gold electrode was hybridized with ferrocene-labeled complementary DNA; an enhanced electrochemical signal provided by the high surface of the CNT-modified electrode was observed.[159] Similarly, the coupling of the CNT nanoelectrode array with the $Ru(bpy)_3^{2+}$-mediated guanine oxidation has facilitated the detection of subattomoles of DNA targets.[160] Using this method, the sensitivity of the DNA detection was improved by several orders of magnitude compared to methods where DNA was immobilized using self-assembled monolayers on conventional electrodes.

Conducting-polymer nanowire biosensors have been shown to be attractive for label-free bioaffinity sensing. Real-time monitoring of nanomolar concentrations of biotin at an avidin-embedded polypyrrole nanowire has been demonstrated.[161] Similarly, label-free conductivity measurements of antibodies associated with human autoimmune diseases have been performed using polyethylene oxide-functionalized CNT.[162]

Functionalized CNT have been used as carriers of multiple enzyme labels for electrochemical DNA and immuno-sensing.[163] For example, CNTs were loaded with about 9600 alkaline phosphatase molecules per CNT and further modified with an oligonucleotide or an antibody. Such a CNT-derived double step amplification pathway (of both the recognition and transduction events) allows the detection of DNA down to the 1.3 zmol level and indicates great promise for PCR-free DNA analysis.

Nanoparticle-based amplification schemes have resulted in an improved sensitivity of bioelectronic assays by several orders of magnitude. The use of colloidal gold tags allowed the highly sensitive detection of the DNA hybridization by anodic stripping voltammetry.[53,164] The same approach was reported for sandwich immunoassays where immunoglobulins G (IgGs) were detected at concentration as low as 3×10^{-12} M, which is competitive with colorimetric ELISA or with immunoassays based on fluorescent europium chelate labels.[165] Subsequent signal amplification of the electro-chemical method for analyzing sequence-specific DNA by gold nanoparticle DNA probes was achieved by silver enhancement where the sensitivity has been increased approximately two orders of magnitude.[166,167] Silver enhancement on the gold nanoparticles was also applied to an electrochemical sandwich type immunoassay.[168] A new strategy for amplifying particle-based electrical DNA detection based on oligonucleotides functionalized with polymeric beads carrying multiple gold nanoparticle tags allowed the detection of DNA targets down to the 300 amol level, and offered great promise for ultrasensitive detection of other biorecognition events.[169]

Inorganic nanocrystals offer an electrodiverse population of electrical tags as needed for designing electronic coding. A novel, sensitive DNA hybridization detection protocol, based on DNA-quantum dot nanoconju-gates coupled with Electrochemical Impedance Spectroscopy (EIS) detec-tion, has been reported.[170] Due to having more negative charges, space resistance and the semiconductor property, CdS nanoparticle labels on target DNA could improve the sensitivity by two orders of magnitude compared with non-CdS tagged DNA sequences. Cu-Au alloy core-shell nanoparticles combine the good surface modification properties of Au with the good electrochemical activity of Cu core and allowed their application as oligonucleotides labels for electrochemical stripping detection of DNA hybridization.[171]

An effective method for amplifying electrical detection of DNA hybridization has been developed using carbon nanotubes carriers loaded with a large number of CdS QDs (see Figure 17).[172] Such use of CNT amplification platforms has been combined with an ultrasensitive stripping-voltammetric detection of the dissolved CdS tags following dual hybridization events of a sandwich assay on a streptavidin modified 96-well microplate. Anchoring of the monolayer-protected quantum dots to the acetone-activated CNT was accomplished via hydrophobic interactions. A substantial (approximately 500-fold) lowering of the detection limit has been obtained compared to conventional single-particle stripping hybridiza-tion assays, reflecting the CdS QDs loading on the CNT carrier.

Nanoparticles hold particular promise as the next generation of barcodes for multiplexing experiments. The labeling of probes bearing different DNA sequences or antibodies with different nanoaprticles enables the simultane-ous detection of more than one target in a sample, as shown in Figures 18[173] and 19.[174] The number of targets that can be readily detected simultane-ously (without using high-level multiplexing) is controlled by the number of voltammetrically distinguishable nanoparticle markers. A multi-target sandwich hybridization assay involving a dual hybridization event, with probes linked to three tagged inorganic crystals and to magnetic beads, has

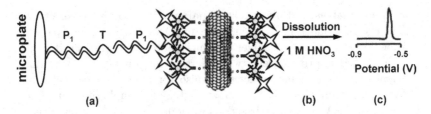

(a)   (b)   (c)

Figure 17.   Schematic representation of the analytical protocol: (a) Dual hybridization event of the sandwich hybridization assay, leading to capturing of the CdS-loaded CNT tags in the microwell; (b) dissolution of the CdS tracer; (c) stripping voltammetric detection of cadmium at a mercury-coated glassy carbon electrode. P1, DNA probe 1; T, DNA target; P2, DNA probe 2. (From Wang et al.,[172] with permission.)

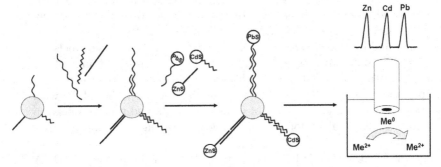

Figure 18.   Multi-target electrical DNA detection protocol based on different colloid nanocrystal tracers (A) Introduction of probe-modified magnetic beads (B) Hybridization with DNA targets (C) Second hybridaztion with the QD-labeled probes (D) Dissolution of QDs and electrochemical detection. (From Wang et al.,[173] with permission.)

been reported (see Figure 18).[173] The DNA-connected QDs yielded well-defined and well-resolved stripping peaks at $-1.12\,V$ (Zn), $-0.68\,V$ (Cd) and $-0.53\,V$ (Pb) at the mercury-coated glassy carbon electrode (versus the Ag/AgCl reference electrode).

A similar approach was used in a multi-analyte electrical sandwich immunoassay for beta(2)-microglobulin, IgG, bovine serum albumin, and C-reactive protein in connection with ZnS, CdS, PbS, and CuS colloidal crystals tracers, respectively as seen in Figure 19.[174] Each biorecognition event yields a distinct voltammetric peak, whose position and size reflects the identity and level, respectively, of the corresponding antigen. Such electrochemical coding could be readily multiplexed and scaled up in multi-well microtiter plates to allow simultaneous parallel detection of numerous proteins or samples. Other attractive nanocrystal tracers for creating a pool of non-overlapping electrical tags for such bioassays are ZnS, PbS, CdS,

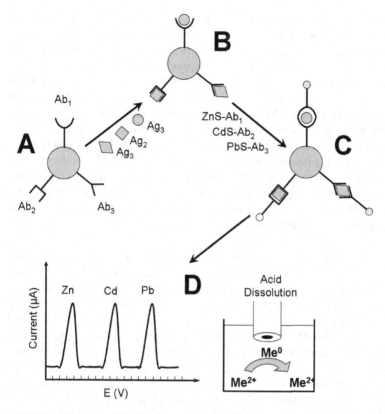

Figure 19. Multiprotein electrical detection protocol based on different inorganic colloid nanocrystal tracers. (A) Introduction of antibody-modified magnetic beads; (B) binding of the antigens to the antibodies on the magnetic beads; (C) capture of the nanocrystal-labeled secondary antibodies; (D) dissolution of nanocrystals and electrochemical stripping detection. (Adapted from Liu *et al.*,[174] with permission.)

InAs, and GaAs semiconductor particles, in view of the attractive stripping behavior of their metal ions.

Magnetic Sensors

Magnetic or paramagnetic materials are being used as labels in a new generation of sensors. Detection of the magnetic labels can be accomplished in a variety of methods. Direct detection of magnetic materials can be achieved by a so-called Superconducting Quantum Interference Device (SQUID) developed by Chemla *et al.*[175] This technique has been introduced for the rapid detection of biological targets through the use of superparamagnetic nanoparticles. In this technique, target proteins are fixed to a mylar film in the well of a micro-titer plate. A suspension of magnetic nanoparticles carrying antibodies is added to the mixture and one-second pulses of magnetic field are applied parallel to the SQUID. This magnetic field magnetizes the nanoparticles; their magnetization relaxes when the field switches off. Nanoparticles that are unbound have the freedom to relax rapidly by Brownian rotation and their magnetization is not detected. Nanoparticles that are specifically bound to the target are restricted to the much slower Néel relaxation and their magnetization (indicative of their presence) is detected by the SQUID. The ability to distinguish between bound and unbound labels makes it possible to run assays which do not require separation and removal of unbound magnetic particles. Although SQUID is a very sensitive detection modality, it is a large and expensive instrument that does not lend itself in broad deployment as a sensor.

One of the methods for detection of magnetic particles is implemented into the Force Amplified BioSensor (FABS) developed by Baselt et al.[176] This biosensor measures the forces that bind DNA-DNA, antibody-antigen, or ligand-receptor pairs together, at the level of single molecules. Another configuration based on magnetic beads detection is the Bead ARray Counter (BARC).[177] It is a multi-analyte biosensor in which DNA probes are patterned onto the solid substrate chip directly above Giant-MagnetoResistive (GMR) sensors. Sample analyte containing complementary DNA hybridizes with the probes on the surface (Figure 20). Labeled, micron-sized magnetic beads are then injected that specifically bind to the sample DNA. A magnetic field is applied, removing any beads that are not specifically bound to the surface.

The beads remaining on the surface are detected by the GMR sensors, and the intensity and location of the signal indicate the concentration and identity of biological targets (such as pathogens) present in the sample. Analyzing millions of transducers offers the possibility to detect or screen thousands of analytes.

A biosensor using magnetic nanoparticles as tags (Figure 21) was developed by Kurlyandskaya et al.[178] The sensor's working principle is based on a magneto-impedemetric response produced by magnetic nanoparticles to an external magnetic field. This response depends on the value of the applied magnetic field and the amplitude and frequency of the flowing current. This configuration is attractive because it offers the possibility of a sensor with enhanced sensitivity and a two-step sensing procedure with separation of the chemical immobilization part from the sensing process, which may have significant potential in the field of medical and pharmaceutical applications.

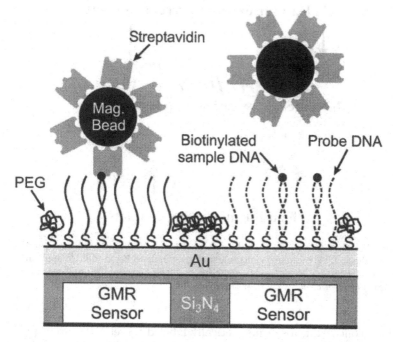

Figure 20. Schematic diagram of the BARC chip surface chemistry and hybridization assay. Thiolated DNA probes are patterned onto a gold layer directly above the GMR sensors on the BARC chip. Biotinylated sample DNA is then added and hybridizes with the DNA probes on the surface when the complementary sequence is present. Unbound sample DNA is washed away. Streptavidin-coated magnetic beads are injected over the chip surface, binding to biotinylated sample DNA hybridized on the BARC chip. Beads that are not specifically bound are removed by applying a magnetic field. Bound beads are detected by the GMR sensors. (From Edelstein *et al.*,[177] with permission.)

Graham *et al.*,[179] developed magnetic field-assisted magnetoresistive-based DNA chips with incorporated spin valve sensors for rapid probe-target hybridisation using low target DNA concentrations. Spin valve sensor is an electronic component, the conductivity of which is highly sensitive to external magnetic field. A schematic representation of the operational principles is presented in Figure 22. When an electric current of 20 mA is passed through both current lines at either side of the sensor (central), a magnetic field gradient is generated resulting in a magnetic force F_1 that attracts the magnetic labels to the narrow regions of the current lines. When a sufficient number of labels have been concentrated on the current lines, the current is switched off and the magnetic labels are then attracted to the nearby sensor by a magnetic force F_2 resulting from the smaller magnetic field generated by the current through the sensor.

Cu lead for exciting current contacts

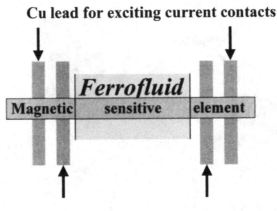

Cu lead for measuring contacts

Figure 21. Principal diagram of the MI sensitive element. Magnetic sensitive element is a horizontal sheet. Lightly shaded central area corresponds to Ferro- fluid® application area. The covered magnetic element is narrower than the Ferrofluid® application area. (From Kurlyandskaya et al.,[178] with permission.)

Thus, high concentrations of magnetically labeled target DNA are rapidly brought into close contact with sensor-bound probe DNA. The detection range of this chip was 140 to 14,000 DNA molecules per sensor, equivalent to 2 to 200 fmole/cm^2.

Grancharov et al.,[180] demonstrated the application of a commercially available Magnetic Tunnel-Junction (MTJ) field sensor to detect specific binding of bio-functionalized manganese ferrite ($MnFe_2O_4$) nanoparticles. The reported detection scheme does not require modulated AC magnetic fields. This fact, combined with the micrometer size of the tunnel junction, suggests that this detection method could potentially be scaled into a multi-analyte "lab-on-a-chip" detector.

Recently, Liu et al.,[181] reported the development of a renewable electrochemical magnetic immunosensor based on the use of magnetic beads and gold nanoparticle labels. Magnetic beads with an anti-IgG antibody-modified surface were attached to a renewable carbon paste transducer surface by a magnet which was incorporated into the sensor. Using a sandwich immunoassay, gold nanoparticles were attached to the surface of magnetic beads. The captured gold nanoparticles were detected by highly sensitive electrochemical stripping which offers a simple and fast quantification, and avoids the use of an enzyme label and substrate. The stripping signal of gold nanoparticles is a measure for the concentration of target

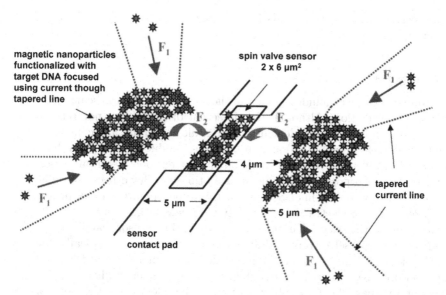

Figure 22. A schematic representation of the operational principles of the tapered current line structures with a magnetoresistive spin-valve sensor. When 20 mA is passed through both current lines at either side of the sensor (central), a magnetic field gradient is generated resulting in a magnetic force F_1 that attracts the magnetic labels to the narrow regions of the current lines. When a sufficient number of labels has been concentrated on the current lines, the current is switched off and the magnetic labels are then attracted to the nearby sensor by a magnetic force F^2 resulting from the smaller magnetic field generated by the current (8 mA) through the sensor. Thus, high concentrations of magnetically labelled target DNA are rapidly brought into close contact with sensor-bound probe DNA. (From Graham et al.,[179] with permission.)

IgG in the sample solution. Such particle-based electrochemical magnetic immunosensors could be readily used for simultaneous parallel detection of multiple proteins by using multiple inorganic metal nanoparticle tracers and are expected to open new opportunities for disease diagnostics and biosecurity.

A piezoelectric immunosensor using magnetic nanoparticles was proposed by Li et al.[182] A goat-anti-IgG antibody (IgGAb) as the model analyte was first covalently immobilized to magnetic nanoparticles, which were surface modified with amino-groups. The magnetic nanoparticles were attached to the surfaces of quartz crystal with the help of a permanent magnet. The specific binding of the analyte (immunoglobulin G) to the IgGAb — coated magnetic nanoparticles was monitored by measuring the shift of the resonance frequency of the quartz crystal. The piezoelectric immunosensor can determine IgG in the range of 0.6 to 34.9 μg/ml with a detection limit of 0.36 μg/ml.

Gas Sensors

For completeness, we include a review of the application of nano-scale materials to gas detection. Although the detection of gases does not rely on the use of bio-recognition elements, nano-scale materials are commonly employed in environmental gas sensors. Semiconductor materials have found wide application as gas sensors, often referred to as electronic noses. Devices that incorporate materials such as tin oxide have been available commercially for some time. Sensors of this type operate on the principal of resistance or electrical conductivity variations as gases adsorb to the surface. Due to the selective nature of this adsorption, sensors can be developed that respond selectively to different gases in a mixture. Typically, the sensing element is made up on a thin film deposited on a substrate. Several investigators have explored the possibility of using nanocrystalline films, or layers of nanoparticles themselves, as the sensing element in the hope of achieving improved sensitivity.

A large variety of nano-scale materials have been examined for application as sensors. Metal oxides constitute a broadly applied family of materials for gas sensing. Tin oxide (SnO_2) has been widely exploited as a suitable semiconductor sensing material, commonly as nanocrystalline films,[35,183] in the form of nanoparticles either by themselves[184] or in conjunction with other nanoparticles such as functionalized copper,[185] or as nanotubes.[186] Tin oxide can be used to detect common air contaminants such as CO, CO_2, NO and NO_2, toxic gases such as H_2S, or ethanol in gasoline. Other nano-scale metal oxides can be used for vapor sensing. Tungsten oxide[187] (WO_3) has been used to detect vapors of H_2S, N_2O and CO; nanostructured thin films of MoO_3 were used by Guidi et al.,[188] to detect CO and NO_2; nanocyrstalline thin films of n-type $MgFe_2O_4$ were used by Liu et al.,[189] to detect vapors of H_2S, CO, liquefied petroleum gas (LPG) and ethanol; α — Fe_2O_3 nanotubes were used by Chen et al.,[103] to detect ethanol and H_2.

Non-oxide materials can also be used for vapor detection — metal nanoparticles have been used as vapor sensors by attaching long chain alkyl-thiols to the particles. The spacing of the array of nanoparticles that is formed by the chain affects the dielectric constant of the assembly and the electrical conductivity. The network of particles and polymer can accept organic vapor molecules that change the particle spacing. The application of Au nanoparticles to vapor sensing is reviewed by Katz et al.[190]

Baraton and Merhari[184] have explored the use of tin oxide nanoparticles as a sensing element for low-cost, deployable, air quality monitoring instruments. The deployment of an array of sensors for air quality monitoring necessitates the development of a technology that is sufficiently cheap and simple to be widely deployed in dispersed arrays. Semiconductor-based gas sensors avoid the difficulties that might be inherent in using optical techniques that require expensive excitation and detection systems. An improvement in the sensitivity and reduction in the cost of production will lead to wider application of sensors of this type.

Baraton and Merhari used laser evaporation of tin oxide starting material to produce nanoparticles of tin oxide with average diameters of about 15 nanometers; the diameters were measured both directly with TEM and indirectly using a BET surface area measurement. The surface reactivity of these nanoparticles is crucial in determining their sensitivity and selectivity to components of a gas mixture that is being analyzed. The authors used FTIR spectroscopy as a screening tool for checking the reactive groups that were attached to the surface of the nanoparticles.

A reduction in the grain size of a semiconductor films, or reduction in the size of semiconductor nanoparticles, confers several advantages in the application of these materials as a gas sensor. Ogawa *et al.*,[191] showed that as the grain size approaches twice the Debye length of the material, there is a significant improvement in sensitivity towards binding of gas molecules at the surface. For example, tin oxide exhibited a maximum sensitivity with grains sizes of about six nanometers. In addition, the increased curvature of very small particles leads to a greater number of surface defects, providing an enhanced number of reactive sites for binding of gas molecules. The combination of increased surface area per unit mass with greater number of reactive sites per surface area to bind yields greater sensitivity for nanoscale materials in gas sensing. The increased sensitivity may also allow the sensor to operate at a lower temperature, thereby reducing power requirements for the instrument.

A Nd:YAG laser was used to evaporate pellets of tin oxide. The powder that was produced was examined with XRD, TEM and FTIR spectroscopy. X-ray diffraction showed that the nanocrystals of tin oxide were formed with average diameters of about 15 nanometers, in agreement with the TEM images (Figure 23).[184] The laser evaporation method provided a production rate that was suitable for the manufacturing of relatively low cost sensors. Other processes, particularly wet methods that are batch processes, may not be simply scalable to useful levels.

FTIR spectroscopy was used to determine the nature of the reactive groups at the surface of nanoparticles. Absorption bands in the infrared spectrum in the region of 4000 to 3000 cm^{-1} were ascribed to the stretch of

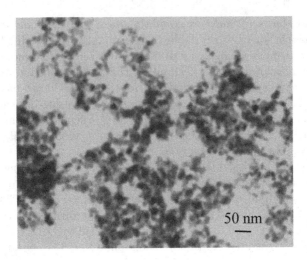

Figure 23. TEM of SnO$_2$ nanoparticles used for gas sensing. (From Baraton and Merhari[184] with permission.)

OH groups there were bonded to the surface of the particle through exposed tin atoms; bending vibrations of the OH group gave rise to absorption features in the 1500 to 1000 cm^{-1} region of the infrared spectrum. The nature of the tin atom at the surface, particularly its valence state, can affect the electron distribution in the absorbed group, thereby changing the infrared spectrum, providing a way to ascertain details of the surface state of the nanoparticles. Baraton and Merhari[184] found that infrared spectroscopy was a very useful tool for rapidly screening the sensing capability of different materials as they were synthesized. This offered a relatively high throughput method for optimizing sensor materials.

The sensitivity of the nano-scale particles to target gases was determined by measuring the infrared absorption spectrum as gases of interest were admitted to a chamber. When carbon monoxide, for example, was admitted to the chamber the absorption spectrum changed as shown in Figure 24. The variation in the absorption peaks arises from the release of free carriers at the surface of the nanoparticles due to reactions of the following type:

$$2CO + O_2^- \rightarrow 2CO_2 + e^-$$
$$CO + O^- \rightarrow CO_2 + e^-$$

(2)

Evacuating the chamber and removing the gases regenerated the sensing material. Reversibility of the reactions at the surface are important so that the sensor is reusable.

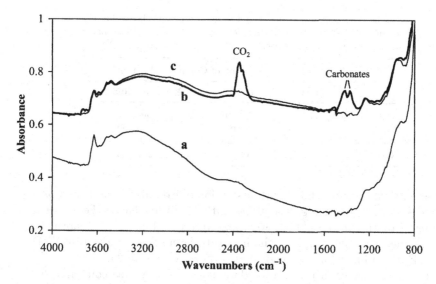

Figure 24. Infrared spectrum of the SnO2 nanoparticles at 300°C: (a) under 50 mbar oxygen; (b, bold line) after addition of 10 mbar CO in presence of oxygen; (c) after evacuation. (From Baraton and Merhari[184] with permission.)

Screen-printing was used to deposit layers of the nanoparticles on a substrate to fabricate a useful device. The particles were dispersed in a suitable solvent and then applied to the substrate through the screen-printing process. Subsequent treatment was found to be important in providing sufficient sintering of the particles to form an intact nanoscale layer without substantial grain growth; a temperature of about 450°C was found to be suitable. Subsequent layers of nanoparticles were applied in a similar process. The authors optimized the screen-printing process for producing multiple layers of nanoparticles.

The nanoparticle sensors were compared with conventional thin film semiconductor devices. The nanoparticle sensor response is compared to conventional devices in Table 1. With the exception of CO, the nanoparticle sensor was competitive with the electrochemical methods but of course can provide the ease of use that comes with a semiconductor device.

Spheres provide the smallest specific surface area of possible geometries of nano-scale particles, i.e. the smallest area per unit mass. Nanotubes offer much greater surface area per unit mass, and this may provide greater sensitivity when the nanotubes are deployed as gas sensors. Both organic and inorganic nanotubes have been investigated for this application. Chen et al.,[103] explored the use of inorganic iron oxide nanotubes for gas detection. They used a templating method for the production of the nanotubes.

Table 1. Comparison between the gas detection thresholds of commercial sensors and the gas detection thresholds of nanoparticle SnO_2 sensor prototypes. (From Baraton and Merhari[184].)

	Typical Detection Limits of Commercial Sensors		
	Semiconductor	Electrochemical	Nanoparticle Sensor
CO	100 ppm	5 ppm	30 ppm
NO_2		600 ppb	500 ppb
No		900 ppb	800 ppb
O_3		200 ppb	200 ppb

An alumina matrix was filled with a solution of iron nitrate and then processed at high temperature. Nanotubes of the stable α-phase of Fe_2O_3 were produced by this process. They were able to create bundles of nanotubes, 60 μm long, with the template; the nanotubes were characterized by XRD and TEM. An image of the nanotubes is shown in Figure 25.

The performance of the iron oxide nanotube sensor was compared for the detection of C_2H_5OH and H_2 at room temperature with nanoparticles

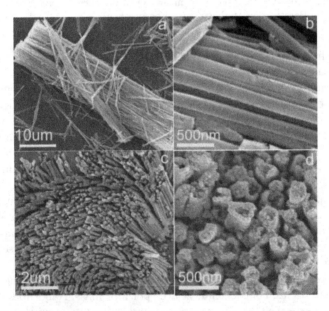

Figure 25. Representative SEM images of the as-prepared alpha Fe_2O_3 nanotubes: (a) Low magnification view of the nanotubes; (b) walls of the nanotube bundles; (c) tips of the tube bundles at low magnification; (d) tips of the tube bundles at high magnification. (From Chen et al.,[103] with permission.)

of the same material that were about 200 nanometers in diameter. The authors measured the change in resistance of the bundle and the nanoparticles when they were exposed to the vapors. The sensitivity is reported as a ratio of resistance in the presence of vapor to the resistance in the absence of the analyte vapor (R_{air}/R_{gas}) where R_{air} is the bundle resistance in the absence of the analyte vapor and R_{gas} is the resistance with the vapor present. The nanotube bundle was about five times more sensitive to ethanol than nanoparticles of the same material (Figure 26); the sensitivity to H_2 was much greater than the sensitivity to ethanol (Figure 27). Furthermore, they demonstrated that the device could be recycled up to 50 times — reusability and repeatability are necessary properties of a useful gas sensor. The change in geometry from sphere to hollow tube improved the sensitivity and suggested that the tube may be the preferred geometry for gas sensors.

Carbon nanotubes can also be used in gas sensors. Since their first application in Field Effect Transistors (FET) in the late 1990s, carbon nanotubes have been found to be sensitive to several gases, including O_2[192] that was able to convert a carbon nanotube from a semiconductor to a metal with a concomitant change in resistance. This work highlighted the problems in dealing with carbon nanotubes in air. Kong *et al.*,[193] found that other gases such as NO_2 and NH_3 could be detected by carbon nanotubes.

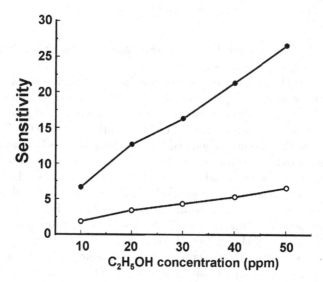

Figure 26. Sensor sensitivity to ethanol: nanotubes (filled symbols); nanoparticles (empty symbols). (From Chen *et al.*,[103] with permission.)

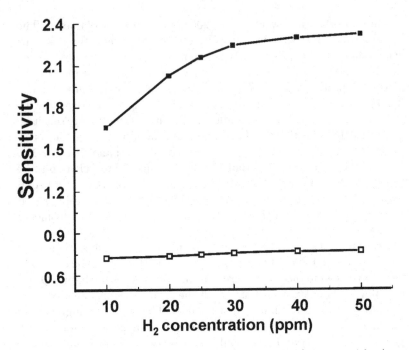

Figure 27. Sensor sensitivity to H_2: nanotubes (filled symbols); nanoparticles (empty symbols) (From Chen *et al.*,[103] with permission.)

Zhang *et al.*,[194] found the multi-walled carbon nanotubes were sensitive to the presence of organic solvent vapor.

The surface of the nanotubes can be functionalized with adducts that provide sensitivity to specific gases. Because a single wall carbon nanotube (SWCNT) behaves as a semiconductor, it can be incorporated into FET devices. Star *et al.*,[195] fabricated a FET that incorporated SWCNT that were fabricated with sequential chemical vapor deposition (CVD) and complementary metal oxide semiconductor (CMOS) processes on an SiO_2 base; the nanotubes served as the conducting source to drain channel. The carbon nanotubes were functionalized with a Na^+, K^+, or Ca^{2+} ion-exchanged Nafion polymer that acted as the chemically sensitive layer. Treatment with the Nafion polymer left a thin layer (< 10 nm) on the nanotubes nanotubes as seen by Atomic Force Microscopy (Figure 28).

Measurements of the electric current between source and drain were made in a flow cell with controllable humidity. The gate voltage was swept from -10 V to $+10$ V in the forward direction and from $+10$ V to -10 V in the reverse direction. The hystersis in source to drain current, I_{SD}, that is the difference in the maximum current in the forward gate voltage sweep

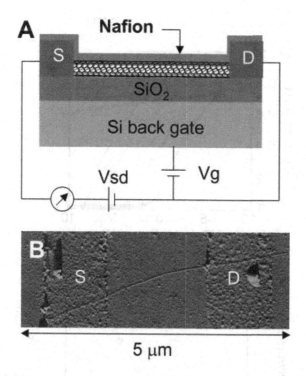

Figure 28. (A) Schematic representation of a Field-Effect Transistor device with a carbon nanotube transducer contacted by two Ti/Au electrodes (source and drain) and a silicon back gate. Carbon nanotube conducting channel is exposed to and covered by the Nafion. (B) Atomic force microscope (tapping mode) topograph of the actual nanotube device coated with Nafion. (From Star *et al.*,[195] with permission.)

versus the maximum current in the reverse sweep is plotted in Figure 29 where the sensitivity to humidity is apparent. The effect of gate voltage on the charge-sensitive nanotube structure was found to depend on the relative humidity over the range of $12 \pm 93\%$ RH with millisecond response time. Although Nafion films are used directly in conductance measurements of humidity, the semiconductor device operated through the interaction of the ionic polymer and the underlying semiconductor SWCNT, offering a new mode of measurement using modulation of the gate voltage with the promise of enhanced operating properties. Star *et al.*,[195] used a similar FET device for CO_2 detection. In this case, the carbon nanotubes were functionalized with PEI/starch polymer. The response of the sensor to CO_2 concentration was rapid with a wide dynamic range from 500 ppm to 10% in air (Figure 30).

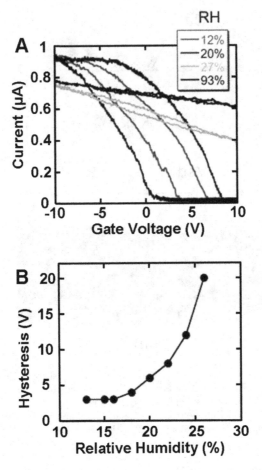

Figure 29.   (A) The device characteristics, Isd vs Vg curves (Vsd 40 mV), of the NTFET device functionalized with Nafion (Na) membrane measured at different relative humidity (RH) values. (B) Humidity dependence of the reversible hysteresis (forward Isd ± reverse Isd) in the device measured in the range of 20 volts (10 V to 10 V) at the sweep rate of 4 Hz. (From Star *et al.*,[195] with permission.)

Conclusion

Nano-scale materials offer new functionalities for sensing. Although the application of this new class of materials in environmental sensors is in its infancy, significant advances in sensor technology can be expected to flow from the application of carbon nanotubes, inorganic nanotubes, nanoparticles and nanostructured surfaces in sensors. Devices based on

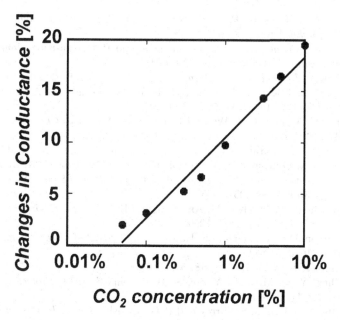

Figure 30.   Calibration curve for the CNT FET sensor for CO_2. (From Star *et al.*,[195] with permission.)

these materials are expected to lead to detection devices that are sufficiently cheap and safe to be widely deployed in environmental sensor arrays.

Acknowledgments

This article was made possible in part by the support of NSF Grant DBI-0102662 and the Superfund Basic Research Program with Grant 5P42ES04699 from the National Institute of Environmental Health Sciences, NIH. This material is based partially upon work supported by the Cooperative State Research, Education, and Extension Service, U.S. Department of Agriculture, under Award No. 05-35603-16280.

References

1. J. Seagrave, J. D. McDonald and J. L. Mauderly, *Experimental and Toxicologic Pathology* **57**, 233–238 (2005).
2. C. Sioutas, R. J. Delfino and M. Singh, *Environmental Health Perspectives* **113**, 947–955 (2005).

3. A. Bossi, S. A. Piletsky, P. G. Righetti and A. P. F. Turner, *Journal of Chromatography A* **892**, 143–153 (2000).
4. R. Carabias-Martinez, E. Rodriguez-Gonzalo and P. Revilla-Ruiz, *Journal of Chromatography A* **1056**, 131–138 (2004).
5. S. D. Harvey, *Journal of Separation Science* **28**, 1221–1230 (2005).
6. S. Piletsky, E. Piletska, K. Karim, G. Foster, C. Legge and A. Turner, *Analytica Chimica Acta* **504**, 123–130 (2004).
7. D. F. Tai, C. Y. Lin, T. Z. Wu and L. K. Chen, *Analytical Chemistry* **77**, 5140–5143 (2005).
8. R. Pon, M. A. Batalin, V. Chen, A. Kansal, D. Liu, M. Rahimi, L. Shirachi, A. Somasundra, Y. Yu, M. Hansen, W. J. Kaiser, M. Srivastava, G. Sukhatme and D. Estrin, *Distributed Computing in Sensor Systems, Proceedings* **3560**, 403–405 (2005).
9. G. C. Windham, P. Mitchell, M. Anderson and B. L. Lasley, *Environmental Health Perspectives* **113**, 1285–1290 (2005).
10. K. Andersson, D. Areskoug and E. Hardenborg, *Journal of Molecular Recognition* **12**, 310–315 (1999).
11. A. Baeumner, *Food Technology* **58**, 51–55 (2004).
12. J. R. Chen, Y. Q. Miao, N. Y. He, X. H. Wu and S. J. Li, *Biotechnology Advances* **22**, 505–518 (2004).
13. P. Fortina, L. J. Kricka, S. Surrey and P. Grodzinski, *Trends In Biotechnology* **23**, 168–173 (2005).
14. M. J. Friedrich, *Jama-Journal of The American Medical Association* **293**, 1965–1965 (2005).
15. K. K. Jain, *Expert Review of Molecular Diagnostics* **3**, 153–161 (2003).
16. P. Kohli, M. Wirtz and C. R. Martin, *Electroanalysis* **16**, 9–18 (2004).
17. T. Vo-Dinh, B. M. Cullum and D. L. Stokes, *Sensors And Actuators B-Chemical* **74**, 2–11 (2001).
18. T. Vo-Dinh and P. Kasili, *Analytical And Bioanalytical Chemistry* **382**, 918–925 (2005).
19. L. J. Yang and Y. B. Li, *Journal of Food Protection* **68**, 1241–1245 (2005).
20. E. R. Goldman, I. L. Medintz, J. L. Whitley, A. Hayhurst, A. R. Clapp, H. T. Uyeda, J. R. Deschamps, M. E. Lassman and H. Mattoussi, *Journal of the American Chemical Society* **127**, 6744–6751 (2005).
21. X. J. Ji, J. Y. Zheng, J. M. Xu, V. K. Rastogi, T. C. Cheng, J. J. DeFrank and R. M. Leblanc, *Journal of Physical Chemistry B* **109**, 3793–3799 (2005).
22. A. R. Clapp, I. L. Medintz, J. M. Mauro, B. R. Fisher, M. G. Bawendi and H. Mattoussi, *Journal of The American Chemical Society* **126**, 301–310 (2004).
23. C. A. Constantine, K. M. Gattas-Asfura, S. V. Mello, G. Crespo, V. Rastogi, T. C. Cheng, J. J. DeFrank and R. M. Leblanc, *Langmuir* **19**, 9863–9867 (2003).
24. A. D. Sheehan, J. Quinn, S. Daly, P. Dillon and R. O' Kennedy, *Analytical Letters* **36**, 511–537 (2003).
25. D. J. Maxwell, J. R. Taylor and S. M. Nie, *Journal of the American Chemical Society* **124**, 9606–9612 (2002).

26. A. P. Alivisatos, W. W. Gu and C. Larabell, *Annual Review of Biomedical Engineering* **7**, 55–76 (2005).
27. R. E. Bailey and S. Nie, In *The Chemistry of Nanomaterials: Synthesis, Properties and Applications.*; C. N. R. Rao, A. Muller, A. K. Cheetam, Eds.; Wiley/VCH: Weinheim, 2004, pp 405–417.
28. D. Gerion, F. Pinaud, S. C. Williams, W. J. Parak, D. Zanchet, S. Weiss and A. P. Alivisatos, *Journal of Physical Chemistry B* **105**, 8861–8871 (2001).
29. W. C. W. Chan, D. J. Maxwell, X. H. Gao, R. E. Bailey, M. Y. Han and S. M. Nie, *Current Opinion in Biotechnology* **13**, 40–46 (2002).
30. D. L. Stokes, Z. H. Chi and T. Vo-Dinh, *Applied Spectroscopy* **58**, 292–298 (2004).
31. M. Seydack, *Biosensors & Bioelectronics* **20**, 2454–2469 (2005).
32. C. R. Yonzon, C. L. Haynes, X. Y. Zhang, J. T. Walsh and R. P. Van Duyne, *Analytical Chemistry* **76**, 78–85 (2004).
33. F. Tam, C. Moran and N. Halas, *Journal of Physical Chemistry B* **108**, 17290–17294 (2004).
34. L. A. DeLouise, P. M. Kou and B. L. Miller, *Analytical Chemistry* **77**, 3222–3230 (2005).
35. J. W. Gong, Q. F. Chen, W. F. Fei and S. Seal, *Sensors And Actuators B-Chemical* **102**, 117–125 (2004).
36. Y. H. Yang, M. M. Guo, M. H. Yang, Z. J. Wang, G. L. Shen and R. Q. Yu, *International Journal of Environmental Analytical Chemistry* **85**, 163–175 (2005).
37. L. Rotiroti, L. De Stefano, N. Rendina, L. Moretti, A. M. Rossi and A. Piccolo, *Biosensors & Bioelectronics* **20**, 2136–2139 (2005).
38. S. Sotiropoulou and N. A. Chaniotakis, *Analytica Chimica Acta* **530**, 199–204 (2005).
39. D. S. dos Santos, P. J. G. Goulet, N. P. W. Pieczonka, O. N. Oliveira and R. F. Aroca, *Langmuir* **20**, 10273–10277 (2004).
40. E. Harlow and D. Lane, *Antibodies: A Laboratory Manual*; Cold Spring Harbor Laboratory, N. Y. (1988).
41. C. P. Price and D. J. Newman, *Principles and Practice of Immunoassay*, 2nd ed., (1997).
42. J. Fotzpatrick, L. Fanning, S. Hearty, P. Leonard, B. M. Manning, J. G. Quinn and R. O'Kennedy, *Analytical Letters* **33**, 2563–2609 (2000).
43. M. C. Estevez-Alberola and M. P. Marco, *Analytical and Bioanalytical Chemistry* **378**, 563–575 (2004).
44. K. A. Fahnrich, M. Pravda and G. G. Guilbault, *Talanta* **54**, 531–559 (2001).
45. C. R. Suri, M. Raje and G. C. Varshney, *Critical Reviews in Biotechnology* **22**, 15–32 (2002).
46. B. Hock, M. Seifert and K. Kramer, *Biosensors and Bioelectronics* **17**, 239–249 (2002).
47. A. J. Baeumner, *Analytical and Bioanalytical Chemistry* **377**, 434–445 (2003).

48. J. O. Mahony, K. Nolan, M. R. Smyth and B. Mizaikoff, *Analytica Chimica Acta* **534**, 31–39 (2005).
49. A. L. Jenkins, R. Yin and J. L. Jensen, *Analyst* **126**, 798–802 (2001).
50. A. L. Jenkins and S. Y. Bae, *Analytica Chimica Acta* **542**, 32–37 (2005).
51. J. H. Zhai, H. Cui and R. F. Yang, *Biotechnology Advances* **15**, 43–58 (1997).
52. M. Campas and I. Katakis, *Trac-Trends in Analytical Chemistry* **23**, 49–62 (2004).
53. F. Lucarelli, L. Authier, G. Bagni, G. Marrazza, T. Baussant, E. Aas and M. Mascini, *Analytical Letters* **36**, 1887–1901 (2003).
54. Y. Liu and B. Danielsson, *Analytical Chemistry* **77**, 2450–2454 (2005).
55. A. L. Simonian, T. A. Good, S. S. Wang and J. R. Wild, *Analytica Chimica Acta* **534**, 69–77 (2005).
56. E. Luzi, M. Minunni, S. Tombelli and M. Mascini, *Trac-Trends in Analytical Chemistry* **22**, 810–818 (2003).
57. V. B. Kandimalla and H. X. Ju, *Analytical Letters* **37**, 2215–2233 (2004).
58. P. D. Patel, *Trac-Trends in Analytical Chemistry* **21**, 96–115 (2002).
59. C. Ziegler, *Fresenius Journal of Analytical Chemistry* **366**, 552–559 (2000).
60. M. P. Marco and D. Barcelo, *Measurement Science & Technology* **7**, 1547–1562 (1996).
61. M. Trojanowicz, *Electroanalysis* **14**, 1311–1328 (2002).
62. H. Schulze, S. Vorlova, F. Villatte, T. T. Bachmann and R. D. Schmid, *Biosensors & Bioelectronics* **18**, 201–209 (2003).
63. D. R. Thevenot, K. Toth, R. A. Durst and G. S. Wilson, *Analytical Letters* **34**, 635–659 (2001).
64. J. H. T. Luong, A. Mulchandani and G. G. Guilbault, *Trends in Biotechnology* **6**, 310–316 (1988).
65. P. Feng, *Journal of Food Protection* **55**, 927–934 (1992).
66. P. Buhlmann, E. Pretsch and E. Bakker, *Chemical Reviews* **98**, 1593–1687 (1998).
67. R. Koncki, S. Glab, J. Dziwulska, I. Palchetti and M. Mascini, *Analytica Chimica Acta* **385**, 451–459 (1999).
68. U. J. Krull, *Chemtech* **20**, 372–377 (1990).
69. S. J. Updike and G. P. HIcks, *Nature* **214**, 986 (1967).
70. A. L. Ghindilis, P. Atanasov and E. Wilkins, *Electroanalysis* **9**, 661–674 (1997).
71. L. Habermuller, M. Mosbach and W. Schuhmann, *Fresenius Journal of Analytical Chemistry* **366**, 560–568 (2000).
72. C. K. O'Sullivan, R. Vaughan and G. G. Guilbault, *Analytical Letters* **32**, 2353–2377 (1999).
73. J. M. Abad, F. Pariente, L. Hernandez and E. Lorenzo, *Analytica Chimica Acta* **368**, 183–189 (1998).
74. K. Bizet, C. Gabrielli and H. Perrot, *Analusis* **27**, 609–616 (1999).
75. D. Ivnitski, I. Abdel-Hamid, P. Atanasov and E. Wilkins, *Biosensors & Bioelectronics* **14**, 599–624 (1999).
76. K. Sapsford, C. R. Taitt and F. S. Ligler, In *Optical Biosensors*; F. S. Ligler, C. R. Taitt, Eds.; Elsevier: Amsterdam, pp 95–121 (2002).

77. C. A. R. Taitt and F. S. Ligler, In *Optical Biosensors*; F. S. Ligler, C. A. R. Taitt, Eds.; Elsevier: Amsterdam, pp 57–94 (2002).
78. D. P. Campbell and C. J. McCloskey, In *Optical Biosensors*; F. S. Ligler, C. A. R. Taitt, Eds.; Elsevier: Amsterdam, pp 277–304 (2002).
79. J. Homola, S. S. Yee and D. Myszka, In *Optical Biosensors*; F. S. Ligler, C. A. R. Taitt, Eds.; Elsevier: Amsterdam, pp 207–251 (2002).
80. J. Homola, *Analytical & Bioanalytical Chemistry* **377**, 528–539 (2003).
81. E. F. Schipper, S. Rauchalles, R. P. H. Kooyman, B. Hock and J. Greve, *Analytical Chemistry* **70**, 1192–1197 (1998).
82. E. F. Schipper, A. J. H. Bergevoet, R. P. H. Kooyman and J. Greve, *Analytica Chimica Acta* **341**, 171–176 (1997).
83. A. M. Tinsley-Bown, L. T. Canham, M. Hollings, M. H. Anderson, C. L. Reeves, T. I. Cox, S. Nicklin, D. J. Squirrell, E. Perkins, A. Hutchinson, M. J. Sailor and A. Wun, *Physica Status Solidi A-Applied Research* **182**, 547–553 (2000).
84. C. S. Ryu, S. M. Cho and B. W. Kim, *Biotechnology Letters* **23**, 653–659 (2001).
85. R. Cush, J. M. Cronin, W. J. Stewart, C. H. Maule, J. Molloy and N. J. Goddard, *Biosensors & Bioelectronics* **8**, 347–353 (1993).
86. M. A. Cooper, *Nature Reviews* **1**, 515–528 (2002).
87. T. Kinning and E. Edwards, In *Optical Biosensors*; C. A. R. Taitt, F. S. Ligler, Eds.; Elsevier: Amsterdam, pp 253–276 (2002).
88. S. Bernad, T. Soulimane and S. Lecomte, *Journal of Raman Spectroscopy* **35**, 47–54 (2004).
89. F. T. Docherty, P. B. Monaghan, R. Keir, D. Graham, W. E. Smith and J. M. Cooper, *Chemical Communications* 118–119 (2004).
90. D. Graham, L. Fruk and W. E. Smith, *Analyst* **128**, 692–699 (2003).
91. A. P. Esposito, C. E. Talley, T. Huser, C. W. Hollars, C. M. Schaldach and S. M. Lane, *Applied Spectroscopy* **57**, 868–871 (2003).
92. J. W. Chan, A. P. Esposito, C. E. Talley, C. W. Hollars, S. M. Lane and T. Huser, *Analytical Chemistry* **76**, 599–603 (2004).
93. M. Bruchez, M. Moronne, P. Gin, S. Weiss and A. P. Alivisatos, *Science* **281**, 2013–2016 (1998).
94. S. Iijima, *Nature* **354**, 56–58 (1991).
95. S. Iijima and T. Ichihashi, *Nature* **363**, 603–605 (1993).
96. M. L. Cohen, *Materials Science & Engineering C-Biomimetic and Supramolecular Systems* **15**, 1–11 (2001).
97. J. J. Davis, K. S. Coleman, B. R. Azamian, C. B. Bagshaw and M. L. H. Green, *Chemistry-a European Journal* **9**, 3732–3739 (2003).
98. R. H. Baughman, A. A. Zakhidov and W. A. de Heer, *Science* **297**, 787–792 (2002).
99. J. W. Mintmire and C. T. White, *Carbon* **33**, 893–902 (1995).
100. E. W. Wong, P. E. Sheehan and C. M. Lieber, *Science* **277**, 1971–1975 (1997).
101. L. R. Hilliard, X. J. Zhao and W. H. Tan, *Analytica Chimica Acta* **470**, 51–56 (2002).

102. R. Fan, M. Yue, R. Karnik, A. Majumdar and P. D. Yang, *Abstracts of Papers of the American Chemical Society* **229**, U141–U142 (2005).

103. J. Chen, L. N. Xu, W. Y. Li and X. L. Gou, *Advanced Materials* **17**, 582–+ (2005).

104. C. N. R. Rao, A. Govindaraj, G. Gundiah and S. R. C. Vivekchand, *Chemical Engineering Science* **59**, 4665–4671 (2004).

105. R. Tenne and C. N. R. Rao, *Philosophical Transactions of the Royal Society of London Series a-Mathematical Physical and Engineering Sciences* **362**, 2099–2125 (2004).

106. M. Remskar, *Advanced Materials* **16**, 1497–1504 (2004).

107. R. Tenne, *Chemistry-a European Journal* **8**, 5297–5304 (2002).

108. J. Q. Hu, Y. Bando, J. H. Zhan, M. Y. Liao, D. Golberg, X. L. Yuan and T. Sekiguchi, *Applied Physics Letters* **87** (2005).

109. S. M. Liu, L. M. Gan, L. H. Liu, W. D. Zhang and H. C. Zeng, *Chemistry of Materials* **14**, 1391–1397 (2002).

110. J. Park, B. Koo, K. Y. Yoon, Y. Hwang, M. Kang, J. G. Park and T. Hyeon, *Journal of the American Chemical Society* **127**, 8433–8440 (2005).

111. J. Perez-Juste, I. Pastoriza-Santos, L. M. Liz-Marzan and P. Mulvaney, *Coordination Chemistry Reviews* **249**, 1870–1901 (2005).

112. R. Tenne, *Angewandte Chemie-International Edition* **42**, 5124–5132 (2003).

113. F. Zhou, X. M. Zhao, H. G. Zheng, T. Shen and C. M. Tang, *Chemistry Letters* **34**, 1114–1115 (2005).

114. Y. C. Zhang, X. Wu, X. Y. Hu and R. Guo, *Journal of Crystal Growth* **280**, 250–254 (2005).

115. K. P. Kalyanikutty, G. Gundiah, C. Edem, A. Govindaraj and C. N. R. Rao, *Chemical Physics Letters* **408**, 389–394 (2005).

116. M. C. Daniel and D. Astruc, *Chemical Reviews* **104**, 293–346 (2004).

117. E. P. Diamandis, *Clin. Biochem* **21**, 139–150 (1988).

118. B. M. Tissue and B. Bihari, *Journal of Fluorescence* **8**, 289–294 (1998).

119. M. Safarikova and I. Safarik, *Chemicke Listy* **89**, 280–287 (1995).

120. Q. A. Pankhurst, J. Connolly, S. K. Jones and J. Dobson, *Journal of Physics D-Applied Physics* **36**, R167–R181 (2003).

121. S. H. Sun, H. Zeng, D. B. Robinson, S. Raoux, P. M. Rice, S. X. Wang and G. X. Li, *Journal of the American Chemical Society* **126**, 273–279 (2004).

122. R. H. Kodama, *Journal of Magnetism and Magnetic Materials* **200**, 359–372 (1999).

123. U. Kreibig and L. Genzel, *Surface Science* **156**, 678–700 (1985).

124. W. H. Yang, G. C. Schatz and R. P. Vanduyne, *Journal of Chemical Physics* **103**, 869–875 (1995).

125. R. Elghanian, J. J. Storhoff, R. C. Mucic, R. L. Letsinger and C. A. Mirkin, *Science* **277**, 1078–1081 (1997).

126. J. J. Storhoff, R. Elghanian, R. C. Mucic, C. A. Mirkin and R. L. Letsinger, *Journal of the American Chemical Society* **120**, 1959–1964 (1998).

127. J. J. Storhoff, A. A. Lazarides, R. C. Mucic, C. A. Mirkin, R. L. Letsinger and G. C. Schatz, *Journal of the American Chemical Society* **122**, 4640–4650 (2000).

128. R. A. Reynolds, C. A. Mirkin and R. L. Letsinger, *Journal of the American Chemical Society* **122**, 3795–3796 (2000).
129. R. A. Reynolds, C. A. Mirkin and R. L. Letsinger, *Pure and Applied Chemistry* **72**, 229–235 (2000).
130. M. Brasuel, R. Kopelman, M. A. Philbert, J. W. Aylott, H. Clark, J. Sumner, H. Xu, M. Hoyer, T. J. Miller, R. Tjalkens, In *Optical Biosensors*; F. S. Ligler and C. A. R. Taitt, Eds.: Amsterdam, pp 497–536 (2002).
131. S. F. Cheng and L. K. Chau, *Analytical Chemistry* **75**, 16–21 (2003).
132. N. Nath and A. Chilkoti, *Analytical Chemistry* **76**, 5370–5378 (2004).
133. A. J. Haes and R. P. Van Duyne, *Journal of The American Chemical Society* **124**, 10596–10604 (2002).
134. L. He, M. D. Musick, S. R. Nicewarner, F. G. Salinas, S. J. Benkovic, M. J. Natan and C. D. Keating, *Journal of the American Chemical Society* **122**, 9071–9077 (2000).
135. J. J. Shi, Y. F. Zhu, X. R. Zhang, W. R. G. Baeyens and A. M. Garcia-Campana, *Trac-Trends in Analytical Chemistry* **23**, 351–360 (2004).
136. W. C. W. Chan and S. M. Nie, *Science* **281**, 2016–2018 (1998).
137. P. T. Tran, E. R. Goldman, G. P. Anderson, J. M. Mauro and H. Mattoussi, *Physica Status Solidi B-Basic Research* **229**, 427–432 (2002).
138. D. Geho, N. Lahar, P. Gurnani, M. Huebschman, P. Herrmann, V. Espina, A. Shi, J. Wulfkuhle, H. Garner, E. Petricoin, L. A. Liotta and K. P. Rosenblatt, *Bioconjugate Chemistry* **16**, 559–566 (2005).
139. R. Tapec, X. J. J. Zhao and W. H. Tan, *Journal of Nanoscience and Nanotechnology* **2**, 405–409 (2002).
140. J. Feng, G. M. Shan, A. Maquieira, M. E. Koivunen, B. Guo, B. D. Hammock and I. M. Kennedy, *Analytical Chemistry* **75**, 5282–5286 (2003).
141. D. Dosev, M. Nichkova, M. Liu, B. Guo, G.-Y. Liu, B. D. Hammock and I. M. Kennedy, *J. Biomed. Optics* **10**, 064006-064001–064006-064007 (2005).
142. W. O. Gordon, J. A. Carter and B. M. Tissue, *Journal of Luminescence* **108**, 339–342 (2004).
143. S.-C. Chen, R. Perron, D. Dosev and I. M. Kennedy, *Proceedings of SPIE* **5275**, 186 (2004).
144. M. D. Rubianes and G. A. Rivas, *Electrochemistry Communications* **5**, 689–694 (2003).
145. S. Hrapovic, Y. L. Liu, K. B. Male and J. H. T. Luong, *Analytical Chemistry* **76**, 1083–1088 (2004).
146. F. Patolsky, Y. Weizmann and I. Willner, *Angewandte Chemie-International Edition* **43**, 2113–2117 (2004).
147. J. Wang, *Analytica Chimica Acta* **500**, 247–257 (2003).
148. M. Musameh, J. Wang, A. Merkoci and Y. H. Lin, *Electrochemistry Communications* **4**, 743–746 (2002).
149. S. H. Lim, J. Wei, J. Y. Lin, Q. T. Li and J. KuaYou, *Biosensors & Bioelectronics* **20**, 2341–2346 (2005).
150. M. Gao, L. M. Dai and G. G. Wallace, *Electroanalysis* **15**, 1089–1094 (2003).
151. Y. H. Lin, F. Lu and J. Wang, *Electroanalysis* **16**, 145–149 (2004).

152. L. A. Bauer, N. S. Birenbaum and G. J. Meyer, *Journal of Materials Chemistry* **14**, 517–526 (2004).

153. T. Y. You, O. Niwa, R. Kurita, Y. Iwasaki, K. Hayashi, K. Suzuki and S. Hirono, *Electroanalysis* **16**, 54–59 (2004).

154. F. F. Zhang, Q. Wan, X. L. Wang, Z. D. Sun, Z. Q. Zhu, Y. Z. Xian, L. T. Jin and K. Yamamoto, *Journal of Electroanalytical Chemistry* **571**, 133–138 (2004).

155. M. G. Li, Y. J. Shang, Y. C. Gao, G. F. Wang and B. Fang, *Analytical Biochemistry* **341**, 52–57 (2005).

156. M. Delvaux, A. Walcarius and S. Demoustier-Champagne, *Analytica Chimica Acta* **525**, 221–230 (2004).

157. J. N. Wohlstadter, J. L. Wilbur, G. B. Sigal, H. A. Biebuyck, M. A. Billadeau, L. W. Dong, A. B. Fischer, S. R. Gudibande, S. H. Jamieson, J. H. Kenten, J. Leginus, J. K. Leland, R. J. Massey and S. J. Wohlstadter, *Advanced Materials* **15**, 1184-+ (2003).

158. H. Cai, X. N. Cao, Y. Jiang, P. G. He and Y. Z. Fang, *Analytical and Bioanalytical Chemistry* **375**, 287–293 (2003).

159. P. G. He and L. M. Dai, *Chemical Communications* 348–349 (2004).

160. J. Li, J. E. Koehne, A. M. Cassell, H. Chen, H. T. Ng, Q. Ye, W. Fan, J. Han and M. Meyyappan, *Electroanalysis* **17**, 15–27 (2005).

161. K. Ramanathan, M. A. Bangar, M. Yun, W. Chen, N. V. Myung and A. Mulchandani, *Journal of the American Chemical Society* **127**, 496–497 (2005).

162. R. J. Chen, S. Bangsaruntip, K. A. Drouvalakis, N. W. S. Kam, M. Shim, Y. M. Li, W. Kim, P. J. Utz and H. J. Dai, *Proceedings of the National Academy of Sciences of the United States of America* **100**, 4984–4989 (2003).

163. J. Wang, G. D. Liu and M. R. Jan, *Journal of the American Chemical Society* **126**, 3010–3011 (2004).

164. J. Wang, D. K. Xu, A. N. Kawde and R. Polsky, *Analytical Chemistry* **73**, 5576–5581 (2001).

165. M. Dequaire, C. Degrand and B. Limoges, *Analytical Chemistry* **72**, 5521–5528 (2000).

166. J. Wang, R. Polsky and D. K. Xu, *Langmuir* **17**, 5739–5741 (2001).

167. H. Cai, Y. Q. Wang, P. G. He and Y. H. Fang, *Analytica Chimica Acta* **469**, 165–172 (2002).

168. H. S. Guo, J. N. Zhang, P. F. Xiao, L. B. Nie, D. Yang and N. Y. He, *Journal of Nanoscience and Nanotechnology* **5**, 1240–1244 (2005).

169. J. Wang, A. N. Kawde and M. R. Jan, *Biosensors & Bioelectronics* **20**, 995–1000 (2004).

170. Y. Xu, H. Cai, P. G. He and Y. Z. Fang, *Electroanalysis* **16**, 150–155 (2004).

171. H. Cai, N. N. Zhu, Y. Jiang, P. G. He and Y. Z. Fang, *Biosensors & Bioelectronics* **18**, 1311–1319 (2003).

172. J. Wang, G. D. Liu, M. R. Jan and Q. Y. Zhu, *Electrochemistry Communications* **5**, 1000–1004 (2003).

173. J. Wang, G. D. Liu and A. Merkoci, *Journal of the American Chemical Society* **125**, 3214–3215 (2003).

174. G. D. Liu, J. Wang, J. Kim, M. R. Jan and G. E. Collins, *Analytical Chemistry* **76**, 7126–7130 (2004).
175. Y. R. Chemla, H. L. Crossman, Y. Poon, R. McDermott, R. Stevens, M. D. Alper and J. Clarke, *Proceedings of the National Academy of Sciences of the United States of America* **97**, 14268–14272 (2000).
176. D. R. Baselt, G. U. Lee, K. M. Hansen, L. A. Chrisey and R. J. Colton, *Proceedings of the Ieee* **85**, 672–680 (1997).
177. R. L. Edelstein, C. R. Tamanaha, P. E. Sheehan, M. M. Miller, D. R. Baselt, L. J. Whitman and R. J. Colton, *Biosensors & Bioelectronics* **14**, 805–813 (2000).
178. G. V. Kurlyandskaya, M. L. Sanchez, B. Hernando, V. M. Prida, P. Gorria and M. Tejedor, *Applied Physics Letters* **82**, 3053–3055 (2003).
179. D. L. Graham, H. A. Ferreira, N. Feliciano, P. P. Freitas, L. A. Clarke and M. D. Amaral, *Sensors and Actuators B-Chemical* **107**, 936–944 (2005).
180. S. G. Grancharov, H. Zeng, S. H. Sun, S. X. Wang, S. O'Brien, C. B. Murray, J. R. Kirtley and G. A. Held, *Journal of Physical Chemistry B* **109**, 13030–13035 (2005).
181. G. D. Liu and Y. H. Lin, *Journal of Nanoscience and Nanotechnology* **5**, 1060–1065 (2005).
182. J. S. Li, X. X. He, Z. Y. Wu, K. M. Wang, G. L. Shen and R. Q. Yu, *Analytica Chimica Acta* **481**, 191–198 (2003).
183. F. Li, L. Y. Chen, Z. Q. Chen, J. Q. Xu, J. M. Zhu and X. Q. Xin, *Materials Chemistry And Physics* **73**, 335–338 (2002).
184. M. I. Baraton and L. Merhari, *Journal of Nanoparticle Research* **6**, 107–117 (2004).
185. R. S. Niranjan, V. A. Chaudhary, I. S. Mulla and K. Vijayamohanan, *Sensors And Actuators B-Chemical* **85**, 26–32 (2002).
186. Y. Wang, J. Y. Lee and H. C. Zeng, *Chemistry of Materials* **17**, 3899–3903 (2005).
187. A. Hoel, L. F. Reyes, P. Heszler, V. Lantto and C. G. Granqvist, *Current Applied Physics* **4**, 547–553 (2004).
188. V. Guidi, D. Boscarino, L. Casarotto, E. Comini, M. Ferroni, G. Martinelli and G. Sberveglieri, *Sensors and Actuators B-Chemical* **77**, 555–560 (2001).
189. Y. L. Liu, Z. M. Liu, Y. Yang, H. F. Yang, G. L. Shen and R. Q. Yu, *Sensors And Actuators B-Chemical* **107**, 600–604 (2005).
190. E. Katz, I. Willner and J. Wang, *Electroanalysis* **16**, 19–44 (2004).
191. H. Ogawa, M. Nishikawa and A. Abe, *J. Appl. Phys.* **53**, 4448–4455 (1982).
192. P. G. Collins, K. Bradley, M. Ishigami and A. Zettl, *Science* **287**, 1801–1804 (2000).
193. J. Kong, N. R. Franklin, C. W. Zhou, M. G. Chapline, S. Peng, K. J. Cho and H. J. Dai, *Science* **287**, 622–625 (2000).
194. B. Zhang, R. W. Fu, M. Q. Zhang, X. M. Dong, P. L. Lan and J. S. Qiu, *Sensors and Actuators B-Chemical* **109**, 323–328 (2005).
195. A. Star, T. R. Han, V. Joshi, J. C. P. Gabriel and G. Gruner, *Advanced Materials* **16**, 2049–+ (2004).

Index